Thermometry at the Nanoscale
Techniques and Selected Applications

RSC Nanoscience & Nanotechnology

Editor-in-Chief:
Professor Paul O'Brien FRS, *University of Manchester, UK*

Series Editors:
Professor Ralph Nuzzo, *University of Illinois at Urbana-Champaign, USA*
Professor Joao Rocha, *University of Aveiro, Portugal*
Professor Xiaogang Liu, *National University of Singapore, Singapore*

Honorary Series Editor:
Sir Harry Kroto FRS, *University of Sussex, UK*

Titles in the Series:

How to obtain future titles on publication:
A standing order plan is available for this series. A standing order will bring delivery of each new volume immediately on publication.

For further information please contact:
Book Sales Department, Royal Society of Chemistry, Thomas Graham House, Science Park, Milton Road, Cambridge, CB4 0WF, UK
Telephone: +44 (0)1223 420066, Fax: +44 (0)1223 420247
Email: booksales@rsc.org
Visit our website at www.rsc.org/books

Thermometry at the Nanoscale
Techniques and Selected Applications

Edited by

Luís Dias Carlos
University of Aveiro, Portugal
Email: lcarlos@ua.pt

Fernando Palacio
CSIC-University of Zaragoza, Spain
Email: palacio@unizar.es

RSC Nanoscience & Nanotechnology No. 38

Print ISBN: 978-1-84973-904-7
PDF eISBN: 978-1-78262-203-1
ISSN: 1757-7136

A catalogue record for this book is available from the British Library

Published by The Royal Society of Chemistry,
Thomas Graham House, Science Park, Milton Road,
Cambridge, CB4 0WF, UK

Registered Charity Number 207890

For further information see our web site at www.rsc.org

Printed in the United Kingdom by CPI Group (UK) Ltd, Croydon, CR0 4YY, UK

Preface

Sensing and measuring temperature is a crucial need for countless scientific investigations and technological developments. Consequently, as technology progresses into the nanoscale an increasing demand for accurate, non-invasive and self-referenced temperature measurements at sub-micrometric length scales has been observed. This is particularly so in microelectronics and micro/nanofluidics, for instance, where the comprehension of heat dissipation, heat transfer and thermal conductivity mechanisms can play a crucial role in areas as diverse as reliability and integration of electronic systems, energy transfer, and cell physiology.

The assortment of luminescent and non-luminescent nanothermometers proposed over the last decade clearly point to an emerging interest in nanothermometry in a large variety of fields, from electronic to photonic devices, from optoelectronic to micro/nanofluidics and nanomedicine. At the same time, nanothermometry is a multidisciplinary and challenging subject requiring new approaches and new techniques, since conventional thermometry is not valid at such scales.

Apart from a few pioneering works published at the very end of the past century, the subject of investigating and developing thermometric devices that work with sub-micrometre space resolution has exploded in the last 10 years. Particularly since 2010, the research on the field has been biased towards luminescent nanothermometry and its applications in photonics, electronics, fluids, and nanomedicine. For instance, luminescent nano-thermometers have already been used to provide thermal readings during photothermal treatments in both culture cells and living organisms.

Several reviews have recently described the progress on high-resolution thermometers operating at the sub-micron scale, including luminescent and non-luminescent nanothermometers, intracellular measurements, ceramic phosphors that can withstand extreme temperatures, multiple optical

RSC Nanoscience & Nanotechnology No. 38
Thermometry at the Nanoscale: Techniques and Selected Applications
Edited by Luís Dias Carlos and Fernando Palacio
© The Royal Society of Chemistry 2016
Published by the Royal Society of Chemistry, www.rsc.org

chemical sensors and temperature-stimuli polymers. Despite the coverage of the subject, we have arrived at the conclusion that this very multidisciplinary subject demands a more solid piece of work, half way between a monograph and a reference book, where specialists in each particular technique can cover it, discussing achievements and limitations as well as future trends and technological possibilities. This was exactly the original purpose of this book.

The book is organized in four quite independent sections (Fundamentals, Luminescence- and Non-Luminescence-Based Thermometry and Applications) comprising sixteen chapters. After a bird's-eye short review on nanoscale thermometry and temperature measurements, the remaining two chapters of the first section encourage the reader to reflect on the basics of nanothermometry, namely the minimal length scales for the existence of local temperature and heat transfer at the nanoscale.

Section II examines in detail luminescent thermometers based on different thermal nanoprobes: Quantum dots (Chapter 4), lanthanide phosphors (Chapter 5), organic dyes (Chapter 6), polymers (Chapter 7), and organic–inorganic hybrids (Chapter 8). Section III provides a comprehensive discussion about three important non-luminescent techniques to measure temperature: Scanning thermal microscopy (Chapter 9), near-field thermometry (Chapter 10) and nanotube thermometry (Chapter 11). Finally, Section IV explores some of the most exciting and intriguing nanothermometry applications, still limited in practice and results, such as cellular thermometers (Chapter 12), thermal issues in microelectronics (Chapter 13), heat transport in nanofluids (Chapter 14), temperature probes in micro/nanofluidics (Chapter 15) and multifunctional platforms for dual-sensing (Chapter 16).

The book renders lucid explorations of a number of significant and difficult problems in nanothermometry providing readers new to the field with a clear overview of this expanding topic, being simultaneously an inspiration to those already well versed in the field. The book is presented in a format that aims to be accessible to postgraduate students and researchers in physics, chemistry, biology and engineering interested in nanothermometry.

This book would have not been possible without the support and contributions from a significant number of people. In the first place it is our pleasure to acknowledge Prof. João Rocha, Series Editor of the _RSC Nanoscience & Nanotechnology Series_, for encouraging us to submit a book proposal to the Royal Society of Chemistry. We are also extremely grateful to the Royal Society of Chemistry for continuous and always stimulating support without which our work would have been much more difficult. Last but not least, we want to deeply thank to all the contributing authors for enthusiastically accepting our invitation and embracing the book's scope and vision. Their time, work and dedication deserve our most profound acknowledgement.

Luís Dias Carlos and Fernando Palacio
Aveiro and Zaragoza

Contents

RSC Nanoscience & Nanotechnology No. 38
Thermometry at the Nanoscale: Techniques and Selected Applications
Edited by Luís Dias Carlos and Fernando Palacio
© The Royal Society of Chemistry 2016
Published by the Royal Society of Chemistry, www.rsc.org

Chapter 3 Introduction to Heat Transfer at the Nanoscale   39
Pierre-Olivier Chapuis

Section II   Luminescence-based Thermometry

Section IV Applications

Section I
Fundamentals

Nanoscale Thermometry and Temperature Measurement

PETER R. N. CHILDS

Faculty of Engineering, Dyson School of Design Engineering, Imperial
College London, Exhibition Road, London SW7 2AZ, UK
Email: p.childs@imperial.ac.uk

1.1 Introduction

Nanoscale temperature measurement concerns the determination of
temperature or temperature difference at the sub-micron scale. Applications
where it is important to be able to measure local temperature at the
nanoscale include microelectronics, optics, microfluidics, chemical reaction
and biochemical processes, such as living cells and nanomedicine. Examples of measurement of intracellular temperature fluctuations include
ref. 1–13, microcircuit temperature mapping,[14–18] and microfluidics.[19–24]
The topic of temperature measurement at the nanoscale has previously been
reviewed.[25–27] The review by the author[28] provides a general overview
of temperature measurement, along with the books in ref. 29 and 30. The
long-standing series *Temperature its Measurement and Control in Science and
Industry*[31–36] is also worth accessing for its archive on the topic.

In conventional temperature measurement a measurement system typically comprises a transducer to convert a temperature-dependent phenomenon into a signal that is a function of that temperature, a method to transmit
the signal from the transducer, some form of signal processing, a display
and method of recording the data. A calibration is then used to convert the
measured quantity into a value of temperature and the significance of the

RSC Nanoscience & Nanotechnology No. 38
Thermometry at the Nanoscale: Techniques and Selected Applications
Edited by Luís Dias Carlos and Fernando Palacio

Published by the Royal Society of Chemistry, www.rsc.org

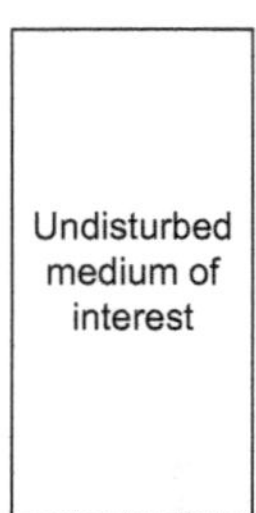

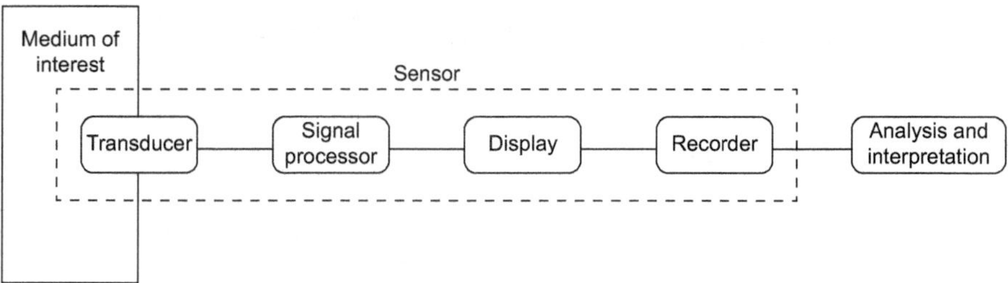

Figure 1.1 A comparison between the undisturbed medium and a typical temperature measurement application.

temperature can then be considered. Temperature measurement systems can be classified into broad themes, according to the physical contact between the transducer and the medium of interest, Figure 1.1, giving invasive, semi-invasive and non-invasive classifications.[28,29] Examples of each type respectively include thermocouples, thermochromic liquid crystals applied to a surface and observed remotely, and infrared pyrometry.

Miniaturisation of electronics and optoelectronics with higher switching speeds and associated energy dissipation has increased the importance of a good knowledge of localised heating in order to resolve manufacture and design issues and improve reliability. An accurate knowledge of living cell temperature is proving significant in monitoring cell health, with cancer cells, for example, exhibiting higher temperature than 'normal tissue' due to the increased metabolic activity.[5,7,8,37,38] Cell temperature alters with cellular activity such as division, gene expression and enzyme reaction, and temperature measurement at this scale can provide valuable diagnostic information.

Temperature can be regarded qualitatively as a measure of hotness or coldness of a body. Quantitatively temperature can be defined from the second law of thermodynamics in terms of the rate of change of entropy with energy.[29] Thermodynamically, temperature is the property that determines whether a system is in thermal equilibrium with another system or systems.

At the nanoscale, issues abound with the definition of temperature, as the scale of the applications means that assumptions of continuum and thermodynamic equilibrium may not hold.[39] For many practical applications, systems are in a non-equilibrium state and it is difficult to determine the conditions for which equilibrium is established at a local level.[40] A length scale has been proposed[39] to identify whether a canonical energy distribution can be assumed. At very low temperatures, *e.g.*, 1 K, a length scale of 0.1 m was found for Si, providing an indication of the challenge for nanoscale measurements. However, the length scale necessary decreases as temperature rises, up to a limit, after which it is reasonably constant. For example a minimum length scale of 100 nm was identified for a temperature of 645 K. The implications of this for nanoscale temperature measurement remain significant, due to non-canonical energy distribution, non-equilibrium and ill-defined material properties. The definition of material properties is particularly important for some transient techniques where knowledge of specific heat capacity is necessary at the temperature of the body concerned.

In order to enable assignment of numerical values to bodies at different temperatures, a temperature scale is necessary. The thermodynamic temperature scale is defined by means of perfect heat engines. The SI unit for the thermodynamic temperature scale is the Kelvin. As perfect heat engines are not realisable in practice, the International Temperature Scale of 1990, denoted by ITS-90, has been developed as a practical approximation to the thermodynamic temperature scale. The ITS-90 is constructed using a number of overlapping temperature ranges and is believed to represent thermodynamic temperature to within ± 2 mK from 2 to 273 K, ± 3.5 mK at 730 K and ± 7 mK at 900 K (one standard deviation limits[41]). The ITS-90 enables traceability between the measurement that is actually made and the temperature scale defined by the ITS-90.

Due to the delicate and intricate nature of accurate measurements to ITS-90, most practical measurements are made using devices and systems that have been calibrated against another device which has itself been calibrated using guidelines specified in ITS-90. Calibration in this context is the process of relating the output value from a measurement system to a known input value. This process is illustrated in Figure 1.2 whereby there is a transfer of calibration between practical and specialist devices. The chain between the thermometer or temperature measurement system in use and

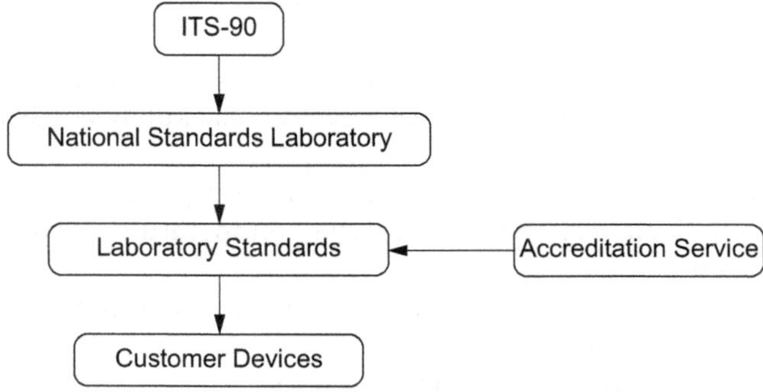

Figure 1.2   A typical chain between practical temperature measurement systems and the International Temperature Scale of 1990.

the ITS-90 may in practice have more links than those shown in Figure 1.2, with each link tending to increase the uncertainty associated with the measurement.

As well as the physical level of contact between a sensor and the application of interest, thermometric systems can be classified according to whether they are characterised by well-established equations of state that directly relate the measured parameter to temperature, and are known as *primary*, or if the system requires calibration then it is known as *secondary*. In addition, some materials exhibit an identifiable transition at a specific temperature, an example being phase change, and these can be useful for determining a particular temperature accurately, or to see if a temperature has been exceeded.[42]

1.2 Methods

A wide range of temperature measurement methods and technologies are available. Conventional technologies such as liquid-in-glass thermometry, thermocouples, and resistance temperature devices (RTDs), have all been adapted to nanoscale applications. Similarly, several of the remote observation approaches such as Raman, fluorescence and optical interferometry have been found to be useful for nanoscale applications The use of biological agents and cellular features to enable the measurement of temperature *in vivo* has provoked particular interest. Although a very wide range of temperatures is experienced in nature and science, here the technologies and temperature ranges considered are limited to those addressed by current technologies, which are generally applicable to cryogenic temperatures and the temperatures associated with electronic circuit operation. Temperature measurement methods relevant to the nanoscale include:

1. Liquid and solid-in-tube nanotube thermometry;
2. Resistance;
3. Thermocouples, based on the Seebeck effect;
4. Coulomb blockade;
5. MEMS (microelectromechanical systems) based resonator quality factor or Fermi-level shift;
6. Themochromic liquid crystals, based on crystal phase transitions;
7. Infrared thermography, based on Planck blackbody emission;
8. Fluorescence;
9. Thermoluminescence or thermographic phosphor thermometry;
10. Thermoreflectance, based on the temperature dependence of reflection;
11. Raman, based on inelastic scattering of monochromatic light;
12. Brownian motion;
13. Near-field scanning optical microscopy;
14. Scanning thermal microscopy, using an atomic force microscope with a thermocouple, thermistor or platinum resistance thermometer tip;

15. Transmission electron microscopy; and
16. Optical interferometry, based on thermal expansion or refractive index change.

In the context of nanoscale applications, the methods can be classified as *luminescent* or *non-luminescent*.[27] An overview of several of these methods is given in Sections 1.2.1 to 1.2.6. The presentation is inevitably brief, but is intended to introduce the principle, provoke interest and offer some relevant references. The chapters in this book also serve as excellent, more in-depth treatments.

1.2.1   Infrared

A body emits energy in the form of thermal radiation due to its temperature, with the quantity of radiation rising with increasing temperature. The energy emitted over the electromagnetic spectrum due to temperature by a black body can be modelled by Planck's law. Infrared thermometers measure the thermal radiation emitted by a body due to its temperature and conventional devices have found applications from cryogenic temperatures to over 6000 K. An infrared temperature measurement system generally consists of three elements, the emitting source of interest, the propagation medium, typically space or air, and the measuring device. An infrared measuring device may comprise an optical system, a detector, processing circuit, and display.[29,43] The purpose of the optical system is to focus the energy emitted by the target onto the infrared detector, typically converting the energy into an electrical signal. Infrared thermometers can be classified depending on whether the device is sensitive to all or a specific band of wavelengths, with those that are sensitive to all wavelengths classed as total radiation or broadband thermometers. Devices sensitive to radiation in a specific band of wavelengths are classed spectral band thermometers, ratio thermometers and thermal imagers. Infrared thermography represents a mature technology. For nanoscale applications the spatial resolution is limited, *ca.* 10 μm, and an estimate of the emissivity is necessary, at the temperature concerned. In addition, careful consideration needs to be given to reflection from other sources of thermal radiation in order to ensure that the measurement is close to the temperature of the undisturbed medium, or that appropriate compensation for this is made in the analysis.

1.2.2   Raman Scattering

Raman scattering refers to the inelastic scatting of light. When a molecule is promoted from the ground state to a higher state by incident radiation, it can either return to the original state, which is called *Rayleigh scattering*, or it can occupy a different vibrational state, which is classed as *Raman scattering*. Raman scattering gives rise to Stokes lines on the observed spectra.

Alternatively if a molecule is already in an excited state, it can be promoted to an even higher unstable state and then subsequently return to its ground state. This process is also called Raman scattering and gives rise to anti-Stokes lines on the observed spectra. Monitoring Raman scattering in order to determine temperature involves shining a monochromatic light source, such as a laser, on a sample and detecting the scattered light with a high-resolution Raman spectrometer. Fibre optics can be used for remote analysers. The process has the advantage that no preparation of the sample is necessary and small probe volumes, <1 μm, are possible. The method is time consuming and the numbers of samples per second is limited and can be as slow as 0.5 points per second. The application of Raman scattering thermometry has been applied to the polymerase chain reaction,[44] thermal transport in wires,[45,46] for strain temperature measurement, temperature and stress,[47] nanotubes,[48] nanowires,[49] and graphene.[50]

1.2.3 Luminescence

The thermal dependence of phosphor luminescence, band shape, peak energy, intensity, and excited states lifetimes can also be used for temperature measurement. The technique is also known as *thermographic phosphor thermometry*. A typical setup requires the phosphor or luminescent probe, applied and bonded to a surface, or incorporated within a medium, a light source such as a laser, and a detection system. Types of luminescent thermal probes include organic dyes, ruthenium complexes, quantum dots, Ln^{3+} ions, polymer and organic–inorganic hybrid matrices incorporating emitting centres. The method has the advantages of remote observation, high sensitivity, good spatial resolution and short acquisition periods.

1.2.3.1 *Organic Complexes*

Organic complexes are readily available and can be selected based on parameters such as the excitation/emission wavelength range required, solubility and facility of vectorisation. An example is nanoparticles of Ru(phen) and PtTFPP embedded in polyacrylonitrile.[51,52] Ru(phen) has a maximum temperature sensitivity of 2.5% K^{-1} at 320 K. Alternatively, a ratio of intensities from the monomer and excimer/exciplex can be analysed.[53] Similarly, the decay of emission lifetimes can be used.

1.2.3.2 *Quantum Dots*

Quantum dots (QDs) are nanocrystals of a semiconducting material with diameters in the range of about 2 to 10 nm. The high surface-to-volume ratios for these particles results in quantum mechanical properties, such as temperature-dependent photoluminescence, which can be exploited for the purpose of temperature measurement. An example is ZnS-coated CdS, which gives a factor of five change in photoluminescence intensity over the range of 100 to 500 K.[54,55] The method involves observing the QDs through a microscope. A series of factors have limited the use of QDs for thermometry to date, including bluing, bleaching and blinking under continuous illumination.[56–58] To overcome some of the practical limitations QDs can be coated with an inert material. The development of a III-nitride QD embedded in a nanowire for room temperature applications has been reported by Holmes *et al.*[59] The design of QDs for *in vivo* applications has also been considered.[60]

1.2.3.3 Phosphors

The luminescent properties of thermographic phosphors are temperature dependent. This dependency can be observed by applying a thin layer of luminescent material to the system of interest, illumination with UV light and observing the subsequent luminescence. There are two commonly used response modes.[61,62] A pulsed excitation source can be used and after each pulse the phosphor exhibits an exponentially decaying emission, the time constant of which is temperature dependent. Alternatively, pulsed or continuous illumination can be used and the ratio of the emission intensities of two distinct spectral lines can be determined, which is temperature dependent.

Ln^{3+} can be incorporated into a polymer binder and sprayed onto a surface.[62] A measurement system will typically comprise the surface of interest, the applied coating, a means of exciting the phosphor, such as a laser, and an optical detector or photomultiplier. The use of the lanthanide coordination polymer $Tb_{0.957}Eu_{0.043}cpda$ as a ratiometric and colorimetric luminescent thermometer for the temperature range 40 to 300 K has been reported.[63]

Thin film deposited chromium-doped aluminium oxide ($Cr–Al_2O_3$, ruby) thermographic phosphors for temperature measurement have been used.[64] Zinc silicate ($Zn_2SiO_4{:}Mn^{2+}$) has been shown to be highly suitable for fluorescence intensity ratio temperature measurement over the range of 30 to 300 °C with a sensitivity of 12.2%.[65] The detailed characteristics of fluorescence spectra measurements have been exploited,[66] in combination with a neural network analysis, in order to determine absolute temperature and this has been demonstrated for Rhodamine B (RhB) and copper chloride dyed glycerol. Polydiacetylene (PDA) embedded PVA film can be used to measure thermal gradients effectively.[67]

1.2.3.4 Biomolecular Intracellular Thermometers

A series of technologies and approaches have been taken to explore possibilities for measuring temperature at the cellular level. These range from the introduction of fluorescent probes and dyes, to the introduction of proteins to control gene expression, the use of materials for the controlled release of chemicals sensitive to temperature, QDs, near-infrared imaging, and heating of nanoparticles to induce cell response at a known temperature. The use of temperature-sensitive green emission of fluorescent $NaYF_4$:Er^{3+},Yb^{3+} nanoparticles over the range 25 to 45 °C with application to HeLa cells has been demonstrated.[68] The use of RhB dye employed as a temperature indicator within a single cell has been reported[69] with a resolution of better than ± 0.3 °C within the cell cytoplasm, but this reduces to ± 1.8 °C as a result of environmental variations. Temperature-sensitive mutants of LacI controlling LacZ expression have been explored.[38] Great care is necessary with such approaches as the signal can be a function of duration at a temperature as well as the magnitude of the temperature. Mesoporous materials, such as silica nanoparticles, offer the potential for the controlled release of chemicals that are responsive to temperature changes and this concept has been explored.[70–72] The heating of nanoparticles of gold injected into a cell, in combination with optical detection of electron spin resonance in single atomic defects in nitrogen vacancy (NV) diamonds, as the sensor, has been demonstrated.[10]

An early example[2] of using luminescence intensity to measure cell temperature used the temperature-dependent luminescence intensity of the $Eu(tta)_3 \cdot 3H_2O$ complex.

A polymer-based thermometry solution has been demonstrated[5] using PNIPAM combined with the water-sensitive fluorophore, DBD-AA. They successfully measured temperatures within the cell over the range 300 to 306 K with a temperature resolution of 0.3 to 0.5 K and a sensitivity of $S_m = 94\%$ K^{-1} at 304 K.

Further consideration of the topic of cell-based temperature measurement is given in Section 1.2.6.4.

1.2.4 Thermoreflectance

Thermoreflectance exploits the temperature dependence of a material's refractive index, using an optical imaging system to observe the surface of interest, with the temperature profile resolution defined by the special resolution of the optical system.[14,26]

1.2.5 Interferometry

Optical interferometry provides both local temperature information, as well as local deformation due to thermal expansion. Interferometers measure differences in optical paths between two light beams, one of which bypasses the test section, while the other passes directly through the test section. A quadriwave lateral shearing interferometer as a wavefront sensor has been used[73] for determining the variation of refractive index for a microwire. Using a water layer on the microwire, and electrical heating, variations in the refractive index generate a distortion of the incident wavefront, enabling mapping of the temperature field.

1.2.6 Non-luminescent

A number of non-luminescent temperature measurement technologies have been developed for nanoscale applications including scanning thermal microscopy, nanolithography, carbon nanotubes, and biomaterials.

1.2.6.1 Scanning Thermal Microscopy

Scanning thermal microscopy uses a small thermocouple with a junction diameter of the order of 20 to 100 nm, formed at a probe tip which is traversed over the surface of interest.[74–76] Alternatively an RTD can be mounted on the probe or a combination of a thermocouple and RTD. This can be useful for determining the thermal conductivity of a sample.[77–79]

1.2.6.2 Nanolithography

Vapour deposition technology[80] can be used to deposit a platinum strip to form a sensor on a surface. Examples include decomposition of organic inks such as $(CH_3)_3Pt$. Combinations of materials can be built up by appropriate layering in order to ensure electrical isolation and lead-outs. US Patent 6905736[81] reports a Pt(Ga)–W(Ga) junction with a temperature coefficient of 5.4 mV$°C^{-1}$. 100 nm^2 gold–nickel thermocouples on silicon and 500 nm^2 thermocouples on quartz have also been produced.[82] A challenge with vapour deposition sensors is ensuring robustness of the sensor, and chemical consistency. It is well known that some materials aggressively leach elements from a substrate, gradually degrading the performance of a sensor with time.

1.2.6.3 Carbon Nanotube Thermometry

Liquid-in-glass thermometry provided the basis for modern temperature measurement.[29,30] Carbon nanotube thermometry was originally proposed[83] using gallium as the thermometric liquid, with a scanning electron microscope to observe the meniscus. The coefficient of thermal expansion of gallium has been observed[83] to vary linearly over the range 323 to 773 K, with values close to the macroscopic quantities. Examples of other thermometric liquids proposed and demonstrated for different temperature ranges include Pb using a ZnO nanotube,[84] Au(Si) using a Ga_2O_3 nanotube,[85] and In using In_2O_3 nanotubes.[86] The use of carbon nanotubes to provide good thermal contact between a microscope and sample has also been reported.[87]

1.2.6.4 Biomolecular-Based Thermometry

Many molecules and biological components exhibit temperature-dependent behaviours that can be exploited for temperature measurement on the nanoscale. Examples include the introduction of proteins to control gene expression, crystal defects, and luminescence.

As noted previously, temperature represents a fundamental parameter for the operation of living cells and a number of mechanisms are apparent that can be exploited or mimicked in the measurement or control of temperature.[88] For histidine kinase, DesK, from Bacillus subtilis, Albanesi *et al.*,[89] report how the plasticity of the helical domain influences the catalytic activities of the protein, through conformational signals transmitted by membrane connections enabling interhelical rearrangements, controlling the alternation between output autokinase and phosphatase activities. Of interest, Albanesi *et al.*[89] suggest that structural comparison of the different DesK variants indicates that incoming signals can take the form of helix rotations and asymmetric helical bends comparable to some other sensing systems, and that similar switching mechanisms could be operational in a wide range of sensor histidine kinases, or indeed exploited in newly designed mechanisms.

Crystal defects, notably the NV centre, in diamond can be exploited to enable the measurement of a series of physical parameters including temperature. The optical excitation of NV nanodiamonds with 532 nm laser light and fluorescence collection with a confocal microscope has been reported by Neumann *et al.*,[10] yielding a noise floor of $130\,\mathrm{mK\,Hz^{1/2}}$ and accuracies as good as 1 mK. Embedded in nanocrystals they can provide robust single-photon sources and fluorescent biomarkers.[13] Nanodiamonds can be injected into a cell, and excited with a laser to alter their spin state and the light emitted measured. By using more than one nanodiamond the temperature difference within a single cell can be evaluated as reported by Kucsko *et al.*,[11] detecting temperature differences as small as $1.8 \pm 0.3\,\mathrm{mK}$. Interestingly Kucsko *et al.*[11] report that by combining nanodiamonds with laser heating of gold particles also injected into a cell, the temperature of the cell can be controlled.

Living cells sense temperature through proteins, nucleic acids and mRNAs that either modify their conformation structure directly[90–92] or undergo complex reactions. Cells respond to harmful temperature changes by inducing heat-shock or cold-shock proteins. RNA temperature-sensing capability is located in specific areas of RNA structure, especially the 5′ untranslated region of certain heat-shock and virulence genes, which shields the ribosome-binding sites at physiological temperatures.[93] Under the onset of heat-shock, destabilisation of RNA commences with associated release of ribosome-binding sites and translation initiation. For cold-shock, cell growth tends to stop, accompanied by a dramatic reduction in bulk gene expression and cold-shock proteins are expressed. The underlying mechanism in cells involves a gradual melting of a weak stem-loop structure rather than a switch between two mutually exclusive conformations. As such, RNA thermometers can be considered as *molecular dimmers* rather than discrete on and off state switches.[93] Naturally occurring thermometers represent highly complex systems, but simpler RNA stem-loop structures,

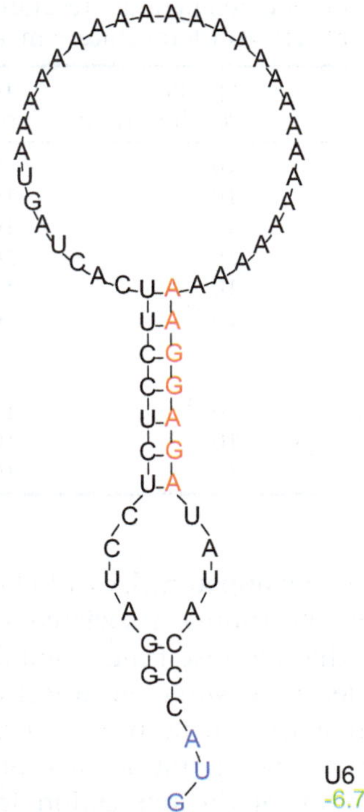

Figure 1.3 Synthetic RNA thermometer example (after Neupert *et al.*[94]).

see for example Figure 1.3, containing the ribosome-binding site can be produced.[93–95] Ideas for the use of the DNA backbone for temperature sensors and switches have also been explored.[12]

1.3 Selection

Conventionally, the selection of a specific sensor or system for temperature measurement involves consideration of a number of parameters including:

(1) Uncertainty;
(2) Temperature range;
(3) Thermal disturbance;
(4) Level of contact;
(5) Size of the sensor;
(6) Transient response;
(7) Sensor protection;
(8) Availability; and
(9) Cost.

In the case of nanoscale applications, however, the small scale and specific requirements, along with available technology may dictate the use of a

Table 1.1 Typical spatial resolution, temperature resolution and response times for a range of nanoscale-relevant temperature measurement methods.

Method	Spatial resolution/μm	Temperature resolution/K	Response time/μs
Infrared thermography	10	10^{-1}	10
Thermoreflectance	10^{-1}	10^{-2}	10^{-1}
Raman	1	10^{-1}	10^{6}
Thermocouple	10^{2}	10^{-1}	10
Fluorescence	10^{-1}	10^{-2}	10
Near-field scanning optical microscopy	10^{-2}	10^{-1}	10
Liquid crystals	10	10^{-1}	10^{2}
Scanning thermal microscopy	10^{-1}	10^{-1}	10^{2}
Transmission electron microscopy	10^{-2}	10^{-1}	10
Optical interferometry	1	10^{-5}	10^{-3}

particular method. Despite any apparent lack of choice, consideration still needs to be given to the uncertainty associated with the measurement. A common assumption is that because fundamental physics is being used, there is no need to consider a measurement uncertainty. However, reasonably simple analysis approaches can be used to evaluate the thermal disturbance associated with the incorporation of a probe into a medium of interest.[29] This may, depending on the application, involve just a conduction equation analysis accounting for any differences in thermal conductivity between the undisturbed medium, and that with the sensor embedded, or coating applied, or may involve a thermal radiation calculation to evaluate whether the presence of a probe has altered the thermal radiation exchanges significantly. It is all too easy to alter the temperature of a sample inadvertently due to the presence of an optical head. In addition, new methods such as the observation of Brownian motion or nanoparticles are emerging,[96] providing new opportunities and technology options to assess.

A summary of the approximate capabilities of a range of methods is given in Table 1.1. This has been adapted from the information given in ref. 26–28 and ref. 97.

In assessing the performance of a luminescent thermometer, a useful parameter is the relative sensitivity, defined by

$$S = \frac{\left(\dfrac{\partial Q}{\partial T}\right)}{Q} \tag{1.1}$$

Where Q, is an experimental parameter and corresponds to the quenching of luminescence with temperature. For example Brites *et al.*[98] determined a sensitivity of 4.9% K^{-1} for an orthosilicate/aminopropyltriethoxysilane (TEOS/APTES) organosilica shell co-doped with [Eu(btfa)$_3$(MeOH)(bpeta)] and [Tb(btfa)$_3$(MeOH)(bpeta)] β-diketonate chelates using the integrated areas I_{Eu} and I_{Tb} of the $^5D_0$-$^7F_2$ (I_{Eu} at 612 nm) and $^5D_4$-$^7F_5$ (I_{Tb} at 545 nm)

Table 1.2 Figure of merit sensitivity ($\%K^{-1}$) and temperature range (K) for a selection of luminescent molecular thermometers. Data adapted from Brites *et al.*[27]

Phosphor	S_m	$\Delta T\ (T_m)$
Ru(phen)	2.5	280–325 (320)
Fluorescein D–Texas Red A in DNA	4.5	278–318 (295)
PtOEP	4.6	290–320 (305)
RhB in PDMS	2.3	293–373 (363)
C_{70} in PtBMA	2.2	293–363 (363)
PNIPAM	10.4	293–318 (308)
Mutant of lacI (Its265)	19.6	308–318 (318)
(CdSe)ZnS QDs	2.2	278–313 (313)
YAG:Ce nanoparticles	0.2	315–350 (350)
$Y_2O_3{:}Eu^{3+}$	2.6	473–973 (973)
$Eu(tta)_3 \cdot 3H_2O$ embedded in PMMA	4.4	293–333 (330)

transitions, in this case, $Q = I_{Eu}^2 - I_{Tb}^2$. Spatial and temporal resolutions of 0.42 µm and 4.8 ms, respectively, have been reported.[99]

Values for the sensitivity of a range of luminescent molecular thermometers are given in Table 1.2, adapted from Brites *et al.*[27]

1.4 Conclusions

The measurement of temperature at the nanoscale represents a practical reality with a wide range of techniques having been demonstrated. A particular challenge remains for systems where thermal gradients exist, the temperature is at cryogenic levels and the thermal mass is low, arising from the fundamental requirement for sufficient continuum to enable a meaningful definition of temperature for the system concerned. Analysis methods have been proposed to enable calculation of the necessary length scales for meaningful measurements of temperature, with the length scale dependent on the material properties and the temperature. The host of technologies available has now led to choice for some applications as well as bespoke solutions emerging, for example the measurement of temperature on microelements in circuitry and within cells. The development of understanding at the cellular level offers opportunities for the transfer of this knowledge to the design of new synthetic devices for both measurement and control. The pioneering work on nanoscale temperature measurement has certainly opened up insight across many domains, as well as posing some fundamental challenges both in terms of the technology and making sense of the data and its validity.

References

1. C. F. Chapman, Y. Liu, G. J. Sonek and B. J. Tromberg, The use of exogenous fluorescent probes for temperature measurements in single living cells, *Photochem. Photobiol.*, 1995, **62**, 416–425.

2. O. Zohar, M. Ikeda, H. Shinagawa, H. Inoue, H. Nakamura, D. Elbaum, D. L. Alkon and T. Yoshioka, Thermal imaging of receptor-activated heat production in single cells, *Biophys. J.*, 1998, **74**, 82–89.
3. M. Suzuki, V. Tseeb, K. Oyama and S. Ishiwata, Microscopic detection of thermogenesis in a single HeLa cell, *Biophys. J.*, 2007, **92**, L46–L48.
4. N. Mohan, C. S. Chen, H. H. Hsieh, Y. C. Wu and H. C. Chang, In vivo imaging and toxicity assessments of fluorescent nanodiamonds in Caenorhabditis elegans, *Nano Lett.*, 2010, **10**, 3692–3699.
5. C. Gota, K. Okabe, T. Funatsu, Y. Harada and S. Uchiyama, Hydrophilic fluorescent nanogel thermometer for intracellular thermometry, *J. Am. Chem. Soc.*, 2009, **131**, 2766–2767.
6. H. Huang, S. Delikanli, H. Zeng, D. M. Ferkey and A. Pralle, Remote control of ion channels and neurons through magnetic-field heating of nanoparticles, *Nat. Nanotechnol.*, 2010, **5**, 602–606.
7. J. M. Yang, H. Yang and L. W. Lin, Quantum dot nano thermometers reveal heterogeneous local thermogenesis in living cells, *ACS Nano*, 2011, **5**, 5067–5071.
8. K. Okabe, N. Inada, C. Gota, Y. Harada, T. Funatsu and S. Uchiyama, Intracellular temperature mapping with a fluorescent polymeric thermometer and fluorescence lifetime imaging microscopy, *Nat. Commun.*, 2012, **3**, 705.
9. C. L. Wang, R. Z. Xu, W. J. Tian, X. L. Jiang, Z. Y. Cui, M. Wang, H. M. Sun, K. Fang and N. Gu, Determining intracellular temperature at single-cell level by a novel thermocouple method, *Cell Res.*, 2011, **21**, 1517–1519.
10. P. Neumann, I. Jakobi, F. Dolde, C. Burk, R. Reuter, G. Waldherr, J. Honert, T. Wolf, A. Brunner, J. H. Shim, D. Suter, H. Sumiya, J. Isoya and J. Wrachtrup, High-precision nanoscale temperature sensing using single defects in diamond, *Nano Lett.*, 2013, **13**, 2738–2742.
11. G. Kucsko, P. C. Maurer, N. Y. Yao, M. Kubo, H. J. Noh, P. K. Lo, H. Park and M. D. Lukin, Nanometre-scale thermometry in a living cell, *Nature*, 2013, **500**, 54–58.
12. A. T. Jonstrup, J. Fredsoe and A. H. Andersen, DNA hairpins as temperature switches, thermometers and ionic detectors, *Sensors*, 2013, **13**, 5937–5944.
13. R. Schirhagl, K. Chang, M. Loretz and C. L. Degen, Nitrogen-vacancy centers in Diamond: nanoscale sensors for physics and biology, *Annu. Rev. Phys. Chem.*, 2014, **65**, 83–105.
14. P. Kolodner and J. A. Tyson, Remote thermal imaging with 0.7 mm spatial resolution using temperature-dependent fluorescent thin films, *Appl. Phys. Lett.*, 1983, **42**, 117–119.
15. L. Aigouy, G. Tessier, M. Mortier and B. Charlot, Scanning thermal imaging of microelectronic circuits with a fluorescent nanoprobe, *Appl. Phys. Lett.*, 2005, **87**(18), 184105.
16. G. Tessier, M. Bardoux, C. Boue and D. Fournier, Back side thermal imaging of integrated circuits at high spatial resolution, *Appl. Phys. Lett.*, 2007, **90**, 171112.

17. W. J. Liu and B. Z. Yang, Thermography techniques for integrated circuits and semiconductor devices, *Sens. Rev.*, 2007, **27**, 298–309.
18. W. Jung, Y. W. Kim, D. Yim and J. Y. Yoo, Microscale surface thermometry using SU8/Rhodamine-B thin layer, *Sens. Actuators, A*, 2011, **171**, 228–232.
19. H. B. Mao, T. L. Yang and P. S. Cremer, A microfluidic device with a linear temperature gradient for parallel and combinatorial measurements, *J. Am. Chem. Soc.*, 2002, **124**, 4432–4435.
20. R. Samy, T. Glawdel and C. L. Ren, Method for microfluidic whole-chip temperature measurement using thin-film poly(dimethylsiloxane)/rhodamine B, *Anal. Chem.*, 2008, **80**, 369–375.
21. C. Gosse, C. Bergaud and P. Low, Molecular probes for thermometry in microfluidic devices, in *Thermal Nanosystems and Nanomaterials, Topics in Advanced Physics*, ed. S. Volz, Springer-Verlag, Berlin, 2009, vol. 118, pp. 301–341.
22. T. Barilero, T. Le Saux, C. Gosse and L. Jullien, Fluorescent thermometers for dual-emission-wavelength measurements: molecular engineering and application to thermal imaging in a microsystem, *Anal. Chem.*, 2009, **81**, 7988–8000.
23. E. M. Graham, K. Iwai, S. Uchiyama, A. P. de Silva, S. W. Magennis and A. C. Jones, Quantitative mapping of aqueous microfluidic temperature with sub-degree resolution using fluorescence lifetime imaging microscopy, *Lab Chip*, 2010, **10**, 1267–1273.
24. J. Feng, K. J. Tian, D. H. Hu, S. Q. Wang, S. Y. Li, Y. Zeng, Y. Li and G. Q. Yang, A Triarylboron-based fluorescent thermometer: sensitive over a wide temperature range, *Angew. Chem., Int. Ed.*, 2011, **50**, 8072–8076.
25. C. Y. Wang and L. J. Chen, Nanothermometers for transmission electron microscopy – fabrication and characterization, *Eur. J. Inorg. Chem.*, 2010, 4298–4303.
26. J. Christofferson, K. Maize, Y. Ezzahri, J. Shabani, X. Wang and A. Shakouri, Microscale and nanoscale thermal characterization techniques, *J. Electron. Packag.*, 2008, **130**, 041101–041106.
27. C. D. S. Brites, P. P. Lima, N. J. O. Silva, A. Millan, V. S. Amaral, F. Palacio and L. D. Carlos, Thermometry at the nanoscale, *Nanoscale*, 2012, **4**, 4799–4829.
28. P. R. N. Childs, J. R. Greenwood and C. A. Long, Review of Temperature measurement, *Rev. Sci. Instrum.*, 2000, **71**, 2959–2978.
29. P. R. N. Childs, *Practical Temperature Measurement*, Butterworth Heinemann, 2001.
30. J. V. Nicholas and D. R. White *Traceable Temperatures*, Wiley, 1994.
31. Temperature, ed. C. M. Herzfield, *Its Measurement and Control in Science and Industry*, Rheinhold, 1962, vol. 3.
32. Temperature, ed. H. H. Plumb, *Its Measurement and Control in Science and Industry*, American Institute of Physics, 1972, vol. 4.
33. Temperature, ed. J. F. Schooley, *Its Measurement and Control in Science and Industry*, American Institute of Physics, 1992, vol. 6.

34. Temperature, ed. J. F. Schooley, *Its Measurement and Control in Science and Industry*, American Institute of Physics, 1982, vol. 5.
35. Temperature, ed. D. C. Ripple, *Its Measurement and Control in Science and Industry*, American Institute of Physics, 2002, vol. 7.
36. Temperature, ed. C. W. Meyer, *Its Measurement and Control in Science and Industry*, American Institute of Physics, 2012, vol. 8.
37. J. B. Weaver, Bioimaging: Hot nanoparticles light up cancer, *Nat. Nanotechnol.*, 2010, **5**, 630–631.
38. K. M. McCabe and M. Hernandez, Molecular thermometry, *Pediatr. Res.*, 2010, **67**, 469–475.
39. M. J. Hartmann, Minimal length scales for the existence of local temperature, *Contemp. Phys.*, 2006, **47**(2), 89–102.
40. H. J. Kreuzer, *Nonequilibrium Thermodynamics and its Statistical Foundation*, Clarandon Press, Oxford, 1981.
41. B. W. Mangum and G. T. Furukawa, Guidelines for realizing the ITS-90, NIST Tech, note 1265, 1990.
42. M. Alaskar, M. Ames, C. Liu, S. Connor, R. Horne, K. Li and Y. Cui, Smart nanosensors for in-situ temperature measurement in fractured geothermal reservoirs, Australian Geothermal Energy Conference, 2011.
43. C. Meola and G. M. Carlomagno, Recent advances in the use of infrared thermography, *Meas. Sci.Technol.*, 2004, **15**, R27–R58.
44. S. H. Kim, J. Noh, M. K. Jeon, K. W. Kim, L. P. Lee and S. I. Woo, Micro-Raman thermometry for measuring the temperature distribution inside the microchannel of a polymerase chain reaction chip, *J. Micromech. Microeng.*, 2006, **16**, 526–530.
45. Y. Yue, G. Eres, X. Wang and L. Guo, Characterization of thermal transport in micro/nanoscale wires by steady-state electro-Raman-thermal technique, *Appl. Phys. A: Mater. Sci. Process.*, 2009, **97**, 19–23.
46. O. Frazão, C. Correia, J. L. Santos and J. M. Baptista, Raman fibre Bragg-grating laser sensor with cooperative Rayleigh scattering for strain-temperature measurement, *Meas. Sci. Technol.*, 2009, **20**, 045203.
47. T. Beechem, S. Graham, S. P. Kearney, L. M. Phinney and J. R. Serrano, Simultaneous mapping of temperature and stress in microdevices using micro-Raman spectroscopy, *Rev. Sci. Instrum.*, 2007, **78**.
48. L. Song, W. Ma, Y. Ren, W. Zhou, S. Xie, P. Tan and L. Sun, Temperature dependence of Raman spectra in single-walled carbon nanotube rings, *Appl. Phys. Lett.*, 2008, **9**.
49. S. Piscanec, M. Cantoro, A. C. Ferrari, J. A. Zapien, Y. Lifshitz, S. T. Lee, S. Hofmann and J. Robertson, Raman spectroscopy of silicon nanowires, *Phys. Rev. B: Condens. Matter Mater. Phys.*, 2003, **68**.
50. I. Calizo, A. A. Balandin, W. Bao, F. Miao and C. N. Lau, Temperature dependence of the Raman spectra of graphene and graphene multi-layers, *Nano Lett.*, 2007, 7, 2645–2649.
51. M. E. Kose, B. F. Carroll and K. S. Schanze, Preparation and spectroscopic properties of a dual luminophore pressure sensitive paint, *Langmuir*, 2005b, **21**, 9121–9129.

52. M. E. Kose, A. Omar, C. A. Virgin, B. F. Carroll and K. S. Schanze, Principal component analysis calibration method for dual luminophore pressure sensitive paints, *Langmuir*, 2005b, **21**, 9110–9120.
53. N. Chandrasekharan and L. A. Kelly, A dual fluorescence temperature sensor based on perylene/exciplex interconversion, *J. Am. Chem. Soc.*, 2001, **123**, 9898–9899.
54. G. Walker, V. Sundar, C. Rudzinski, A. Wun, M. Bawendi and D. Nocera, Quantum-dot optical temperature probes, *Appl. Phys. Lett.*, 2003, **83**, 3555–3557.
55. H. Sakaue, A. Aikawa, Y. Iijima, T. Kuriki and T. Miyazaki, Quantum dots as global temperature measurements, in *Quantum Dots – A Variety of New Applications*, ed. A. Al-Ahmadi, InTech ch. 7, 2012.
56. W. G. J. H. M. van Sark, P. L. T. M. Frederix, A. A. Bol, H. C. Gerritsen and A. Meijerink, Blueing, bleaching, and blinking of single CdSe/ZnS quantum dots, *Chem. Phys. Chem.*, 2002, **3**, 871–879.
57. K. H. Isnaeni Kim, D. L. Nguyen, H. Lim, T. N. Pham and Y. H. Cho, Shell layer dependence of photoblinking in CdSe/ZnSe/ZnS quantum dots, *Appl. Phys. Lett.*, 2011, **98**, 012109–012112.
58. M. Nakamura, S. Ozaki, M. Abe, T. Matsumoto and K. Ishimura, One-pot synthesis and characterization of dual fluorescent thiol-organosilica nanoparticles as non-photoblinking quantum dots and their applications for biological imaging, *J. Mater. Chem.*, 2011, **21**, 4689–4695.
59. M. J. Holmes, K. Choi, S. Kako, M. Arita and Y. Arakawa, Room-temperature triggered single photon emission from a III-nitride site-controlled nanowire quantum dot, *Nano Lett.*, 2014, **14**(2), 982–986.
60. B. Somogyi and A. Gali, Computational design of in vivo biomarkers, *J. Phys.: Condens. Matter*, 2014, **26**, 143202.
61. A. L. Heyes, Thermographic phosphor thermometry for gas turbines, in *Advanced Measurement Techniques for Aero and Stationary Gas Turbines*, ed. C. H. Sieverding and J. F. Brouckaert, VKI LS 2004-04, 2004.
62. A. L. Heyes, On the design of phosphors for high temperature thermometry, *J. Lumin.*, 2009, **129**, 2004–2009.
63. Y. Cui, W. Zou, R. Song, J. Yu, W. Zhang, Y. Yang and G. Qian, A ratiometric and colorimetric luminescent thermometer over a wide temperature range based on a lanthanide coordination polymer, *Chem. Commun.*, 2014, **50**, 719–721.
64. W. Li, Z. J. Coppens, D. Greg Walker and J. G. Valentine, Electron beam physical vapor deposition of thin ruby films for remote temperature sensing, *J. Appl. Phys.*, 2013, **113**, 163509.
65. V. Lojpur, M. G. Nikolic, D. Jovanovic, M. Medic, Z. Antic and M. D. Dramicanin, Luminescence thermometry with Zn_2SiO_4:Mn^{2+} powder, *Appl. Phys. Lett.*, 2013, **103**, 141912.
66. L. Liu, S. Creten, Y. Firdaus, J. Jesus, A. Flores Cuautle, M. Kouyate, M. Van der Auweraer and C. Glorieux, Fluorescence spectra shape based dynamic thermometry, *Appl. Phys. Lett.*, 2014, **104**, 031902.

67. O. Yarimaga, S. Lee, D. Y. Ham, J. M. Choi, S. G. Kwon, M. Im, S. Kim, J.-M. Kim and Y.-K. Choi, Thermofluorescent conjugated polymer sensors for nano- and microscale temperature monitoring, *Macromol. Chem. Phys.*, 2011, **212**, 1211–1220.

68. F. Vetrone, R. Naccache, A. Zamarron, A. J. De La Fuente, F. Sanz-Rodriguez, L. M. Maestro, E. M. Rodriguez, D. Jaque and J. Garcia, Temperature sensing using fluorescent nanothermometers, *ACS Nano*, 2010, **4**, 3254–3258.

69. C. Paviolo, A. H. A. Clayton, S. L. McArthur and P. R. Stoddart, Temperature measurement in the microscopic regime: a comparison between fluorescence lifetime- and intensity-based methods, *J. Microsc.*, 2013, **250**, 179–188.

70. Q. Fu, G. V. R. Rao, L. K. Ista, Y. Wu, B. P. Andrzejewski, L. A. Sklar, T. L. Ward and G. P. Lopez, Control of molecular transport through stimuli-responsive ordered mesoporous materials, *Adv. Mater.*, 2003, **15**, 1262–1266.

71. A. Schlossbauer, S. Warncke, P. M. E. Gramlich, J. Kecht, A. Manetto, T. Carell and T. Bein, A Programmable DNA-Based Molecular Valve for Colloidal Mesoporous Silica, *Angew. Chem., Int. Ed.*, 2010, **49**, 4734–4737.

72. E. Aznar, L. Mondragon, J. V. Ros-Lis, F. Sancenon, M. D. Marcos, R. Martinez-Manez, J. Soto, E. Perez-Paya and P. Amoros, Finely tuned temperature-controlled cargo release using paraffin-capped mesoporous silica nanoparticles, *Angew. Chem., Int. Ed.*, 2011, **50**, 11172–11175.

73. P. Bon, N. Belaid, D. Lagrange, C. Bergaud, H. Rigneault, S. Monneret and G. Baffou, Three-dimensional temperature imaging around a gold microwire, *Appl. Phys. Lett.*, 2013, **102**, 244103.

74. G. Binnig, H. Rohrer, C. Gerber and E. Weibel, Tunneling through a controllable vacuum gap, *Appl. Phys. Lett.*, 1982, **40**, 178–180.

75. A. Majumdar, J. P. Carrejo and J. Lai, Thermal imaging using the atomic force microscope, *Appl. Phys. Lett.*, 1993, **62**, 2501–2503.

76. G. Mills, H. Zhou, A. Midha, L. Donaldson and J. M. R. Weaver, Scanning thermal microscopy using batch fabricated thermocouple probes, *Appl. Phys. Lett.*, 1998, **72**, 2900–2902.

77. A. Hammiche, M. Reading, H. M. Pollock, M. Song and D. J. Hourston, Localized thermal analysis using a miniaturized resistive probe, *Rev. Sci. Instrum.*, 1996, **67**, 4268–4274.

78. P. G. Royall, V. L. Kett, C. S. Andrews and D. Q. M. Craig, Identification of crystalline and amorphous regions in low molecular weight materials using microthermal analysis, *J. Phys. Chem. B*, 2001, **105**, 7021–7026.

79. S. Lefevre and S. Volz, 3ω-scanning thermal microscope, *Rev. Sci. Instrum.*, 2005, **76**, 033701.

80. J. R. Greenwood, P. R. N. Childs and P. Chaloner, Gold leads to PRTs for monitoring high temperatures, *Gold Bull.*, 1999, **32**(3), 85–89.

81. *US Pat.*, No. 6905736, Fabrication of nano-scale temperature sensors and heaters, 2005.

82. D. C. Chu, W. K. Wong, K. E. Goodson and R. F. W. Pease, Transient temperature measurements of resist heating using nanothermocouples, *J. Vac. Sci. Technol., B*, 2003, **21**, 2985–2989.

83. Y. H. Gao and Y. Bando, Carbon nanothermometer containing gallium, *Nature*, 2002, **415**, 599.

84. C. Y. Wang, N. W. Gong and L. J. Chen, High sensitivity solid state Pb(core)/ZnO(shell) nanothermometers fabricated with a facile galvanic displacement method, *Adv. Mater.*, 2008, **20**, 4789–4792.

85. N. W. Gong, M. Y. Lu, C. Y. Wang, Y. Chen and L. J. Chen, Au(Si)-filled Beta-Ga$_2$O$_3$nanotubes as wide range high temperature nanothermometers, *Appl. Phys. Lett.*, 2008, **92**, 073101–073103.

86. Y. B. Li, Y. Bando and D. Golberg, Single-crystalline In$_2$O$_3$ nanotubes filled with In, *Adv. Mater.*, 2003, **15**, 581–585.

87. P. D. Tovee, M. E. Pumarol, M. C. Rosamond, R. Jones, M. C. Petty, D. A. Zezeb and O. V. Kolosov, Nanoscale resolution scanning thermal microscopy using carbon nanotube tipped thermal probes, *Phys.Chem.Chem.Phys.*, 2014, **16**, 1174–1181.

88. B. Klinkert and F. Narberhaus, Microbial thermosensors, *Cell. Mol. Life Sci.*, 2009, **66**, 2661–2676.

89. D. Albanesi, M. Martin, F. Trajtenberg, M. C. Mansilla, A. Haouz, P. M. Alzari, D. de Mendoza and A. Buschiazzo, Structural plasticity and catalysis regulation of a thermosensor histidine kinase, *Proc. Natl. Acad. Sci. U. S. A.*, 2009, **106**, 16185–16190.

90. J. Kortmann, S. Sczodrok, J. Rinnenthal, H. Schwalbe and F. Narberhaus, Translation on demand by a simple RNA-based thermosensor, *Nucleic Acids Res.*, 2011, **39**, 2855–2868.

91. S. Chowdhury, C. Maris, F. H. T. Allain and F. Narberhaus, Molecular basis for temperature sensing by an RNA thermometer, *EMBO J.*, 2006, **25**, 2487–2497.

92. F. Narberhaus, T. Waldminighaus and S. Chowdhury, RNA thermometers, *FEMS Microbiol. Rev.*, 2006, **30**, 3–16.

93. T. Waldminghaus, J. Kortmann, S. Gesing and F. Narberhaus, Generation of synthetic RNA-based thermosensors, *Biol. Chem.*, 2008, **389**, 1319–1326.

94. J. Neupert, D. Karcher and R. Bock, Design of simple synthetic RNA thermometers for temperature-controlled gene expression in Escherichia coli, *Nucleic Acids Res.*, 2008, **36**, e124.

95. J. Neupert and R. Bock, Designing and using synthetic RNA thermometers for temperature-controlled gene expression in bacteria, *Nat. Protocol*, 2009, **4**, 1262–1273.

96. J. Millen, T. Deesuwan, P. Barker and J. Anders, Nanoscale temperature measurements using non-equilibrium Brownian dynamics of a levitated nanosphere, *Nat. Nanotechnol.*, 2014.

97. M. Asheghi and Y. Yang Micro- and nano-scale diagnostic techniques for thermometry and thermal imaging of microelectronic and data storage

devices, in *Microscale Diagnostic Techniques*, ed. K. S. Breuer, Springer-Verlag, 2005, pp. 155–196.

98. C. D. S. Brites, P. P. Lima, N. J. O. Silva, A. Millan, V. S. Amaral, F. Palacio and L. D. Carlos, Thermometry at the nanoscale using lanthanide-containing organic–inorganic hybrid materials, *J. Lumin.*, 2013a, **133**, 230–232.

99. C. D. S. Brites, P. P. Lima, N. J. O. Silva, A. Millan, V. S. Amaral, F. Palacio and L. D. Carlos, Organic-Inorganic Eu^{3+}/Tb^{3+} codoped hybrid films for temperature mapping in integrated circuits, *Front. Chem.*, 2013b, **1**, 9.

Minimal Length Scales for the Existence of Local Temperature

MICHAEL J. HARTMANN

Institute of Photonics and Quantum Sciences (IPaQS),
School of Engineering and Physical Sciences, Heriot-Watt University,
Edinburgh EH14 4AS, UK
Email: m.j.hartmann@hw.ac.uk

2.1　Introduction

Macroscopic systems in a thermal equilibrium state may, despite their very large number of degrees of freedom, be characterized by only a small set of quantities such as pressure, volume, particle number and temperature. In physics, one refers to this kind of description as a *thermodynamical* description. This description works very well for equilibrium states because, with increasing number of particles, an overwhelming majority of the system's microstates have the same macroscopic properties. *Microstate* here refers to a description that includes all degrees of freedom of the system under study. As a result of the above tendency, thermodynamical behaviour becomes 'typical'. In more technical terms this fact is called the existence of the *thermodynamic limit*, meaning that intensive quantities, such as the energy per particle, approach a limiting value that does no longer depend on the detailed configuration of the system as its size increases. As an example, the energy per particle of a very large piece of solid is independent of whether this piece is surrounded by air or water, provided it is in an equilibrium state, *i.e.*, has the same temperature as its surrounding.

RSC Nanoscience & Nanotechnology No. 38
Thermometry at the Nanoscale: Techniques and Selected Applications
Edited by Luís Dias Carlos and Fernando Palacio
© The Royal Society of Chemistry 2016
Published by the Royal Society of Chemistry, www.rsc.org

In the thermodynamic limit, differences between particles inside the solid and on the surface, where they interact with the surrounding, obviously become negligible. Hence the size of the solid is important. With increasing transverse dimensions of an object, the number of particles sitting on its surface quickly becomes vanishingly small compared to the number of particles in the bulk. For example, a sphere with radius r has a surface area $4\pi r^2$ and a volume $(4\pi/3)r^3$, so that for a uniform particle density, the ratio of particles sitting on the surface over the total number scales as $1/r$ and thus becomes negligible as r goes to infinity.

To analyze the existence of local temperatures, small parts of large systems are of interest. These parts do inevitably interact with their surrounding. Yet, provided interactions between the constituent particles are short range, only those particles in close vicinity to the boundary of the considered part interact with the environment. Thus, the described scaling properties immediately give rise to the following question: How large do parts of a system need to be to permit a local thermodynamical description, that is a thermodynamical description of the part alone?

For a long time, this fundamental question might have been of purely academic interest since thermodynamics was only used to describe macroscopic systems, where deviations form the thermodynamic limit may safely be neglected. However, with the advent of nanotechnology, the microscopic limit of the applicability of thermodynamics became relevant for the interpretation of experiments, and may in the near future even have technological importance.

In recent years, amazing progress in the synthesis and processing of materials with structures on nanometre length scales has been made.[1–4] Experimental techniques have improved to such an extent that the measurement of thermodynamic quantities like temperature with a spatial resolution on the nanoscale has become accessible.[5–10] To provide a better basis for the interpretation of present day and future experiments in nanoscale thermometry and for a better understanding of the limits of the validity of thermodynamics, it is thus of crucial importance to examine the applicability of thermodynamical concepts on small length scales, starting from the most fundamental theory at hand, *i.e.*, quantum mechanics. In this context, we here discuss the particularly important and interesting question: Can temperature be meaningfully defined on nanometre length scales?

The existence or non-existence of local temperature has practical implications. Local physical properties may show different behaviour depending on whether the local state is thermal or not. Moreover, there may be a limit for the spatial resolution on which a temperature profile can be meaningful.[11] However a spatially varying temperature calls for non-equilibrium—a complication which we will exclude here. The existence of thermodynamical quantities, *i.e.*, the existence of the thermodynamic limit strongly depends on the correlations between the considered parts of a system. As mentioned above, with increasing diameter, the volume of a region in space grows faster than its surface, and effective interactions between two regions, provided they are short ranged, become less relevant.[12,13] This scaling behaviour is used to show that the thermodynamic limit exists[14–16] since correlations between a region and its environment become negligible in the limit of infinite region size.

The scaling of interactions between parts of a system as compared to the energy contained in the parts themselves thus sets a minimal length scale on which correlations are still small enough to permit the definition of local temperatures. Here we review an approach to study this connection quantitatively.[17–21]

2.2 Definition of Temperature

Temperature is one of the central concepts of thermodynamics and statistical mechanics. Yet, it is not a direct observable and in quantum mechanics there is no operator associated to it. There are two standard ways of defining it.

2.2.1 Definition in Thermodynamics

Since thermodynamics is an empirical description for systems whose macroscopic physics can be sufficiently characterized by a few variables such as volume, energy and particle number, the definition of temperature within this framework is empirical as well. In thermodynamics, a system is said to be in equilibrium if the few variables describing it are stationary for given constraints. An important, special case of such constraints, is a bi-partite (or multipartite) system with a fixed total energy, where the parts may exchange energy among themselves. These parts are then said to be in *thermal equilibrium*. In thermodynamics, temperature is thus defined by the property:

> *Two systems that can exchange energy and are in thermal equilibrium, have the same temperature.*

To fix a temperature scale, one needs a reference system. A canonical choice for this reference system is the ideal gas, where temperature may be defined by

$$T = \frac{pV}{nk_{\mathrm{B}}} \tag{2.1}$$

here, p is the pressure, V the volume and n the number of particles of the gas. k_{B} is Boltzmann's constant. The above definition is unambiguous only if the states of thermal equilibrium form a one-dimensional manifold,[22] where a single parameter suffices to characterize them. This parameter can then be identified with the temperature T.

2.2.2   Definition in Statistical Mechanics

In statistical mechanics, temperature is defined *via* the derivative of the entropy S with respect to the internal energy $\bar{E}$,

$$\frac{1}{T} = \frac{\partial S}{\partial \bar{E}}. \tag{2.2}$$

Hence this temperature T exists as long as the entropy S is a function of the internal energy $\bar{E}$. In quantum mechanics the entropy can be defined as the von Neumann entropy

$$S = k_{\mathrm{B}} Tr(\rho \ln \rho), \tag{2.3}$$

and is a measure of the number of possible pure states the system could be in. The entropy as defined in eqn (2.3) always exists, but it shows its standard properties, such as extensivity, only in the thermodynamic limit.[12] In statistical mechanics, an equilibrium state is defined to be the state with the maximal entropy. For systems that have a fixed energy expectation value but can exchange energy with their surrounding, this equilibrium state is the so-called *canonical state*

$$\rho = \frac{\exp{(-\beta H)}}{Z} \tag{2.4}$$

with the partition sum $Z = Tr\{\exp(-\beta H)\}$. In quantum mechanics the internal energy is given by

$$\bar{E} = Tr(H\rho), \tag{2.5}$$

where H is the Hamiltonian of the considered system proper. This means, it does not contain any interactions of the system with its environment and the internal energy is therefore a property of the system itself, as it does neither depend on the state nor Hamiltonian of the environment. However, the notion of temperature, as defined in eqn (2.2), just like entropy, shows its characteristic thermodynamical properties (see above) only for equilibrium states.[12,22,23]

2.2.3   Local Temperature

Local temperature is the temperature of a part of a larger system. This part can therefore not be perfectly isolated but is able to exchange energy with its surrounding. As we here limit our consideration to cases without particle exchange, the following convention appears to be reasonable:

> *Local temperature exists if the considered part of the system is in a canonical state.*

Besides its roots in statistical mechanics there are important practical reasons for this definition. The canonical distribution (2.4) is an exponentially decaying function of energy that is characterized by one single parameter only. This parameter can be identified with temperature. These properties imply that there exists a one-to-one mapping between temperature and the expectation values of observables, which are usually employed to measure it. Whereas for distributions with several parameters, temperature measurements *via* different observables could yield different results, they are unambiguous for distributions characterized by a single parameter.

Let us emphasize that this is a basic property of systems that are amenable to a thermodynamic description. Provided it exists, temperature describes a system in a sufficiently complete way such that several properties of it can be predicted if one only knows its temperature.

Furthermore, the strong decay of the distribution with increasing energy is also crucial. Large systems typically have a modular structure so that their density of states strongly grows as a function of energy.[23] Hence the product of the density of states times a strongly (even exponentially) decaying distribution of occupation probabilities will result in a very pronounced narrow peak at the internal energy $\bar{E}$, see Figure 2.1. If on the other hand, the distribution does not decay fast enough, the product of the density of states times the distribution would not have this pronounced peak and physically relevant quantities like energy could not have 'sharp' values.

According to the above definition, temperature exists locally, *i.e.*, for a given part of the system, if the respective part is in a thermal equilibrium state. We thus define temperature to exist on a certain length scale, if all possible partitions of the corresponding size are simultaneously in an

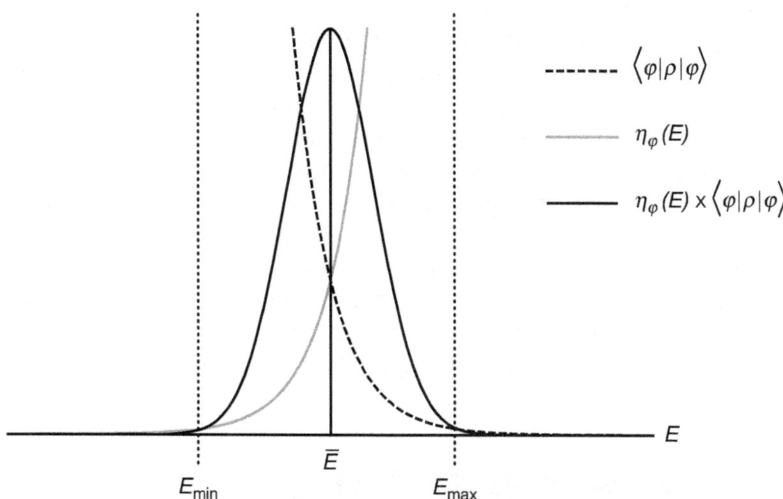

Figure 2.1 The product of the density of states $\eta_\varphi(E)$ and the occupation probabilities $\langle \varphi | \rho | \varphi \rangle$ form a strongly pronounced peak at the energy expectation value $\bar{E}$.

equilibrium state. This requirement for simultaneous local thermal states is motivated by demanding that it should not make a difference whether a temperature profile is scanned by one single thermometer, which is moved in small steps across the sample, or whether the profile is measured by several closely spaced thermometers simultaneously.

After discussing the criteria, when local temperature is defined to exist locally, we now turn to describe the approach for analyzing these criteria.

2.3   General Theory for the Existence of Local Temperature

For systems, which are globally in a non-equilibrium state it is very difficult to decide, under what conditions equilibrium states show up locally[24] and only very few exact results are known.[25] Yet, whenever thermal equilibrium exists locally, one would expect the macroscopic temperature gradient to be small. In a first approximation to this scenario, we derive here conditions for the local thermal state in systems which are in a global equilibrium state, *cf.* eqn (2.4). In such situations, one expects sub-units of the total system to be in an equilibrium state whenever their mutual interactions are weak enough and hence correlations between them are small so that the global thermal state approximately factorizes into a product of local thermal states.

To explore on which length scales local temperature can exist, we investigate how small the considered parts can be while still permitting a description with local thermal states. To this end we consider parts of different sizes and explore the scaling behaviour of boundary *versus* bulk energies, which ensures the existence of the thermodynamic limit (*cf.* Section 2.2).[15,16] Since modular structures are typical in the description of condensed matter and solid-state systems, we consider systems that are composed of elementary units with mutual short-range interactions. For simplicity we focus here on nearest-neighbour interactions. If we then form parts of n adjacent sub-systems, as represented in Figure 2.2, the energy of the part is n times the average energy per sub-system and thus grows approximately linearly with n. Since sub-systems only interact with their nearest neighbours, two adjacent parts interact *via* the two sub-systems at the respective boundaries, only. As a consequence, the coupling between two parts is independent of their sizes n and the interaction energy of this coupling becomes less relevant compared to the energy contained in the parts as n increases.

Figure 2.2   We form groups of n adjacent sub-systems.

2.3.1 The Model

We consider here homogeneous (*i.e.*, translation-invariant) systems with nearest-neighbour interactions, which we divide into identical parts. The Hamiltonian of the system can thus be written as

$$H = \sum_i (H_i + I_{i,i+1}), \tag{2.6}$$

where the index i labels the elementary sub-systems, H_i is the Hamiltonian of sub-system i, $I_{i,i+1}$ the interaction between sub-systems i and $i+1$, and periodic boundary conditions are assumed.

Now N_G groups of n sub-systems each (index $i \to (\mu-1)n+j$; $\mu=1,\ldots,N_G$; $j=1,\ldots,n$) are formed and the Hamiltonian is split into two parts,

$$H = H_0 + I, \tag{2.7}$$

where H_0 contains the Hamiltonians of the isolated groups,

$$H_0 = \sum_{\mu=1}^{N_G} \mathscr{H}_\mu \quad \text{with}$$
$$\mathscr{H}_\mu = \sum_{j=1}^{n} H_{n(\mu-1)+j} + \sum_{j=1}^{n} I_{n(\mu-1)+j,n(\mu-1)+j+1} \tag{2.8}$$

and I contains the interactions of each group with its neighbouring groups,

$$I = \sum_{\mu=1}^{N_G} I_{\mu n,\mu n+1}. \tag{2.9}$$

The eigenstates $|a\rangle$ of the Hamiltonian H_0, $H_0|a\rangle = E_a|a\rangle$ are direct products of the eigenstates of the group Hamiltonians,

$$|a\rangle = \prod_{\mu=1}^{N_G} |a_\mu\rangle \quad \text{with} \quad \mathscr{H}_\mu|a_\mu\rangle = E_\mu|a_\mu\rangle, \tag{2.10}$$

where E_μ is the energy of one sub-group and $E_a = \sum_{\mu=1}^{N_G} E_\mu$.

2.3.2 Thermal State in the Product Basis

To test whether a part H_{μ_0} is in a thermal state, we need to calculate its reduced density matrix and compare it to a local canonical state. For performing the necessary trace over all other parts, $\mu \neq \mu_0$, we need to represent the global equilibrium state in the basis formed by the product states, $|a\rangle$, see eqn (2.10). Denoting the eigenstates and energy eigenvalues of the global Hamiltonian by $|\varphi\rangle$, $|\psi\rangle$ and E_φ, E_ψ, the global equilibrium state ρ reads

$$\langle \varphi \mid \rho \mid \psi \rangle = \frac{e^{-\beta E_\varphi}}{Z} \delta_{\varphi,\psi} \tag{2.11}$$

in the global eigenbasis. In turn, the diagonal elements in the product basis are,

$$\langle a \mid \rho \mid a \rangle = \int_{E_0}^{E_1} \omega_a(E) \frac{e^{-\beta E}}{Z} \, \mathrm{d}E, \tag{2.12}$$

where sums over global eigenstates have been replaced by an integral over energy, and $\omega_a(E)$ is the probability of obtaining an energy value between E and $E + \Delta E$ in a measurement of the total energy for a system in the state $|a\rangle$,

$$w_a(E) = \frac{1}{\Delta E} \sum_{\{|\varphi\rangle \, : \, E \leq E_\varphi < E+\Delta E\}} | \langle a \mid \varphi \rangle |^2. \tag{2.13}$$

Here the sum runs over all states $|\varphi\rangle$ with energy eigenvalues E_φ in the respective energy range and ΔE is small. Moreover, E_0 is the energy of the ground state and E_1 the (potential) upper limit of the energy spectrum.

For computing the reduced density matrices of the groups and analyzing whether they are of canonical form, we thus need the explicit form of $\omega_a(E)$. Under quite general conditions, this can be shown to be a Gaussian[26,27] of the form,

$$\lim_{N_{\mathrm{G}} \to \infty} \omega_a(E) = \frac{1}{\sqrt{2\pi}\,\Delta_a} \exp\left(-\frac{(E - \bar{E}_a)^2}{2\Delta_a^2}\right), \tag{2.14}$$

where $\bar{E}_a$ is the expectation value of H in the state $|a\rangle$, and Δ_a^2 its variance, *i.e.*,

$$\begin{aligned} \bar{E}_a &= \langle a \mid H \mid a \rangle \quad \text{and} \\ \Delta_a^2 &= \langle a \mid H^2 \mid a \rangle - \langle a \mid H \mid a \rangle^2. \end{aligned} \tag{2.15}$$

Note that the limit of an infinite number of groups is taken here while the size of the individual groups remains finite.

The expectation value $\bar{E}_a$ of the entire Hamiltonian H in the state $|a\rangle$ is the sum of the energy eigenvalue of the isolated groups E_a and a term that contains the interactions, $\bar{E}_a = E_a + \varepsilon_a$. Therefore, the two quantities ε_a and Δ_a^2 can also be expressed in terms of the interaction only,

$$\begin{aligned} \varepsilon_a &= \langle a \mid I \mid a \rangle \quad \text{and} \\ \Delta_a^2 &= \langle a \mid I^2 \mid a \rangle - \langle a \mid I \mid a \rangle^2. \end{aligned} \tag{2.16}$$

Hence, ε_a is the expectation value and Δ_a^2 the squared width of the interactions in the state $|a\rangle$. Note that whereas ε_a has a classical counterpart, Δ_a^2 is purely quantum mechanical as it only appears if the commutator $[H, H_0]$ is non-zero and, as a consequence, the distribution $\omega_a(E)$ has non-zero width. Applying eqn (2.14) to calculate the integral in eqn (2.12) yields for $N_{\mathrm{G}} \gg 1$,

$$\begin{aligned} \langle a \mid \rho \mid a \rangle = \frac{1}{Z} \exp\left(-\beta(E_a + \varepsilon_a) + \frac{\beta^2 \Delta_a^2}{2}\right) \\ \times \frac{1}{2}\left[\mathrm{erfc}\left(\frac{E_0 - E_a - \varepsilon_a - \beta\Delta_a^2}{\sqrt{2}\Delta_a}\right) - \mathrm{erfc}\left(\frac{E_1 - E_a - \varepsilon_a + \beta\Delta_a^2}{\sqrt{2}\Delta_a}\right)\right], \end{aligned} \tag{2.17}$$

where erfc is the conjugate Gaussian error function,[28] and the second error function appears only if the energy has an upper bound. Note that the arguments of the error function grow as $\sqrt{N_G}$ or faster so that an asymptotic expansion[28] may be used for $N_G \gg 1$. Moreover, the off-diagonal elements $\langle a|\rho|b\rangle$ vanish for $|E_a + E_b| > \Delta_a + \Delta_b$ because the overlap of the two Gaussian distributions becomes negligible. In turn, for $|E_a + E_b| < \Delta_a + \Delta_b$, the transformation involves an integral over frequencies and thus these terms are significantly smaller than the entries on the diagonal.

2.3.3   Conditions for Local Thermal States

We now analyze under what conditions a product of canonical density matrices with temperature β_{loc} for each sub-group μ can approximate the density matrix ρ. Since the trace of a matrix is invariant under basis transformations, it suffices to verify the correct energy dependence of the product density matrix. Since we assume periodic boundary conditions, all reduced density matrices are equal and if they were canonical their product would be of the form $\langle a|\rho|a\rangle \propto \exp(-\beta_{\text{loc}}E_a)$. We thus have to verify whether the logarithm of the right-hand side of eqn (2.17) is a linear function of the energy E_a, that is, whether

$$\ln(\langle a|\rho|a\rangle) \approx -\beta_{\text{loc}}E_a + c, \tag{2.18}$$

where β_{loc} and c are constants. Applying the asymptotic expansion of the conjugate error function to eqn (2.17) shows that eqn (2.18) can only hold if

$$\frac{E_a + \varepsilon_a - E_0}{\sqrt{N_G}\Delta_a} > \beta \frac{\Delta_a^2}{\sqrt{N_G}\Delta_a} \quad \text{and} \tag{2.19}$$

$$-\varepsilon_a + \frac{\beta}{2}\Delta_a^2 \approx c_1 E_a + c_2 \tag{2.20}$$

where c_1 and c_2 are constants. These two conditions constitute the general result of our approach. Temperature becomes intensive, if the constant c_1 becomes negligible,

$$|c_1| \ll 1 \Rightarrow \beta_{\text{loc}} = \beta \tag{2.21}$$

Yet, even if this was not the case, temperature might still exist locally. For a local thermal description to be accurate, one should only require a canonical distribution of the diagonal elements (2.12) for an appropriate energy range, $E_{\text{min}} < E_a < E_{\text{max}}$. As discussed in the previous section, the density of states $\eta(E)$ is, for large modular systems, a fast-growing function of energy and its product with the exponentially decaying canonical distribution $\langle \varphi|\rho|\varphi\rangle$ forms a strongly pronounced peak at the expectation value of the global energy $\bar{E}$. Provided that the diagonal elements (2.12) are canonically distributed in an energy range, that is centred at this peak and large enough to entirely cover it, all observables with non-vanishing matrix

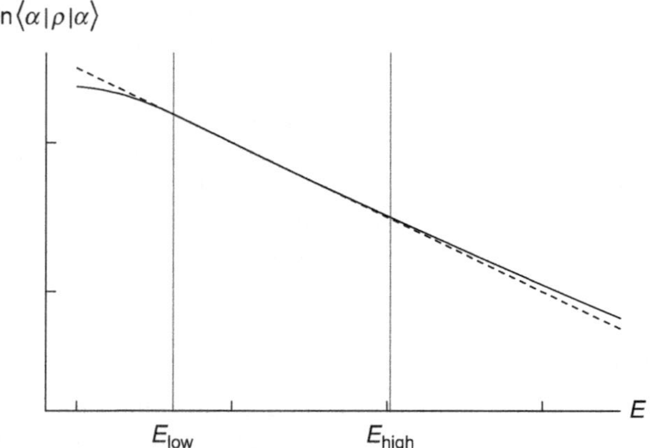

Figure 2.3 $\ln\langle a|\rho|a\rangle$ for ρ as in eqn (2.17) (solid line) and for a canonical density matrix (dashed line) for the case of a harmonic chain (figure adapted from ref. 20).

elements in that range show the same behaviour as for a canonical distribution on all energy scales. Observables which are not of that kind are in general not of interest.

A suitable and safe choice for the considered energy range $E_{\text{min}} < E_a < E_{\text{max}}$ should thus be

$$E_{\text{min}} = \max([E_a]_{\text{min}}, \alpha^{-1}\bar{E} + E_0)$$
$$E_{\text{max}} = \min([E_a]_{\text{max}}, \alpha\bar{E} + E_0), \tag{2.22}$$

where $\alpha \gg 1$ and $\bar{E}$ in general depends on the global temperature β. In eqn (2.22), $[E_x]_{\text{min}}$ and $[E_x]_{\text{max}}$ denote the minimal and maximal values E_x can take on.

Figure 2.3 shows the logarithm of eqn (2.17) and the logarithm of a canonical distribution with the same β for the example of a harmonic chain. The actual density matrix is more mixed than the canonical one. In the interval between the two vertical lines, both criteria (2.19) and (2.20) are satisfied. For $E < E_{\text{low}}$, criterion (2.19) is violated and criterion (2.20) for $E > E_{\text{high}}$. To allow for a description by means of local canonical density matrices, the group size needs to be chosen such that $E_{\text{low}} < E_{\text{min}}$ and $E_{\text{max}} < E_{\text{high}}$.

For a model of the class considered here, the two conditions (2.19) and (2.20) must both be satisfied. In the following section, we will apply them to a concrete model.

2.4   Harmonic Chain

We consider an harmonic chain of $N_{\text{G}} \times n$ particles of mass m and spring constant $\sqrt{m}\omega_0$, where the respective terms of the Hamiltonian (2.6) read

$$H_i = \frac{m}{2}p_i^2 + \frac{m}{2}\omega_0^2 q_i^2$$
$$I_{i,i+1} = -m\omega_0^2 q_i q_{i+1}. \tag{2.23}$$

Here p_i is the momentum of the particle at site i and q_i the displacement from its equilibrium position $i \times a_0$ with a_0 being the distance between neighbouring particles at equilibrium. We divide the chain into N_G groups of n particles each and thus get a partition of the type considered in Section 2.3.

The Hamiltonian of each group can be diagonalized by a Fourier transform and we get

$$E_a = \sum_{\mu=1}^{N_G} E_\mu \quad \text{with} \quad E_\mu = \sum_k \omega_k \left(n_k^a(\mu) + \frac{1}{2} \right), \tag{2.24}$$

where $k = \pi l/(a_0(n+1))$ for $l = 1,2,\ldots,n$ and the frequencies ω_k are given by $\omega_k^2 = 4\omega_0^2 \sin^2(ka_0/2)$. $n_k^a(\mu)$ is the occupation number of mode k of group μ in the state $|a\rangle$. We here choose units, where $\hbar = 1$. This model has the required properties so that the central limit theorem (2.14) is applicable.

Here, the expectation values of the group interactions, see eqn (2.16), vanish, $\varepsilon_a = 0$. The widths Δ_μ^2 in turn depend on the occupation numbers $n_k(\mu)$, see eqn (2.16), and therefore also on the energies E_μ. To analyze conditions (2.19) and (2.20) we make use of the continuum or Debye approximation,[29] which requires $n \gg 1$, $a_0 \ll l$ (where $na_0 = l$), and a finite length of the chain. The applicability of the Debye approximation will be corroborated by our results below, where the minimal group sizes $n_{\min}$ we find are larger than 10^3 for all temperatures.

In this approximation we have $\omega_k = vk$ with the constant velocity of sound $v = \omega_0 a_0$ and the energy width of the group–group interactions obey the relation

$$\Delta_\mu^2 = \frac{4}{n^2} E_\mu E_{\mu+1}, \tag{2.25}$$

where we have used $n + 1 \approx n$.

The relevant energy scale is introduced by the thermal energy expectation value of the entire chain

$$\bar{E} = E_0 + N_G n k_B \Theta \left(\frac{T}{\Theta} \right)^2 \int_0^{\Theta/T} \frac{x}{e^x - 1} \, dx, \tag{2.26}$$

and the ground-state energy E_0 is given by

$$E_0 = N_G n k_B \Theta \left(\frac{T}{\Theta} \right)^2 \int_0^{\Theta/T} \frac{x}{2} \, dx = \frac{N_G n k_B \Theta}{4}, \tag{2.27}$$

where Θ is the so-called *Debye temperature*,[29] which characterizes the considered material and can be found tabulated.

Inserting eqn (2.26) and (2.27) into eqn (2.19) and (2.20), and taking into account eqn (2.22), one can now calculate the minimal n for given α, Θ and T. For obtaining a specific value of $n_{\min}$, one needs to introduce another accuracy parameter δ that quantifies, for the right-hand side of eqn (2.20),

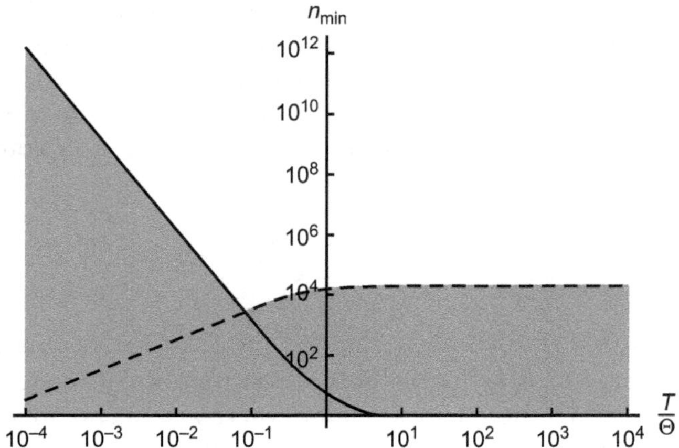

Figure 2.4 $n_{\min}$ as a function of T/Θ for a harmonic chain as determined by eqn (2.19) (solid line) and by eqn (2.20) (dashed line). $\alpha = 10$ and $\delta = 1/1000$. Local temperature does not exist in the shaded area.

how much smaller terms of quadratic and higher order in E_a are compared to the zero- and linear-order ones. δ is thus the ratio of the higher order terms to the (at most) linear ones.

Figure 2.4 shows $n_{\min}$ for $\alpha = 10$ and $\delta = 1/1000$ as a function of T/Θ. Moreover, for high (low) temperatures, $n_{\min}$ can be estimated by

$$n_{\min} \approx \begin{cases} \dfrac{2\alpha}{\delta} & \text{for} \quad T > \Theta \\ \dfrac{3\alpha}{2\pi^2} \dfrac{\Theta^3}{T^3} & \text{for} \quad T < \Theta \end{cases} \tag{2.28}$$

In addition, local temperatures are equal to the global one whenever they exist, $\beta_{\text{loc}} = \beta$, which implies that temperature is intensive [see eqn (2.21)]. In the following section the results obtained above will be applied to real materials.

2.5 Estimates for Real Materials

Thermal properties of insulating solids are successfully described by harmonic lattice models, where for example the temperature dependence of the specific heat is correctly predicted by the Debye theory.[29] Therefore one would expect the present approach to provide reasonable estimates for real materials. We thus take the results obtained in Section 2.4 for the harmonic chain and insert the corresponding parameters (in particular the Debye temperature) for the materials we apply it to. One can obtain a length scale by multiplying $n_{\min}$ with the corresponding lattice constant. Hence our approach predicts the minimal length scale $l_{\min}$ on which intensive temperatures exist in insulating solids

$$l_{\min} = n_{\min} a_0, \tag{2.29}$$

where a_0 is the lattice constant, *i.e.*, the distance between neighbouring atoms. Since $n_{\min}$ has been calculated for a one-dimensional model the results we obtain here should apply for one-dimensional or at least quasi-one-dimensional structures of the respective materials. We consider the following two examples.

Silicon, which is used in many branches of technology, has in its crystalline form a Debye temperature of $\Theta = 645$ K and a lattice constant of $a_0 = 2.4$ Å. Figure 2.5 shows the minimal length scale on which temperature can exist in a one-dimensional silicon wire as a function of global temperature. Here, the accuracy parameters α [see eqn (2.22)] and δ [see paragraph below eqn (2.27)] are chosen to be $\alpha = 10$ and $\delta = 1/1000$. Local temperature does not exist in the shaded area.

Recently, carbon has emerged as a very successful material in several branches of nanotechnology and nanofabrication.[30,31] In particular, carbon nanotubes feature in many applications. They have diameters of only a few nanometres, and measurements of their specific heat have shown that their thermal properties can be accurately modelled with one-dimensional harmonic chains. Our approach can thus be expected to provide accurate estimates for them. Carbon nanotubes have a Debye temperature of $\Theta = 1100$ K and a lattice constant of $a_0 = 1.4$ Å.

Figure 2.6 shows the minimal length scale on which temperature can exist in a carbon nanotube as a function of global temperature. It provides a good estimate of the maximal accuracy, with which temperature profiles in such tubes can be meaningfully discussed.[1] Again, the accuracy parameters α and δ are chosen to be $\alpha = 10$ and $\delta = 1/1000$. Local temperature does not exist in the shaded area.

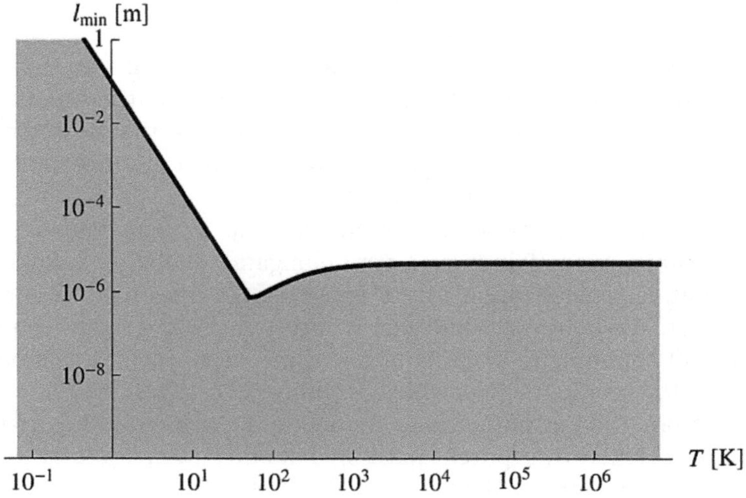

Figure 2.5　$l_{\min}$ as a function of T for a very elongated crystalline silicon wire. Local temperature does not exist in the shaded area.

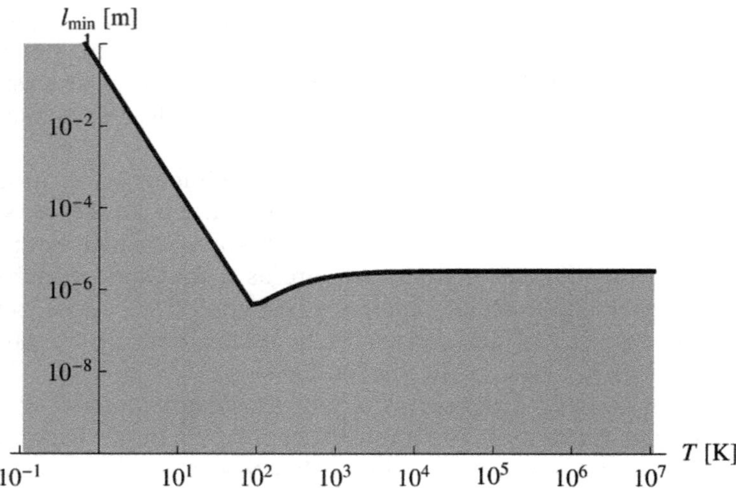

Figure 2.6 $l_{\min}$ as a function of T for a carbon nanotube. Local temperature does not exist in the shaded area.

Of course the validity of the harmonic lattice model will eventually break down at high but finite temperatures. The estimates drawn from the considered approach, in particular the results presented in Figures 2.5 and 2.6, will then no longer apply.

2.6   Discussion of the Resulting Length Scales

The length scales obtained here are, in particular for low temperatures, surprisingly large. As one might thus wonder whether the approach really captures the relevant physics, we finish our presentation with a discussion of some of its possible limitations.

Firstly, one may argue that taking the limit of an infinite number of groups, as required for the central limit theorem, will not correspond to the physically relevant situations. However, having in mind that we intended to analyze when a small part of a larger system can be in a thermal state, taking this limit should be a good approximation.

Secondly, we only considered one-dimensional models, whereas a real physical system, even if it is of a very elongated shape, is always three-dimensional. A generalization of the approach to those models is thus of high interest. We however would like to stress that the general conditions (2.19) and (2.20) apply to systems of any dimension, so that it is only the application to specific models, which requires generalization.

Furthermore, the harmonic chain model does not include any scattering of phonons. Hence this purely harmonic model does, for example, not predict any expansion or shrinking of the material caused by heating or cooling. It is therefore possible that it may fail to give reliable results for our present investigation, in particular at low temperatures.

Finally, one might speculate whether the length scales we find could significantly change if the assumption of a global equilibrium state was relaxed. Whereas this is certainly possible, one would expect our estimates to still apply as long as temperature gradients are small.

References

1. D. Cahill, W. Ford, K. Goodson, G. Mahan, A. Majumdar, H. Maris, R. Merlin and S. Phillpot, *J. Appl. Phys.*, 2003, **93**, 793.
2. C. C. Williams and H. K. Wickramasinghe, *Appl. Phys. Lett.*, 1986, **49**, 1587.
3. J. Varesi and A. Majumdar, *Appl. Phys. Lett.*, 1998, 72, 37.
4. J. K. Schwab, E. A. Henriksen, J. M. Worlock and M. L. Roukes, *Nature*, 2000, **404**, 974.
5. G. Kucsko, P. C. Maurer, N. Y. Yao, M. Kubo, H. J. Noh, P. K. Lo, H. Park and M. D. Lukin, *Nature*, 2013, **500**, 54.
6. P. Neumann, I. Jakobi, F. Dolde, C. Burk, R. Reuter, G. Waldherr, J. Honert, T. Wolf, A. Brunner, J. H. Shim, D. Suter, H. Sumiya, J. Isoya and J. Wrachtrup, *Nano Lett.*, 2013, **13**, 273.
7. X. Wang and Y. Yue, *Nano Rev.*, 2013, **3**, 11586.
8. Y. Gao and Y. Bando, *Nature*, 2002, **415**, 599.
9. H. Pothier, S. Gueron, N. O. Brige, D. Esteve and M. H. Devoret, *Phys. Rev. Lett.*, 1997, **79**, 3490.
10. J. Aumentado, J. Eom, V. Chandrasekhar, P. M. Baldo and L. E. Rehn, *Appl. Phys. Lett.*, 1999, 75, 3554.
11. P. Kim, L. Shi, A. Majumdar and P. L. McEuen, *Phys. Rev. Lett.*, 2001, **87**, 215502.
12. M. Schmidt, R. Kusche, B. von Issendorf and H. Haberland, *Nature*, 1998, **393**, 238.
13. M. Hartmann, J. Gemmer, G. Mahler and O. Hess, *Euro. Phys. Lett.*, 2004, **65**, 613.
14. M. E. Fisher, *Arch. Ratl. Mech. Anal.*, 1964, **17**, 377.
15. D. Ruelle, *Statistical Mechanics*, W.A. Benjamin Inc., New York, 1969.
16. J. L. Lebowitz and E. H. Lieb, *Phys. Rev. Lett.*, 1969, **22**, 631.
17. M. Hartmann, G. Mahler and O. Hess, *Phys. Rev. Lett.*, 2004, **93**, 080402.
18. M. Hartmann, G. Mahler and O. Hess, *Phys. Rev. E*, 2004, **70**, 066148.
19. M. Hartmann and G. Mahler, *Europhys. Lett.*, 2005, **70**, 579.
20. M. Hartmann, *On the Microscopic Limit for the Existence of Local Temperature*, PhD dissertation, Universität Stuttgart, 2005.
21. M. Hartmann, *Contemp. Phys.*, 2006, **47**, 89.
22. G. Adam and O. Hittmair, *Wärmetheorie*, Vieweg, Braunschweig, Wiesbaden, 4th Edn, 1992.
23. R. C. Tolman, *The Principles of Statistical Mechanics*, Oxford Univ. Press, London, 1967.
24. H. J. Kreuzer, *Nonequilibrium Thermodynamics and Its Statistical Foundation*, Clarandon Press, Oxford, 1981.

25. J. Meixner, *Ann. Phys.*, 1941, **39**, 333.
26. M. Hartmann, G. Mahler and O. Hess, *Lett. Math. Phys.*, 2004, **68**, 103.
27. M. Hartmann, G. Mahler and O. Hess, *J. Stat. Phys.*, 2005, **119**, 1139.
28. M. Abramowitz and I. Stegun, *Handbook of Mathematical Functions*, Dover Publ., New York, 9th edn, 1970.
29. C. Kittel, *Quantum Theory of Solids*, Wiley, New York, 1963.
30. M. S. Dresselhaus, G. Dresselhaus, and P. C. Eklund, *Science of Fullerenes and Carbon Nanotubes*, Academic Press, San Diego, 1996.
31. J. Hone, M. C. Llaguno, M. J. Biercuk, A. T. Johnson, B. Batlogg, Z. Benes and J. E. Fischer, *Appl. Phys. A: Mater. Sci. Process.*, 2002, **74**, 339.

Introduction to Heat Transfer at the Nanoscale

PIERRE-OLIVIER CHAPUIS

Centre d'Energétique et de Thermique de Lyon (CETHIL), CNRS-INSA Lyon-UCBL Campus La Doua-Lyon Tech., 69621 Villeurbanne, France
Email: olivier.chapuis@insa-lyon.fr

3.1   Introduction

Heat transfers from hot areas to cold ones, in matter or through whatever medium including vacuum. Heat conduction and radiative heat transfer are described by well-established theories at the macroscopic scale, known respectively as *Fourier's heat diffusion* and *incoherent thermal radiation* (the latter involves radiation in transparent and semitransparent media). Because the characteristic lengths associated with these two phenomena, such as the mean free path and the wavelength, lie in the micro to nanometre-scale ranges, the theories had to be revisited in these size regimes. The consequence is that the laws describing the heat transfer are different at these scales than at the macroscopic scale. Nanoscale samples or thermal sensors should be analyzed in light of this novel physics.

In this chapter, we will mostly deal with heat conduction but we will also briefly address thermal radiation. We will first recall the heat equation and the main variables that it involves, such as the thermal conductivity and the heat capacity. We will then analyze the imperfections of this well-known equation, due to stationary and transient effects. We will then see how to take these effects into account for the various types of materials that one can find: solid—crystalline, metallic ones, as well as amorphous

RSC Nanoscience & Nanotechnology No. 38
Thermometry at the Nanoscale: Techniques and Selected Applications
Edited by Luís Dias Carlos and Fernando Palacio
© The Royal Society of Chemistry 2016
Published by the Royal Society of Chemistry, www.rsc.org

ones—and polymers. Liquids are not tackled in this chapter, because the stratification or confinement effects they undergo may not appear at spatial scales much larger than 1 nm. We will briefly address thermal radiation, because additional dramatic phenomena such as near-field heat transfer come into play when thermometry is performed with near-field sensors. The last part of the chapter describes experiments that are used to measure nanoscale thermal properties or to observe the heat-transfer dynamics. We note that there are several reliable books that can introduce to the topic, such as ref. 1–4. First approaches are presented here, and the reader is referred to these advanced monographs or to references therein for more details.

3.2   Heat Conduction Equations

The main historical work for heat transfer in solids was done at the beginning of the 19th century by J. Fourier, in his treatise[5] *Théorie analytique de la chaleur* (*Analytical theory of heat*) published in 1822. Since then, we have known that the conducted heat flux follows a diffusion equation. As a consequence, the *density of heat flux* $\vec{q}$ (the modulus of which is a power per unit surface [W m^{-2}]) is proportional to the gradient of *temperature* [K]:

$$\vec{q} = -\kappa \vec{\nabla} T \tag{3.1}$$

where κ is a coefficient called the *thermal conductivity* [W m^{-1} K^{-1}] and $\vec{\nabla}$ represents the spatial gradient. This expression known as Fourier's law has been useful over the two last centuries and remains valid provided that the involved dimensions and time scales are not too small. It was realized in the 1950s by a few researchers such as Cattaneo[6] and Vernotte[7] that Fourier's law was suffering from some deficiencies. In this section, we will review these deficiencies and analyze how they can be avoided by modifying or replacing Fourier's law.

3.2.1   Establishing the Heat Equation

3.2.1.1   *Energy Conservation*

In the majority of everyday life applications, the heat conduction equation can be applied. This equation is demonstrated by considering the equation of conservation for energy

$$\frac{\partial u}{\partial t} = -\vec{\nabla} \cdot \vec{q} + \dot{Q} \tag{3.2}$$

where u is the local energy per unit volume [J m^{-3}], t the time [s] and $\dot{Q}$ the term associated with local energy generation per unit volume [W m^{-3}] that can be due to chemical reactions or to a non-local mode of energy deposition such as thermal radiation. Thermodynamics teaches us that

$\dfrac{\partial u}{\partial t} = \rho c_\text{v} \dfrac{\partial T}{\partial t}$, where ρ is the density [kg m^{-3}] and c_v the heat capacity at constant volume [J kg^{-1} K^{-1}]. As a consequence,

$$\rho c_\text{v} \frac{\partial T}{\partial t} - \vec{\nabla}(\kappa \vec{\nabla} T) = \dot{Q} \tag{3.3}$$

which is known as the *heat equation*. Note that it can be useful to consider the local variation of enthalpy instead of the one of local energy, and then one should use the heat capacity at constant pressure c_p instead of c_v. We do so in the following.

3.2.1.2 Thermal Conductivity

The thermal conductivity is particularly useful for describing the ability of a material to transfer heat in the stationary regime. In particular, the heat flux $-\kappa \vec{\nabla} T$ is constant in such case in the absence of additional heat sources and κ will then determine the shape of the temperature profile. Over small temperature ranges, eqn (3.3) reduces to

$$\rho c_\text{p} \frac{\partial T}{\partial t} - \kappa \Delta T = \dot{Q}, \tag{3.4}$$

where Δ represents the second-order spatial derivative known as the *Laplacian*. Figure 3.1 presents the thermal conductivity of various representative materials as a function of temperature: a metal (Al), two crystalline materials (Si and graphite), an amorphous material (glass), a polymer (polystyrene) and two liquids (water and ethanol). It can be observed that the thermal conductivity depends strongly on temperature in general and for large temperature differences eqn (3.4) cannot be used.

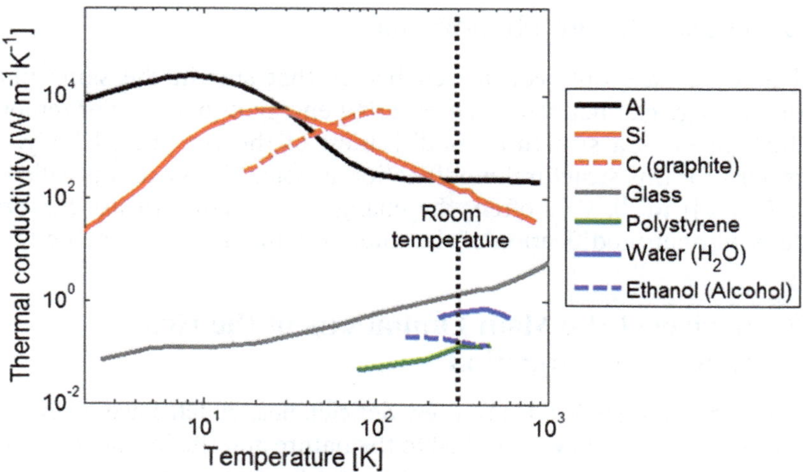

Figure 3.1 Thermal conductivity as a function of temperature. The data are taken from ref. 8–10.

3.2.1.3 *Heat Capacity, Diffusivity and Effusivity*

The heat capacity is required for the analysis of transient problems. It also depends on temperature, strongly at low temperature but weakly at room temperature (see ref. 11 for solids and ref. 12 for liquids). Bodies with large heat capacities are the ones which are able to store more energy. Very often, the heat capacity is combined with the thermal conductivity to define the thermal diffusivity a [m^2 s^{-1}] which is the parameter that appears naturally in eqn (3.4): $a = \kappa/\rho c_p$. The characteristic time τ of thermalization for a body of characteristic length L is given by $\tau = L^2/a$. In addition to the thermal diffusivity, it is useful to define the thermal effusivity $b = \sqrt{\kappa \rho c_p}$ [J s$^{-1/2}$ K^{-1} m^{-2}] which is the quantity that evaluates the ability of a body to exchange heat with another body by means of heat diffusion when they come into contact.

3.2.1.4 *Green's Function Solution to the Heat Equation*

In an infinite material initially at T_0, the increase of temperature at a given point M due to a modification of the temperature θ during an infinitesimal time δt at a point O of infinitesimal volume δV is

$$T(\vec{r}, t - t_0) = T_0 + \frac{\theta \delta V}{8[\pi a(t - t_0)]^{3/2}} e^{-\frac{|\vec{r} - \vec{r}_0|}{2\sqrt{a(t - t_0)}}} \tag{3.5}$$

where $|\vec{r} - \vec{r}_0|$ is the distance between O located at $\vec{r}_0$ and $M(\vec{r})$, and t the time after the temperature modification that happened at t_0. The response to the perturbation θ is sometimes called *Green's function*. If the perturbation is different or longer, one can make a convolution between the perturbation and Green's function to determine the temperature field.

3.2.2 Phase Change Phenomena

Until now we have only considered bodies that stay in the same phase. Another change can happen when bringing energy into a body or removing it, which involves a structural modification of the considered body. The energy cost for this structural modification is usually large and is called the *latent heat l* [J m^{-3}]. Very often, the change of structure (*phase transition*) happens without modification of the internal temperature of the body.

3.2.3 Review of the Main Limitations of the Heat Conduction Equation

The heat equation suffers from a few deficiencies, which have been recognized over the years. They are linked to the nature and the interactions of the heat carriers. Table 3.1 summarizes the main heat carriers in various states of matter, ranging from very dilute ones such as gases to dense ones (crystalline solids).

Table 3.1 Media and heat carriers.

Medium	Main heat carrier
Gas	Molecule
Solid	
Dielectric crystal (electrically insulating or semiconducting)	Phonon = collective vibration of the atoms
Amorphous dielectric	Localized vibration (extending over a few atoms)
Metal	Electron
Liquid	Molecule
Polymer	Extended vibration inside a molecular chain; almost localized vibration mode when involving at least two molecular chains

One can associate characteristic lengths with the heat carriers that depend on their nature.

3.2.3.1 Stationary Effects

3.2.3.1.1 Thermal Boundary Resistance. Let us consider the heat flow in a mono-dimensional configuration across a material of thermal conductivity κ_0 [see Figure 3.2(a)]. The material is sandwiched between two other materials. It appears that the total thermal resistance $R_{th,tot} = (T_A - T_B)/q$ between the materials at each side of interfaces A and B can be written as:

$$R_{th,tot} = R_{tb,A} + \frac{e}{\kappa_0 S} + R_{tb,B}, \tag{3.6}$$

where $R_{tb,i}$ is the thermal boundary resistance of interface i, e is the depth of the material under study and S a cross-section. Usually the resistance is considered to be $R_{th,tot} = \dfrac{e}{k_0 S}$ and the thermal boundary resistances are neglected, because e is so large that the main contribution to the overall resistance is the one related to the bulk of the material. However, when size e is decreased, the thermal boundary resistances will enter into play and can become the leading terms of eqn (3.6) [see Figure 3.2(b)]. As a consequence, it is important to determine these thermal boundary resistances for sub-micrometric scales.

3.2.3.1.2 Sub-diffusive Regime for the Heat Carriers. The energy transferred through heat diffusion is carried by heat carriers such as air molecules, electrons, phonons (quasi-particles associated with collective vibrations of the atoms in crystals), *etc.* In the diffusive regime, the heat carriers can be assimilated to particles that undergo many collisions in the whole volume, and one can define a *local thermal equilibrium* (LTE) everywhere: at each point of the volume, the distribution of energy of

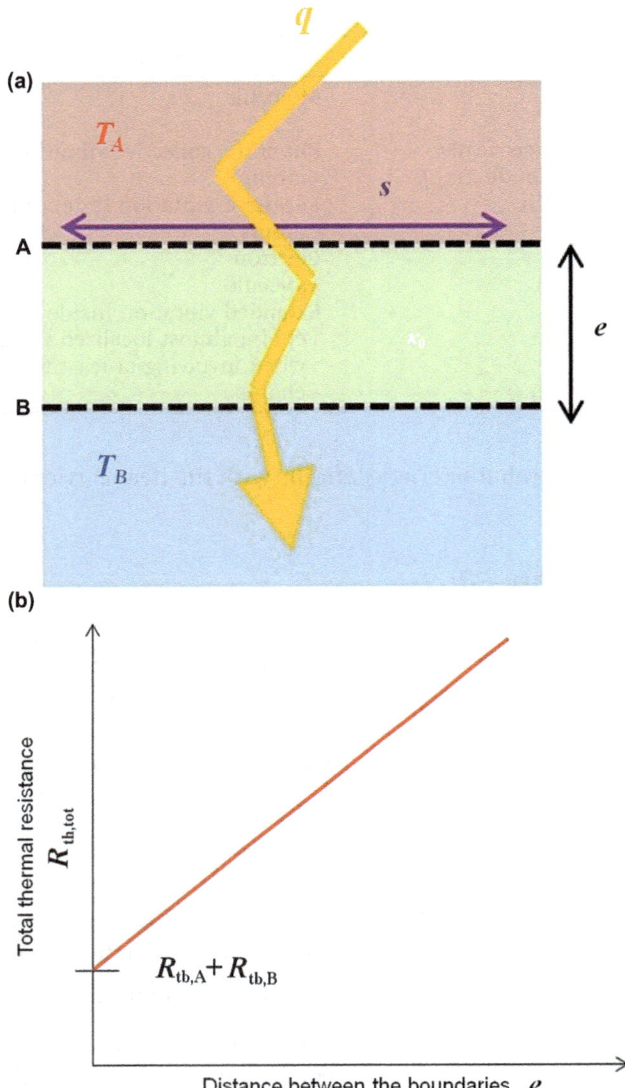

Figure 3.2 (a) Heat flow across a sandwiched material. (b) Total thermal resistance measured as a function of the thickness e.

the carriers is very close to an equilibrium distribution given by thermodynamics.[13] The mean free path of the heat carriers, here denoted Λ, is much smaller than the characteristic size of the medium, represented by e on Figure 3.3(a). If the mean free path is larger than e, no LTE can be defined. The heat carriers are no longer scattered in the volume of the material and will travel from one side of the material to the other one. The heat is then carried in a totally different way from one side to the other one. This regime is called *ballistic* and appears when the size e is small. $\text{Kn} = \Lambda/e$ is called the *Knudsen number* and allows us to determine if the transport regime is *diffusive* ($\text{Kn} \rightarrow 0$) or ballistic ($\text{Kn} \rightarrow \infty$).

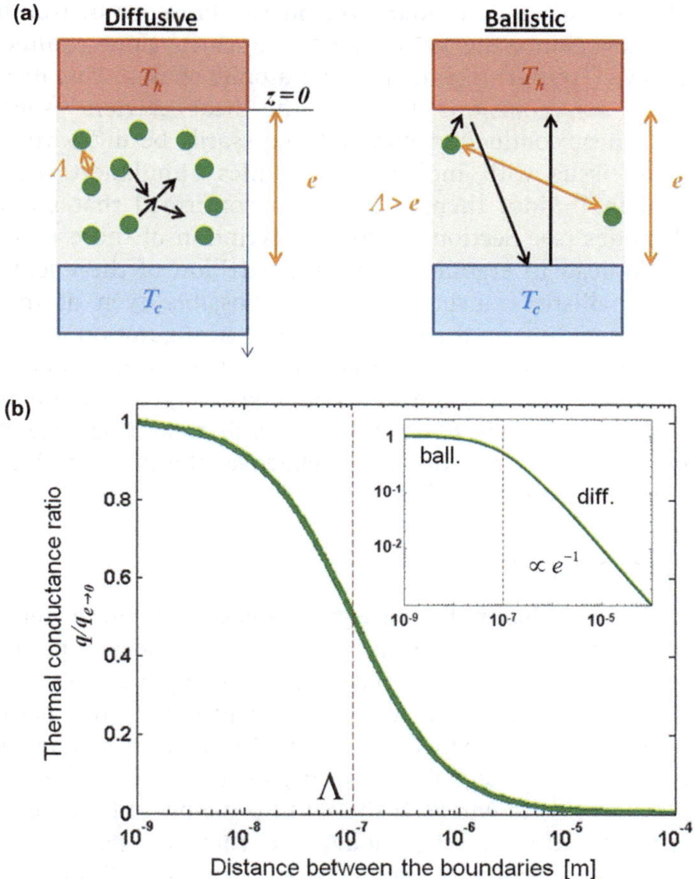

Figure 3.3 (a) Difference between the diffusive and the ballistic regimes of heat carriers' motion. (b) Thermal conductance in the ballistic and diffusive regimes in a mono-dimensional configuration. Inset: the same on a logarithmic scale.

In the ballistic regime the Fourier law is not valid anymore and more advanced strategies have to be considered, especially for complex geometries. They can be treated with the *Boltzmann transport equation* (BTE). In 1D, a simple analysis can be done as follows. We observe that Fourier's law diverges when $e \to 0$: $q = \kappa \dfrac{\Delta T}{e}$. Since the flux q cannot become infinite, there should be a spatial cut-off. It is provided by the mean free path, and one writes:

$$q = \kappa \frac{\Delta T}{e + C\Lambda} \tag{3.7}$$

where C is a constant. There is a maximal flux that can be transferred, which is the ballistic heat flux. As a consequence, the thermal conductance $G = q/\Delta T$ behaves in the same way, as shown in Figure 3.3(b).

3.2.3.1.3 Non-diffusive Behaviour Even in the Presence of Non-linearity.
It has been shown since the 1950s that non-linearity alone cannot guarantee the diffusive (Fourier) regime in linear atomic chains. This means that, even if there are collisions between the heat carriers [such as in Figure 3.3(a)], heat conduction may not necessarily be diffusive. The first to detect this issue with molecular dynamics simulation were Fermi, Pasta and Ulam.[14] Since then, it has been confirmed that scattering of vibrational modes (see Section 3.3 for a description of these modes) does not necessarily lead to ergodicity and equipartition of the energy.[15] As a consequence, ballistic transport would be possible even at the macroscopic scale, depending on the dimension and configuration.

It has been speculated in the past that the effective thermal conductivity of carbon nanotubes[16,17] and graphene[18] could be divergent *i.e.*, the measured conductivities would only be due to the contact thermal resistance. Recently, length-dependent (*i.e.*, non-diffusive) effective thermal conductivity of nanowires has been observed.[19,20]

3.2.3.2 Transient Effects

3.2.3.2.1 Infinite Velocity for Energy Transfer. When a temperature step is applied at a point O, Fourier's law tells us that any point M feels the increase of temperature. Indeed, eqn (3.5) implies that there is an increase of temperature in the whole space just after the perturbation appears, also for point M which is very far from O. Of course this is an approximation since no signal can propagate faster than the velocity of light $c \approx 3 \times 10^8$ m s^{-1}, so eqn (3.5) should in principle at least be restricted to $r \ll ct$: either the time is long enough (r/c) that heat has propagated to M, or the considered point M should be in a region delimited by ct. We will see in the following that one way to remedy this issue is to consider the nature of the heat carriers.

3.2.3.2.2 Non-equilibrium between Heat Carriers. Heat can be transported by more than one type of energy carrier. Typically electrons and phonons can be responsible for the heat flux in some semiconductors. However these two types of (quasi)particle may not be excited in the same way. As an example, one can mention that visible light may excite (heat up) the electrons without doing the same to the phonons. In this case two temperatures have to be defined: one for the electrons and one for the phonons (lattice). A coupling constant G_{ep} links these two temperatures,[21] as energy can be transferred from one heat reservoir (*e.g.*, the electron-based one) to the other one (the phonon-based one). This model is called the *two-temperature model*:[22]

$$\rho_e c_{p,e} \frac{\partial T_e}{\partial t} = \dot{Q}_e + \kappa_e \Delta T + G_{ep}(T_p - T_e) \tag{3.8a}$$

$$\rho_p c_{p,p} \frac{\partial T_p}{\partial t} = \dot{Q}_p + \kappa_p \Delta T + G_{ep}(T_e - T_p) \tag{3.8b}$$

where all the quantities are defined per type of particle (e for electrons and p for phonons). The usual Fourier law is not sufficient since it considers only one temperature, with all the particles being in local thermal equilibrium. We should notice that this model reduces to the classical one when the time is long enough, *i.e.*, when $t \gg \rho_i c_{\mathrm{p},i}/G_{\mathrm{ep}}$.

3.2.4 Non-local Heat Conduction Equations

3.2.4.1 Inertial Effects

There have been many attempts to establish a new heat equation, which does not suffer from the previously mentioned deficiencies. In particular, non-locality in time was first considered. The Cattaneo–Vernotte assumption [eqn (3.9a)][6,7] adds an inertial term in the Fourier law that removes the infinite-velocity paradox. With the energy conservation [eqn (3.2)], it leads to a hyperbolic equation:

$$\tau \frac{\partial \vec{q}}{\partial t} + \vec{q} = -\kappa \vec{\nabla} T \tag{3.9a}$$

$$\rho c_{\mathrm{p}} \left(\tau \frac{\partial^2 T}{\partial t^2} + \frac{\partial T}{\partial t} \right) = \kappa \Delta T \tag{3.9b}$$

Let us rewrite eqn (3.9a) in the Fourier space, where ω is the frequency:

$$\vec{q} = -\frac{\kappa}{1 + i\omega\tau} \vec{\nabla} T \tag{3.10}$$

We find that there is an apparent thermal conductivity, $\frac{\kappa}{1 + i\omega\tau}$, that will decay at high frequency.[23] Such an expression shows that a thermal excitation at high frequency will not be dissipated as easily as low-frequency ones.

Despite such interesting insight, it has been realized since the 1950s that eqn (3.9a) is also limited, and that it can lead to some unphysical results.[24,25] Some authors tried to develop other versions of the heat equation, on heuristic physical grounds. Tzou produced a body of work with the *dual-phase lag equation*,[26] which is based on inertia in time with two different time scales, one for the flux (linked to transport τ_q) and another one for local temperature (linked for example to a time τ_T required to equilibrate when there are two types of heat carrier):

$$\vec{q}(\vec{x}, t + \tau_q) = -\kappa \vec{\nabla} T(\vec{x}, t + \tau_T). \tag{3.11}$$

At first order in both τ_q and τ_T, one can derive a Jeffreys-like equation:

$$\tau_q \frac{\partial \vec{q}}{\partial t} + \vec{q} = -\kappa \left(\vec{\nabla} T + \tau_T \frac{\partial \vec{\nabla} T}{\partial t} \right) \tag{3.12a}$$

$$\rho c_{\mathrm{p}} \left(\tau_q \frac{\partial^2 T}{\partial t^2} + \frac{\partial T}{\partial t} \right) = \kappa \left(\Delta T + \tau_T \frac{\partial \Delta T}{\partial t} \right) \tag{3.12b}$$

It is interesting to note that eqn (3.12b) is similar to the equation that was studied by Guyer and Krumhansl[27] in the 1960s, but derived from another modified temperature–flux relationship. One can also expand eqn (3.11) at the second order, *etc.* However, all these equations suffer from the fact that: (1) inertia is limited to one or two relaxation times; and (2) no spatial dispersion is included. Point (1) can be removed when writing the Fourier law in the frequency space:

$$\vec{q}(\vec{x}, \omega) = -\kappa(\vec{x}, \omega)\vec{\nabla}T(\vec{x}, \omega),$$

$$\vec{q}(\vec{x}, t) = -\int_{-\infty}^{t} \kappa(\vec{x}, t' - t)\vec{\nabla}T(\vec{x}, t')\mathrm{d}t' \tag{3.13}$$

which is equivalent to considering that there is a *thermal conductivity kernel* that can involve more complicated dynamics in the flux–temperature relationship. Eqn (3.10)–(3.12) are particular cases of eqn (3.13), with especially simple kernels. Joseph and Preziosi[28] have listed such types of equations in the past.

3.2.4.2   *Effect of the Nanoscale Dimension: Spatial Dispersion*

Since at the nanoscale we are mostly dealing with spatial effects, it is important to note that all these equations reduce to the Fourier law in the stationary regime. However, it was shown that the nanoscale can also have a strong effect in this regime. As a consequence, spatial non-locality should also be considered. The most general case is therefore to write Fourier's law in the Fourier spatial space and in the frequency domain:

$$\vec{q}(\vec{k}, \omega) = -\kappa(\vec{k}, \omega)\vec{\nabla}T(\vec{k}, \omega), \tag{3.14}$$

where $\vec{k}$ is the wave vector (see Section 3.3.1.4). To find the kernel $\kappa(\vec{k}, \omega)$, one should rely on physical arguments. This can be done by both accounting for the spatial dimensions of the samples and by analyzing carefully the behaviour of the energy carriers, with the help of the BTE.[29] At this stage, it is easiest to change scale and directly compute the evolution of the heat carriers instead of trying to analyze the variation of ensembles, which may not distinguish between those that are impacted by the nanoscale and those that are not.

3.3   Beyond the Heat Equation

3.3.1   Characteristic Scales

3.3.1.1   *Wavelength*

Since a crystalline solid is ordered with a periodic spatial arrangement, it is natural to introduce periodic lengths, *i.e.*, wavelengths (λ). For an electron

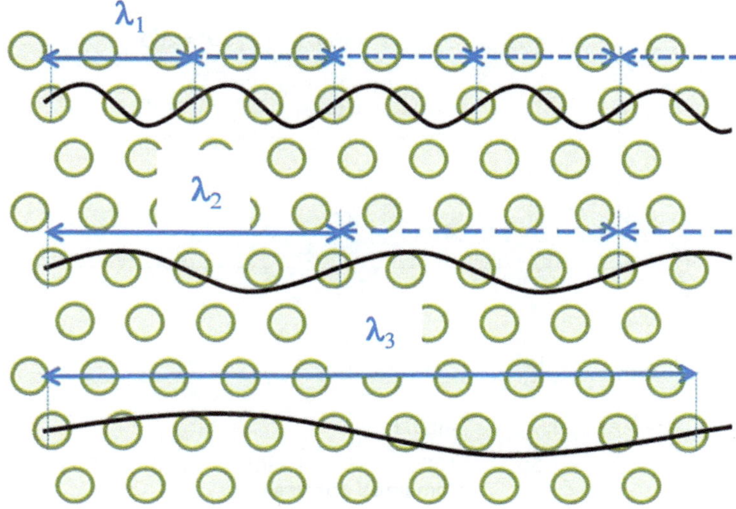

Figure 3.4 Crystalline solid and waves linked to the periodicity of the arrangement of the atoms.

that experiences a periodic electric potential due to the positively charged atomic lattice in a metal or a semiconductor, it is also natural to consider a wavelength. More generally, ordered materials can be described with wavelengths (see Figure 3.4).

3.3.1.2 Coherence Length

A key concept linked to waves is the coherence length (ℓ_c): it tells how long the extension of the wave is. Indeed, a wave can interact so that its amplitude will progressively decay due to exchange of energy, *e.g.*, with the medium, with another wave, *etc.* It is of critical importance to know how far the amplitude will stray from its initial value. Figure 3.5 explains what ℓ_c is: the extension of the wave packet is on the order of ℓ_c.

3.3.1.3 Mean Free Path

In a gas, molecules fly until they collide with another molecule. The average distance between two collisions is called the *mean free path* (Λ). In solids, one can consider gases of heat carriers that interact with each other or with other collision partners. Electrons may collide with another electron or can be scattered by impurities that modify the electrical potential they feel. The average distance between two modifications of the path of the heat carrier can be defined as the mean free path by analogy. For a travelling wave, the mean free path is the distance after which the amplitude will have vanished (see Figure 3.5).

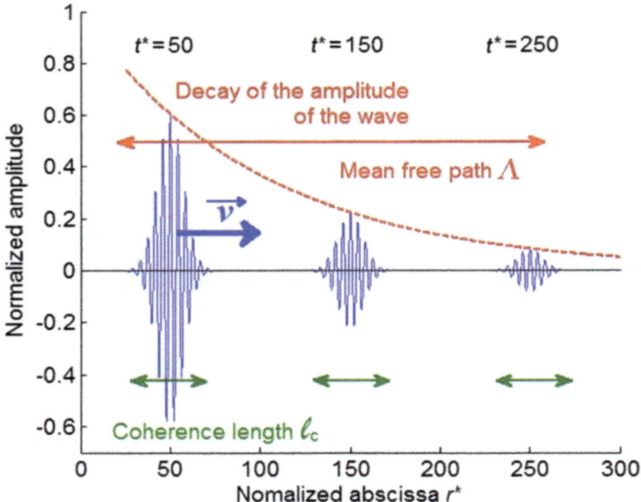

Figure 3.5 A wave packet propagating in a medium for a few normalized times t^*: graphical representation of the coherence length and the mean free path.

3.3.1.4   *Vibration Frequency*

While in a monoatomic gas the velocity of the atoms (*i.e.*, their energy) is the key ingredient, atoms in solids are mainly described by the frequency of oscillation around their equilibrium position. In crystalline solids, the atomic oscillations can be collective due to the ordering of the atoms. As a consequence, waves are possible and the propagation of the oscillations is possible. The wavelength (λ) already mentioned is linked to this oscillation frequency (ω) by the wave speed v: $\omega = 2\pi v/\lambda$. Plane waves are also defined by their wave vector of modulus $k = 2\pi/\lambda$ (see Section 3.2.4.2). In disordered solids such as amorphous materials (glass, *etc.*), the absence of order does not allow us to define proper waves (*i.e.*, with wave vectors and wavelengths) at the atomic scale. However, the atoms still oscillate and the frequency then becomes the essential parameter in the usual absence of detailed knowledge of the real structure of the material.

3.3.2   Heat Carriers

Heat carriers in ordered solids such as electrical insulators and semi-conductors are mostly phonons, which are modes of collective vibrations of atoms. In electrically conducting materials (metals), the heat is carried by electrons, which also possess an electrical charge.

3.3.2.1   *Phonons*

As is well known, the motion of atoms in an ordered solid can be decomposed in a sum of collective vibration modes of the total solid. These

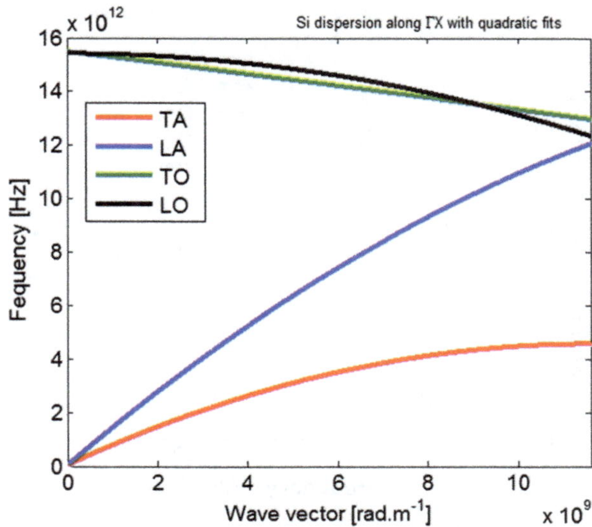

Figure 3.6　Bulk phonon dispersion curves in the ΓX direction (one of the principal directions of the crystal) fitted from experimental data.

modes are called *phonons* and are acoustic plane waves. The phonon modes are described by their dispersion relation such as the one represented in Figure 3.6, which connects their momentum $p = \hbar k$, where $\hbar = h/2\pi$ is the reduced Planck constant and k is the wave vector, to their energy $E = \hbar\omega$: only some of the pairs (ω, k) are allowed. When the atomic oscillation is parallel to the direction of propagation of the wave, the phonon is termed *longitudinal*; and when the oscillation is perpendicular, the phonon is *transverse*. When there is more than one atom that can be distinguished in the primitive cell (minimal volume of the lattice that can produce the whole solid by being reproduced and translated), *optical* branches can appear in the spectrum. These are the ones at higher frequencies. The *acoustic* phonons are those for which $\omega \to 0$ when $k \to 0$. In the example of silicon shown in Figure 3.6, there are two transverse acoustic (TA) branches which are degenerate (same ω and same k) and one longitudinal acoustic (LA) branch. In addition, there are two transverse optical (TO) branches and one longitudinal optical (LO) branch.

The heat capacity and the thermal conductivity due to phonons can be calculated once the dispersion curves are known. We recall how to do so in bulk. The heat capacity is calculated as follows if the medium is large enough to be considered as continuous:

$$c_{\mathrm{p}} = \sum_{\substack{\text{all branches}}} \iiint_{\substack{\theta \in [0,\pi] \\ \varphi \in [0,2\pi]}} \hbar\omega \frac{\mathrm{d}n}{\mathrm{d}T} \frac{\mathrm{d}k \, k\mathrm{d}\theta \, k\sin\theta\mathrm{d}\varphi}{(2\pi)^3} \tag{3.15}$$

where $n(\omega, T)\hbar\omega$ is the mean energy of a Planck oscillator, and $n(\omega, T) = 1/(e^{\frac{\hbar\omega}{k_{\mathrm{B}}T}} - 1)$ is the Bose–Einstein distribution. Assuming that the solid is isotropic, the integration over the angles can be performed

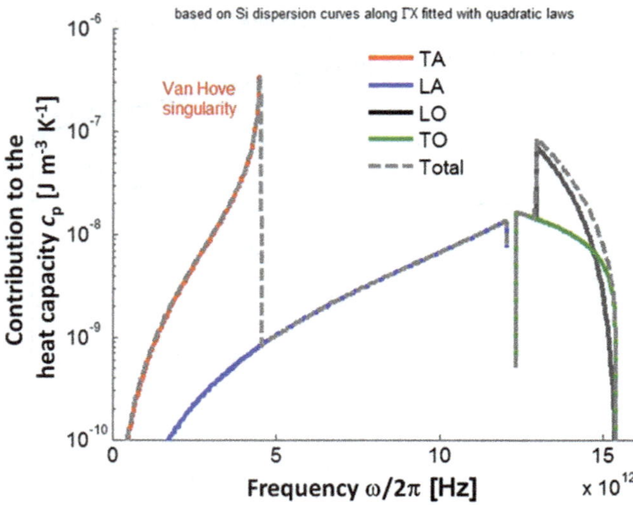

Figure 3.7   Spectral contribution to the heat capacity in the isotropic model. One can notice strong resonances at Van Hove singularities.

immediately. One can then introduce the density of states $g(\omega)$ per unit volume so that $g(\omega)\mathrm{d}\omega = 4\pi k^2 \mathrm{d}k/(2\pi)^3$. The heat capacity is then written as a function the oscillation frequency: $c_p = \displaystyle\sum_{\text{all branches}} \int \hbar\omega \frac{\mathrm{d}n}{\mathrm{d}T} g(\omega)\mathrm{d}\omega$. Figure 3.7 shows the contribution of a given frequency to the heat capacity as a function of the circular frequency at $T = 300$ K, for the approximation of quadratic fits to the dispersion curves. It can be observed that some modes contribute strongly, especially those which have a null group velocity $v_g = \mathrm{d}\omega/\mathrm{d}k$: since in the isotropic approximation $g(\omega) = \dfrac{\omega^2}{2\pi^2 v_\phi} \cdot \dfrac{1}{v_g}$, this model shows a divergence (Van Hove singularity) at given frequencies.

It is intriguing to know which phonon modes contribute to the heat capacity. Figure 3.8 indicates the percentage of the contribution of a specific branch to the mono-chromatic heat capacity at $T = 300$ K. In addition, the total contribution to the heat capacity of a branch is indicated. One can see that at ambient temperature the energy will be shared between both acoustic and optical modes, with TA modes being allowed to store close to 40% of the energy. TO modes may also store double the energy of LO modes. As a consequence, one can notice that more than two thirds of the heat capacity is due to transverse modes.

The thermal conductivity κ from the phonon contribution in the isotropic approximation can be deduced from the BTE for the heat carriers. The Boltzmann equation is a conservation equation for the number of phonons in the medium:

$$\left(\frac{\partial n_p}{\partial t}\right) + \vec{v}_{g,p} \cdot \vec{\nabla}_{\vec{r}} n_p = \frac{n_p(\vec{r}, \vec{k}) - n^0(\vec{r}, \vec{k})}{\tau_p(\vec{r}, \vec{k})} \tag{3.16}$$

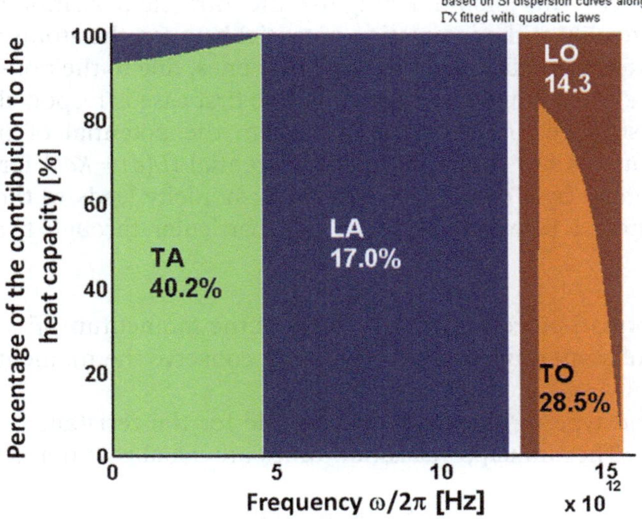

Figure 3.8 Spectral location of the contribution to the heat capacity in the quadratic approximation; indication of the total contribution of the branch to the total heat capacity.

where p stands for the polarization, and the right-hand side of the equation is the collision term for the phonons. Here this term is expressed under the relaxation time approximation, which is always assumed for phonon transport. τ is the *relaxation time*, which is the time during which the phonon can vibrate. After τ, some interaction will happen, leading to a new state for the phonon. Note that $v_{g,p}\tau$ is the mean free path (Λ) of the phonon. In the stationary regime, one can deduce the thermal conductivity by writing that

$$n_{\mathrm{p}}(\vec{r},\vec{k}) = n^0(\vec{r},\vec{k}) - \tau_{\mathrm{p}}(\vec{r},\vec{k})\vec{v}_{g,\mathrm{p}} \cdot \vec{\nabla}_{\vec{r}} n_{\mathrm{p}} \tag{3.17}$$

In the LTE hypothesis, one can write that $\vec{\nabla}_{\vec{r}} n_{\mathrm{p}} = \dfrac{\mathrm{d}n}{\mathrm{d}T}\vec{\nabla}_{\vec{r}} T$. Assuming without loss of generality that $\vec{\nabla}_{\vec{r}} T = \dfrac{\mathrm{d}T}{\mathrm{d}z}\vec{e}_z$ and that θ is the angle with the z-axis, we find finally that

$$\kappa = \sum_{\substack{\text{all branches}}} \iiint_{\substack{\theta \in [0,\pi] \\ \varphi \in [0,2\pi]}} \hbar\omega v_g^2(\omega)\tau(\omega)\cos^2\theta \frac{\mathrm{d}n}{\mathrm{d}T}\frac{\mathrm{d}k\,k\mathrm{d}\theta\,k\sin\theta\mathrm{d}\varphi}{(2\pi)^3} \tag{3.18}$$

Here we have left the index p out for simplicity. In the isotropic approximation, we can perform the integration over the angles and shift to the frequency dependent expression which can be written as follows:

$$\kappa = \frac{1}{3}\sum_{\substack{\text{all branches}}} \int \hbar\omega \frac{\mathrm{d}n}{\mathrm{d}T}g(\omega)v_g^2(\omega)\tau(\omega)\mathrm{d}\omega \tag{3.19}$$

They key ingredients to the calculation of the thermal conductivity are the relaxation times $\tau(\omega)$. There has been a body of work since the 1950s to determine them. One should distinguish the intrinsic relaxation times, due to the non-ideality of the interaction potential between the atoms that exists even for perfect structures, and the extrinsic ones, due to the non-ideality of the structure of the considered material. The first case is responsible for the intrinsic resistance to the flux propagation: the potential of interaction between atoms is not a pure harmonic potential $(U(d) = Kd^2)$ for a pair of atoms separated by a distance d. The anharmonicity leads to the fact that phonons interact between themselves, in particular through two types of processes:

(1) N (Normal) processes, which conserve the momentum $(\vec{p} = \hbar\vec{k})$; and
(2) U (Umklapp) processes, which do not conserve the momentum.

The second type of process is responsible for the resistance to the flux propagation. The Umklapp relaxation times are usually written as:

$$\frac{1}{\tau} = A\omega^n T^m e^{C/T} \tag{3.20}$$

where A and C are real constants and n and m are integers. At low temperatures, one can very often take $C = 0$, and $n + m = 5$ (ref. 30–32). At medium temperatures, often $n + m = 3$. The reader is referred to classical textbooks[33] or key articles[30–34] for a review of the expressions of the intrinsic relaxation times as a function of frequency polarization at a given temperature.

The extrinsic scattering events are due to defects in the structure: interstitial atoms (such as dopants), isotopes with different mass, grain boundaries or crystal dislocations. The relaxation times are written as a function of the frequency:

$$\frac{1}{\tau} = A\omega^q \tag{3.21}$$

where $q = 4$ for some of the processes such as those involving isotopes or impurities. In some cases, the inverse relaxation times can be added in a so-called *Mathiessen rule*, which can be justified in the case of independent scattering processes. Analytical expressions like the Callaway model[34] or the Holland model[31] allow reproduction of the trend of the thermal conductivity over a large temperature range. More advanced models have also been developed since the 1960s.[33] Figure 3.9 shows that the impact of the impurities is significant at a few tens of K. Around room temperature, the thermal conductivity decreases with temperature as $k \propto T^{-m}$, with $m \sim 1$ if the dominant scattering mechanism is phonon–phonon interaction.

At low temperature, the mean free paths $\Lambda = v_g\tau$ are so large that they exceed the size of the medium. The mean free path can then be set to the size D of the sample. Since the group velocity is essentially equal to a constant at low frequencies, the thermal conductivity is proportional to the heat capacity

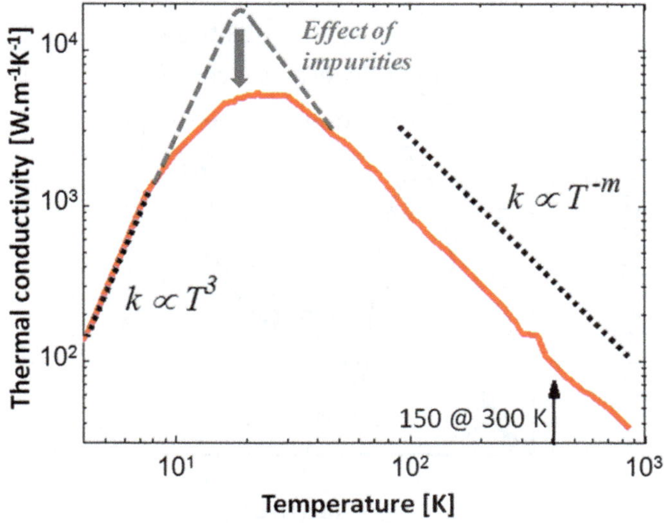

Figure 3.9 Thermal conductivity of silicon and regions of a predominant scattering mechanism.

and shows a typical $k \propto T^3$ behaviour (see Figure 3.9), as does the heat capacity in the Debye model.[11]

We note that optical modes do not carry much energy because of their low group velocities $v_g = \dfrac{\mathrm{d}\omega}{\mathrm{d}k}$.

3.3.2.2 Other Vibrational Modes in Solids

In amorphous materials there is no long-range order for the locations of the atoms. As a consequence, collective modes of vibration of the ordered atoms cannot be defined over large distances, especially at high frequencies (ω). This means that the concept of a phonon is ill-defined. However, it has been shown with molecular dynamics that heat can be transferred through various types of vibration modes, some of them being long-range even though they do not show a dispersion relation like waves. Allen and colleagues[35] have developed a terminology to split the contributions of various types of vibration modes. In particular, they term *extendons* the vibration modes that have some long-range interaction (see Figure 3.10). In this category they rank *propagons*, which are close to phonons and have a wave behaviour (lowest frequencies), and *diffusons*, which are long-range order modes that do not show a wave behaviour. At the highest frequencies, localized vibration modes are called *locons*.

To determine the thermal conductivity of amorphous materials, one writes:

$$k = \sum_i c_{\mathrm{p}_i} D_i \qquad (3.22)$$

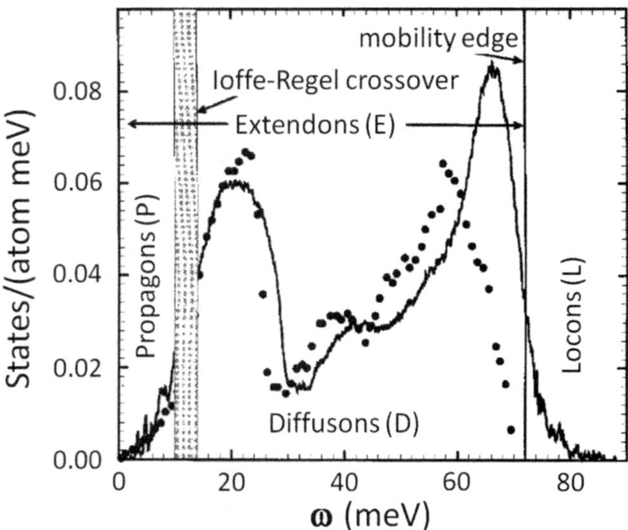

Figure 3.10 Density of states in amorphous silicon and associated vibration modes. Reproduced with permission from ref. 35. © Taylor & Francis 1999.

where the discrete sum is over all modes of vibrations denoted by i, $c_{p_i} = \hbar\omega_i \dfrac{\partial}{\partial T}(n(\omega_i, T_i))$ with $n(\omega_i, T_i) = 1/(e^{\hbar\omega_i/k_B T_i} - 1)$ being the heat capacity of the mode i and D_i a diffusion coefficient associated with the mode. Note that it is straightforward to transform the discrete sum into an integral. One can also notice that the expression for crystals [eqn (3.22)] can be retrieved with $D_i = \dfrac{1}{3}v_i\Lambda_i$, which leads to the well-known expression of the thermal conductivity similar to the kinetic expression in the isotropic approximation: $k = \dfrac{1}{3}\sum_i c_{p_i} v_i \Lambda_i$. In amorphous materials, since what would be the mean free path can be smaller than half the wavelength (to the right of the Ioffe–Regel limit on Figure 3.10), another framework has been developed and one has to trace back to the matrix element describing the interaction between the vibration modes to calculate the *mode diffusivity*:

$$D_i = \sum_{j \neq i} \frac{\pi V}{3(\hbar\omega_i)^2} |S_{ij}|^2 \delta(\omega_i - \omega_j) \tag{3.23}$$

where $S_{ij} = \langle i|S|j\rangle$ is the intermode matrix element of the heat flow operator. A temperature-independent mode diffusivity proposed by various authors (cited by ref. 35) is on the order of $D_i = a\omega_D^2$, where a is the interatomic distance and ω_D the Debye circular frequency. As shown by Figure 3.1, the temperature dependence of the thermal conductivity of amorphous solids (in this case, glass) is very different to that of crystalline solids.[36,37]

3.3.2.3   Heat Transfer in Polymers

Polymers are made of long molecular chains that can be intertwined or not (see Figure 3.11). If they are not, one speaks about *crystalline polymers*. Polymer thermal conductivity is usually in the range 0.1–0.5 W m^{-1} K^{-1} close to ambient temperature, so polymers are usually used as insulators. The very peculiar nature of polymers allows their thermal conductivity to be modified when altering their structure. It has been shown that when stretched, polymers detangle. The matter may then appear in an aligned way. Figure 3.12(a) presents a recent result for a polyethylene fibre. Its equivalent thermal conductivity along the chain increases strongly with its draw ratio. It is also interesting to observe the evolution of the thermal conductivity in the perpendicular direction, as shown in Figure 3.12(b). It appears that the thermal properties become strongly anisotropic, in contrast to the initial case the anisotropy ratio may reach 20, which is much more than for usual anisotropic solids.

Note that polymer organization is an interesting topic which is currently under investigation, especially since the boom in self-assembled molecules (SAMs) (see Figure 3.11).[40]

It is interesting to note that the glass-transition temperature is modified when the size of a polymer film is decreased to the nanoscale.[41] Despite the fact that the size $\mathcal{R}$ becomes, in principle, comparable to the one between two solids sandwiching the polymer [e in Figure 3.11(b)], the impact of percolation on the thermal conductivity is not clear yet.[42] Intrachain heat conduction could be responsible for the majority of the heat transfer, since heat conduction along chains is more favourable than interchain conduction, but experiments have not confirmed this picture.

3.3.2.4   Electrons

Electrons are the main hear carriers in metals, while their contribution is usually modest in semiconductors. Since it is difficult to split the contributions of the electrons and the phonons when measuring the heat transfer,

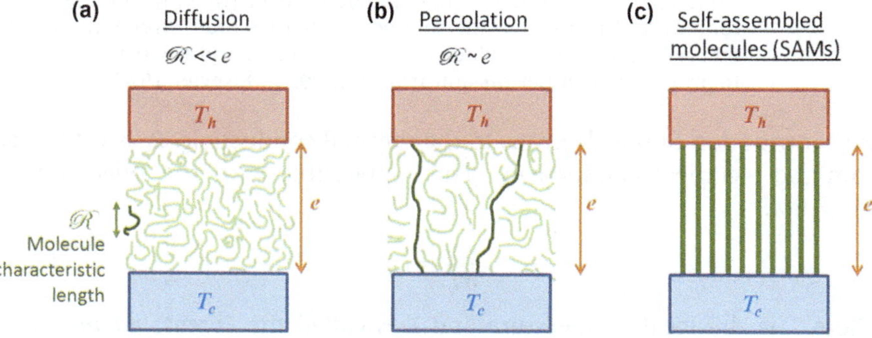

Figure 3.11   Three types of polymer arrangement. (a) Standard, (b) percolated, and (c) self-assembled molecules.

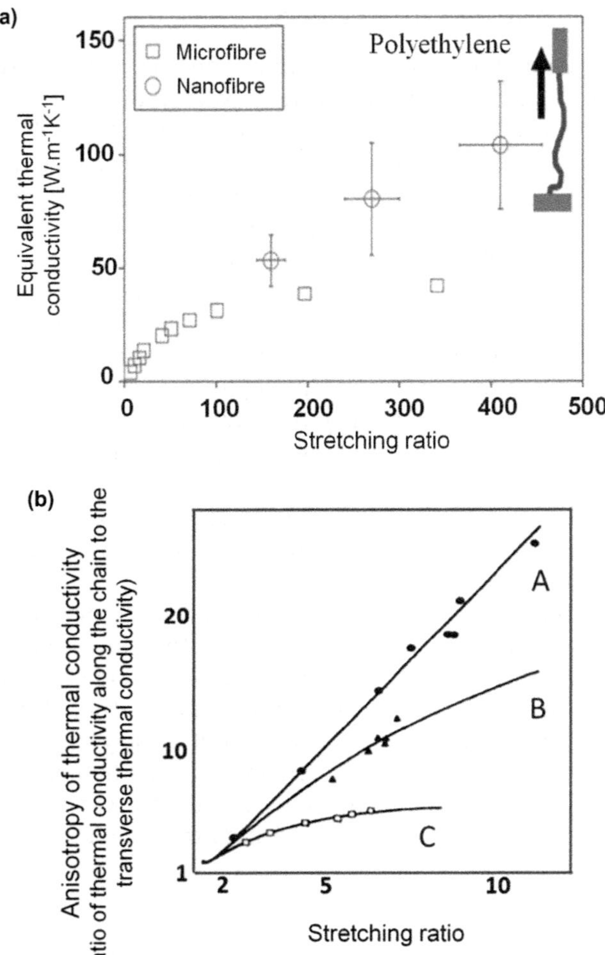

Figure 3.12 Thermal conductivity of polymers. (a) Equivalent thermal conductivity of at stretched polyethylene single fibre. Reproduced with permission from ref. 38. © Macmillan Publishers 2010. (b) Ratio of the longitudinal and transverse thermal conductivities after stretching a polymer with crystallinity values $A = 0.74$, $B = 0.67$, and $C = 0.45$. Reproduced with permission from ref. 39. © Elsevier 1977.

the contribution of the electrons to the thermal conductivity k_e is estimated from the electrical conductivity. This is possible with the Wiedemann–Franz law:

$$\frac{k_e}{\sigma T} = L \tag{3.24}$$

where T is the local temperature and L is called the *Lorentz number*. It is usually close to $L° = \frac{\pi^2}{3}\left(\frac{k_B}{e}\right)^2 = 2.45 \times 10^{-8}\,\mathrm{W\,\Omega\,K^{-1}}$. This relation is due to the fact that each electron carries a unit charge e and some energy linked to the temperature T.

The heat transport due to electrons is usually described with a BTE in the relaxation time approximation.[43] Heat conduction can be computed with the same tools as used for phonons. An important difference is that the spectrum is not broad, since all the electrons have an energy close to the Fermi level.[11]

3.4 Heat Conduction at the Nanometre Scale

Here, we will successively analyze the effects of the nanometre scale on the thermal transport of phonons. These are linked to thermal boundary issues and confinement.

3.4.1 Thermal Boundary Resistances

Eqn (3.6) showed that the thermal resistance at boundaries between materials can lead to the overall thermal resistance. The consequence is that the temperature field is not necessarily continuous at these boundaries: a discontinuity can appear, such as the one represented on Figure 3.13. These thermal boundary resistances (sometimes called *Kapitza resistances*) can exist for various reasons:

1. Macroscale irregularities at interfaces [Figure 3.14(a)];
2. A transmission factor at the interface due to dissimilarity of the materials [Figure 3.14(b)];
3. Atomic-scale roughness at the interface [Figure 3.14(c)]; and
4. Non-perfect transmission from one type of particle to another one at the interface.

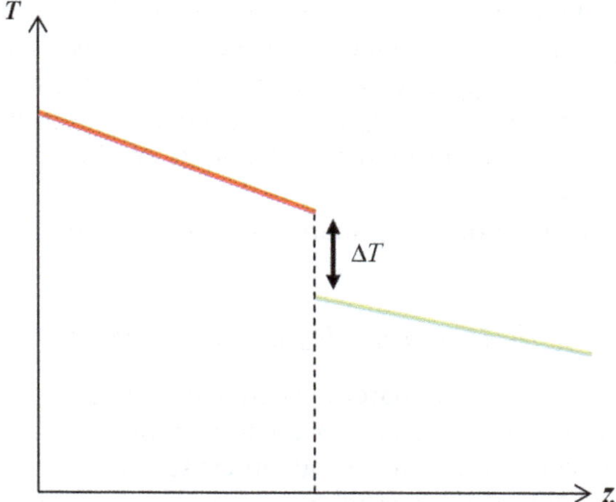

Figure 3.13 Temperature profile close to a boundary and discontinuity due to a thermal boundary resistance.

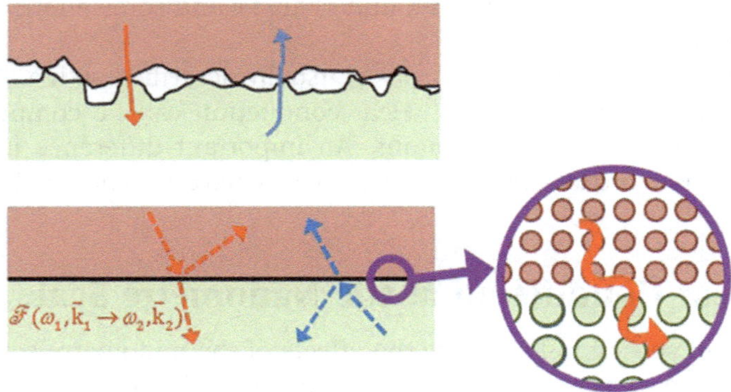

Figure 3.14   Imperfect heat transfer at interfaces for low-quality contact (top) or perfect contact (bottom).

The thermal boundary conductance associated with this discontinuity of temperature is defined as $G_C = |q/(T(\vec{r}^A) - T(\vec{r}^B))|$, where q is the net flux that is transmitted through the boundary. This flux is due to the energy flowing from one side $\vec{r}^A$ of the boundary to the other side $\vec{r}^B$, and can be written as a function of the energy transmission coefficient $\mathscr{F}(k, \theta, \varphi)$:

$$q_z = \sum_p \iiint_{\substack{\theta \in [0,\pi/2] \\ \varphi \in [0,2\pi]}} \left(\hbar\omega_A(\vec{k}) \cdot v_{gA}(\vec{k}) \cos\theta \cdot n_A(\vec{k}) \cdot \mathscr{F}_{AB}(k, \theta, \varphi) \right.$$

$$\left. - \hbar\omega_B(\vec{k}) \cdot v_{gB}(\vec{k}) \cos\theta \cdot n_B(\vec{k}) \cdot \mathscr{F}_{BA}(k, \theta, \varphi) \right) \frac{dk \cdot k d\theta \cdot k \sin\theta \, d\varphi}{(2\pi)^3} \tag{3.25}$$

where θ is the angle between the normal direction to the boundary and the direction of incidence of the wave, φ the second angle of the spherical coordinates, v_g is the group velocity respectively in each medium, $\hbar\omega$ is the vibration mode energy and $n(\omega, T)$ the Bose–Einstein distribution. At thermal equilibrium $(T_A = T_B = T_0)$ there is no net flux. For example, if both sides are made of the same material (A and B are identical), one should retrieve $\mathscr{F}_{AB} = \mathscr{F}_{BA}$.

Let us consider now how to evaluate the $\mathscr{F}$ coefficients.

3.4.1.1   *Macroscale Irregularities at the Contact*

The bad contact between the materials can lead to constrictions of the heat flux lines, a phenomenon which has been known for a very long time.[44] Since the materials involved may have various microstructures and hardnesses, it is difficult to evaluate the effect when the actual geometry of the contact is not known (almost always the case). Theories aim at establishing a dependence between the pressure applied to the materials to get them into better contact and the thermal boundary conductance.[45,46] Usually, the thermal boundary conductance G_C is then written as:

$$G_C \propto F^n \qquad (3.26)$$

where F is the applied load and n an exponent. Models link the *apparent area of contact* A_a to F, an average H of the hardnesses H_i of the two materials and to the roughness parameters such as the mean deviation to an average plane σ_i, the characteristic distance between two asperities η_i, or the fractal dimensionality associated with the roughness d_i. Here we do not need the microscopic expression [eqn (3.25)] to determine G_C.

3.4.1.2 Perfect Acoustic Contact

In crystalline solids, the heat carriers are the phonons, which are acoustic waves. As a consequence, phonons behave like other types of acoustic waves and their propagation is linked to the acoustic impedance $Z = \rho v$, where v is the acoustic velocity of the phonon for the given polarization and direction of propagation, and ρ is the material density. When matter is assimilated to a continuous medium, the reflection coefficient at the planar interface between two dissimilar media can be computed based on usual elastic-continuum acoustics. The condition of continuity of the stress $\overline{\overline{T}} \cdot \vec{n}$, where $\overline{\overline{T}}$ is the stress tensor and $\vec{n}$ the unit vector perpendicular to the surface, and the condition of continuity of the displacement field $\vec{u}$ at the boundary allows us to write these transmission coefficients:

$$\vec{u}(\vec{r}^A) = \vec{u}(\vec{r}^B) \qquad (3.27)$$

$$(\overline{\overline{T}} \cdot \vec{n})(\vec{r}^A) = (\overline{\overline{T}} \cdot \vec{n})(\vec{r}^B) \qquad (3.28)$$

where $\vec{r}^A$ and $\vec{r}^B$ denote locations at each side of the boundary. The angles of the incident, reflected and transmitted waves can be deduced from the acoustic Snell laws:

$$\frac{\omega}{v_i} \sin \theta_i = \frac{\omega}{v_r} \sin \theta_r = \frac{\omega}{v_t} \sin \theta_t \qquad (3.29)$$

where of course $v_r = v_i$ is the acoustic velocity in the medium before the interface and v_t is the one after transmission. If we make the assumption of two isotropic solids, we find the acoustic Fresnel coefficients. For a shear wave which is horizontally polarized (SH) so that it oscillates in the direction perpendicular to the incident plane, one finds:

$$t_{\text{SH} \to \text{SH}} = \frac{2Z^A \cos \theta_i}{Z^A \cos \theta_i + Z^B \cos \theta_t} \qquad (3.30)$$

The index SH $\to$ SH means that the wave is transmitted and that there is no change in polarization. In the general case, waves that are incident on a surface with a given polarization can be transmitted in more than one polarization state. For a vertically polarized shear (SV) wave, which oscillates in the plane of incidence, one finds that the transmission will lead to two waves after the interface, both having their displacement field in the plane of incidence: $t_{SV \to SV}$ and $t_{SV \to L}$ will be defined.[47] Similarly, a longitudinal wave (also denoted P for *primary wave* because the longitudinal velocity is larger than the transverse one, so it travels faster) will also be transmitted through two components: $t_{L \to SV}$ and $t_{L \to L}$.[47] In the end, we are able to determine the $\mathscr{F}$ energy transmission coefficients from the acoustic amplitude transmission coefficients $t_{X \to Y}$. Some approximations can be made to simplify the expressions.[48] The technique to evaluate $\mathscr{F}$ that we have described here is called the *Acoustic Mismatch Model* (AMM).

3.4.1.3 *Diffuse Acoustic Contact*

The previous section is valid provided the wavelength (λ) is much larger that the characteristic lengths associated with the microscopic structure of the medium. If the wavelength is close to the lattice constant, the contact between the two media will not be a totally flat interface. In addition, if one considers a little roughness at this interface or intermixing of atoms, it is absolutely sure that it will not be possible to apply the procedure of the previous section. To estimate the $\mathscr{F}$ coefficients, another model called the *Diffuse Mismatch Model* (DMM) can then be applied.[49] It consists of saying that when a phonon is leaving an interface, it is unknown if it has been transmitted through the interface or if it has been reflected by this interface. This means that:

$$\mathscr{F}_{AB}(\omega) = 1 - \mathscr{F}_{BA}(\omega) \tag{3.31}$$

However, a drawback of this model is that the transmission factor is never one for the case of an interface of similar materials: $\mathscr{F}_{AB}(\omega) = \mathscr{F}_{BA}(\omega) = \dfrac{1}{2}$ (ref. 50).

Note that the phonons are out of equilibrium close to the interface and their populations n_A and n_B in eqn (3.25) should be determined according to the BTE close to the interface, for both the cases of acoustic and diffuse contact.[50]

3.4.2 Confinement in Nanoscale Objects

Nanoscale heat conduction effects appear when a characteristic size of the medium becomes comparable to the heat carrier mean free path (Λ). Unfortunately, the mean free paths are not well known. Measurements with novel techniques[51,52] and determinations through simulations[53,54] are currently under way, at least for common materials such as Si or Ge. In the following, we describe the nanoscale effects without detailing how the mean free paths are determined. The reader is referred to the fast-growing literature for this topic (see, for example, ref. 55). One key point is that the

Table 3.2 Gross estimates of average mean free paths.

Material	Typical average mean free path (Λ)/nm
Silicon	300
Gold	30
Glass	1
Water	1

distribution of mean fee paths spans over orders of magnitude, from the very nanometric scale up to tens of microns for some room-temperature crystals. It is useful to write the thermal conductivity as a function of the mean free path distribution: $\kappa = \int_0^{+\infty} \kappa(\Lambda)\mathrm{d}\Lambda$. As one wants to define a single Knudsen number, it is practical to define an average mean free path. It provides a characteristic distance for the change of regime, from diffusive to ballistic, but it does not tell us about the extension of the intermediate regime, that can be very large (for example, three orders of magnitude of Kn around $\mathrm{Kn} = 1$ can be involved if $\kappa(\Lambda)$ possess significant values over three orders of magnitude). Table 3.2 presents gross estimations of the average mean free path for various media.

3.4.2.1 *Heat Conduction in Nanowires and Membranes*

When the mean free path is larger than the size of the solid, the plane waves are reflected at the boundaries. This impacts the heat conduction in the medium. In free-standing nanowires or membranes (supposed to be in contact with other materials only through an area A much smaller than the surface S at the boundary of these materials), the confinement induced by the walls of the solids leads to a modification of the effective thermal conductivity. This is taken into account in considering an effective mean free path $\Lambda_S(k)$ which can be written as:

$$\Lambda_S(k) = F(k, D, p) \cdot \Lambda_V(k) \tag{3.32}$$

where $\Lambda_V(k)$ is the bulk mean free path and $F(k, D, p)$ is the correction factor linked to the confinement. D is the characteristic size linked to the confinement, such as the diameter in a cylindrical nanowire or the thickness in a thin film. $F(k, D, p)$ is a function which depends on:

1. The geometry of the nano-object since it depends on the characteristic size D; and
2. The surface roughness through the reflection parameter p ($p \in [0;1]$), which tells us about the probability of a phonon being specularly reflected by the boundary, with $p = 1$ if the reflection is specular or $p = 0$ if it is diffuse.[56]

Table 3.3 shows us how to calculate the factor $F(k, D, p)$[57] for free-standing cylindrical nanowires and thin films. Note that the specularity parameter p

Table 3.3 The function $F(k, D, p)$ as a function of the specularity parameter p and the Knudsen number $\mathrm{Kn}(k) = \Lambda_V(k)/D$.

Configuration	Function $F(k, D, p)$	Approximation of F for large $\mathrm{Kn}\ (p<1)$
Membrane of thickness D	$1 - \dfrac{3}{2}\mathrm{Kn}\,(1-p)\displaystyle\int_1^\infty \left(\dfrac{1}{t^3} - \dfrac{1}{t^5}\right)\dfrac{1 - e^{-\frac{t}{\mathrm{Kn}}}}{1 - pe^{-\frac{t}{\mathrm{Kn}}}}\,\mathrm{d}t$	$\dfrac{3}{4}\dfrac{1+p}{1-p}\dfrac{\ln(\mathrm{Kn})}{\mathrm{Kn}}$
Nanowire of characteristic dimension D	$1 - \dfrac{12}{\pi}(1-p)^2\displaystyle\sum_{n=1}^{+\infty} n\,p^{n-1}\int_0^1 (1-t^2)^{1/2} S_4\left(\dfrac{t}{n\cdot\mathrm{Kn}}\right)\mathrm{d}t$ with $S_n(u) = \displaystyle\int_1^\infty e^{-ut}(t^2-1)^{1/2}t^{-n}\,\mathrm{d}t$	Cylindrical: $\dfrac{1+p}{1-p}\dfrac{1}{\mathrm{Kn}}$ Square cross-section: $0.897\dfrac{1+p}{1-p}\dfrac{1}{\mathrm{Kn}}$

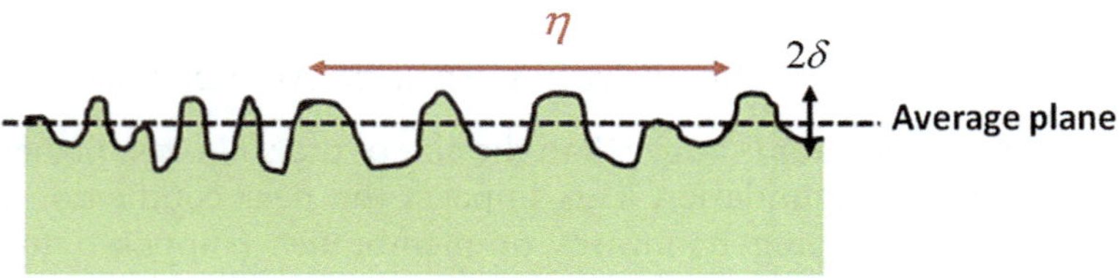

Figure 3.15 Rough surface, roughness correlation length η and mean deviation δ to the mean surface.

can be traced back to the surface roughness[58] (see Figure 3.15). Indeed, for a transverse correlation length of surface roughness η, which represents the distance over which there is some regularity on the surface, we can write that $p(k) = e\exp(-4k^2\delta^2)$, where root mean square deviation to the mean surface δ is much smaller than η.

3.4.2.2 Constrictions

At the nanoscale heat can be forced to go through small mechanical contacts lying between macroscale bodies, which leads to a constriction of the heat flux lines. This is represented in Figure 3.16(a), where two regimes are depicted: the diffusive regime (left), where the heat carriers mostly interact inside the volume of the material, and the ballistic regime (right), where they mostly interact (collide) with the boundaries of the medium. This last regime is of course similar to the previous section, but here the heat conduction from the bulk to the thin constriction is also to be considered. For the

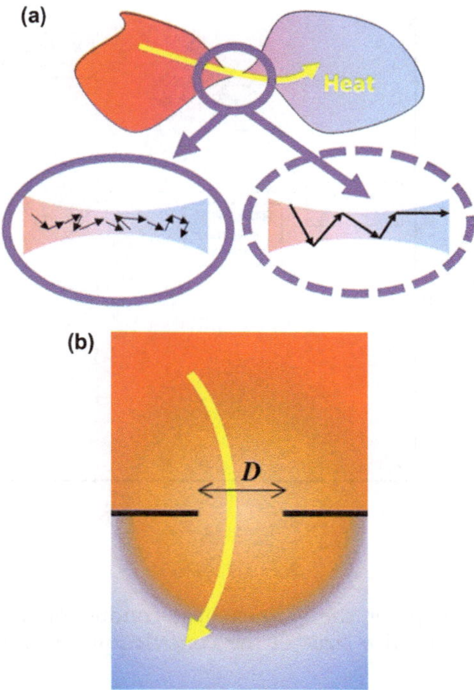

Figure 3.16 Heat transfers from the hot side to the cold side through a constriction. (a) Diffusive regime and ballistic regimes in a smooth constriction. (b) Abrupt constriction.

diffusive case, Fourier's law can be applied and the famous Maxwell formula can be used:

$$R_{\mathrm{M}} = \frac{1}{Dk} \tag{3.33}$$

where D is the characteristic size of the constriction and depends of course on its shape. For a circular constriction such as the one of Figure 3.16(b), D is the disk diameter.

If the size D is smaller than the mean free path of the heat carriers, Fourier's law cannot be applied. For electrons, this limiting case is known as the *Sharvin constriction resistance*. If we assume that the kinetic theory for heat carriers is correct and consider the case of two macroscopic bodies contacted through a circular hole, the thermal resistance is then equal to:

$$R_{\mathrm{S}} = \frac{16}{\pi c_{\mathrm{p}} v D^2} \tag{3.34}$$

where v is the velocity of the heat carriers. Here the thermal resistance does not depend on the thermal conductivity because this quantity is

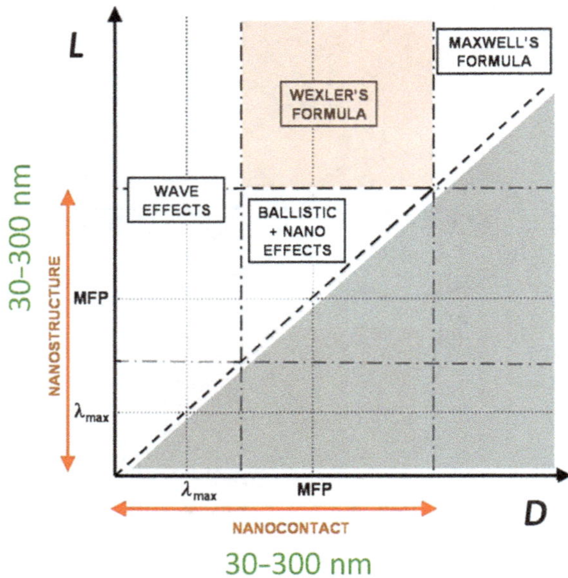

Figure 3.17   Regime map for the various heat conduction regimes that depend on the respective sizes of the constriction and the connected bodies (see ref. 59).

not relevant anymore. This approach is particularly valid for electrons, since their distribution of mean free paths is narrower than that of phonons. For phonons, eqn (3.34) should be regarded as approximate.

Interestingly, an interpolation formula can often be found in the literature, which is known as the *Wexler formula* and reads $R_W = R_M + R_S$. The principle is that each formula has its domain of validity, where the other resistance is negligible. Such interpolation works very well in a large number of cases. Wexler's formula can also be rewritten as follows:

$$R_W = R_M(1 + \beta Kn) \tag{3.35}$$

where β is a geometric parameter depending on the shape of the constriction.

Recently, it was proposed that a formula like eqn (3.35) can only be valid when carriers from one side cannot be backscattered from the other side before being thermalized. As a consequence, a regime map depending on the shape of the constriction, but also on the characteristic size of the side bodies can be plotted.[59] This is presented in Figure 3.17. Experimental confirmation should be provided in the near future.

3.4.2.3   Nanoscale Heat Sources

The heat source can also be smaller than the mean free path. Figure 3.18 shows that this configuration is close to the previous one

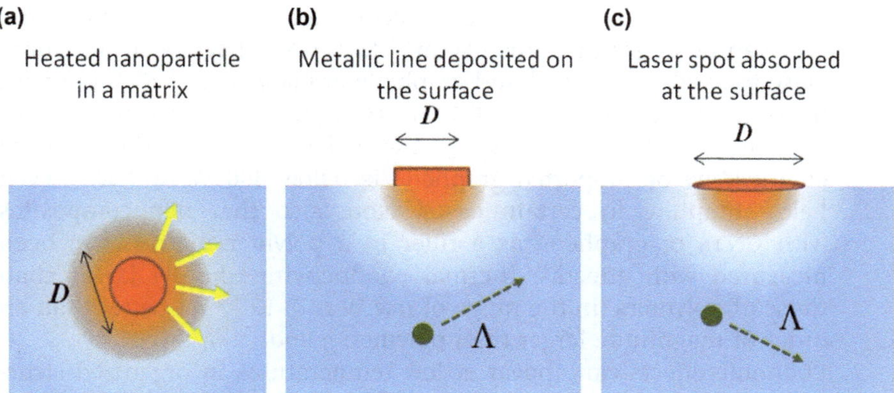

Figure 3.18 (a) A spherical heat source embedded in a host medium. (b) Cross-section of a configuration where a metallic line is deposited on top of a substrate. (c) Schematic of a hot spot due to laser absorption on top of the substrate.

[see Figure 3.18(b)], but with only one half of the geometry. It has been shown in ref. 60 that energy is not transferred as well as predicted by macroscopic Fourier heat conduction when the heat source is smaller than the mean free path. Knowing the actual profile of the heat source is extremely important, especially for optical experiments [the case in Figure 3.18(c)] where the lateral extension of the spot is not step-like (often it is Gaussian). A set of experiments involving optical excitation[61–63] or electrical heaters,[64] where the source also acts as the sensor, showed that the *apparent* thermal conductivity measured with sub-mean free path devices is indeed smaller than the real thermal conductivity of the material. The error that is made in measuring the apparent thermal conductivity with a source of given size is accounted for with a *suppression function S(Λ)* associated with the source profile, and one can write:

$$\kappa(S) = \int_0^{+\infty} S(\Lambda)\kappa(\Lambda)\,\mathrm{d}\Lambda \tag{3.36}$$

where S is a *transfer function* associated with the source/detector and can be treated as a filter function.

3.4.3 Other Effects

In addition to confinement and boundary effects, various other effects that appear at the nanoscale should also be mentioned:

1. Large intrinsic effective thermal conductivity of low-dimensional materials has been observed in free-standing few-atom thick nanostructures such as carbon nanotubes (CNTs)[16] and graphene.[18,65] These

materials possess the best equivalent thermal conductivities, and they can be larger than 1000 W m^{-1} K^{-1} (CNTs have been measured close to 3000 W m^{-1} K^{-1} and graphene at 6000 W m^{-1} K^{-1}, but with uncertainties on the order of 50%). However, they are very sensitive to contacts that can destroy these large values: the thermal conductivity of supported graphene is below 500 W m^{-1} K^{-1}, even below graphite in certain cases. Note also that nanocomposites with CNTs or graphene as a filler in a polymer matrix have been measured with effective thermal conductivities hardly larger than those of polymers (in the range of few W m^{-1} K^{-1}, not better than an order of magnitude larger than polymers); and

2. Phononic effects can appear at low temperatures in organized structures at the nano- to microscale[66] or in very small (~ 1 nm) superlattices.[67]

3.5   Thermal Radiation at the Nanoscale

3.5.1   At the Macroscale

Thermal radiation is the way heat transfers without the need of a medium. We restrict ourselves in this section to situations where opaque bodies are immersed in a non-absorbing medium. For the sake of simplicity the medium will be assimilated to vacuum but other non-absorbing media can be considered in principle. At equilibrium, thermal radiation is described by Planck's law, which states that the equilibrium spectral radiance is:[65]

$$I^{\circ}(\omega, T) = \frac{c}{4\pi} \cdot \frac{\omega^2}{\pi^2 c^3} \cdot \hbar\omega \cdot \frac{1}{e^{\hbar\omega/k_B T} - 1} \tag{3.37}$$

where c is the speed of light in vacuum, and ω is the circular frequency. This expression is written as the product of four factors: the second one is the *density of states* $\rho(\omega) = \dfrac{\omega^2}{\pi^2 c^3}$, the third one is the energy of a harmonic oscillator $E(\omega) = \hbar\omega$, and the last one is the Bose–Einstein distribution $n^{\circ}(\omega, T) = \dfrac{1}{e^{\hbar\omega/k_B T}}$ which gives the probability for a photon (electromagnetic mode) to be populated. The first factor is the ratio of the specific intensity $I^{\circ}(\omega, T)$ to the density of electromagnetic energy $u_{EM}(\omega) = \rho(\omega)n^{\circ}(\omega, T)E(\omega)$. A surface emits thermal radiation in a given solid angle per unit surface according to the product of the spectral angular emissivity $\varepsilon(\omega, \vec{\Omega})$ with the equilibrium spectral radiance:

$$I(\omega, \vec{\Omega}) = \varepsilon(\omega, \vec{\Omega})I^{\circ}(\omega, T) \tag{3.38}$$

If one considers two parallel opaque bodies denoted respectively by 1 and 2 separated by a distance e, it can be shown that the net heat exchange per unit surface does not depend on the distance:

$$q = \iiint_{\substack{\omega \in [0;\infty] \\ \theta \in [0;\frac{\pi}{2}] \\ \phi \in [0;2\pi]}} \frac{\varepsilon_1(\omega,\vec{\Omega})\varepsilon_2(\omega,\vec{\Omega})}{1 - [(1 - \varepsilon_1(\omega,\vec{\Omega}))(1 - \varepsilon_2(\omega,\vec{\Omega}))]} I^\circ(\omega,T)\,\mathrm{d}\theta \sin\theta\,\mathrm{d}\varphi \quad (3.39)$$

where q is the net het flux. However, this frame is valid only provided that the radiation wavelength is much larger than the characteristic size of the geometrical configuration. Here, this characteristic size is given by e and Wien's wavelength $\lambda_{\max}$ is known to be around 10 μm at ambient temperature (infrared radiation). As a consequence, this formalism cannot be used for smaller distances and in particular in the nanoscale regime.

3.5.2 Interbody Distances Smaller than the Thermal Wavelength

One needs to come back to the physical origin of thermal radiation, *i.e.*, the motion of charges in matter that radiate electromagnetic waves, to be able to compute the energy transfer. The wave picture is now required, which will lead to the appearance of the following phenomena:

1. Interferences, which may lead to an increase (constructive interference) or to a decrease (destructive interference) in the heat flux exchanged through thermal radiation;[68,69]
2. Near-field effects, which lead to a very strong increase (orders of magnitude!) in the thermal radiation exchanged when the bodies are separated by a distance which is much smaller than $\lambda_{\max}$.[70,71]

Note in particular that the wave formalism is able to reproduce the results of the far-field regime mentioned earlier, at the price of larger computational expense.

3.5.2.1 *Computation of Thermal Radiation in the Wave Regime*

Crystalline solids are made of periodic arrangements of atoms that can be partially ionized. In metals, electrons move in the lattice of positively charged ions. In dielectric crystals, atoms are partially charged. In both cases, there are positive and negative charges that move randomly due to thermal motion. It is well known that moving charges radiate an electromagnetic field: as a consequence, the motion of all these charges is responsible for thermal radiation. At equilibrium, the temperature is fixed and despite the fact that these charges radiate, which means that they lose energy, no cooling can be observed. This is due to the fact that some energy is also brought to them by other charges: indeed, a charge in an electromagnetic field may start to move according to the field and therefore acquire some energy. Finally, we understand that the equilibrium is guaranteed by the balance between the energy which is radiated and the energy which is acquired (absorbed).

The fluctuation–dissipation theorem (FDT) allows us to determine the energy radiated by the charges. It links the fluctuations of the electric and magnetic polarizations to the ability to absorb an electromagnetic field, represented by the imaginary part of the dielectric function. The FDT can be written as:

$$\langle P_i(\vec{r}_1,\omega)P_j^*(\vec{r}_2,\omega')\rangle = \frac{4}{\pi\omega}\Theta(\omega,T)\mathrm{Im}(\epsilon_r)\epsilon_0\delta_{ij}\delta(\omega-\omega')\delta(\vec{r}_1-\vec{r}_2) \qquad (3.40a)$$

$$\langle M_i(\vec{r}_1,\omega)M_j^*(\vec{r}_2,\omega')\rangle = \frac{4}{\pi\omega}\Theta(\omega,T)\,\mathrm{Im}(\mu_r)\mu_0\delta_{ij}\delta(\omega-\omega')\delta(\vec{r}_1-\vec{r}_2) \qquad (3.40b)$$

where P and M are respectively, the electric and magnetic polarizations, i and j being (polarization) directions, $\vec{r}_1$ and $\vec{r}_2$ locations, ω and ω' circular frequencies, ϵ_r and μ_r respectively the relative dielectric permittivity and the magnetic permeability, ε_0 and μ_0 the vacuum permittivity and permeability, and δ_{ij} and δ the Kronecker and Dirac symbols. $\Theta(\omega,T)=n°(\omega,T)E(\omega)$ is the mean energy of a Planck oscillator.

Let us make few comments about these equations: First, one can notice that they state that the polarizations at different locations or in different directions are totally incoherent (only $\langle X_i(\vec{r}_1,\omega)X_i^*(\vec{r}_1,\omega)\rangle$ may be non-zero). Second, there is only radiation when the temperature is non-zero: $\Theta(\omega,T)\to 0$ for $T\to 0$. Finally, many materials are non-magnetic (in particular, $\mathrm{Im}(\mu_r)=0$) and for them only the dielectric function may play a role [eqn (3.40a)].

Once the FDT is known, it is possible to calculate the heat flux exchanged between the two surfaces. The net heat flux can be calculated by taking the difference between the energy flux due to the Poynting vector generated by one of the media and the energy flux due to the Poynting vector generated by the other medium. A Green's function approach allows us to determine the electromagnetic field received at a given point $\vec{r}_2$: $\vec{E}(\vec{r}_2)=\iiint_{\vec{r}_1}\overline{\overline{G^{EP}}}(\vec{r}_2,\vec{r}_1)\cdot\vec{P}(\vec{r}_1)dV$, where $\vec{r}_1$ is the location where the fluctuation $\vec{P}$ takes place and is of volume dV. $\vec{P}(\vec{r}_1)\delta V$ can be considered as the electric dipole moment associated with the charges in the volume δV. It is well known that the energy deposited is proportional to the mean quadratic field $\langle\vec{E}(\vec{r}_2)\vec{E}(\vec{r}_2)\rangle$; the application of the FDT allows us to explicitly calculate the Poynting vector, which is also a quadratic quantity:

$$\langle\vec{E}(\vec{r}_2)\vec{H}^*(\vec{r}_2)\rangle = \left\langle \iiint_{\vec{r}_1}\overline{\overline{G^{EP}}}(\vec{r}_2,\vec{r}_1)\cdot\vec{P}(\vec{r}_1)dV \cdot \iiint_{\vec{r}_1'}\overline{\overline{G^{EP}}}^*(\vec{r}_2,\vec{r}_1')\cdot\vec{H}^*(\vec{r}_1')dV'\right\rangle$$

$$(3.41)$$

3.5.2.2 Heat Flux between Two Flat Surfaces

For the case of two semi-infinite media (media 1 and 3) separated by a vacuum gap (medium 2) of size e, the computation of the integral in eqn (3.41) leads to the following expression:[70,71]

$$q_{\mathrm{net},13}^{\mathrm{tot}} = \int_0^\infty \mathrm{d}\omega \, \frac{\Theta(\omega, T_1) - \Theta(\omega, T_3)}{4\pi^2}$$

$$\times \left\{ \int_0^{k_0} k_\parallel \mathrm{d}k_\parallel \left[\frac{\left(1 - |r_{21}^{\mathrm{TE}}|^2\right)\left(1 - |r_{23}^{\mathrm{TE}}|^2\right)}{\left|1 - r_{21}^{\mathrm{TE}} r_{23}^{\mathrm{TE}} e^{2ik_{2\perp}d}\right|^2} + \frac{\left(1 - |r_{21}^{\mathrm{TM}}|^2\right)\left(1 - |r_{23}^{\mathrm{TM}}|^2\right)}{\left|1 - r_{21}^{\mathrm{TM}} r_{23}^{\mathrm{TM}} e^{2ik_{2\perp}d}\right|^2} \right] \right.$$

$$\left. + \int_{k_0}^\infty 4k_\parallel \mathrm{d}k_\parallel e^{-2\mathrm{Im}(k_{\perp 2})e} \left[\frac{\mathrm{Im}\left(r_{21}^{\mathrm{TE}}\right)\mathrm{Im}\left(r_{23}^{\mathrm{TE}}\right)}{\left|1 - r_{21}^{\mathrm{TE}} r_{23}^{\mathrm{TE}} e^{-2\mathrm{Im}(k_{2\perp})e}\right|^2} + \frac{\mathrm{Im}\left(r_{21}^{\mathrm{TM}}\right)\mathrm{Im}\left(r_{23}^{\mathrm{TM}}\right)}{\left|1 - r_{21}^{\mathrm{TM}} r_{23}^{\mathrm{TM}} e^{-2\mathrm{Im}(k_{2\perp})e}\right|^2} \right] \right\}$$

$$(3.42)$$

which contains various terms involving the Fresnel optical reflection coefficients $r_{ij}^{\mathrm{TE,TM}}$. Here $k_\parallel$ is the projection of the wave vector along the direction of the interface, $k_\parallel = \frac{\omega}{c}\sin\theta$, and $k_z^2 = \left(\frac{\omega}{c}\right)^2 - k_\parallel^2$. TE and TM are the Transverse Electric and the Transverse Magnetic polarization of the electromagnetic fields. The first integral is the sum of all the propagating waves ($k_\parallel < \omega/c$) which are also considered in the far-field expression. Under various assumptions, it may be shown that it is equal to eqn (3.39). The existence of the polarizations as well as the Fabry–Pérot factors involving resonances in the denominator are the principal differences. The second term is due to the evanescent waves that are usually locked at interfaces. Here more flux will be transmitted since the presence of a material in the vicinity of the surfaces where the evanescent waves are bound allows them to be scattered, opening an additional channel of transmission for the heat flux. Figure 3.19 shows that the net exchanged heat flux becomes much larger at small separation distances e: this is due to the additional contribution of evanescent waves.

This case of two parallel plates exemplifies the importance of accounting for the near-field effect of thermal radiation. Recent measurements have verified the theory with good accuracy.[72–75]

3.5.2.3 Radiative Flux Spectrum

Another important point is that the spectrum of thermal radiation can be very different. Figure 3.20 shows the spectral density of energy $u_{\mathrm{EM}}(\omega)$ (the quantity of energy per unit volume in a given frequency band which is located close to the surface, in J m^{-3}/(rad s^{-1})), close to a flat surface of glass. It can be seen that the spectra are very different in the far-field, where one observes the modulation of the Planck spectrum by the emissivity of the

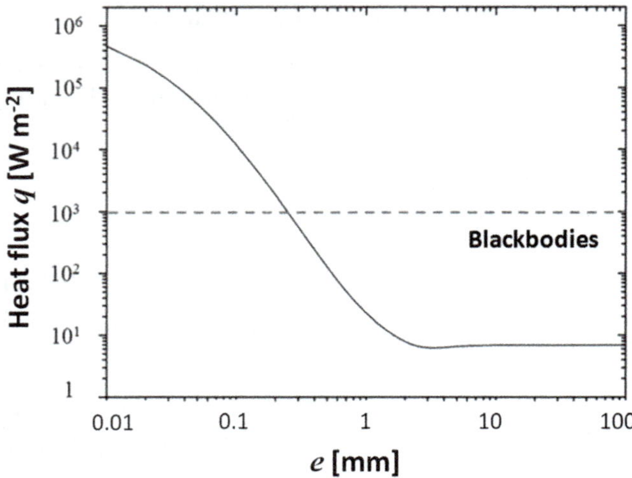

Figure 3.19 Total heat flux exchanged by two parallel flat media in aluminium ($T_1 = 400$ K, $T_3 = 300$ K) as a function of the distance d between them.

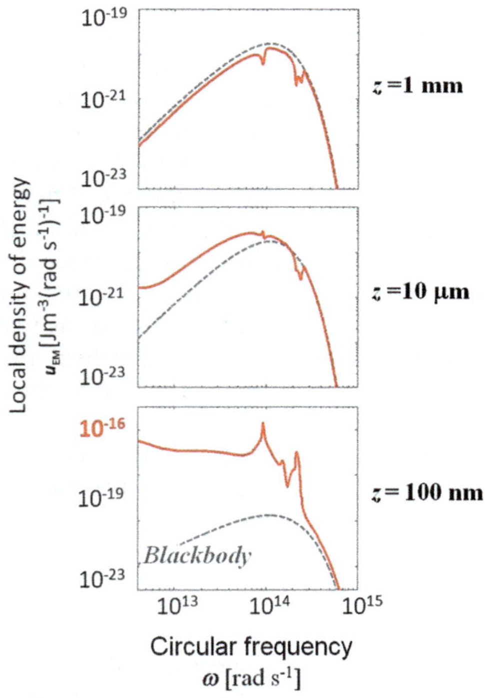

Figure 3.20 Local density of energy (DOE) u_{EM} close to a surface of glass (SiO_2): While in the far-field it is always smaller than the equilibrium value, it can exceed it in the near-field. Note also that the low-frequency DOE does not decrease when ω decreases. The far-field blackbody DOE (equilibrium value) is represented with a dashed grey line. Inspired by ref. 76.

surface (see eqn 3.38), and in the near-field, where the density of energy exceeds Planck's limit due to the additional surface electromagnetic (evanescent) modes. Note in particular that the dips in the far-field spectra are 'replaced' by peaks in the near-field, and the density of energy does not become zero at low frequencies.

3.5.3 Bodies Smaller than the Thermal Wavelength

When the size of a radiating body is on the order of the thermal wavelength or smaller, radiation also becomes different to when the size is much larger. Wave phenomena enter into play, modifying the properties of thermal emission and thermal absorption of the body. If we consider a small sphere, that can be assimilated to the sum of a dipole of electric polarizability

$$\alpha_{\mathrm{E}} = 4\pi R^3 \mathrm{Im}\left(\frac{\epsilon_{\mathrm{r}} - 1}{\epsilon_{\mathrm{r}} + 2}\right) \quad \text{and} \quad \text{magnetic polarizability} \quad \alpha_{\mathrm{M}} = \frac{2\pi}{15} R^5 \mathrm{Im}\,(\epsilon_{\mathrm{r}} - 1),$$

where R is the sphere radius, one can show that the radiated power is:[77]

$$P = \int_0^{+\infty} \frac{\omega^3}{\pi^2 c^3} \theta(\omega, T)[\mathrm{Im}(\alpha_{\mathrm{E}}) + \mathrm{Im}(\alpha_{\mathrm{M}})]\mathrm{d}\omega \tag{3.43}$$

Assuming that $\mathrm{Im}(\alpha_{\mathrm{E}}) \gg \mathrm{Im}(\alpha_{\mathrm{M}})$, we find that the radiative power emitted per unit surface is

$$P(\omega) = \frac{\hbar\omega^4}{\pi^2 c^3} \frac{1}{\mathrm{e}^{\hbar\omega/k_{\mathrm{B}}T}} R\,\mathrm{Im}\left(\frac{\epsilon_{\mathrm{r}} - 1}{\epsilon_{\mathrm{r}} + 2}\right) \tag{3.44}$$

which is different to the result obtained from the macroscopic theory [eqn (3.37)]: $\int_{\theta \in [0,\frac{\pi}{2}],\varphi \in [0,2\pi]} \varepsilon(\omega, \theta, \varphi) \cos\theta\, I^\circ(\omega, T)\mathrm{d}\theta \sin\theta\, \mathrm{d}\varphi = \frac{\hbar\omega^3}{4\pi^2 c^2} \cdot \frac{1}{\mathrm{e}^{\hbar\omega/k_{\mathrm{B}}T}} \bar{\varepsilon}(\omega),$ where $\bar{\varepsilon}$ does not depend on the radius and is termed *hemi-spherical emissivity*. Even without specifying the value of $\bar{\varepsilon}$, we see that the dependence of the radius is different to the one of eqn (3.44). Since the total power radiated by the small sphere is proportional to R^3, it is tempting to say that we observe a change from surface radiation (which depends on R^2) to volume radiation. However, if we had decided that the magnetic dipole contribution was predominant in eqn (3.43), we would have found a total radiated power proportional to R^5, which cannot be predicted by macroscopic thermal radiation.

The normalized radiative power emitted by a sphere as a function of its radius has been computed by Krüger *et al.*[78] As shown by eqn (3.43) with the expressions of the polarizabilities, it is much smaller than the macroscopic incoherent theory prediction for nanoscale objects since it depends on power laws of $\frac{\omega}{c}R$ which become small when R decreases.

We finally note that all optical effects involved in plasmonics and nano-optics may play a role in thermal radiation. Some of these effects can be 'averaged' due to the broad spectrum of the radiated heat and may not be apparent when analyzing at the integrated scale (energy or power, without looking at the spectrum) but this is not necessarily the case. It is generally considered that this field is an avenue for new interesting research.

3.6 Characterization of Heat-transfer Properties in Nanomaterials

There are many ways to characterize the heat-transfer properties of materials. In the stationary regime, thermal properties of materials of interest, such as the apparent thermal conductivity, can be directly measured. Once the temperature at a given location or the full temperature field has been measured, a model of the sample is used to retrieve the parameter to be measured. *Usually the measurement chain is based on an analysis using macroscopic heat transfer.* It is important to account for the effects of the previous sections if one wants to avoid mistakes. To improve the signal-to-noise ratio, modulated measurements may be preferred. Finally, there is a set of experiments that allows direct probing of the heat-transfer dynamics. In this section, we focus on heat conduction. We do not address scanning probe experiments, as they are tackled elsewhere.

3.6.1 Electrical Measurements

We start by considering electrical measurements of the thermal conductivity. The techniques are based on the Joule heating of an electric wire and the monitoring of its temperature as a function of the studied sample. To do so, a thin metallic wire is deposited on top of a flat sample (see Figure 3.21). An electric current I that flows in the wire will undergo Joule heating RI^2. As the electrical resistance is inversely proportional to the cross section, this power can be significant if the cross section of the wire is small enough. Since the electrical resistance of the wire depends linearly on its temperature at first order

$$R = R_0 \left(1 + \alpha \Delta T\right), \tag{3.45}$$

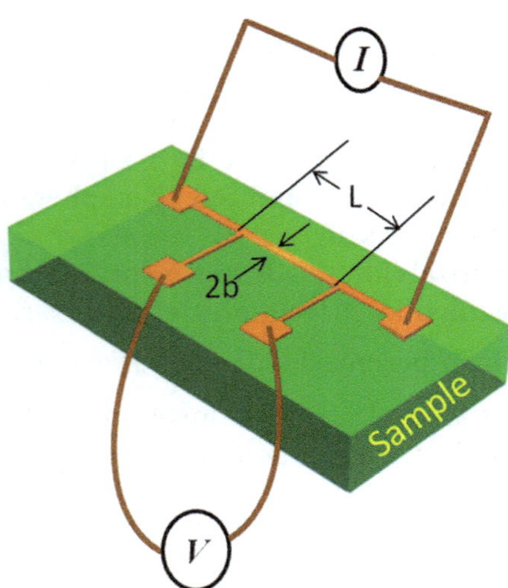

Figure 3.21 Setup of electrical measurements such as the 3ω method.

the measurement of the wire voltage $V = RI$ is a way to probe the sample. Indeed, if the sample is thermally conductive, the wire will not heat a lot because the thermal flux will spread into the sample. Conversely, if the sample is thermally insulating, the temperature of the wire will increase, because the heat will not be allowed to flow into the sample.

The peculiarity of the *3ω method*[79–81] is that the current is harmonic: $I = I_0 \cos \omega t$. The probed voltage is then analyzed thanks to lock-in detection at the third harmonic: it can be demonstrated that this component of the signal can be used as a thermometer. Indeed, it is directly proportional to the 2ω component, $\theta_{2\omega}$, of the wire temperature $T = T_0 + \theta_{\mathrm{dc}} + \theta_{2\omega} \cos(2\omega t + \phi_{2\omega})$ where T_0 is the ambient temperature and θ_{dc} is the increase of the continuous temperature due to the Joule heating:

$$V_{3\omega} = \alpha/2 \; \theta_{2\omega} \; R_0 \; I_0 \; \cos(3\omega t + \phi_{2\omega}). \tag{3.46}$$

These measurements require generally knowing with very good accuracy the coefficient of temperature α of the electrical resistance of the wire. Since this depends strongly on the conditions of fabrication, it is absolutely impossible to use a value from literature. As a consequence, such 3ω setups are usually built on temperature-dependent stages, which enable one to directly measure α. The most efficient setups are located in cryostats, which also allow for the measurement of the sample's thermal properties as a function of temperature. Note that it is worth measuring various widths and lengths with various 3ω devices on a single sample since the ratio of the thermal penetration depth to the wire width is the parameter that may depart from the Cahill's conditions.[79] Measuring with a few devices allows one to perform a double-check of this condition.

Other types of electrical measurements, in particular those mixing harmonics, are also possible.[82] We also note that heat pulse experiments[83,84] were performed in the 20th Century but this type of experiment does not appear to be currently used.

3.6.2 Optical Measurements

3.6.2.1 Stationary and Modulated Regimes

A typical way to probe thermal properties of materials is *thermoreflectance*,[4] which allows for temperature measurements with an uncertainty that can be below 1 K. The principle is as follows (see Figure 3.22): a light beam illuminates a sample and is partially reflected. The intensity and/or the phase of the reflected light is collected. The reflection can be analyzed in light of the Fresnel reflection coefficients, which depend on the refractive index of the sample n. The refractive index depends on temperature. At first order, one can write:

$$n_\lambda(T) = n_\lambda(T_0) + \mathrm{d}n_\lambda/\mathrm{d}T \, (T - T_0). \tag{3.47}$$

It is important to note that n_λ depends also on the wavelength: this allows optimization of the detected signal so that the reflection is maximal.

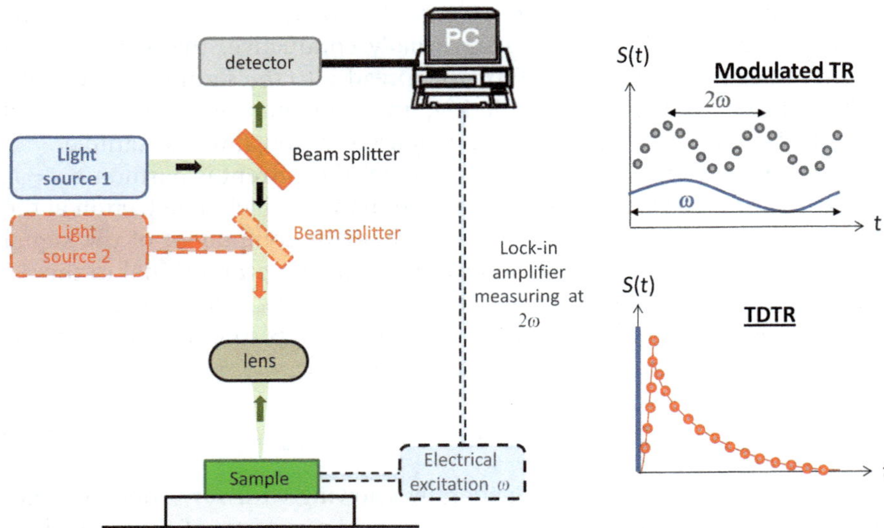

Figure 3.22 Schematic of thermoreflectance setups, in the modulated and time-domain (TDTR) configurations.

To improve the sensitivity, the incident light beam can be modulated and the reflected light is then detected with a lock-in amplifier (see Figure 3.22 in the case of a setup with the grey-dashed elements).

Another optical method that allows thermal properties to be measured in the stationary regime is based on *Raman spectroscopy*.[85] It appears that the position of a particular phonon peak (such as the LO phonon mode) can depend on temperature. At first order, on can write:

$$\sigma(T) = \sigma(T_0) + a(T - T_0). \tag{3.48}$$

Note that the temperature dependence of the shift σ can be more complicated than linear (here we have noted a the coefficient). By increasing the power of the incident light beam, more energy is deposited on the surface of a sample, which heats up. If a calibration of the spectral shift has been done previously, one can trace back to the surface temperature with this setup. This allows the thermal conductivity of samples to be determined.

If one wants to measure the full temperature field, a setup which decouples the heating and the measurement, with two lasers, enables one to scan the sample.[86]

3.6.2.2 Transient Regime

Optical techniques are limited by the lateral resolution of few hundreds of nanometres (for visible light), which may appear large in comparison to some probe techniques. However, a striking point of interest of optical techniques is that they allow probing of the dynamics of samples at very high frequencies, by means of ultrafast pump–probe spectroscopy. *Time-domain thermoreflectance* (TDTR) is currently the main technique. Here two laser pulses are used,[87] which are slightly time-lagged: the first one heats the sample up, while the second measures it by being reflected toward the detector (see Figure 3.22, for the case where the setup involves the red-dashed areas). This can be done by splitting a single beam into two beams and delaying one of them, or by using two lasers with slightly shifted frequencies in a heterodyne configuration. By repeating the experiments with increasing time delays, the evolution of the sample temperature as a function of the time can be plotted. The heat propagation is then monitored, with typical electron excitation at first (sub-picosecond time scale) and then lattice thermalization [see eqn (3.8)] on the picosecond time scale, and then thermal decay associated with heat diffusion is often observed.

In the nanosecond regime, heat diffusion takes place. The transient thermal grating (TTG) technique,[88] which does not require an ultrafast pulsed laser, can be used in this regime.

In this section we have presented a snapshot of a few techniques that allow thermal dynamics in nanoscale samples to be measured, either in the modulated regime or in the transient regime. The observation of the dynamics of heat transfer is currently an expanding field and novel exciting results should appear soon.

3.7 Conclusions

In this chapter, we have seen that heat can be transferred in various ways, through oscillations of atoms in a coherent manner or in a completely disorganized way. Since the characteristic sizes of the heat carriers fall in the nanoscale range, the impact of nanostructuring on the heat-transport properties can be very strong. We have discussed the impact of confinement on the heat flux for both heat conduction and thermal radiation. We have seen that, in addition to confinement, transmission issues can take place at boundaries. We have underlined the effect of constrictions of the heat flux lines and size effects for heat sources or thermal sensors. We have also highlighted that the modifications undergone by thermal radiation can already take place in the micrometric regime. Finally, we have given a brief overview of techniques that can be used to measure thermal transport properties or to directly observe the heat transfer. In thermometry, sensors can be either based on optical principles or on electrical detection. All the physical phenomena mentioned have to be taken into account when analyzing experiments involving nanoscale heat sources, sensors or samples.

Acknowledgements

I thank S. Merabia and K. Termentzidis for useful discussions, as well as Y. Tsurimaki for Figure 3.19, W. Jaber for Figure 3.21, R. Vaillon and S. Gomes. This work has been partly supported by ANR RPDOC NanoHeat, INSA BQR MaNaTherm and EU FP7 QuantiHeat.

References

1. G. Chen, *Nanoscale Energy Transport and Conversion: A Parallel Treatment of Electrons, Molecules, Phonons, and Photons*, MIT-Pappalardo Series in Mechanical Engineering ed., Oxford, 2005.
2. M. Kaviany, *Heat Transfer Physics*, Cambridge University Press, Cambridge, 2008.
3. S. Volz, *Thermal Nanosystems and Nanomaterials*, Springer-Verlag, Berlin, Heidelberg, 2009.
4. S. Volz, *Microscale and Nanoscale Heat Transfer*, Topics in Applied Physcis, Springer, 2007, vol. 107.
5. J. Fourier, *Theorie Analytique de la Chaleur*, Firmin Didot, Paris, 1822.
6. C. Cattaneo, *C. R. Acad. Sci., Sér. I*, 1958, **247**, 432.
7. P. Vernotte, *C. R. Acad. Sci.*, 1958, **246**, 3154.
8. Y. Touloukian, *Thermal Conductivity*, Nonmetallic Solids, FI/Plenum, 1970, vol. 2.
9. Y. Touloukian, R. Powell, C. Ho and P. Klemens, *Thermal Conductivity: Metallic Elements and Alloys*, Plenum Press, New York, 1970.
10. Y. Touloukian, P. Liley and S. Saxena, *Thermal Conductivity: Nonmetallic Liquid and Gases*, IFI/Plenum, New-York and Washington, 1970.
11. C. Kittel, *Introduction to Solid State Physics*, Wiley, 8th edn, 2005.
12. D. Bolmatov, V. Brazhkin and K. Trachenko, *Sci. Rep.*, 2012, **2**, 421.
13. J. M. G. Vilar and J. M. Rubi, *Proc. Natl. Acad. Sci. U. S. A.*, 2001, **98**, 11081.
14. E. Fermi, J. Pasta and S. Ulam, *Studies of Nonlinear Problems*, Document LA-1940, 1955.
15. S. Lepri, R. Livi and A. Politi, *Phys. Rep.*, 2003, **377**, 1.
16. P. Kim, L. Shi, A. Majumdar and P. McEuen, *Phys. Rev. Lett.*, 2001, **87**, 201502.
17. E. Pop, D. Mann, Q. Wang, K. Goodson and H. Dai, *Nano Lett.*, 2006, **6**, 96.
18. S. Ghosh, W. Bao, D. Nika, S. Subrina, E. Pokatilov, C. Lau and A. A. Balandin, *Nat. Mater.*, 2010, **9**, 555.
19. N. Yang, G. Zhang and B. Li, *Nano Today*, 2010, **5**, 85.
20. T. Hsiao, H. Chang, S. Liou, M. Chu, S. Li and C. Chang, *Nat. Nanotechnol.*, 2013, **8**, 534.
21. A. Arbouet, N. Del Fatti, F. Vallée, M. Pellarin and M. Pileni, *Phys. Rev. Lett.*, 2003, **90**, 177401.
22. J. Hohlfeld, S. Wellershoff, J. Gudde, U. Conrad, V. Jahnke and E. Matthias, *Chem. Phys.*, 2000, **251**, 237.
23. S. G. Volz, *Phys. Rev. Lett.*, 2001, **87**, 074301.
24. G. Chen, *Phys. Rev. Lett.*, 2001, **86**, 2297.
25. G. Chen, *ASME J. Heat Transfer*, 2002, **124**, 320.
26. D. Tzou, *Macro- To Micro-Scale Heat Transfer: The Lagging Behavior*, CRC Press, 1996.
27. R. Guyer and J. Krumhansl, *Phys. Rev.*, 1964, **5A**, 1411.

28. D. Joseph and L. Preziosi, *Rev. Mod. Phys.*, 1989, **61**, 41.
29. G. Mahan and F. Claro, *Phys. Rev. B: Condens. Matter Mater. Phys.*, 1988, **38**, 1963.
30. P. Carruthers, *Rev. Mod. Phys.*, 1961, **33**, 92.
31. M. G. Holland, *Phys. Rev.*, 1963, **132**, 2461.
32. M. G. Holland, *Phys. Rev.*, 1964, **A134**, 471.
33. G. Srivastava, *The Physics of Phonons*, Taylor and Francis, 1990.
34. J. Callaway, *Phys. Rev.*, 1959, **113**, 1046.
35. P. Allen, J. Feldman, J. Fabian and F. Wooten, *Philos. Mag. B*, 1999, **79**, 1715.
36. D. G. Cahill, S. K. Watson and R. O. Pohl, *Phys. Rev. B: Condens. Matter Mater. Phys.*, 1992, **46**, 6131.
37. D. G. Cahill and R. O. Pohl, *Phys. Rev. B: Condens. Matter Mater. Phys.*, 1987, **15**, 4067.
38. S. Shen, A. Henry, J. Tong, R. Zheng and G. Chen, *Nat. Nanotechnol.*, 2010, **5**, 251.
39. C. L. Choy, *Polymer*, 1977, **18**, 984.
40. R. Wang, R. Segalman and A. Majumdar, *Appl. Phys. Lett.*, 2006, **89**, 173113.
41. J. Forrest, K. Dalnoki-Veress and J. Dutcher, *Phys. Rev. E: Stat. Phys., Plasmas, Fluids, Relat. Interdiscip. Top.*, 1997, **56**, 5705.
42. M. Losego, L. Moh, K. Arpin, D. G. Cahil and P. Braun, *Appl. Phys. Lett.*, 2010, **97**, 011908.
43. S. Datta, *Electronic Transport in Mesoscopic Systems*, Cambridge Studies in Semiconductor Physics and Microelectronic Engineering, 1997.
44. M. M. Yovanovich, Conduction and thermal contact resistance (conductance), in *Handbook of Heat Transfer*, McGraw-Hill, New York, 1998.
45. A. Majumdar and C. L. Tien, *J. Heat Transfer*, 1991, **113**, 516.
46. M. Williamson and A. Majumdar, *J. Heat Transfer*, 1992, **114**, 802.
47. B. Auld, *Acoustic Fields and Waves in Solids*, R.E. Krieger, 1990, vol. 1 and 2.
48. W. Little, *Can. J. Phys.*, 1959, **37**, 334.
49. E. T. Swartz and R. O. Pohl, *Rev. Mod. Phys.*, 1989, **61**, 605.
50. S. Merabia and K. Termentzidis, *Phys. Rev. B: Condens. Matter Mater. Phys.*, 2012, **86**, 094303.
51. B. C. Daly, K. Kang, Y. Wang and D. G. Cahill, *Phys. Rev. B: Condens. Matter Mater. Phys.*, 2009, **80**, 174112.
52. A. J. Minnich, J. A. Johnson, A. Schmidt, K. Esfarjani, M. Dresselhaus, K. Nelson and G. Chen, *Phys. Rev. Lett.*, 2011, **107**, 095901.
53. A. Henry and G. Chen, *J. Comput. Theor. Nanosci.*, 2008, **5**, 141.
54. A. Ward and D. Broido, *Phys. Rev. B: Condens. Matter Mater. Phys.*, 2010, **81**, 085205.
55. A. J. Minnich, 2014, Towards a microscopic understanding of phonon heat conduction, arXiv:1405.0532.
56. H. B. G. Casimir, *Physica*, 1938, **5**, 595.
57. E. Sondheimer, *Philos. Mag.*, 1952, **1**, 1.
58. J. M. Ziman, *Electrons and Phonons*, Clarendon Press, Oxford, 1960.

59. S. Volz and P. O. Chapuis, *J. Appl. Phys.*, 2008, **103**, 034306.

60. G. Chen, *ASME J. Heat Transfer*, 1996, **118**, 539.

61. J. A. Johnson, A. Maznev, J. Cuffe, J. Eliason, A. J. Minnich, T. Kehoe, C. M. Sotomayor Torres and K. Nelson, *Phys. Rev. Lett.*, 2013, **110**, 025901.

62. A. J. Minnich, *Phys. Rev. Lett.*, 2012, **109**, 205901.

63. K. Regner, D. Sellan, Z. Su, C. Amon, A. McGaughey and J. Malen, *Nat. Commun.*, 2013, **4**, 1640.

64. P. O. Chapuis, M. Prunnila, A. Shchepetov, L. Scheider, S. Laakso, J. Ahopelto and C. M. Sotomayor Torres, Effect of phonon confinement on heat dissipation in ridges, in *Proceedings of THERMINIC 16*, Barcelona, Spain, 2010.

65. M. F. Modest, *Radiative Heat Transfer*, Academic Press, 2003.

66. N. Zen, T. Puurtinen, T. Isotalo, S. Chaudhury and I. Maasilta, *Nat. Commun.*, 2014, **5**, 3435.

67. J. A. Ravichandran, R. Yadav, P. Cheaito, A. Rossen, S. Soukiassian, J. Duda, B. M. Foley, C. H. Lee, Y. Zhu, A. Lichtenberger, J. Moore, D. Muller, D. Schlom, P. Hopkins, A. Majumdar, R. Ramesh and M. Zurbuchen, *Nat. Mater.*, 2014, **13**, 168.

68. E. Blandre, P. O. Chapuis, M. Francoeur and R. Vaillon, *AIP Adv.*, 2015, **5**, 057106.

69. J. Mayo and A. Narayanaswamy, Mimimum radiative transfer between two metallic surfaces, in *11th ASME-AIAA Joint Thermophysics and Heat Transfer Conference*, Atlanta, 2014.

70. D. Polder and M. Van Hove, *Phys. Rev. B: Solid State*, 1971, **4**, 3303.

71. K. Joulain, J. Mulet, F. Marquier, R. Carminati and J. J. Greffet, *Surf. Sci. Rep.*, 2005, **57**, 59.

72. S. Shen, A. Narayanaswamy and G. Chen, *Nano Lett.*, 2009, **9**, 2909.

73. E. Rousseau, A. Siria, G. Jourdan, S. Volz, F. Comin, J. Chevrier and J. J. Greffet, *Nat. Photonics*, 2009, **3**, 514.

74. R. Ottens, V. Quetschke, S. Wise, A. Alemi, R. Lundock, G. G. Mueller, D. Reitze, D. Tanner and B. Whiting, *Phys. Rev. Lett.*, 2011, **107**, 014301.

75. T. Kralik, P. Hanzelka, M. Zobac, V. Musilova, T. Fort and M. Horak, *Phys. Rev. Lett.*, 2012, **109**, 224302.

76. A. V. Shchegrov, K. Joulain, R. Carminati and J. J. Greffet, *Phys. Rev. Lett.*, 2000, **85**, 1548.

77. K. Joulain, Rayonnement thermique aux courtes échelles, in *Micro et Nanothermique*, ed. S. Volz, CNRS edn, 2006, p. 119.

78. M. Krüger, G. Bimonte, T. Emig and M. Kardar, *Phys. Rev. B: Condens. Matter Mater. Phys.*, 2012, **86**, 115423.

79. D. G. Cahill, *Rev. Sci. Instrum.*, 1990, **61**, 802.

80. T. Borca-Tasciuc, A. Kumar and G. Chen, *Rev. Sci. Instrum.*, 2001, **72**, 2139.

81. J. Y. Duquesne, D. Fournier and C. Frétigny, *J. Appl. Phys.*, 2011, **108**, 086104.

82. C. Dames and G. Chen, *Rev. Sci. Instrum.*, 2005, **76**, 124902.

83. R. Gereth and K. Hubner, *Phys. Rev.*, 1964, **134**, A235.

84. T. F. McNelly, S. J. Rogers, D. J. Channin, R. J. Rollefson, W. M. Goubau, G. E. Schmidt, J. A. Krumhansl and R. O. Pohl, *Phys. Rev. Lett.*, 1970, **24**, 100.

85. S. Perichon, V. Lysenko, B. Remaki, B. Champagnon and D. Barbier, *J. Appl. Phys.*, 1999, **86**, 4700.

86. J. S. Reparaz, E. Chavez-Angel, M. R. Wagner, B. Graczykowski, J. Gomis-Bresco, F. Alzina and C. M. S. Torres, *Rev. Sci. Instrum.*, 2014, **85**, 034901.

87. S. Dilhaire, G. Pernot, G. Calbris, J. Rampnoux and S. Grauby, *J. Appl. Phys.*, 2011, **110**, 114314.

88. J. A. Johnson, A. Maznev, M. Bulsara, E. Fitzgerald, T. Harman, S. Calawa, C. Vineis, G. Turner and K. Nelson, *J. Appl. Phys.*, 2012, **111**, 023503.

Section II
Luminescence-based Thermometry

Quantum Dot Fluorescence Thermometry

DANIEL JAQUE GARCÍA* AND JOSÉ GARCÍA SOLÉ

Fluorescence Imaging Group, Departamento de Física de Materiales, Facultad de Ciencias, Universidad Autónoma de Madrid, Madrid 28049, Spain
*Email: daniel.jaque@uam.es

4.1 Quantum Dots: An Introduction

In general the colour of a given crystal does not depend on its size or shape. For instance, the typical pink colour of ruby, the first laser crystal, is given by the energy level diagram of Cr^{3+} dopant ions in the Al_2O_3 matrix, and so independent of the crystal size or shape.[1] However the fluorescence properties of semiconductor crystals, that have their absorption edge in the visible or near-infrared spectral region, are strongly modified when they are confined to the nanoscale, in such a way that the emission colour becomes size-dependent. For instance CdSe bulk crystals have a typical grey colour, which is mostly due to their energy gap at 1.74 eV (712 nm).[2] However when these crystals are synthesized as spheres of nanometric diameter (*quantum dots, QDs*) the optical properties of CdSe change dramatically.[3] As an example, Figure 4.1 shows how the emission colour of a solution containing QDs of CdSe gradually shifts from red to blue as the dot diameter is reduced from 8 to 2 nm, respectively. This amazing size-induced change in the emission colour provides a variety of applications in different fields, such as laser technology, display devices, highly

RSC Nanoscience & Nanotechnology No. 38
Thermometry at the Nanoscale: Techniques and Selected Applications
Edited by Luís Dias Carlos and Fernando Palacio
© The Royal Society of Chemistry 2016
Published by the Royal Society of Chemistry, www.rsc.org

Figure 4.1 Digital pictures of the fluorescence produced by water solutions containing CdSe QDs of different size, ranging from 2 to 8 nm. Excitation was performed by UV radiation.
Reproduced with permission from ref. 71. © Elsevier 2009.

efficient solar cells and high-brightness and high-resolution fluorescence imaging.[4–6] In particular, the ability of QDs to generate size-controlled well-defined emission colours makes it possible to use these dots as highly sensitive fluorescent thermal nanosensors.

In this chapter we first describe the basic mechanisms leading to the ability of QDs as size-controllable fluorescent probes (Section 4.2.1). Then we analyze how different fluorescence features (emission peak, emission intensity and emission lifetime) can be affected by temperature changes and, therefore, how QDs can be used as fluorescent nanothermo-meters (Section 4.2.2). We will describe how the thermal response of the optical properties of QDs is extremely dependent on their environment, (surface coating plus liquid or solid medium). This strong environmental dependence is described in detail in Sections 4.2.2 and 4.2.3. The last part of the chapter (Section 4.3) is dedicated to highlighting relevant recent applications of these spectacular fluorescent thermal sensors. Certainly, QD fluorescence thermometry is an emerging new field and the application of these probes as thermal sensors is evolving dramatically. In particular, here we describe what for us are the most relevant applications in biomedicine (Sections 4.3.1 and 4.3.3) and in microelectronics (Section 4.3.2). These have propelled the development of novel complex QD-based nanoparticles specifically engineered for accurate thermal sensing. Thus, the last section of this chapter (Section 4.3.4) is devoted to describing the structure and operating methods of novel QD-based nanoheterostructures.

4.2 Fluorescent Properties of Quantum Dots

4.2.1 Basic Principles: Quantum Confinement Effects

When a semiconductor crystal is excited with photons of energy equal to or larger than the energy gap, electrons (and holes) became free to move in the crystal. It is also well known that when free particles (electrons or holes) are spatially confined, for instance in a box, their energy states become restricted, leading to size-dependent discrete energy levels. As a general rule, the separation between these energy levels increases with the degree of confinement. For small enough systems (nanocrystals) these quantum confinement effects start to be manifested leading to strong size-dependent fluorescence properties. In bulk semiconductors excited electrons (and holes) can move as free particles with an associated characteristic wavelength, the so-called *de Broglie wavelength*, $\lambda_B = h/p_x$ (h being the Plank constant and p_x the linear momentum in the confinement direction, x). Let us, for instance, consider that the thickness of a semiconductor bulk crystal is reduced down to about the de Broglie wavelength, λ_B, corresponding to the thermal motion of electrons.[7] Such one-dimensional (1D) confined structures are usually denoted *quantum wells* (QWs). We can roughly estimate how short the thickness of a QW in a semiconductor crystal must be in order to observe quantum confinement effects by simply considering that the thermal energy associated with a 1D degree of motion, $\frac{1}{2}kT$ (k being the Boltzmann constant and T the absolute temperature), is the kinetic energy of the electron, $p_x^2/2m_e^*$ (m_e^* being the effective mass of the electron). Taking into account that confinement effects will be manifested when $x \approx h/p_x$, this condition can be written as:

$$\frac{(h/x)^2}{2m_e^*} \approx \frac{1}{2}kT \tag{4.1}$$

and solving for the thickness

$$x \approx \sqrt{\frac{h^2}{m_e^* kT}}. \tag{4.2}$$

For the case of CdSe, the semiconductor crystal host of the QDs included in Figure 4.1, $m_e^* = 0.13m_e$ (m_e being the electronic mass) and so we find that the thickness must be reduced to about 30 nm in order to observe confinement effects. This rough calculation indicates the importance of confining to the *nanoscale* in order to observe the size-dependent effects caused by quantum confinement. QWs are 1D confined structures but the spatial confinement of charge carriers can also be extended to two dimensions (structures called *quantum wires*) or three dimensions (QDs). QDs are very small spheres and so correspond to a spatial three-dimensional (3D) spherical confinement of the carrier particles. To understand the spectroscopic features and optical properties associated with a QD, we have to first

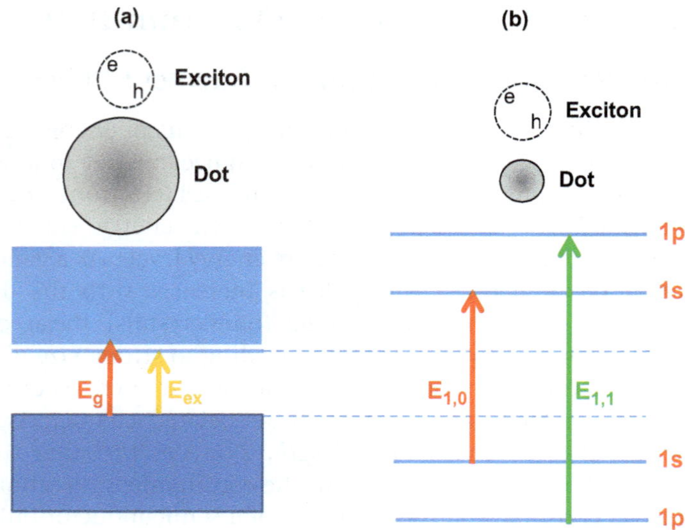

Figure 4.2 Schematics showing the energy level structure corresponding to (a) large and (b) small QDs, as defined with respect to exciton size.

think about the optical transitions that are produced when a bulk semiconductor is excited with photons near the bandgap energy E_g. Excitation just below the bandgap leads to the formation of electron–hole pair structures [see Figure 4.2(a)]. These electron–hole pairs, which can somehow move together in the semiconductor host crystal, are called *excitons*. The exciton size is typically characterized by the so-called *exciton Bohr radius*, a_B^e, given by $a_B^e = \varepsilon_r(m_e/\mu)a_B$, where μ is the effective exciton mass $\left(\mu = \dfrac{m_e^* + m_h^*}{m_e^* \cdot m_h^*}\right)$, m_e^* and m_h^* are the effective masses of the electron and hole (respectively), ε_r the dielectric constant of the semiconductor crystal and a_B^e the Bohr radius (0.053 nm).[8] At this point we can say that quantum confinement effects in QDs are observed when the dot size is on the order of or smaller (strong confinement) than the exciton size in the bulk crystal [see Figure 4.2(b)].

Considering again the case of CdSe QDs ($\varepsilon_r = 10.1$, $m_e^* = 0.13 m_e$ and $m_h^* = 0.45 m_e$), a bulk exciton radius $a_B^e \approx 5.3$ nm is estimated. Thus, quantum confinement effects should be observed for dot sizes (diameters) shorter than about 10 nm. This is in fact corroborated by Figure 4.3, which shows absorption spectra of CdSe QDs of different sizes.[9,10] Indeed the absorption spectrum of the 20 nm size dots (*i.e.*, dot radius 10 nm) still resembles that of the bulk crystal, and so no confinement effects are observed. However, a blue-shift together with the appearance of defined absorption peaks are both observed for dots smaller than about 10 nm. In particular, it can be seen how for QDs smaller than about 5 nm, a series of individual peaks is resolved. These peaks correspond to different transitions

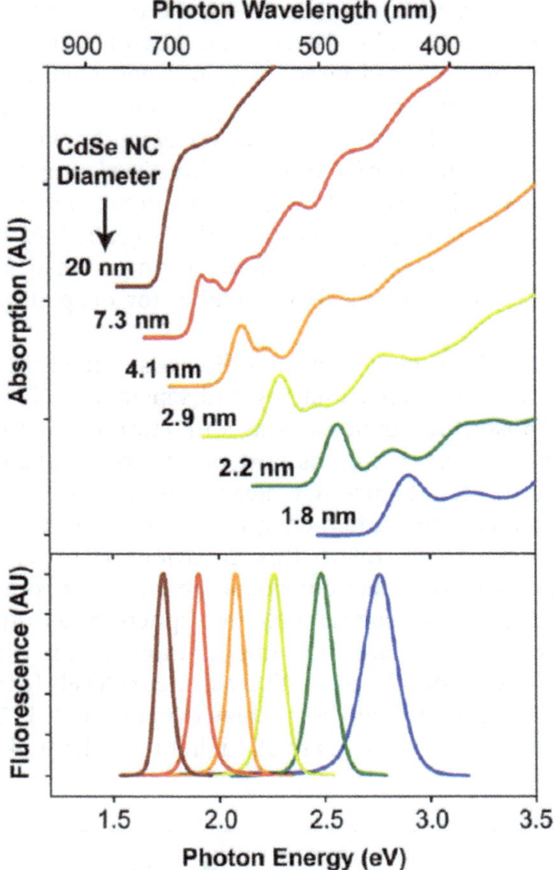

Figure 4.3 Normalized room temperature absorption (top) and emission (bottom) spectra of CdSe QDs of different diameters. Data extracted from ref. 9.

Table 4.1 Energy gaps and Bohr radii of II–VI and IV–VI semiconductors for relevant QD applications.

Material	Bandgap/eV	Bohr radius/nm
CdS	2.42	2.8
CdSe	1.74	5.3
CdTe	1.49	7.5
PbS	0.37	20
PbSe	0.27	46

in the dots [see Figure 4.2(b)]. Typically, the lowest energy absorption peak (transition energy $E_{0,1}$ in Figure 4.2) is denoted as the *first exciton transition*.

In Table 4.1 the Bohr radii and energy gaps of some II–VI and IV–VI semiconductor bulk crystals, typically used for the synthesis of fluorescent QDs, are listed. It should be noted that the IV–VI Pb-chalcogenides have much larger Bohr radii than the II–VI Cd-based nanomaterials, so that confinement effects can be observed for much larger dots. In addition, the bandgaps of the Pb-based QDs are also much smaller than those of the Cd QDs and so the former are very suitable for tuneable optical detectors and sources in the infrared (wavelengths larger than about 800 nm). In fact, a number of PbS QDs with sizes ranging from 3–7 nm are now commercially available and considered as very promising infrared fluorescence probes working in the so-called *biological window* (BW, 800–1600 nm). Since in this spectral range tissues are partially transparent, these QDs are especially suitable for deep tissue biomedical imaging.[11–13]

Let us follow our discussion with the CdSe QDs, one of the most popular QDs in the field of fluorescence imaging and sensing. In Figure 4.4 we have plotted the first absorption band peak as a function of the CdSe dot radius for dot radii shorter than 10 nm, as obtained from the absorption spectra displayed in Figure 4.3. Figure 4.4 clearly shows how the first exciton absorption peak shifts to higher energies as the dot radius is reduced. The simplest approach to explain this behaviour is to use the so-called *effective mass approximation*, that considers the quantum confinement of an isolated electron and of an isolated hole in a sphere by assuming that their effective masses in the dot are the same as those in the bulk.[14] Solution of the Schrödinger equation leads to discrete energy levels (denoted by the n and l quantum numbers), like those shown Figure 4.2(b). The transition energies between these levels, $E_{n,l}$, are all higher than the bulk E_g energy gap

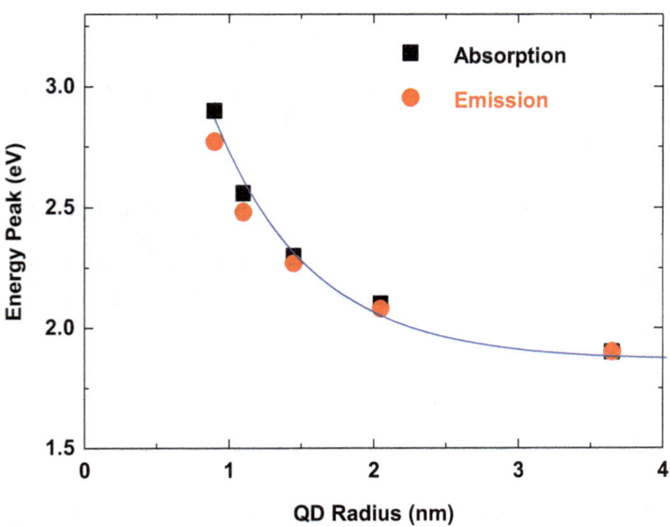

Figure 4.4 First exciton absorption and emission energies as a function of dot radius for CdSe QDs.
Data obtained from the analysis of Figure 4.3. Solid line is a guide for the eyes.

by a size-dependent energy shift, $\Delta E_{n,l}$, so that $E_{n,l} = E_g + \Delta E_{n,l}$ and the $\Delta E_{n,l}$ energies are given by:

$$\Delta E_{n,l} = \frac{h^2 \alpha_{n,l}^2}{8\pi^2 R^2 \mu} \tag{4.3}$$

where $\alpha_{n,l}$ are dimensionless values to account for the different transitions. For instance for $n = 1$ and $l = 0$ (s states) $\alpha_{1,0} = \pi$.[15] Thus, the first observed absorption transition ($E_{1,s} \rightarrow E_{1,s}$) should scale such that $E_{1,s \rightarrow 1,s} - E_g \propto \dfrac{1}{R^2}$. Indeed, this $\dfrac{1}{R^2}$ trend does not fit with the behaviour observed in Figure 4.4. In fact, in this first approach, we have considered that electron and hole are independent, but the exciton nature of the first excitations involves electron–hole pairs, in which both particles attract each other by Coulomb interaction. Therefore, when the dot size of a semiconductor crystal is smaller than the exciton Bohr radius, the dependence of the electron–hole Coulomb interaction with dot radius must be considered, so that the first exciton absorption transition ($E_{1,s \rightarrow 1,s} = E_{1,0}$) should appear at an energy given by:

$$E_{1,0} = E_g + \Delta E_{1,0} - E_{exc} \tag{4.4}$$

where $E_{exc} = \dfrac{1}{R}\left(\dfrac{1.8e^2}{4\pi\varepsilon_0\varepsilon_r}\right)$ describes the effective Coulomb electron–hole interaction.[15]

This fact makes the problem more difficult to solve. Indeed, in a more realistic calculation, aspects such as the size-induced variation of the dielectric constant ε_r and real boundary conditions of the wave functions (not being an infinite potential at the QD surface) should be also considered. To elucidate this problem Meulenberg *et al.* experimentally determined the exciton binding energy ($E_{1,0}$–E_g) of CdSe QDs as a function of dot size by using X-ray absorption and photoemission spectroscopy.[16] They estimated a complex dependence due to different scaling factors arising from the valence ($1/R^{1.6}$) and conduction bands ($1/R^{0.6}$), which are in good agreement with direct pseudopotential calculations of exciton Coulomb and exchange energies in semiconductor QDs.[17] Certainly, at the present time, a proper theoretical model to scale the first exciton absorption peak with the dot radius in order to provide a direct QD size readout from the extinction spectra is still a subject of intense research. Nevertheless, in this respect it is important to mention that a systematic experimental study of the absorption spectra of high-quality CdTe, CdSe and CdS QDs has been performed by Yu *et al.*[18] This study can be used to get a direct readout of the dot size of these semiconductors from both their first exciton absorption peak spectral position and extinction coefficient. As an example, Figure 4.5 shows the *sizing curves* for three Cd-based QDs. These curves are very useful for determining the dot diameter D from the wavelength λ of the first exciton absorption peak. In fact, empirical fitting $D(\lambda)$ functions have been given for direct numerical estimation.

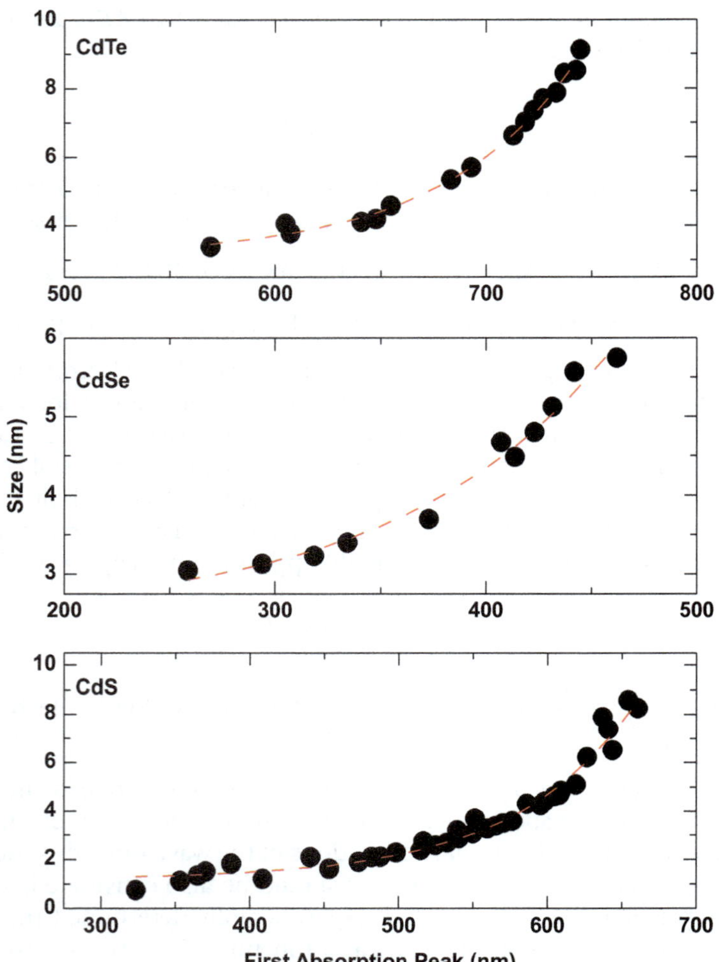

Figure 4.5 Sizing curves determined by Yu *et al.* for CdTe, CdSe and CdS QDs. The circles are experimental data and the dashed lines correspond to the best fitting curves (see text).
Data extracted from ref. 18.

Figure 4.3 also includes the emission spectra of the CdSe QDs with different sizes. For each dot size a single and symmetric emission band is obtained at different spectral locations, thus explaining the different emission colours observed in Figure 4.1 for the different QD sizes. These single emission bands are, in general, produced by the radiative de-excitation from to the lowest electronic excited state of the dot (1s in Figure 4.2(b) to the highest ground state. In fact, excitation at wavelengths shorter than the first exciton produces essentially the same single-emission spectrum due to non-radiative depopulation down to the first excited electron level of the exciton [1s in Figure 4.2(b)]. In Figure 4.4 we have also included the emission peak positions for each dot radius, which can be compared to the first exciton absorption peak. The emission peak also follows quite a similar trend with dot size, as the one observed for the first absorption peak. However, the existence of a slight red-shift of emission with respect to the first exciton absorption peak should be noted. This shift, usually called *Stokes shift*, is a general feature observed for localized states and, in semiconductors, occurs because the excitation and emission processes involve different energy levels. In fact, the fine structure of both the excited and ground-state first exciton levels is not as simple as that shown in Figure 4.2(b). Indeed, the first exciton levels are split by crystal field and electron–hole exchange interactions, leading to forbidden transitions (*dark excitons*) and allowed transitions (*bright excitons*). This fine structure accounts for the Stokes shift of QD emission spectra as the excitation/emission steps involve different pairs of levels and the emission is also mediated by non-radiative de-excitations.[19] Moreover, surface defects can also produce additional red-shift.[20] It is important to note (see Figure 4.4) that this Stokes shift decreases with increasing dot size, being negligible for radii larger than 3.5 nm.

The emission efficiency is a key parameter when using QDs for different applications. Bright QDs (QDs with large emission efficiencies) are especially suitable for fluorescence bio-imaging, lasers, white luminescent emission diodes, electroluminescent devices, and solar concentrators. The luminescence efficiency is usually given by the quantum yield (QY), defined as the ratio of the number of photons emitted to those absorbed per unit time. The QY can be estimated if the radiative (k_r) and non-radiative (k_{nr}) de-excitation rates from the excitonic excited state are both known, so that; $QY = k_r/(k_r + k_{nr})$. Here the problem of fluorescence blinking, random fluctuations in the emission of single QDs, constitutes an important problem to overcome.[21] Using a sufficiently large number of CdTe QDs, to avoid these fluctuations, Maestro *et al.* have shown that an optimum dot size can be designed in order to enhance the QY [see Figure 4.6(a)].[22] In that work, the largest QY (above 40%) was observed for a 3.8 nm dot diameter. This dependence was mostly attributed to a minimum non-radiative rate for this specific size, as shown in Figure 4.6(b); that includes the contributions of both k_r and k_{nr} to the total de-excitation rate, $k = k_r + k_{nr}$. While the radiative rate, k_r, displays a monotonic increasing behaviour with emission frequency peak (with decreasing dot size),[23] the non-radiative, k_{nr}, rate shows two-well defined regions: It decreases with dot size down to about a CdTe QD size of 3.8 nm, then reverses this tendency; for QD sizes smaller than 3.8 nm, k_{nr} displays an increasing tendency as the QD size is reduced. This latter tendency was related to a dramatic increase in the density of surface trap states (that act as fluorescence quenching centres) as result of the large surface/volume ratios arising for small QDs. On the other hand, the trend observed for dots larger than 3.8 nm, can be explained as a result of the quantum confinement effects that lead to larger separation between energy levels and then reduce the non-radiative probability. Data included in Figure 4.6 show evidence of the critical role played by the surface traps in the final QY of QDs. Indeed,

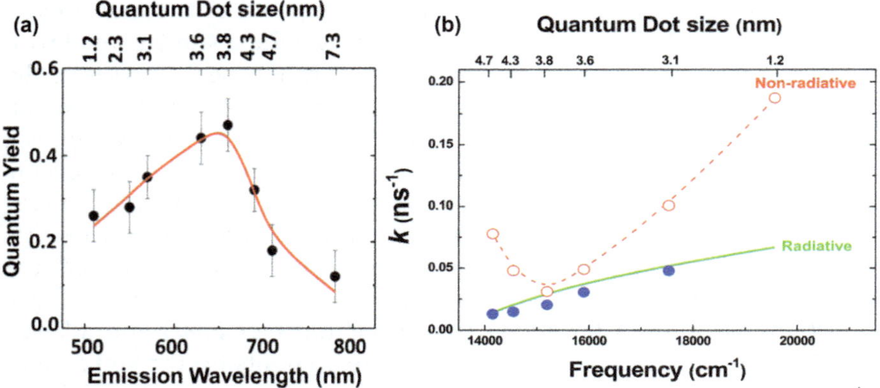

Figure 4.6   (a) Quantum yield as a function of the emission wavelength peak (quantum dot diameter) for CdTe QDs (dispersed in water). (b) Radiative and non-radiative rates as a function of the emission peak (dot size). Reproduced with permission from ref. 22. © AIP Publishing 2012.

the synthesis of QDs with QYs larger than 90% has been demonstrated based on core–shell architectures that reduce the density of trap states.[24]

4.2.2   Temperature Dependence of Quantum Dot Luminescence

A number of features (emission intensity, fluorescence decay time, emission peak position, emission bandwidth and Stokes shift) of QD fluorescence are affected by temperature variations. As will be demonstrated in Section 4.3, this thermal dependence of the fluorescence properties of QDs has been used by many research groups to obtain sub-micrometric thermal images of a great variety of systems. In the following we will focus on describing the fundaments of this thermal dependence. Figure 4.7(a) shows, as a relevant example, the emission spectra of CdSe QDs dispersed in water at three different temperatures within the physiological temperature range. As the temperature increases, the fluorescence intensity decreases and the emission is red-shifted. In some cases these two effects are accompanied by a remarkable change in the QD fluorescence lifetime. The temperature variations of these three parameters (peak position, luminescence intensity and lifetime) as obtained for the particular case of CdSe QDs are also included in Figure 4.7 [(b)–(d), respectively]. As will be demonstrated in the next section, the temperature-induced changes in these three emission features can be used for thermometry at the nanoscale. In the following we will focus on describing the basic physical principles at the origin of these temperature-induced changes.

4.2.2.1   *Temperature-induced Spectral Shift*

The temperature shift of the emission peak position is mostly related to the size-dependent temperature variation of the first exciton absorption peak.

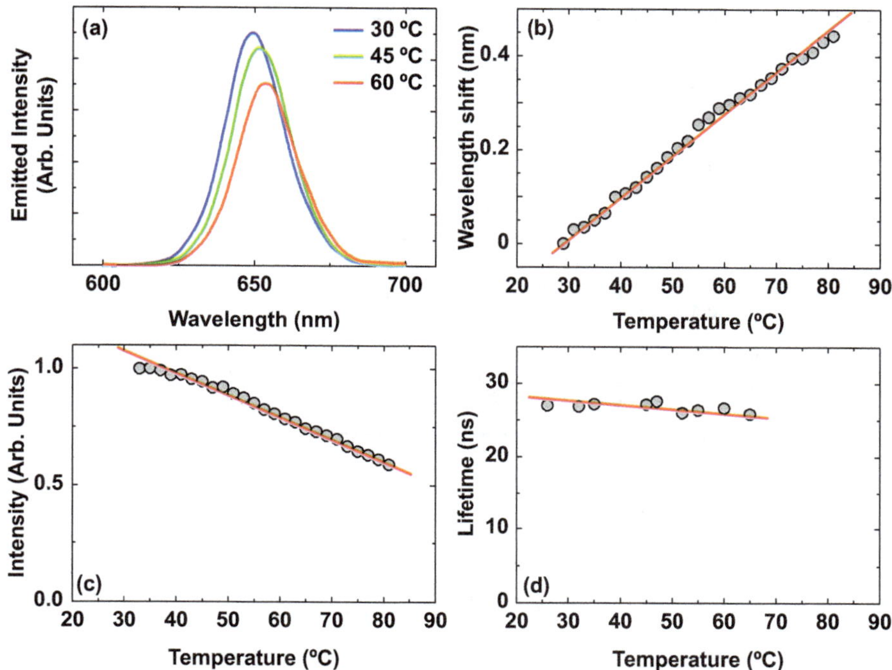

Figure 4.7 (a) Emission spectra of CdSe–SZn (core–shell) QDs (core of 4 nm, dispersed in water) at three different temperatures. The temperature dependence (in the physiological temperature range) of the peak displacement, integrated intensity and lifetime is shown in parts (b), (c) and (d), respectively.
Data were taken from the PhD thesis of Laura Martinez Maestro (Universidad Autónoma de Madrid, Spain).

Indeed, as shown in Figure 4.4, the emission peak is very close to the first exciton absorption peak, except for the weak Stokes shift, that will obviously also depend on temperature. The peak position of the first exciton absorption depends on a number of temperature-related effects, such as thermal expansion of the dot host lattice, temperature-induced changes in the confinement energy, temperature-induced mechanical strength and electron–phonon coupling. Olkhovets *et al.* investigated in great detail these effects in lead salt QDs.[25] Indeed IV–VI semiconductors, such as PbS QDs and PbSe QDs, are excellent systems for investigating size-quantized effect as they have large exciton Bohr radii (see Table 4.1) and so provide easy access to strong quantum confinement. Figure 4.8 shows, as a relevant example, the temperature variation (for low temperatures) of the absorption spectra of two PbS QDs embedded in oxide glasses with different radii, 8.5 and 4.5 nm. In both cases a blue-shift is observed in the first exciton peak as the temperature is increased.

However the temperature sensitivity of the first exciton peak ($dE_{1,0}/dT$) is clearly dependent on the dot size. Olkhovets *et al.* studied these

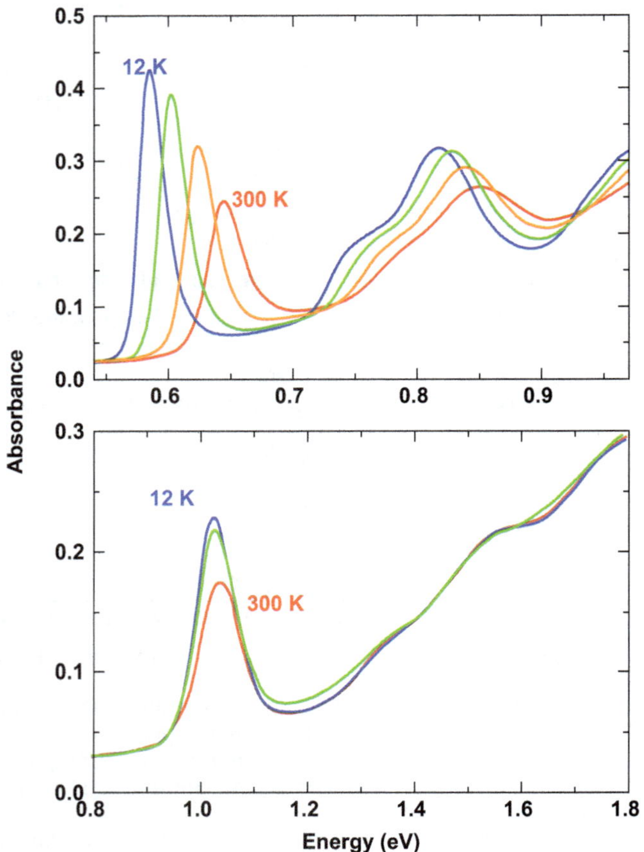

Figure 4.8 Absorption spectra of PbS QDs of two different sizes (8.5, upper figure, and 4.5 nm, bottom figure) in oxide glasses recorded at different temperatures.
Data extracted from ref. 25.

sensitivities for PbS QDs of different sizes embedded in various hosts (phosphate glass, oxide glass and polymers) (see Figure 4.9) and explained the observed behaviour by considering three main contributions to $dE_{1,0}/dT$, such that:[25]

$$\frac{dE_{1,0}}{dT} \approx \left(\frac{\partial E_{1,0}}{\partial T}\right)_{\text{lattice}} + \left(\frac{\partial E_{1,0}}{\partial T}\right)_{\text{conf.}} + \left(\frac{\partial E_{1,0}}{\partial T}\right)_{\text{el−ph coupling}} \tag{4.5}$$

where the first term on the right-hand side accounts for the thermal expansion coefficient of the PbS QD (lattice term), the second one for the temperature-induced change in the quantum confined energy, and the third

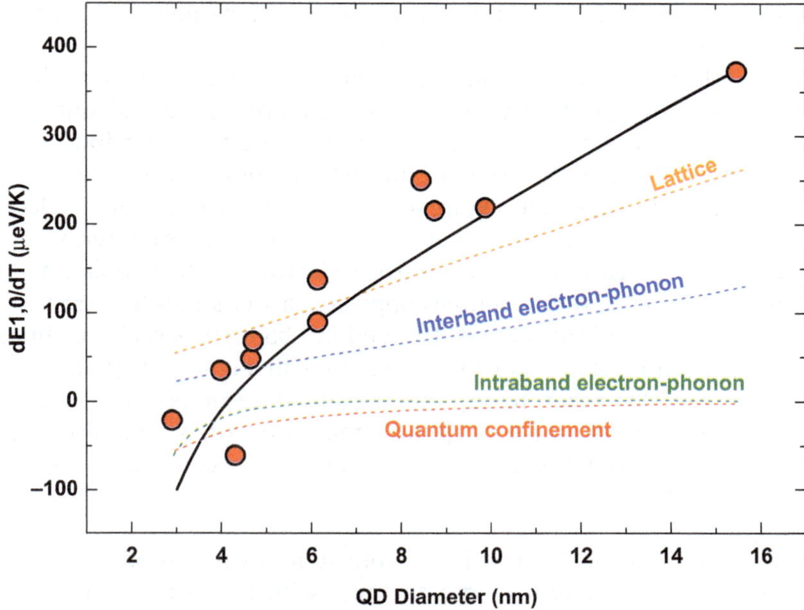

Figure 4.9 Thermal sensitivity of PbS QDs embedded in different hosts as a function of dot diameter. Dotted lines represent the different contributions to the thermal sensitivity. The solid line is the sum of all contributions. Data extracted from ref. 25.

one for temperature-related changes in the electron–phonon coupling. We now briefly describe the origin of these terms separately:

(1) The lattice dilation term takes into account the effect on the exciton energy as a result of thermal expansion of the lattice constant, a, of the dot. Thus it can be expressed as $\left(\partial E_{1,0}/\partial T\right)_{\text{lattice}} = \partial E_{1,0}/\partial a \cdot \partial a/\partial T$. In PbS, this term increases with the dot size up to the bulk value ($\approx 340\ \mu\text{e V K}^{-1}$). Its contribution to the overall thermal sensitivity is shown in Figure 4.9 (orange dashed line).

(2) Thermal expansion (QD size increment) implies less confinement, and so it affects the term due to quantum confinement [$\Delta E_{1,0}$ in eqn (4.4)], so that we can write:

$$\left(\frac{\partial E_{1,0}}{\partial T}\right)_{\text{conf.}} = \frac{\partial E_{1,0}}{\partial R}\cdot\frac{\partial R}{\partial T} = \frac{\partial E_{1,0}}{\partial R}\cdot\alpha R \tag{4.6}$$

where α is the thermal expansion coefficient of the dot and R is the dot radius. According to eqns (4.3) and (4.6) we can write that $\left(\partial E_{1,0}/\partial T\right)_{\text{conf.}} \propto -\dfrac{\alpha}{\mu}\cdot\dfrac{1}{R^2}$, so that this term accounts for a negative thermal sensitivity; the sensitivity decreasing as the dot size increases.

Indeed the effect of this term is only relevant for small (*ca.* <4 nm) PbS dot sizes (see red-dashed line in Figure 4.9).

(3) The electron–phonon coupling term arises from the non-discrete nature of the QD energy levels, which are broadened and shifted as a result of the vibrating lattice. In particular, two contributions are considered when electrons are promoted from occupied levels [hole levels, ground states in Figure 4.2(b)] to non-occupied levels [electron levels, excited states in Figure 4.2(b)]: An intraband term, which couples like carrier states (electron–electron or hole–hole), and an interband term, which couples opposite carrier states (electron–hole). The latter term is always positive and in PbS increases almost linearly with the dot size,[25] as can be observed in Figure 4.9 (blue dashed line). On the other hand, the intraband term is always negative and proportional to the coupling strength factor, S, that also slightly depends on the dot radius for small dots (see Figure 4.9, green dashed line).

The sum of all the above-mentioned contributions (see solid black line in Figure 4.9) is in reasonably good agreement with the experimental results obtained (dots in Figure 4.9). Thus, it can be realized that for large PbS QDs (>5 nm) positive values of $dE_{1,0}/dT$ occur so that QD fluorescence experiences a blue-shift with temperature. Moreover, for these dots the thermal sensitivity increases with the PbS dot size as the interband electron–phonon and lattice dilation terms are dominant. For very small dot sizes $dE_{1,0}/dT$ becomes negative, in such a way that for these 'small' QDs the QD fluorescence band is red-shifted as the temperature increases. This is due to the dominant character of the intraband electron–phonon coupling and confinement terms for these small dot sizes. This sign inversion in the thermal sensitivity is clearly observed for PbSe QDs (see Figure 4.10).[25,26] This figure shows the temperature dependence of the first exciton absorption peak for PbSe QDs of different sizes; 3.9 and 6.9 nm. As can be observed, the 3.9 nm QDs suffer from a linear red-shift with increasing temperature. In contrast, for the 6.9 nm PbSe dots a blue-shift is observed as the temperature increases.

4.2.2.2 Temperature-induced Fluorescence Quenching

As is clearly evidenced in Figure 4.7, temperature does not only cause a spectral shift in the QD fluorescence but also a remarkable fluorescence intensity reduction that, for the case of CdSe QDs, follows a linear relation with temperature [see Figure 4.7(c)]. However, this fluorescence reduction is not accompanied by any marked decrease in the average fluorescence lifetime, as can be observed in Figure 4.7(d), so that the assignment of the intensity decrease to a purely thermal quenching mechanism may not be totally true, as we will discuss below. As for every luminescent centre, the quantum efficiency of QDs is temperature dependent.[27] Generally speaking, above a given temperature (quenching temperature) the fluorescence

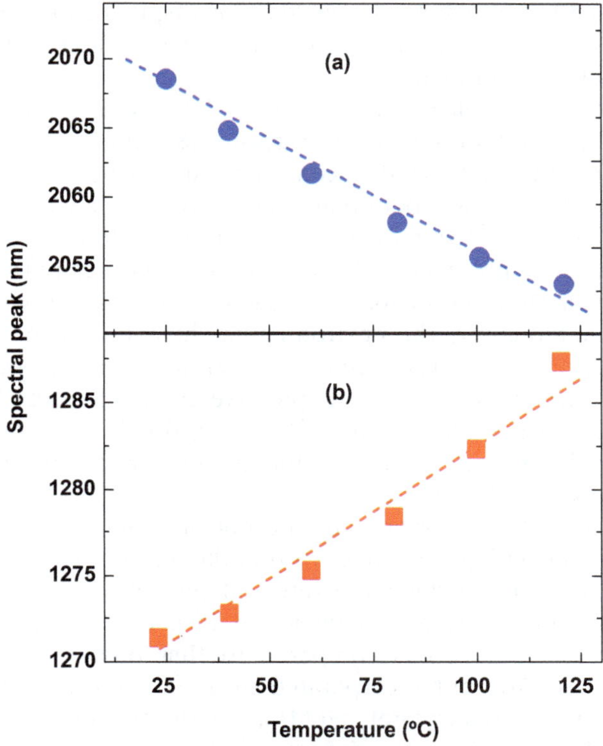

Figure 4.10 Temperature dependence of the first absorption peak of PbSe QDs of two different sizes; (a) 6.9 nm and (b) 3.9 nm.
Data extracted from ref. 26.

efficiency starts to decrease. For localized emitting centres, such as optical dopant ions in insulating crystals, two main mechanisms are responsible for thermal quenching:

(1) A very different electron–phonon coupling of the excited and ground states that accounts for large Huang–Rhys factors; and
(2) A multiphonon relaxation in centres with high-energy effective phonons so that the gap between the excited and ground states can be bridged by about five or fewer phonons.[1,27]

However, the thermal quenching of semiconductor materials is not related to these mechanisms. It is mostly due to thermally activated photo-ionization of the carrier charges (electron or holes), so that they can be more delocalized and reach a nearby trap non-fluorescent state. For instance, we can think of an exciton (electron–hole pair) where the temperature increase allows carriers to escape from the Coulomb attraction until a trap quenching state is reached. In semiconductor QDs the trap quencher states are usually surface states and, consequently, the dot environment has a tremendous influence on the thermal fluorescence quenching. Indeed, the fluorescence thermal quenching of organically coated QDs displays complex behaviours even from very low temperatures and is strongly dependent on the QD surface nature and environment.[28–30]

Surface traps and defects have been used to explain the fluorescence quenching above 50 K of organically capped CdSe and CdTe QDs.[28–30] Zhao *et al.* have recently published a comprehensive paper on thermal fluorescence quenching above room temperature for a variety of core–shell QD based nanostructures.[29] As a matter of fact, in order to reduce surface trap states related to the presence of defects and/or impurities in the immediate surroundings of QDs, the majority of practical devices are based on core–shell systems. Moreover, the environmental host in which the dots are deposited plays an important role. Zhao *et al.* tried several core–shell environment strategies in order to preserve high quantum fluorescence efficiencies at elevated temperatures.[29] Indeed, this is quite important not only for nanothermometry, but also for the application of QDs as colour converters in warm-white LEDs.[31]

In Figure 4.11 the importance of the dot environment in the thermal fluorescence quenching of CdSe (core)/CdS/ZnS (double-shell) QDs is manifest. In this figure the temperature dependence of the normalized emission intensity of these dots dispersed in a high-boiling-point liquid solvent (octadecene, ODE) is compared to that obtained for the same core/double-shell QDs but incorporated in a solid polymer matrix cross-linked poly(lauryl metha-crylate) (cPLMA). The thermal quenching is clearly different for the two different environments. The fluorescence quenching is significantly faster for the QDs dispersed in the liquid than those allocated in the solid environment. Indeed, as shown in the digital picture [Figure 4.11(b)], at 150 °C the fluorescence signal of the liquid solution has completely disappeared, while that of the dots in the solid environment is still clearly observed even at 200 °C. At this point it can be seen how the linear decrease in intensity [Figure 4.11(a)], for liquid-dispersed CdSe–ZnS core–shell QDs, is in qualitative agreement with that observed (Figure 4.7) in the physiological temperature range for similar dots dispersed in water. Another important aspect that can be extracted from Figure 4.11 is that the temperature dependences of the emitted intensity obtained during the cooling procedure are completely different to those observed during the heating process. Indeed, once the QDs/liquid solution has been sufficiently heated, the fluorescence cannot be recovered, so the process becomes irreversible. In contrast, this irreversibility is only partial for the QDs/solid environment system.

By performing an exhaustive experimental study, Zhao *et al.* were able to differentiate between reversible and irreversible fluorescence quenching processes. In general, the reversibility of a thermal quenching process occurs within a certain temperature range. Above this temperature, irreversible quenching takes place and the fluorescence is partially or completely lost. For instance, the thermal intensity decrease of the core/double shielded dots in the solid environment [see Figure 4.11(a)] is only reversible below 100 °C.

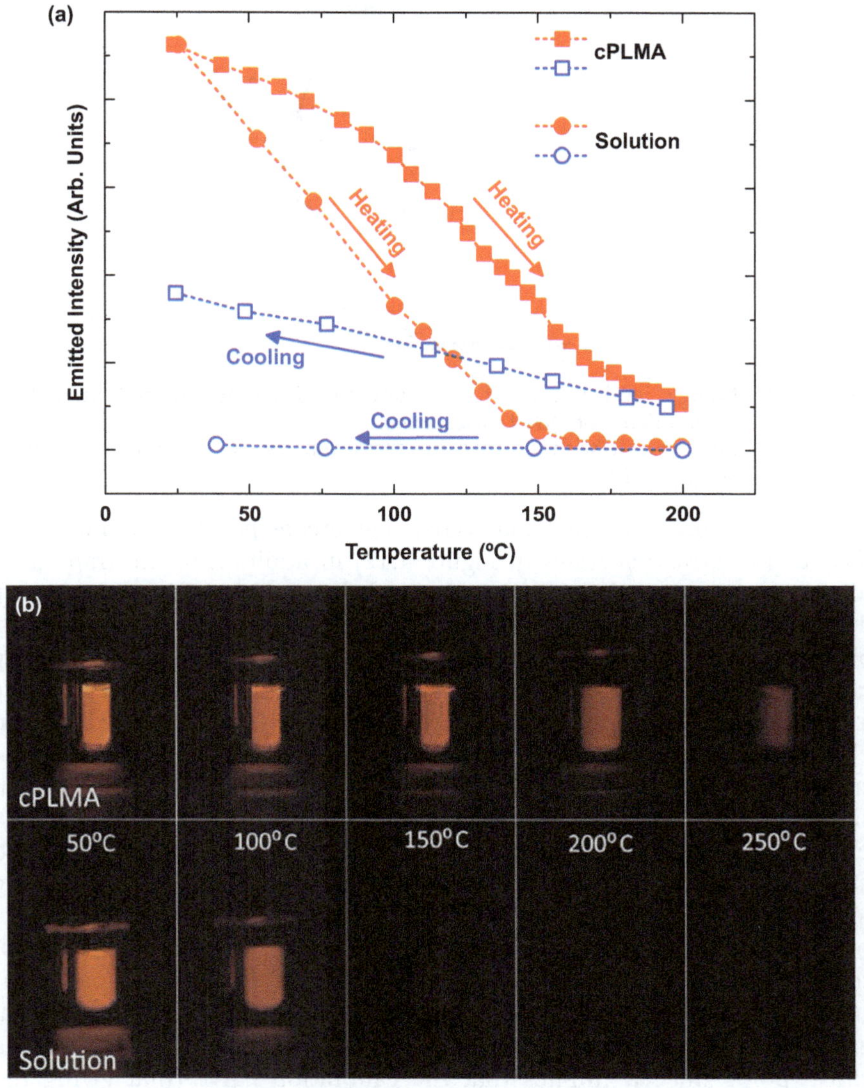

Figure 4.11 (a) Normalized emission intensity as a function of temperature for CdSe–CdS/ZnS (core–shell/shell) QDs (core diameter of 3.8 nm and full particle size of 7 nm) in two different environments (ODE, squares, and cPLMA, circles). The solid symbols correspond to heating cycles, while the open symbols correspond to cooling cycles. Pictures in (b) correspond to the QD solutions as obtained under UV excitation at different temperatures.
Reproduced with permission from ref. 29. © American Chemical Society 2012.

However, if the samples are heated above 100 °C the initial intensity and average lifetime are not recovered because of the creation of thermally induced structural changes (for instance defects induced by a mismatch

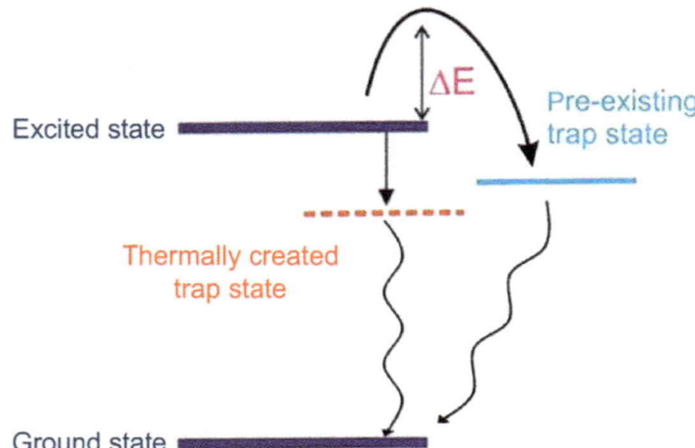

Figure 4.12   Energy level scheme to account for different thermal quenching mechanisms of QDs fluorescence.
Reproduced with permission from ref. 29. © American Chemical Society 2012.

between the core and shell materials) that create permanent trap states (*thermally created trap states* in Figure 4.12) in addition to the trap states existing prior to any thermal cycle (*pre-existing trap states* in Figure 4.12). These thermally induced permanent traps are responsible for the irreversible fluorescence quenching. In contrast, the reversible thermal quenching was explained by either thermally activated escape of carriers to the pre-existing trap states or by new thermally created trap states which can relax upon cooling. It is also important to note that, for many core–shell structures, the average lifetime does not follow the same trend with temperature as the emitted intensity. For instance, for the CdSe–CdS/ZnS solid-environment dots, the decreasing trend observed in the emitted intensity for temperatures below 100 °C is accompanied by an increasing trend in the average lifetime.[29] Thus, the thermal quenching processes are complicated and depend to a great extent not only on the particular dot structure (core, core–shell or core multishell) but also on the particular environment where the dot structures are hosted. For the purpose of thermal sensing based on the analysis of fluorescence intensity variations, this is a critical point since it implies that the calibration curve (that giving the temperature variation of fluorescence intensity) could be modified during sensing due to undesirable changes in the QD environment.

4.2.2.3   Temperature-induced Lifetime Variations

The emission decay curve is another interesting fluorescence feature that, in some cases, can be used for thermal sensing. Indeed, fluorescence lifetime is not affected by either the QD concentration or by slight intensity variations of the excitation source. As a consequence, QD-based lifetime thermal sensing appears to be an interesting method for fluorescence nanothermometry. Recently Haro-González *et al.* have demonstrated the possibility of using CdTe QDs for thermal sensing.[32] They first demonstrated that the fluorescence time decay is strongly dependent on the dot size. In general the fluorescence decay curves display a non-exponential shape and the average lifetime decreases with decreasing size. This lifetime decrease is particularly relevant for dot sizes shorter than the optimum one that we previously mentioned (about 4 nm, Figure 4.6) for fluorescence imaging. Importantly, the average lifetime of small dots is sensitive to changes in the temperature of their environment, as shown in Figure 4.13(a). In this figure, emission decay time curves at 30 and 50 °C are displayed for 1 and 7.8 nm CdTe QDs dispersed in water. It can be seen [Figure 4.13(b)] that, for 1 nm size dots, the average lifetime decreases with temperature, displaying a linear behaviour in the physiological range of temperatures. In contrast, the average lifetime of the 7.8 nm CdSe QDs/water dispersion is almost temperature-insensitive in this physiological range. In fact, the smaller the dot size, the larger the thermal sensitivity, as is shown in Figure 4.14(a). In this figure the normalized lifetime thermal coefficient $\alpha_\tau = d\tau^{nor}(T)/dT$, where $\tau^{nor}(T) = \tau(T)/\tau(25\,°C)$ is the average lifetime at temperature T normalized to 25 °C), is shown as a function of the CdTe dot size. It can be seen how α_τ monotonically increases by more than one order of magnitude as the dot size is shortened from 7.8 to 1 nm. In Figure 4.14(b) the temperature dependence of the normalized lifetime of CdTe dots (3.8 nm, a size designed for optimum fluorescence imaging) is compared to that of CdSe–ZnS (core–shell) QDs of similar size (4 nm). Indeed, the thermal sensitivity of the CdTe QDs is about five times that of the core–shell CdSe QDs and makes the former dots

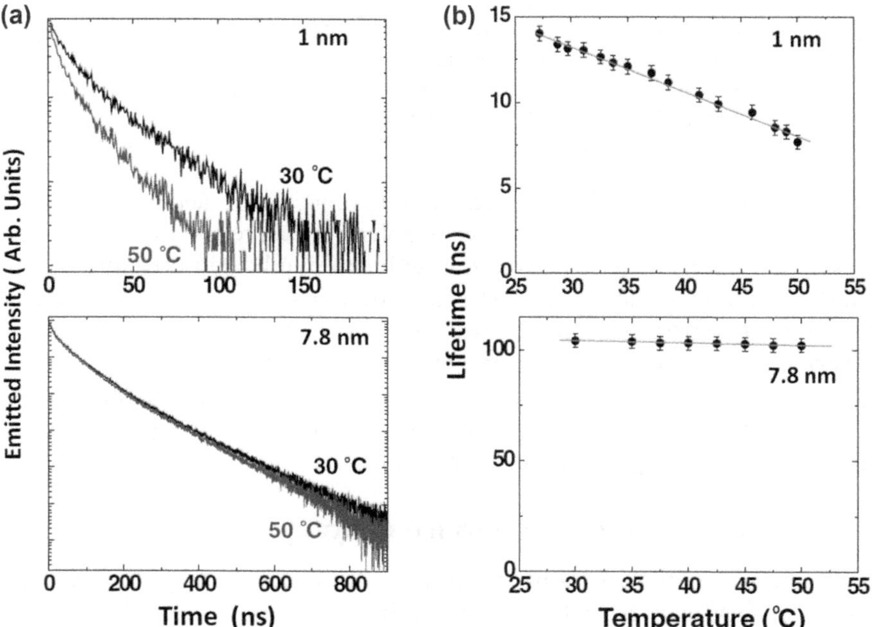

Figure 4.13 (a) Fluorescence decay time curves at two different temperatures for CdTe QDs of two different sizes. (b) Temperature dependence of the average lifetime for these dots.
Reproduced with permission from ref. 32. © Wiley-VCH 2012.

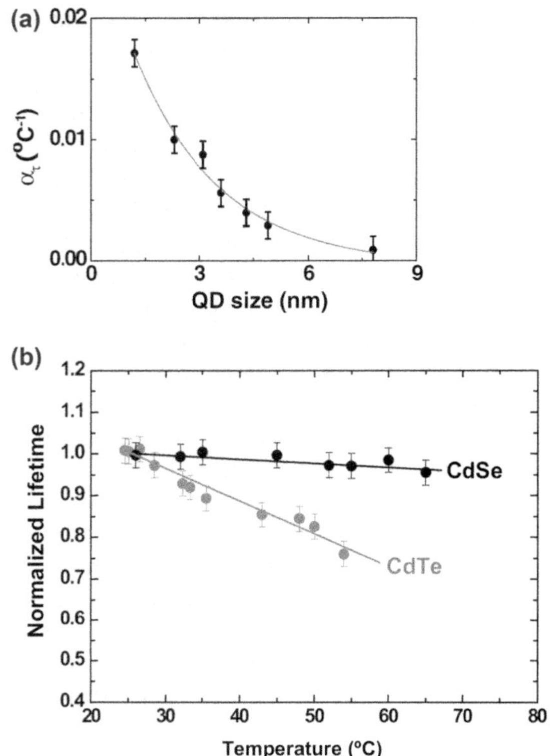

Figure 4.14 (a) Normalized thermal sensitivity of CdTe QDs (dispersed in water) as a function of the dot size. Line is to guide the eyes. (b) Normalized lifetime as a function of temperature for CdTe QDs (3.8 nm) and CdSe–SZn (core–shell, 4 nm core) QDs.
Reproduced with permission from ref. 32. © Wiley-VCH 2012.

excellent candidates for fluorescence lifetime imaging. In fact, based on CdTe (3.8 nm) QDs, Haro-González *et al.* were able to measure the thermal loading caused by a tightly focussed infrared laser beam that heated up a micro water solution containing these dots.[32] The high lifetime thermal sensitivity of these dots must be strongly related to the thermal quenching induced by their surface coating and, very likely, by their solvent environment too. Thus, each particular experiment would require an *in situ* calibration of the thermal sensitivity. Nevertheless, these results have provided a first step towards lifetime thermal imaging based on fluorescent QDs.

4.2.3 The Importance of Surface Coating

As described in the previous section, fluorescence temperature sensing based on QDs can be achieved either by monitoring their fluorescence features *i.e.*, intensity and average lifetime, or from analysis of the temperature-induced spectral shift of their fluorescence band (first exciton emission). We mentioned in the previous section how the QDs' environment can affect their thermally induced fluorescence quenching. As a matter of fact, the shift in the emission peak is probably the most relevant feature for thermal sensing, but it is also affected by the dot coating or environment. In this section we briefly discuss the importance of surface in this thermally induced spectral shift.

We have seen that the spectral shift displays a linear dependence with temperature in most cases. As has been already explained in Section 4.2.2, this thermally induced spectral shift is mostly caused by the temperature dependence of the first exciton absorption (excitation) peak, which varies due to a combination of different processes, including the effect of lattice dilation, electron–phonon coupling and the temperature-induced changes in the quantum confinement, due to QD lattice dilation and constriction (see Figure 4.9). In some cases, such as II–IV semiconductor QDs, the QD size variation due to temperature changes is small due to their small lattice thermal expansion coefficients. Nevertheless the assumption that the thermal expansion coefficient of a given QD is that of its host is only correct for large (>10 nm) QDs. Indeed, in these dots, only a small fraction of atoms is located on the surface and, thus, most of them form the 'bulk' core. On the other hand, for 'small' QDs (<3 nm) this is not fully correct. For instance, for 3 nm CdTe QDs it is estimated that about 35% of atoms are located on the surface. For these dots the 'effective' thermal expansion coefficient is strongly influenced by the properties of these 'surface atoms'. In this case, the QD surface treatment could have a strong influence on the thermal expansion of the QD and, hence, on the temperature-induced change in the quantum confinement. Indeed, this fact has been recently demonstrated by Zhou *et al.* who analyzed in detail the fluorescence thermal sensitivity of CdTe QDs with different surface treatments (coatings).[33] In particular [see Figure 4.15(a)] Zhou *et al.* decorated aqueous dispersible CdTe QDs with *per*-6-thio-β-cyclodextrin (so-called β-CD). This complex is a quite well-known macrocyclic compound that it is composed of six or more glucopyranose rings that undergo relevant rotation around C–O bonds as a consequence of temperature changes.[34–36] When β-CD complexes are linked to a given dot surface these temperature-induced rotations (that are, indeed, conformational changes) create considerable torsional forces at the surface. When linked to the QD surface, the torsional forces generated by these conformational changes are conducted to the inner lattice of the QD. It has been demonstrated that the combination of the intrinsic lattice expansion/contraction of the QD, and the additional contribution generated by the coating stress enhances the thermally induced spectral shift of CdTe QDs. This is clearly shown in Figure 4.15(b) which includes the temperature-induced fluorescence spectral shift as obtained from β-CD and 3-mercatopropionic acid (MPA)-decorated CdTe QDs. It is shown that the existence of these conformational changes at the surface coating doubles the thermal sensitivity of CdTe QDs. By comparing the temperature-induced spectral shifts in both cases, and also taking into account the thermal coefficients of the CdTe energy gap, Zhou *et al.* claimed that the external

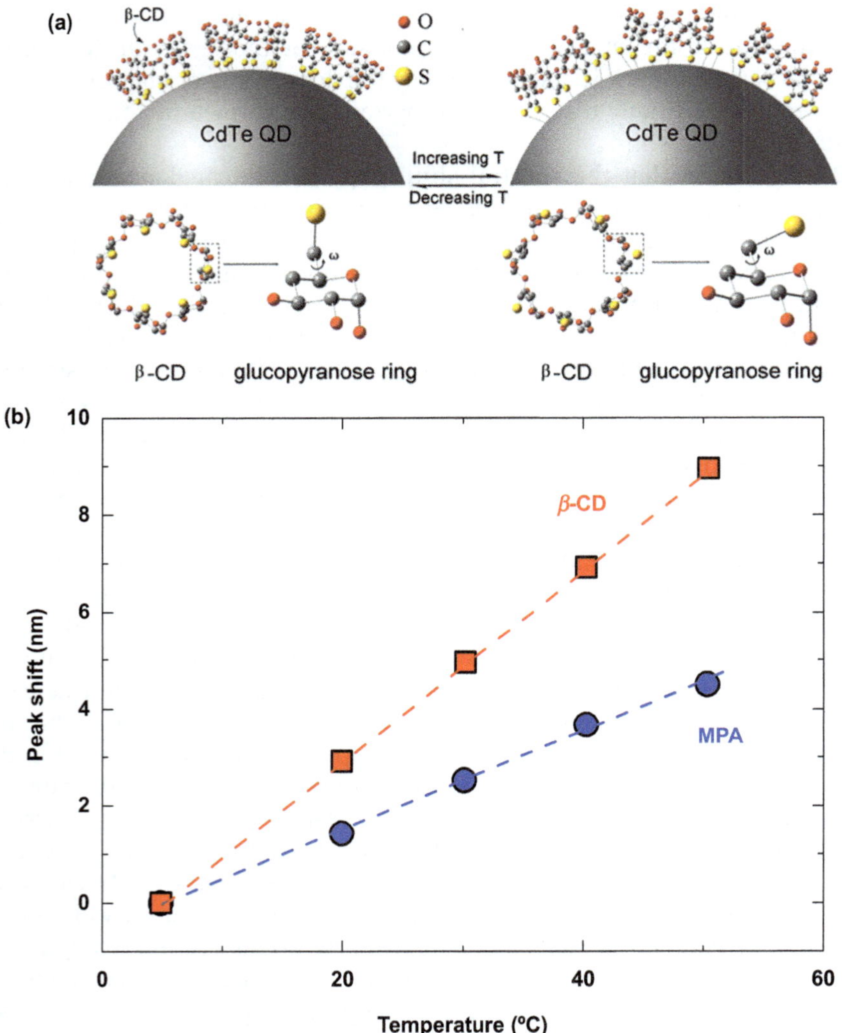

Figure 4.15　(a) Scheme showing the temperature-induced conformation change of β-CD. For better visibility, the hydrogen atoms are not shown. (b) Temperature-induced fluorescence shift of β-CD and MPA-decorated CdTe QDs. Circles and squares are experimental data and dashed lines are the best linear fits.
Reproduced with permission from ref. 33. © American Chemical Society 2013.

pressure induced by the surface coating acting on the QD surface could be as large as 8.4 MPa per degree.[33]

The existence of coating-related surface stress, shown in the work of Zhou *et al.*, could also explain the controversy about the size-dependence of the thermal spectral sensitivity of CdTe QDs. Some authors have found that the smaller the CdTe QD, the larger its spectral thermal sensitivity (as shown in Figure 4.14(a)).[37,38] In contrast, other authors have reported a size-independent thermal sensitivity for CdTe QDs.[39] The explanation for this discrepancy could be the role played by the surface coating in the 'effective' thermal expansion/contraction of QDs. If QDs are affected by conformational changes of complexes at their surface, then the relative contribution of these changes to the effective thermal expansion coefficient increases as the QD size is reduced, due to the larger surface-to-volume ratio. On the other hand, if the surface coating does not create additional stress then the thermal sensitivity should be size-independent. Thus, the different QD coatings used by different authors could explain the different size-dependent thermal sensitivities found in the literature.

Finally, it should be noted that an adequate choice of the complexes used for QD coating could not only enhance the thermal sensitivity of QDs but also their stability during heating/cooling cycles. For the particular case of CdTe, spontaneous QD growth has been observed in colloidal solutions when heating above the so-called *critical growth temperature* ($T_{critical}$), which strongly depends on the surface coating. When a QD solution is heated above this temperature the QD size increases and, consequently, a red-shift in fluorescence is produced.[40] If this happens, then the experimentally measured spectral shift cannot be unequivocally related to temperature change, but may also be due to an unwanted size increase during the heating process. Hence, in this situation the QDs become non-valid fluorescent nanothermometers. In fact, these QDs would display thermal hysteresis, undesirable for a suitable nanothermometer. This undesirable effect can be avoided by using surface coating complexes with large $T_{critical}$ values. This is illustrated in Figure 4.16 which displays the variation of the emission peak position of CdTe QDs decorated with β-CD [Figure 4.16(a)] and MPA [Figure 4.16(b)] during a temperature cycle between 65 and 70 °C. As can be observed, since MPA-decorated CdTe QDs have a critical temperature below 65 °C, a continuous red-shift is observed. On the other hand, the relatively high $T_{critical}$ of β-CD decorated CdTe QDs avoids temperature-induced dot growth and the fluorescence properties of these QDs are not affected by consecutive thermal cycles.[33]

4.3 Quantum Dots as Thermal Sensors: Applications

Up to this point we have discussed a variety of features that affect the thermal sensitivity of fluorescent QDs. Hereafter we describe the most relevant applications developed to date.

4.3.1 Intracellular Thermal Sensing

One of the most relevant advances in the synthesis of QDs is the almost total control over their surface properties and composition. This control has allowed, in turn, the preparation of colloidal suspensions of QDs not only in water but also in biocompatible liquids, such as phosphate buffered saline

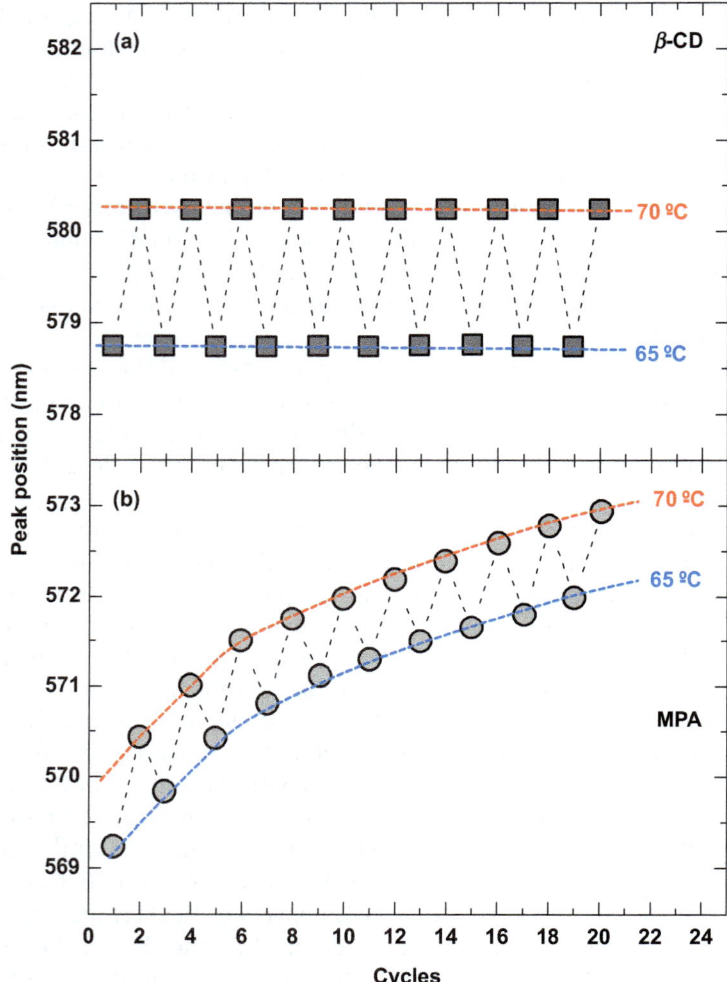

Figure 4.16   Variation in the fluorescence peak position of CdTe QDs decorated with β-CD and MPA during a temperature cycle between 65 and 70 °C. Data extracted from ref. 33.

(PBS), required for cell culture and manipulation during *in vivo* experiments. Based on this possibility, it has been possible to incubate living cells in media containing QDs and, in this way, to force either their incorporation *in* or their adhesion *to* living cells.[41,42] Initially this possibility was used for *in vitro* fluorescence imaging by using conventional fluorescence or modern multiphoton fluorescence microscopes.[41,42] During recent years, researchers have taken advantage of these techniques to get information about intracellular temperature changes in order to understand different intracellular processes that could be affected or triggered by temperature (such as cell division),[43–45] or that could cause a modification in the intracellular temperature. In general, variations in cell temperature are not larger than a few degrees and take place around the normal body temperature, typically 37 °C. Thus, intracellular thermal sensing requires high-resolution temperature determination in the so-called *biophysical range* (30–60 °C). The general procedure for QD-based intracellular thermal sensing is as follows: Cells to be studied are incubated in a PBS solution containing fluorescent QDs. After incubation, the presence of QDs, inside the cells or at their membrane, is corroborated by either epifluorescence or multiphoton fluorescence microscopy. Once the presence of QDs is confirmed, a visible laser beam or a tightly focused infrared beam (for multiphoton microscopy), is used to stimulate the QD fluorescence (*i.e.*, to stimulate the cellular fluorescence). The subsequent spectral analysis of this fluorescence provides intracellular thermal readings. This simple procedure is shown schematically in Figure 4.17 which includes fluorescence images of NIH/3T3 murine fibroblast cells after incubation with 655 nm emitting CdSe QDs, as reported by Yang *et al.*[46] Red spots in Figure 4.17(a) show the intracellular incorporation of QDs inside the cells. In this case, the fluorescence of intracellular QDs was excited with a mercury lamp and the spectral properties of the QD fluorescence were analyzed by using a conventional monochromator [Figure 4.17(b)]. A typical fluorescence spectrum obtained under these experimental conditions is included in Figure 4.17(c) where dots are experimental data and the red solid line corresponds to the best fit of the central part of the spectrum. From this fit it is possible to determine the peak wavelength and so the QD temperature, *i.e.*, the intracellular temperature. In this case the spatial resolution of the temperature measurement is determined by the numerical aperture of the microscope objective used for lamp focusing and fluorescence collection. In general, and assuming that spatial filters were not used in the confocal regime, the spatial resolution, Δr, could be assumed to be:[47]

$$\Delta r = \frac{0.5 \times \lambda_{\text{em}}}{NA}. \tag{4.7}$$

where NA is the numerical aperture of the microscope objective and λ_{em} is the QD emission peak wavelength. For conventional microscope objectives (with $NA \approx 0.6$) a sub-micrometric spatial resolution can be easily obtained. Regarding the thermal resolution achievable by this method, its evaluation is quite challenging and depends on a great variety of factors, including the spectral resolution of the dispersing systems as well as the signal-to noise ratio of the luminescence spectrum. It also depends on the accuracy of the fitting procedure and the influence of background signals. Giving an accurate number for it is complicated and unrealistic. Nevertheless, sub-degree thermal resolution can be easily obtained. Indeed, Yang *et al.* reported thermal resolutions well below 0.5 °C.[46] As a critical comment to this approach, it should be mentioned that this is only valid in the complete absence of QD aggregation. When QD aggregation occurs, side bands appear in the fluorescence spectrum and this could lead to spectral shifts that are not related to temperature change. In any case, Yang *et al.* demonstrated that

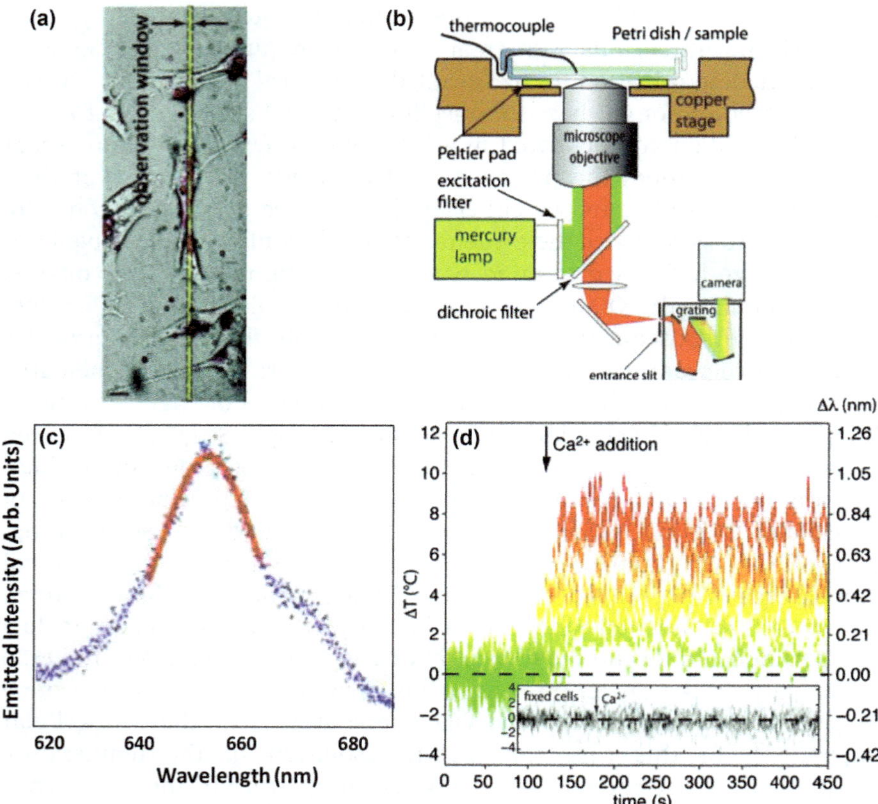

Figure 4.17 (a) Merged optical and fluorescence images of NIH/3T3 murine fibroblast cells after incubation with 655 nm emitting CdSe QDs. (b) Schematic diagram of the experimental setup used for QD-mediated intracellular thermal sensing. (c) Characteristic intracellular CdSe QD fluorescence. Dots are experimental data and the solid red line is the peak fit used for temperature determination. (d) Time evolution of intracellular temperature after a Ca^{2+} shock as obtained at different intracellular locations.
Data reproduced with permission from ref. 46. © American Chemical Society 2011.

their thermal sensing approach was sensitive enough for real-time monitoring of the intracellular temperature change during a chemical shock. This is manifested in Figure 4.17(d), which shows the local temperature change as a function of time as obtained at different points inside the cell (where QDs had been incorporated) after the external addition of Ca^{2+} ions.[46] As can be observed, the intracellular temperature change depends on the location inside the cell. This unequivocally reveals the heterogeneous cell response when subjected to a chemical shock. This, in turns, reveals the presence of subcellular temperature gradients caused by highly localized heat sources.

Intracellular thermal sensing using fluorescent QDs has also been demonstrated by Maestro *et al.* inside HeLa cancer cells during an externally induced heat-shock.[37] CdSe–ZnS (core–shell) QDs were used and, again, thermal readings were obtained from the spectral analysis (peak wavelength determination) of the fluorescence generated by these QDs. This showed a thermal spectral shift of 0.1 nm per °C, leading to a thermal resolution better than 1 °C. QD fluorescent intracellular thermal sensors have also been used to achieve real-time thermal readings of cell temperature during photo-thermal therapy. The idea behind this is that efficient and controlled photothermal therapy for cancer cells requires high-resolution and real-time thermal sensing at the cellular level. Such thermal knowledge is required to limit the intracellular heating to a pre-designed treatment temperature range.[45,48–50] For so-called *hyperthermia treatments*, the cell temperature should be increased from normal temperature up to the 40–50 °C range. Higher temperatures lead to direct and fast cell ablation. This makes it impossible to access the transient cell state that follows a heat-shock, which is the basis of the reduced resistance of heated cells to traditional treatments, including radiotherapy and chemotherapy.[51] Real-time thermal sensing during *in vitro* thermal therapy was first demonstrated by Han *et al.*, who designed and developed a smart but simple experimental setup that is schematically represented in Figure 4.18(a).[52] The experiments were performed on prostate cancer cells (PC-3) that were incubated with a mixed saline solution containing gold nanoshells (GNSs) and CdSe–ZnS (core–shell) QDs. The GNSs acted in these experiments as heating nano-particles. When the cells were excited with a 820 nm laser beam, a collective motion of electrons was induced at the surface of the GNSs. Relaxation of these surface electron currents caused a remarkable heat production, increasing the temperature of the PC-3 cells.[52] Simultaneously, a second light beam [the lamp in Figure 4.18 (a)] was used to optically excite the QDs. A simple CCD camera was used to record the fluorescence images of treated cells, based on the QDs taken up by these cells. A proper analysis of these fluorescence images pro-vided information, in real time, on the intracellular temperature. In this case thermal reading was not achieved from spectral analysis of the QD emission band (*i.e.*, by recording the spectral shift of peak emission wavelength) but from the analysis of the integrated emitted intensity of the QDs. As re-ported by Han *et al.*, this is possible thanks to the fact that the integrated emitted intensity of CdSe–ZnS QDs suffers from a linear thermal quenching in the 10–70 °C temperature range [Figure 4.18(b)].[52] Regardless of the exact origin of this fluorescence quenching, Han *et al.* found that the normalized intensity change at temperature T ($\Delta I(T)$) with respect to the fluorescence intensity at a given reference temperature (T_{ref}) varied linearly in the physio-logical range, so that it can be written as:[52]

$$\Delta I(T) = \frac{I(T) - I(T_{\mathrm{ref}})}{I(T_{\mathrm{ref}})} = b(T - T_{\mathrm{ref}}) \tag{4.8}$$

where b should be experimentally obtained in each case and is strongly dependent on the particular QDs used. For the particular case of the CdSe–ZnS QDs used by Han *et al.*, this was found to be close to 0.006 °C^{-1}.[52] Based

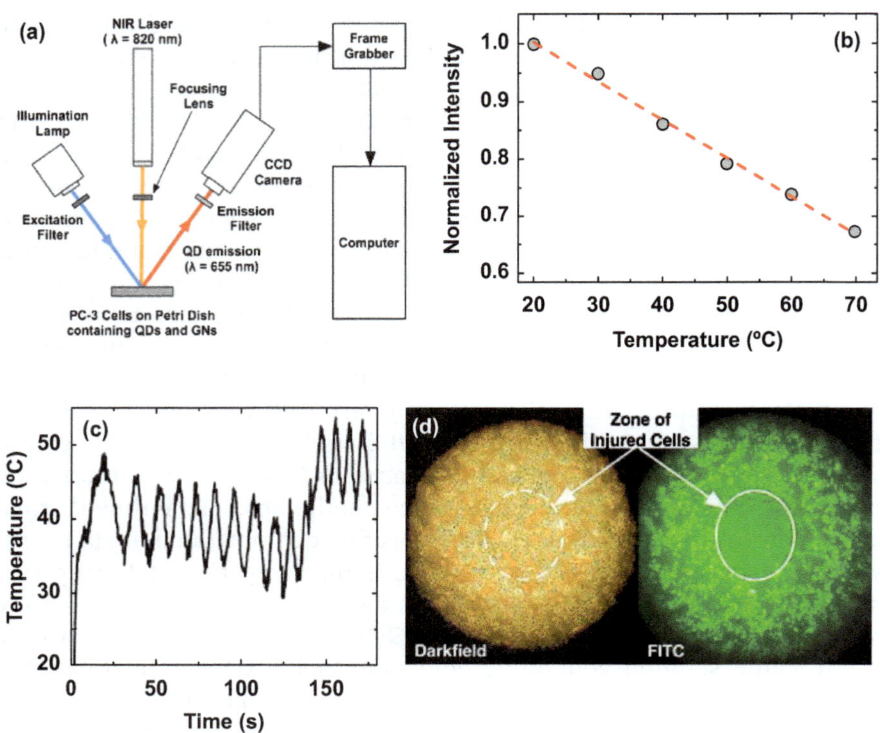

Figure 4.18 (a) Schematic representation of the experimental setup used for QD-mediated real-time thermal sensing during *in vivo* thermal therapy of prostate cancer cells. (b) Temperature dependence of the CdSe QD fluorescence intensity. Dots are experimental data and the dashed line is the best linear fit. (c) Time evolution of the average cell temperature after switching on the 820 nm heating laser. (d) Micrographs of darkfield (left) and luminescent field (right) indicating that the cells are still attached to the dish after heat treatment. The luminescence image reveals cell death in the surroundings of the laser heating focus area. Data reproduced with permission from ref. 52. © Springer 2009.

on eqn (4.8), the temperature at the QD location can be easily estimated by measuring the QD fluorescence intensity and comparing it with the fluorescence intensity obtained at a reference temperature. A typical result, obtained by Han *et al.* is shown in Figure 4.18(c) which displays the time evolution of the temperature of PC-3 cells after 820 nm laser irradiation with a 0.9 W cm^{-2} power density. Han *et al.* found that the cell's temperature rose rapidly within the first 20 s and fluctuated periodically with a $\pm 5\,^{\circ}$C amplitude afterwards, this fact very likely due to fluctuations in the laser source. At this point it should be noted that the intracellular temperature determined in this way should be considered a first-order approximation, since it does not take into account the temperature effects of any other external stimulus which may affect the electronic configuration of the QDs, such as pH.[53] The presence of these additional external stimuli should be carefully delineated to precisely measure temperature in a physiological environment. Note that Han *et al.* reported on cell temperatures well above 40 °C, and such large temperatures are expected to cause severe cell damage. Indeed, this was corroborated by the optical and fluorescence images of the treated cells [Figure 4.18(d)] which revealed cell injury in the 820 nm irradiated zone. The pioneering work of Han *et al.* inspired later works also hoping to achieve real thermal control during photothermal treatment at the single-cell level. Maestro *et al.* reported on intracellular thermal sensing during gold nanorod mediated hyperthermia treatment.[54] In this case, the authors also used CdSe–ZnS (core–shell, 4 nm core size) QDs as thermal sensors but, in contrast with the work of Han *et al.*, intracellular temperature was determined from the spectral analysis of QD fluorescence. This approach is more suitable, as problems due to instabilities in the excitation source and redistribution of QDs inside the cells are avoided. Indeed thermal resolutions below 0.5 °C have been reported.[55]

4.3.2 Thermal Characterization of Microelectronic Devices

The first demonstration of the potential use of luminescent QDs as high-resolution nanothermometers for thermal imaging of micro/nano-electronic devices was provided by Li *et al.*[56] The authors used the luminescence of CdSe QDs to determine the temperature around an electrical microheater, consisting of a high-quality aluminium microwire with dimensions $1200 \times 40 \times 0.1$ μm^3 fabricated on top of a Pyrex wafer [schematic representation in Figure 4.19(a)]. CdSe QDs were placed on the top of the microheater by just depositing a microdroplet of an aqueous suspension containing QDs and drying it in air. Then, the QDs were optically excited by visible laser light and the subsequent red fluorescence (650 nm) was collected by a microscope objective and spectrally analyzed by using a monochromator and a high-sensitivity camera. Any local change in the microheater temperature was then detected by the appearance of a red-shift in the CdSe QD emission peak that shifted at a rate of 0.1 nm per °C [Figure 4.19(b)]. By scanning the focusing/collecting microscope objectives, the authors were able to measure the temperature profiles along the microheater with sub-micrometric resolution. Representative results are shown in Figure 4.19(c), where the temperature profiles obtained for different applied voltages are shown. The experimentally obtained profiles account well for the predictions made based on Joule's law, *i.e.*, the temperature rises with the square of the applied voltage. From the thermal profiles the temperature uncertainty can be estimated to be close to 1 °C.

An important point highlighted by Li *et al.* is the fact that the peak emission wavelength varied from dot to dot. When performing statistical analysis over about 100 dots, Li *et al.* observed a standard deviation in the peak emission wavelength of almost 3 nm (around a central emission wavelength of 660 nm). This remarkable uncertainty in the peak wavelength was tentatively explained in terms of the existence of a broad dot size

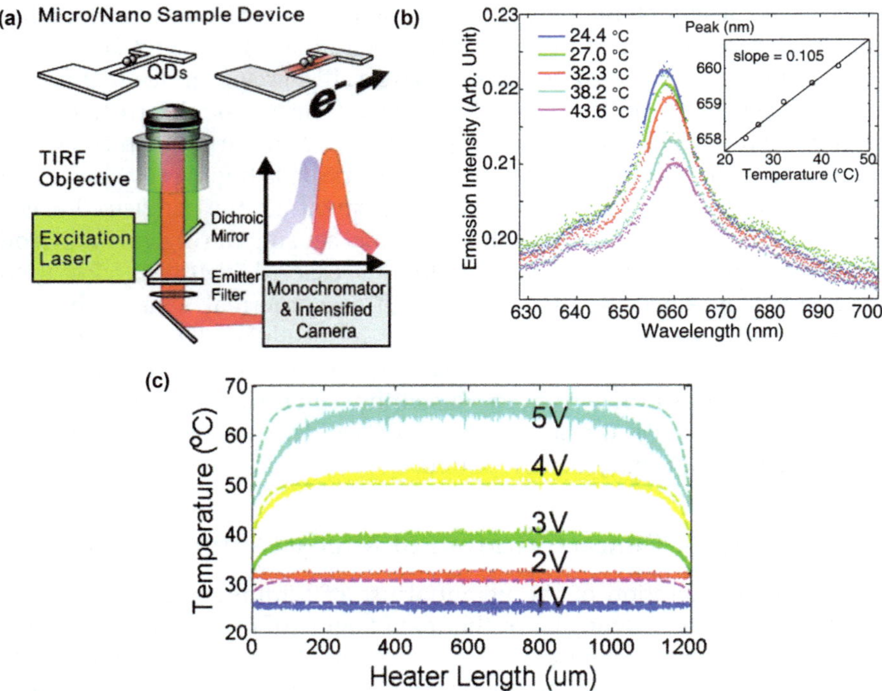

Figure 4.19 (a) Schematic representation of the experimental setup used for QD-mediated thermometry of an integrated microheater. (b) Emission spectra as obtained from CdSe QDs at different temperatures. Dots are experimental data and solid lines are the fits used for determination of the peak position. The calibration curve obtained is included as an inset. (c) Temperature profiles obtained along the microheater for different applied voltages.
Data reproduced with permission from ref. 56. © American Chemical Society 2007.

distribution (close to 20%). As explained in Section 4.2 the emission wavelength of QDs is determined by their size, in such a way that the presence of different sizes leads to the appearance of different peak emission wavelengths.[57] This is a critical point for the purpose of thermal measurements. As discussed by Li *et al.*, this fact constitutes a serious limitation for thermal measurements since different sizes could also lead to different temperature responses (different thermally induced spectral shifts). At the same time, size dispersion limits the capability of QDs to give absolute values of temperature and only the measurement of relative temperature changes seems to be possible. Obviously this limitation disappears if measurements are performed not on single QDs but over a relative large number of QDs. Li *et al.* state that the resolution of QD-based absolute thermal measurements depends critically on the number of QDs contributing to the fluorescence signal.[56] Based on simple statistics, the authors concluded that absolute temperature determination with an accuracy better than 1 °C requires analysis of the fluorescence generated by at least 1200 QDs. This implies a minimum area in which temperature can be defined: The area in which it is possible to ensure that at least 1200 QDs are present. Li *et al.* concluded that, under their experimental conditions, such a number of QDs was only achieved if a surface area of about 350 nm in diameter was analyzed.[58] Of course, both the minimum number of analyzed QDs, as well as the corresponding minimum analyzed area, are needed for accurate temperature measurements, and both depend on the particular quality of the QDs. The use of higher quality QDs suspensions (with a narrower size distribution curves) would reduce the number of QDs required for sub-degree thermal resolution. This, in turn, would reduce the required measurement area, and so would enhance the spatial resolution of the thermal images.

4.3.3 Sub-tissue Thermal Sensing

As stated in Section 4.2, one of the main advantages of using QDs is that their fluorescence (excitation and emission) bands can be pre-designed by an appropriate selection of either the constituent material and/or the QD size. This fact is being presently explored in order to synthesize QDs whose emission lies in one of the BWs.[11,59] These are spectral ranges in which optical extinction of tissues is strongly reduced, in such a way that large optical penetration depths can be achieved. BWs are located in the infrared and most efforts are being devoted to the development of QDs emitting in the so-called *first* and *second* BWs that correspond to the 700–950 nm and 1000–1400 nm spectral ranges, respectively.[12,13,60,61] With such infrared-emitting QDs the acquisition of deep tissue fluorescence images has been proved to be possible.[60] In principle, the combination of these large optical penetration depths with the thermal sensitivity of QDs allows for remote sub-tissue thermal sensing, which is of great interest for many biomedical applications including real-time monitoring during *in vivo* hyperthermia treatments. QD-based sub-tissue thermometry was proposed by Ghosh *et al.*[62] For this purpose, the authors monitored the emitted intensity generated by sub-tissue allocated QDs (see Figure 4.20). Under this approach, sub-tissue temperature was estimated by using a pre-determined intensity–temperature calibration curve. This simple operational procedure becomes much more complicated in real applications. Ghosh *et al.* realised that sub-tissue temperature determination on the basis of a calibration curve (QD intensity *vs.* temperature curve) obtained in the absence of tissue is not valid.[63] The first experiments were performed by using commercially available water-dispersible CdTe–ZnS (core–shell) QDs, with an emission band close to 610 nm. These QDs were placed below a phantom tissue and their fluorescence intensity was analyzed through the tissue. QDs were positioned on a temperature-controlled base, so that their temperature could be varied in the 10–80 °C range.

Figure 4.20 shows the intensity *vs.* temperature calibration curve as obtained without tissue. A linear relationship is clearly observed with a slope

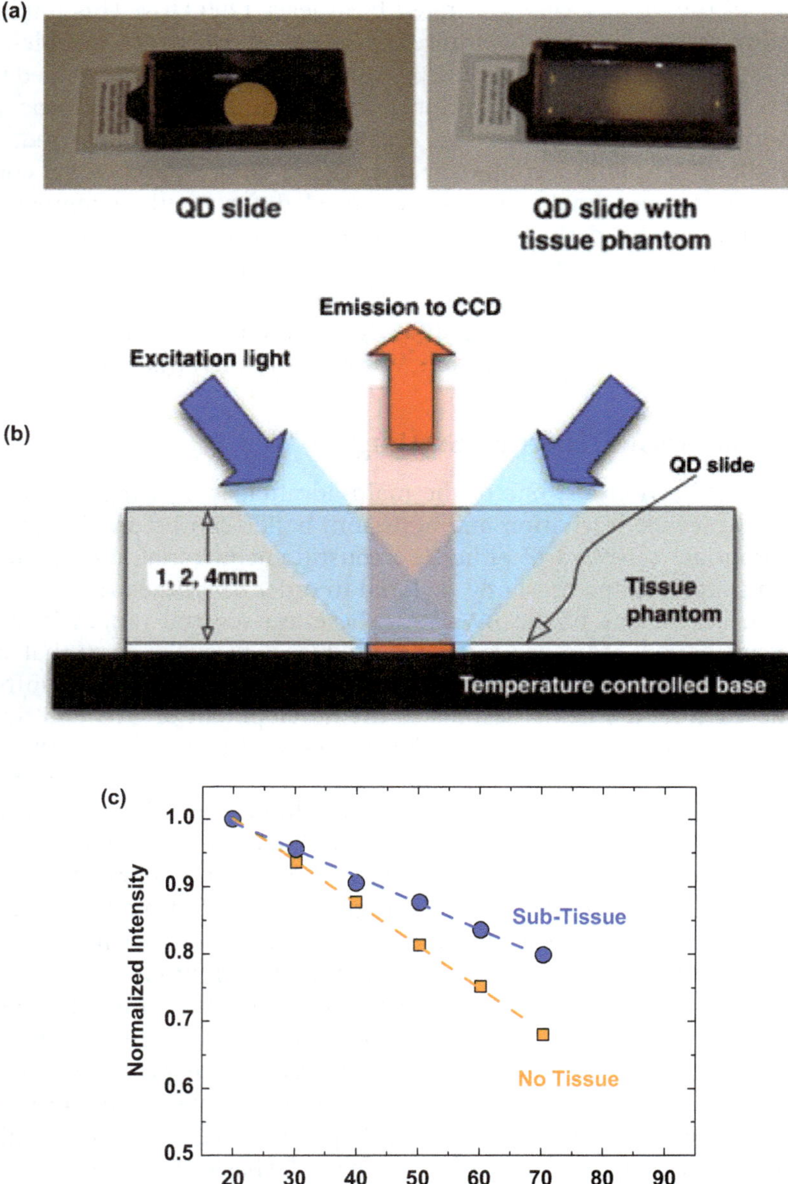

Figure 4.20 (a) QD slides with and without phantom tissues designed to investigate the ability of visible-emitting QDs for sub-tissue thermal sensing. (b) Schematic representation of the experimental setup used for QD-mediated sub-tissue thermometry. (c) Calibration (emitted intensity *vs.* temperature) curves as obtained in the presence and absence of a 4 mm phantom tissue. Dots are experimental data and dashed lines are the best linear fits.
Data extracted from ref. 62.

close to $6.4 \times 10^{-3}\,°C^{-1}$. When the same experiments were performed by placing a 4 mm phantom tissue on top of the QDs, a linear relationship was again obtained but with a significantly reduced slope: $4 \times 10^{-3}\,°C^{-1}$. This surprising result was initially explained in terms of the light–tissue interactions that resulted in a strong scattering of the luminescence radiation. This scattering leads to marked differences between the emission intensity outcoming from the tissue surface and that generated by sub-tissue QDs (underneath the tissue).[63] Thus, for real sub-tissue thermal sensing it would be necessary to correlate the measured fluorescence intensity (at the surface of the tissue) with that generated from the QDs. Indeed, this is possible by means of an *inverse solution* algorithm; that is to inversely solve the radiative transfer equation in the diffusion approximation. Ghosh *et al.* stated that, by using this algorithm, and if the geometry (depth and shape) of the treated volume (a tumour, for instance) is known from prior diagnosis, the spatiotemporal change of sub-tissue QD fluorescence could be obtained through this inverse procedure. Thus, it is possible to determine sub-tissue temperature from the experimental intensity data at the skin surface. Despite the good results obtained by Ghosh *et al.*, several factors should be considered before applying QDs for sub-tissue *in vivo* thermal imaging. First, noise during imaging should be minimized, as the accuracy of the inverse solution is significantly affected by noise. The proof of concept, provided by Ghosh, was performed by using phantom tissues, but the situation could be much more complex when dealing with real tissues due to the rather heterogeneous characteristics of optical parameters and other complications (inhomogeneity of tissue composition and anisotropy of physical and optical properties). Finally, real *in vivo* applications require full reversibility of the thermal response of QDs for dynamic adjustment of the therapy parameters. However, reversibility in QD fluorescence properties in the wide thermal range required during photothermal therapy (20–60 °C) still needs to be tested as many processes, such as temperature-induced agglomeration, could take place.

4.3.4 Heterostructures

Up to now, we have described how single QDs can be used as fluorescence nanothermometers. In those cases previously described, thermal readings were achieved from spectral analysis of the luminescence generated by single a QD or by a collection of them. This is a valid approximation when temperature is extracted from the emission spectral position, as it gives, after proper calibration, an absolute temperature value without requiring any additional reference. On the other hand, when temperature readings are obtained from the thermally induced fluorescence quenching, temporal and spatial variations of the concentration of QDs could cause intensity variations not related to temperature, and so leading to erroneous measurements and interpretation.[64,65] To mitigate this error source related to intensity measurements, a fluorescence reference should be used simultaneously, so that these intensity fluctuations can be corrected. This could be achieved by developing complex luminescent nanostructures composed of a QD and another luminescent material that is used for reference purposes. Indeed, this was the scheme adopted by Albers *et al.* for the development of dual-emitting fluorescence nanothermometers, shown schematically in Figure 4.21.[66] This dual-emitting nanothermometer consisted of a red-emitting CdSe–CdS Quantum Dot–Quantum Rod (QD–QR) semi-conductor nanocrystal heterostructure passivated with an amphiphilic polymer shell, the shell appended with far-red-emitting cyanine dyes.[66] This QD–QR

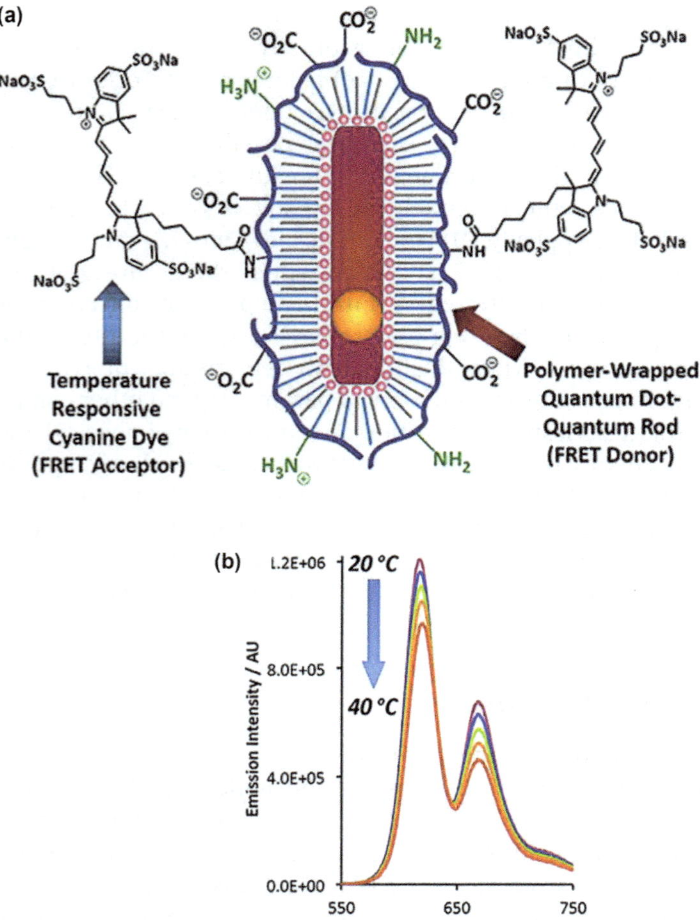

Figure 4.21   (a) Schematic representation of the dual-emitting hybrid nano-thermometer consisting of a QD–QR core labelled with a tempera-ture-responsive dye. (b) Fluorescence spectra as generated by the hybrid structure, obtained at different temperatures.
Reproduced with permission from ref. 66. © American Chemical Society 2012.

heterostructure has been proven to display a large extinction coefficient together with a strong temperature-dependent broad emission band. At the same time, cyanine dyes show temperature-dependent fluorescence quantum yields but not wavelength shifts. Albers *et al.* demonstrated that when both are put together in the same nanostructure, resonant energy transfer (FRET) processes are activated.[67] As a consequence, the overall fluorescence generated by the nanostructure consists of both the QD–QR emission band (at around 620 nm) and the dye emission broad band (at around 670 nm). When the temperature of the heterostructure is increased above 20 °C the intensity of both bands decreases monotonically. Due to the different mechanisms involved in each case, the intensity ratio between these two bands varies monotonically with temperature, as can be appreciated in Figure 4.21(b). In fact, a pseudolinear response was observed, which displayed fully reversible behaviour. Indeed, from the data included in Figure 4.21, it was concluded that the sensitivity of the QD–QR nanosized heterostructure was as large as 2.4% per °C (in the 20–40 °C temperature range), which is enough to achieve thermal measurements with a precision better than 0.2 °C.[66] Such precision would allow for the observation of thermal signatures of different intracellular processes. Motivated by this possibility, Albers *et al.* introduced these QD–QR heterostructures into living cells. Very surprisingly, the authors found that the intracellular thermal sensitivity of the heterostructures was enhanced compared to that obtained in aqueous buffers. The reason for this is not clear at present, although the authors claimed that it could be related to the specific response of the dyes to cytosolic constituents. Several explanations were proposed, including either cyanine dye deterioration in the cell, or an increase in the average distance to the QD–QR core in the cytosol.[66] Whatever the particular cause, the dramatic differences observed between cellular and cuvette measurements highlight the importance of calibrating the probes within a cell rather than using the calibration curve obtained in the buffer. This is a critical point, as many of the works published up to now, related to QD-based intracellular thermal sensing, have omitted this previous calibration step.

An alternative approach for fluorescence nanothermometry using QD heterostructures was that proposed by Lee *et al.*[68,69] In this approach the authors described reversible nanothermometers built from QDs interconnected with metallic nanoparticles by a polymer that was acting as a molecular spring.[68,69] The nanosized heterostructure consisted of a core Au nanosphere covered by a poly(ethylene glycol) (PEG) film with a thickness of few nanometres. The outer side of the PEG film was conjugated with 3.7 nm CdTe–QDs. The Au nanosphere showed a surface plasmon resonance wavelength at around 633 nm. Thus, when the heterostructure was optically excited at this wavelength, field enhancement was induced at the surface of the Au nanosphere, and, due to plasmon–exciton interactions, QD fluorescence (at around 550 nm) was activated. The efficiency of this plasmon–excitation energy transfer is strongly dependent on the distance between the Au surface and the QDs, *i.e.*, on the PEG film thickness. PEG is known to undergo a drastic expansion in the 20–50 °C range, so that a relevant change in the QD luminescence intensity is produced in this temperature range. Lee *et al.* demonstrated the applicability of this kind of heterostructure for thermal sensing with sensitivities close to 0.6% per °C.[68,69]

4.4 Conclusions and Perspectives

Because of their unique combination of outstanding properties, QDs have emerged as reliable fluorescent nanothermometers based on their temperature-dependent fluorescence features. Temperature variations cause changes in fluorescence intensity, fluorescence lifetime and spectral shape, so a multiparameter spectral analysis could provide high-resolution thermal readings. This approach has been proven to work nicely in a great variety of systems, ranging from electronic microdevices to living cells with sub-degree resolution. These outstanding results are accompanied by some concerns about their possible biotoxicity and about the possible presence of photo-damage. All these questions should be faced and addressed in the near future. We are convinced that solving these problems will imply the design and synthesis of novel QD heterostructures. In addition, for the particular purpose of thermal sensing in biosystems, the fact that the thermal response of QDs can be dependent on the environment could be a severe limitation. It will be necessary to know the exact cause of this variable sensitivity in order to control it. Again, the design of novel heterostructures for enhanced isolation of QDs from their environment would be, very likely, a reliable alternative.

Finally, it should be noted that most of the reviewed applications of QDs report on single-point measurements. Thermal imaging implies a step forward and would require either a homogeneous distribution of QDs over the whole system to be imaged or, on the other hand, the remote manipulation of a single QD that would be scanned all over this system. In many applications, such as those related to biosystems, the number of QDs incorporated into the system studied should be kept at a minimum, thus QD thermal imaging would require remote manipulation. Such 3D remote manipulation of QDs has already been demonstrated by the use of optical traps acting on either single QDs or QD aggregates.[70] Despite the good results obtained up to now on this optical manipulation of QDs, much work is still to be done in order to clarify the trapping mechanisms as well as determining how the optical forces could affect the thermal response of QD fluorescence.

References

1. B. H. a. G. F. Imbusch, *Optical Spectroscopy of Inorganic Solids*, Oxford Science, New York, 1989.
2. Y. S. Park and D. C. Reynolds, *Phys. Rev.*, 1963, **132**, 2450–2457.

3. D. J. Norris and M. G. Bawendi, *Phys. Rev. B: Condens. Matter Mater. Phys.*, 1996, **53**, 16338–16346.

4. A. P. Alivisatos, W. Gu and C. Larabell, *Annu. Rev. Biomed. Eng.*, 2005, 7, 55–76.

5. H. Arya, Z. Kaul, R. Wadhwa, K. Taira, T. Hirano and S. C. Kaul, *Biochem. Biophys. Res. Commun.*, 2005, **329**, 1173–1177.

6. D. Bera, L. Qian, T.-K. Tseng and P. H. Holloway, *Materials*, 2010, **3**, 2260–2345.

7. A. M. Fox, *Optical Properties of Solids*, Oxford University Press, 2001.

8. W. J. Parak, L. Manna and T. Nann, in *Nanotechnology*, Wiley-VCH Verlag GmbH & Co. KGaA, 2010, DOI: 10.1002/9783527628155.nanotech004.

9. A. M. Smith, S. Dave, S. Nie, L. True and X. Gao, *Expert Rev. Mol. Diagn.*, 2006, **6**, 231–244.

10. A. M. Smith, H. W. Duan, A. M. Mohs and S. M. Nie, *Adv. Drug Delivery Rev.*, 2008, **60**, 1226–1240.

11. R. R. Anderson and J. A. Parrish, *J. Invest. Dermatol.*, 1981, 77, 13–19.

12. G. H. Au, W. Y. Shih, S. J. Tseng and W. H. Shih, *Nanotechnology*, 2012, **23**, 275601.

13. L. M. Maestro, J. E. Ramirez-Hernandez, N. Bogdan, J. A. Capobianco, F. Vetrone, J. G. Sole and D. Jaque, *Nanoscale*, 2012, **4**, 298–302.

14. G. Dresselhaus, *J. Phys. Chem. Solids*, 1956, **1**, 14–22.

15. H. Haug and S. W. Koch, in *Quantum Theory of the Optical and Electronic Properties of Semiconductors*, 3rd edn, World Scientific Publishing Co., River Edge, NJ, 1990, pp. 386–399.

16. R. W. Meulenberg, J. R. Lee, A. Wolcott, J. Z. Zhang, L. J. Terminello and T. Van Buuren, *ACS Nano*, 2009, **3**, 325–330.

17. A. Franceschetti and A. Zunger, *Phys. Rev. Lett.*, 1997, **78**, 915–918.

18. W. W. Yu, L. Qu, W. Guo and X. Peng, *Chem. Mater.*, 2003, **15**, 2854–2860.

19. A. L. Efros, M. Rosen, M. Kuno, M. Nirmal, D. J. Norris and M. Bawendi, *Phys. Rev. B: Condens. Matter Mater. Phys.*, 1996, **54**, 4843–4856.

20. N. O. Dantas, F. Y. Qu, R. S. Silva and P. C. Morais, *J. Phys. Chem. B*, 2002, **106**, 7453–7457.

21. X. Y. Wang, X. F. Ren, K. Kahen, M. A. Hahn, M. Rajeswaran, S. Maccagnano-Zacher, J. Silcox, G. E. Cragg, A. L. Efros and T. D. Krauss, *Nature*, 2009, **459**, 686–689.

22. L. M. Maestro, C. Jacinto, U. Rocha, M. C. Iglesias, F. De La Cruz, A. Sanz-Rodriguez, J. G. Juarranz, Sole and D. Jaque, *J. Appl. Phys.*, 2012, 111.

23. A. F. van Driel, G. Allan, C. Delerue, P. Lodahl, W. L. Vos and D. Vanmaekelbergh, *Phys. Rev. Lett.*, 2005, **95**, 236804.

24. S. Coe-Sullivan, *Nat. Photonics*, 2009, **3**, 315–316.

25. A. Olkhovets, R. C. Hsu, A. Lipovskii and F. W. Wise, *Phys. Rev. Lett.*, 1998, **81**, 3539–3542.

26. Q. Dai, Y. Zhang, Y. Wang, M. Z. Hu, B. Zou, Y. Wang and W. W. Yu, *Langmuir*, 2010, **26**, 11435–11440.

27. J. Solé, L. Bausa and D. Jaque, *An Introduction to the Optical Spectroscopy of Inorganic Solids*, John Wiley & Sons, 2005.

28. C. de Mello Donega, M. Bode and A. Meijerink, *Phys. Rev. B: Condens. Matter Mater. Phys.*, 2006, **74**, 085320.

29. Y. Zhao, C. Riemersma, F. Pietra, R. Koole, C. de Mello Donegá and A. Meijerink, *ACS Nano*, 2012, **6**, 9058–9067.

30. P. Jing, J. Zheng, M. Ikezawa, X. Liu, S. Lv, X. Kong, J. Zhao and Y. Masumoto, *J. Phys. Chem. C*, 2009, **113**, 13545–13550.

31. S. Nizamoglu, T. Erdem, X. W. Sun and H. V. Demir, *Opt. Lett.*, 2010, **35**, 3372–3374.

32. P. Haro-González, L. Martínez-Maestro, I. R. Martín, J. García-Solé and D. Jaque, *Small*, 2012, **8**, 2652–2658.

33. D. Zhou, M. Lin, X. Liu, J. Li, Z. Chen, D. Yao, H. Sun, H. Zhang and B. Yang, *ACS Nano*, 2013, 7, 2273–2283.

34. D. J. Wood, F. E. Hruska and W. Saenger, *J. Am. Chem. Soc.*, 1977, **99**, 1735–1740.

35. K. B. Lipkowitz, *Chem. Rev.*, 1998, **98**, 1829–1873.

36. T. Kozar and C. A. Venanzi, *J. Mol. Struct.:THEOCHEM*, 1997, **395**, 451–468.

37. L. M. Maestro, E. M. Rodriguez, F. S. Rodriguez, M. C. I. la Cruz, A. Juarranz, R. Naccache, F. Vetrone, D. Jaque, J. A. Capobianco and J. G. Sole, *Nano Lett.*, 2010, **10**, 5109–5115.

38. L. M. Maestro, C. Jacinto, U. R. Silva, F. Vetrone, J. A. Capobianco, D. Jaque and J. G. Solé, *Small*, 2013, **9**, 3198–3200.

39. D. Zhou and H. Zhang, *Small*, 2013, **9**, 3195–3197.

40. N. Khlebtsov and L. Dykman, *Chem. Soc. Rev.*, 2011, **40**, 1647–1671.

41. X. Michalet, F. F. Pinaud, L. A. Bentolila, J. M. Tsay, S. Doose, J. J. Li, G. Sundaresan, A. M. Wu, S. S. Gambhir and S. Weiss, *Science*, 2005, **307**, 538–544.

42. K. T. Yong, J. Qian, I. Roy, H. H. Lee, E. J. Bergey, K. M. Tramposch, S. He, M. T. Swihart, A. Maitra and P. N. Prasad, *Nano Lett.*, 2007, 7, 761–765.

43. J. L. Roti Roti, *Int. J. Hyperthermia*, 2008, **24**, 3–15.

44. V. Tseeb, M. Suzuki, K. Oyama, K. Iwai and S. Ishiwata, *HFSP J.*, 2009, **3**, 117–123.

45. G. Hegyi, G. P. Szigeti and A. Szasz, *J. Evidence-Based Complementary Altern. Med.*, 2013, **2013**, 672873.

46. J. M. Yang, H. Yang and L. Lin, *ACS Nano*, 2011, **5**, 5067–5071.

47. M. Gu, *Advanced Optical Imaging Theory*, Springer, 2000.

48. L. E. Gerweck, *Cancer Res.*, 1985, **45**, 3408–3414.

49. P. Wust, B. Hildebrandt, G. Sreenivasa, B. Rau, J. Gellermann, H. Riess, R. Felix and P. M. Schlag, *Lancet Oncol.*, 2002, **3**, 487–497.

50. R. H. Donkol and A. Al Nammi, 2013.

51. J. van der Zee, D. González, G. C. van Rhoon, J. D. P. van Dijk, W. L. J. van Putten and A. A. M. Hart, *Lancet*, 2000, **355**, 1119–1125.

52. B. Han, W. L. Hanson, K. Bensalah, A. Tuncel, J. M. Stern and J. A. Cadeddu, *Ann. Biomed. Eng.*, 2009, **37**, 1230–1239.
53. V. Pilla, L. P. Alves, J. F. de Santana, L. G. da Silva, R. Ruggiero and E. Munin, *J. Appl. Phys.*, 2012, **112**, 104704.
54. L. M. Maestro, P. Haro-González, M. C. Iglesias-de la Cruz, F. SanzRodríguez, Á. Juarranz, J. G. Solé and D. Jaque, *Nanomedicine*, 2013, **8**, 379–388.
55. D. Choudhury, L. M. Maestro, W. T. Ramsay, L. Paterson, D. Jaque and A. K. Kar, 2012.
56. S. Li, K. Zhang, J.-M. Yang, L. Lin and H. Yang, *Nano Lett.*, 2007, **7**, 3102–3105.
57. E. Alipieva, A. S. Zlatov, V. A. Polischuk, A. P. Briukhovetskiy and D. E. Grigoriev, 2013.
58. O. Zohar, M. Ikeda, H. Shinagawa, H. Inoue, H. Nakamura, D. Elbaum, D. L. Alkon and T. Yoshioka, *Biophys. J.*, 1998, **74**, 82–89.
59. A. M. Smith, M. C. Mancini and S. Nie, *Nat. Nanotechnol.*, 2009, **4**, 710–711.
60. Y. Zhang, G. Hong, Y. Zhang, G. Chen, F. Li, H. Dai and Q. Wang, *ACS Nano*, 2012, **6**, 3695–3702.
61. G. Hong, J. T. Robinson, Y. Zhang, S. Diao, A. L. Antaris, Q. Wang and H. Dai, *Angew. Chem., Int. Ed.*, 2012, **51**, 9818–9821.
62. S. Ghosh, W. Hanson, N. Abdollahzadeh and B. Han, *Meas. Sci. Technol.*, 2012, **23**, 045704.
63. H. Yi, D. Ghosh, M.-H. Ham, J. Qi, P. W. Barone, M. S. Strano and A. M. Belcher, *Nano Lett.*, 2012, **12**, 1176–1183.
64. D. Jaque and F. Vetrone, *Nanoscale*, 2012, **4**, 4301–4326.
65. C. D. Brites, P. P. Lima, N. J. Silva, A. Millan, V. S. Amaral, F. Palacio and L. D. Carlos, *Nanoscale*, 2012, **4**, 4799–4829.
66. A. E. Albers, E. M. Chan, P. M. McBride, C. M. Ajo-Franklin, B. E. Cohen and B. A. Helms, *J. Am. Chem. Soc.*, 2012, **134**, 9565–9568.
67. T. Förster, *Ann. Phys.*, 1948, **437**, 55–75.
68. J. Lee and N. A. Kotov, *Nano Today*, 2007, **2**, 48–51.
69. J. Lee, A. O. Govorov and N. A. Kotov, *Angew. Chem., Int. Ed.*, 2005, **44**, 7439–7442.
70. L. Jauffred and L. B. Oddershede, *Nano Lett.*, 2010, **10**, 1927–1930.
71. N. Tomczak *et al.*, *Prog. Polym. Sci.*, 2009, **34**, 393.

Luminescent Nanothermometry with Lanthanide-doped Nanoparticles

MARTA QUINTANILLA, ANTONIO BENAYAS,
RAFIK NACCACHE AND FIORENZO VETRONE*

Institut National de la Recherche Scientifique – Énergie, Matériaux et
Télécommunications, Université du Québec, 1650 Boulevard Lionel-Boulet,
Varennes, Québec J3X 1S2, Canada
*Email: vetrone@emt.inrs.ca

5.1 Introduction

In recent years, nanomaterials have come to light with the potential of
revolutionizing current approaches to the development of novel materials
and tools. It is in this regard that there has been a large body of work
dedicated to the study and integration of these materials at the nanoscale
with the goal of improving existing devices or further, to develop altogether
new nanoscale devices and technologies.[1-7] A specific focus has been on
optical nanoparticles such as semiconductor quantum dots (QDs) and noble
metal nanomaterials (silver nanoparticles, gold nanoparticles/nanorods) for
their implementation in biological applications and nanomedicine.[8-15] This
small (or nano) revolution is bringing forth the possibility of personalized
nanomedicine with the simultaneous ability to target specific biological
sites, detect disease by providing diagnostic information by one or more
techniques (fluorescent imaging, magnetic resonance imaging, thermal

RSC Nanoscience & Nanotechnology No. 38
Thermometry at the Nanoscale: Techniques and Selected Applications
Edited by Luís Dias Carlos and Fernando Palacio

Published by the Royal Society of Chemistry, www.rsc.org

sensing, *etc.*) and deliver a localized therapeutic modality (photodynamic therapy, photothermal therapy, drug release, or combinations thereof). This theranostic approach to medicine has the potential to bring forth a paradigm shift in the biomedical landscape. Not only will such an approach result in the reduction of the vast and mounting costs associated with healthcare, but will also ensure a more rapid intervention for patients and will eventually translate to a greater probability of therapeutic success.

While a large body of research has focused on the above-mentioned optical materials, significant interest has focused on lanthanide-doped nanoparticles, which exhibit highly interesting luminescent properties including narrow bandwidth, stability with respect to photobleaching, flexible surface chemistry, *etc.*[16] Moreover, these nanoparticles are also capable of (up)converting near-infrared (NIR) excitation light to higher energies spanning the ultraviolet (UV), visible or even NIR wavelengths *via* a multiphoton excitation process known as *upconversion* (*vide infra*).[17–21] This renders these upconverting nanoparticles (UCNPs) interesting for biological applications where NIR excitation offers several advantages,[22] and as a result are attractive alternatives, relative to the conventional organic dye based fluorophores or semiconductor QDs typically excited with high-energy UV or blue light. For example, NIR radiation offers greater tissue penetration depths relative to UV excitation. In fact, tissue penetration of up to several centimetres has been recently reported following NIR excitation (of lanthanide-doped nanomaterials),[23] in comparison to the micron range penetration depths afforded by UV sources.[22] Furthermore, previous studies have demonstrated the relative safety of NIR radiation on biological systems in contrast to its UV counterpart, which is known to induce significant photodamage.[24,25] Moreover, these lanthanide-doped UCNPs possess a lower chemical toxicity relative to semiconductor QDs, which are typically prepared using elements such as cadmium, tellurium and lead, to name a few.[26,27]

In light of their interesting optical properties and the advances in preparation methods including controlled and application-tailored synthetic procedures, surface modification and functionalization techniques, these lanthanide-doped UCNPs have been intensively studied for integration into biological and medical applications ranging from diagnostics, optical and/or magnetic imaging, therapeutics and drug delivery, as well as photodynamic therapy for example.[28–37] More recently, scientists have recognized the importance of temperature sensing in biological applications and as such have looked towards the development of a sensitive temperature probe in order to relay diagnostic information following the use of the above-mentioned applications. Knowing that some of the lanthanide ions possess certain temperature-sensitive transitions[38–45] (detailed explanations and theory in the following section), a new arm of research in the field of lanthanide spectroscopy was dedicated to the development of non-invasive optical thermometer probes with a specific emphasis on thermometry at the nanoscale. Through the study of their luminescent properties (intensity or intensity ratios, or even luminescence lifetimes), it is possible to establish a calibrated correlation with temperature that allows the optical probe to act as a sensitive thermometer. Owing to their ability to upconvert NIR light, lanthanide-doped UCNPs can potentially be used in a biological milieu providing a method for real-time monitoring of the sample under study.

While thermometry at the nanoscale has been demonstrated in semiconductor QDs and various other materials,[46] this chapter is solely devoted to nanothermometry development based on lanthanide-doped UCNPs and discusses the advances achieved in this area to date as well as providing a perspective on the future of this research within the field of nanomaterials, and their application to biology and nanomedicine.

5.2 Some Insight into the Properties of the Lanthanides

5.2.1 Lanthanide Ions as Luminescent Probes

Lanthanides (Ln) are the elements of the periodic table whose atomic numbers range from 58 to 71, *i.e.*, from cerium to lutetium, although often lanthanum is also included in the group (atomic number 57). These elements also form, together with scandium and yttrium (atomic numbers 21 and 39), the group of elements traditionally known as the *Rare Earths* (RE).[47] Lanthanides belong to the sixth row of the periodic table. However, they do not follow the general periodic rules governing it. This fact is related to the electronic configuration of the lanthanide ions: while most of the elements have the valence electrons located in the outermost electron shell, the valence electrons of the lanthanides are associated with the 4f orbital which is defined by a closer distance to the nucleus than the already filled orbitals 5s, 5p and 6s.[48,49]

The electronic configuration of the lanthanides results thus in the shielding of valence electrons, therefore, they are only weakly affected by the environment and can be accurately described by the approximation of the free ion.[50] In this context, the 4f orbital is split into several allowed energy positions, or energy states, which electrons occupy. The splitting will depend first on the field created by electrons in the surrounding unfilled layers (electron–electron term), and then on the effect of spin–orbit coupling (the spin–orbit term). From Laporte's selection rule, it can be deduced that optical transitions between 4f states are normally forbidden. Nevertheless, when the ions are incorporated into a crystalline matrix, the effect of the crystal field generated by a non-symmetric local environment can make them partially allowed. Moreover, although the effect of the environment is weak, some splitting of the energy states related to the crystal field can be observed. To further clarify the energy level structure of lanthanides, a scheme showing the effect of each contribution to the splitting and its magnitude is presented in Figure 5.1.

It is clear that the crystal field contribution, although relevant, is weaker than any of the previous contributions. Therefore, the energy states of the lanthanides can be precisely calculated to obtain a plethora of states that will

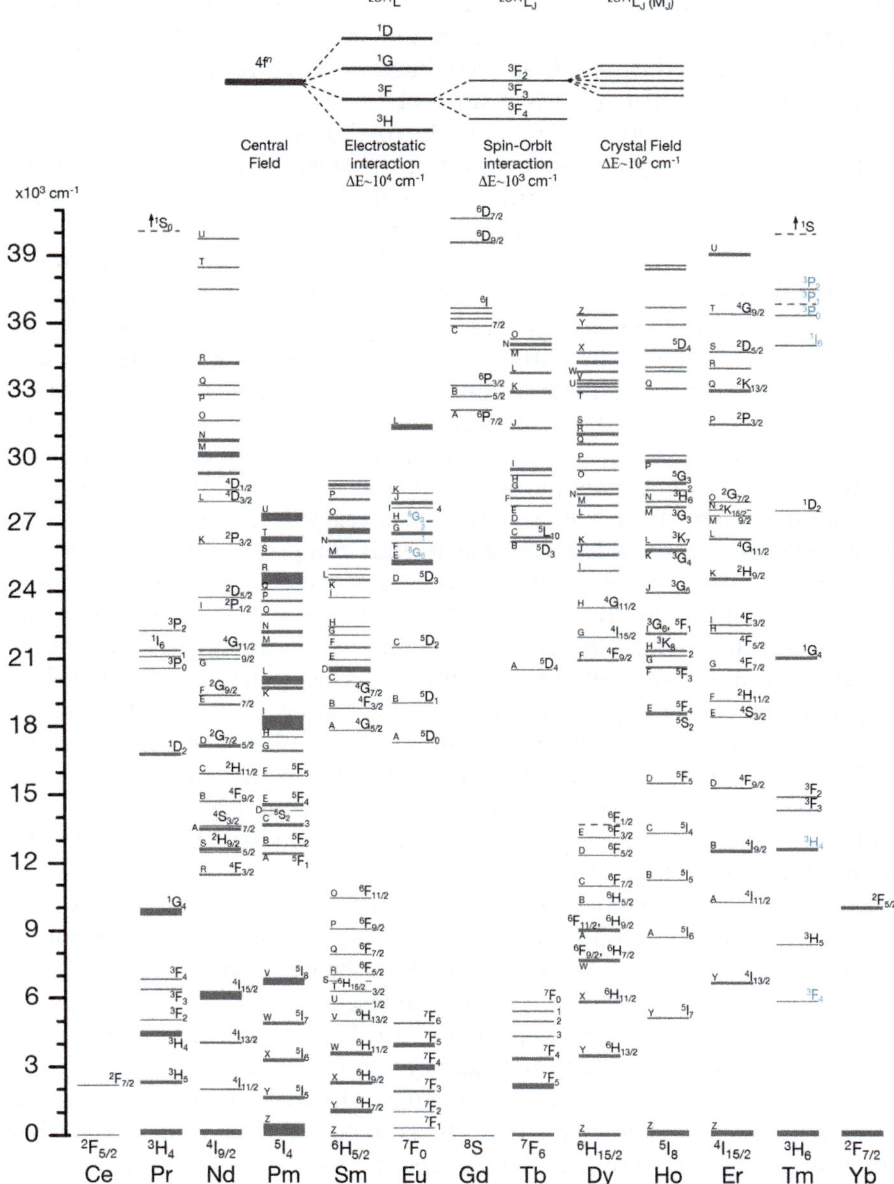

Figure 5.1 Top: Scheme of the different perturbations and resulting level splitting in lanthanides in a crystal environment. Bottom: The Dieke diagram shows the expected positions of the energy levels of each lanthanide ion. The levels are labelled according to the standard nomenclature. The labels in blue account for recalculations and consequent relabelling subsequent to Dieke's work.[51–54]

remain unaffected to a large extent regardless of the crystal host. In the case of tri-valent ions (Ln^{3+}), the most common valence state,[48,50] the expected levels are shown in the broadly applied diagram of Figure 5.1, which was first obtained by Dieke[51] using $LaCl_3$ as host. In the diagram the energy states are labelled as $^{2S+1}L_J$, where L, S and J are the total orbital momentum, the total spin and the total angular momentum of the state, respectively. It must be noted that after Dieke's work, further recalculations considering L and S state mixing gave rise to some differences in the labelling compared to Dieke's original work.[52–54] These new labels are highlighted in the figure in blue.

The absorption and emission bands associated with Ln^{3+} correspond to transitions between the energy states in Figure 5.1 that, as depicted, are typically narrow due to the partial screening of the valence electrons. Additionally, the electronic shielding contributes to enlarging the characteristic lifetimes that are normally in the range from milliseconds to microseconds. However, they can become as fast as nanoseconds in some circumstances, particularly related to nanoparticles and the activation of alternative de-population routes of the energy states.[55] Generally speaking, when it comes to the relaxation of the states and thus to emission probabilities, a small gap between two consecutive states would favour the relaxation of the upper state following a multiphononic route, *i.e.*, *via* the release of energy in the form of heat to the host lattice. On the other hand, larger gaps account for larger probabilities of photonic emissions. It can be calculated from Dieke's diagram that the energy gaps, and corresponding electronic transitions between the different states of the lanthanides, lie between the ultraviolet and the NIR ranges of the electromagnetic spectrum, covering thus the whole visible range. Therefore, each Ln^{3+} will be defined by a different set of emission bands in this range that, like a signature, will be added to the transparent crystal host in which they are doped.

Due to the electromagnetic range in which the energy states lie, tri-valent lanthanide ions are applied in different fields of science and technology related to light generation. A good example is the widespread use of Nd^{3+} to obtain laser action in the common 1064 nm lasers ($^4F_{3/2} \rightarrow {}^4I_{11/2}$); or the use of Er^{3+} in optical communications, where it excels because of its characteristic NIR emissions matching the transparency window of silica ($^4I_{13/2} \rightarrow {}^4I_{15/2}$) at around 1550 nm. In these two key examples, the ions are generally excited using photons of higher energy than the emitted ones. Nevertheless, further applications become possible thanks to the possibility of, under certain conditions, obtaining emitted photons of higher energy than the excitation photons. This effect is not exclusive to lanthanide-doped materials, though it is particularly relevant in them, as will be discussed in the following section.

5.2.2 Cooperative Processes between Lanthanide Ions: Energy Transfer and Upconversion

A careful analysis of Dieke's diagram in Figure 5.1 reveals that there are several Ln^{3+} ions with excited energy states defined by the same or very close energy values. This energy matching between two ions, A and B (for example), allows them to transfer energy to each other, as long as one of them (A) has its excited energy state populated and the other one (B) does not. In this case, the second ion (B) can use the transferred energy to promote an electron from its ground state to the excited state as shown in Figure 5.2. This process can take place through the emission and later re-absorption of a photon or through an exchange interaction if the two ions are in close proximity (at distances typically around or below 10 Å), as is often the case of dopants inside a crystal host. The first process is known as *radiative energy transfer*, and in contrast, the second one is known as *non-radiative energy transfer*, and the ions involved are called the *sensitizer* (or *donor*) and *activator* (or *acceptor*), respectively, for the initially excited ion (A) and the ion that receives the energy (B). It could appear that both cases are equivalent in the sense that the outcome is the same: the acceptor ion is populated. However, there are strong differences concerning the entire dynamics of the donor ions, since the non-radiative process is an alternative depopulation route that can compete with the emission of photons and the multiphonon relaxation, while the radiative process takes place after the emission of the sensitizer, and therefore, no further modifications of its emission probabilities are expected.

Non-radiative energy transfer is a very common phenomenon between Ln^{3+} ions as it occurs very efficiently between ions of the same species, since they have identical energy states and therefore there exists a great overlap between the energy gaps. This mechanism is often referred as *energy migration* as the energy can hop from ion to ion, all of them equivalent, and thus travel inside the material.[56] However, the case in which sensitizer and activator are not of the same species, or at least the energy levels involved are not the same, is more relevant in the systems used for nanothermometry. This is, for instance, the case of the widely applied pair of ions Er^{3+}/Yb^{3+}. In this case the only excited state of Yb^{3+} ($^2F_{5/2}$) is resonant with the $^4I_{11/2}$

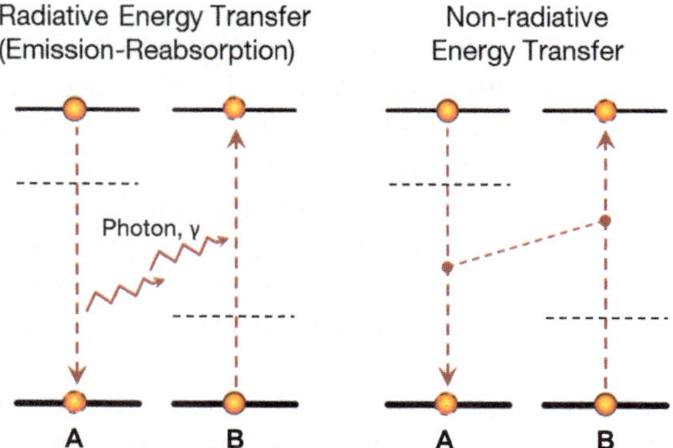

Figure 5.2 Energy transfer processes between lanthanide ions, radiative or mediated by photons (left) and non-radiative or mediated by exchange interaction (right).

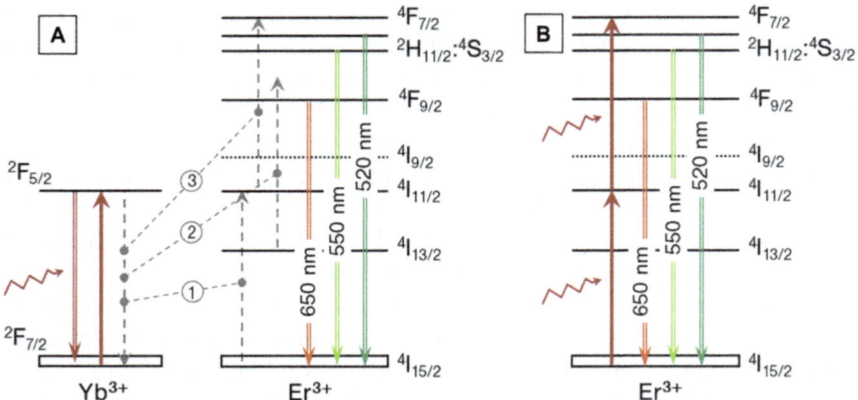

Figure 5.3 Typical upconversion processes in Er^{3+}/Yb^{3+} systems following (A) energy transfer processes and (B) excited state absorption in singly doped Er^{3+} systems.

excited state of Er^{3+}, as can be seen in Dieke's diagram and in Figure 5.3, and therefore energy transfer is possible.

Up to this point, we have considered transitions that involve the ground states of the ions; however, it is possible to find transitions between two excited states in two different lanthanides with the same energy gap, where energy transfer is also possible. For instance, in the Er^{3+}/Yb^{3+} system apart from the already mentioned possibility involving the $^2F_{5/2}$ of Yb^{3+}, $^4I_{11/2}$ of Er^{3+} excited states and their respective ground states ($^2F_{7/2}$ of Yb^{3+}, $^4I_{15/2}$ of Er^{3+}, respectively), two more energy transfer processes are possible (Figure 5.3A):$^2F_{5/2} \rightarrow {}^2F_{7/2}$ (Yb^{3+}):$^4I_{13/2} \rightarrow {}^4F_{9/2}$ (Er^{3+}) or $^2F_{5/2} \rightarrow {}^2F_{7/2}$ (Yb^{3+}):$^4I_{11/2} \rightarrow {}^4F_{7/2}$ (Er^{3+}), provided that the Er^{3+} ion involved has been previously excited to the starting excited state in each case ($^4I_{13/2}$ or $^4I_{11/2}$). In that regard the lifetime of the starting state is highly relevant, since the longer the state remains populated, the higher the probability of promoting an electron from it.

It must be noted that in some cases the energy match of the sensitizer and activator gaps is not absolute, as is the case in the process labelled as 2 in Figure 5.3A. In these situations energy transfer can still be possible, but a small amount of (excess) energy must be taken in or released in the form of heat (phonons) to the host lattice.[57] Logically, the larger the required external contribution; the lower the probability of the process to occur. In any case, these processes are generally very sensitive to temperature changes in the environment, since the energy required to bridge the gap is directly linked to it.

Following the energy diagram for the Er^{3+}/Yb^{3+} system in Figure 5.3A, it is clear that both of the energy transfers labelled as 2 and 3 in which Er^{3+} is not originally in the ground state will finish with a populated state that is higher in energy than any of the starting states. Therefore a radiative transition from this state back to the ground state will create a photon of higher energy than the excitation photons. The group of processes triggering emissions of larger energy than the excitation are generally called *upconversions*.[20] In the studied case of Er^{3+}/Yb^{3+}, for instance, the sensitizer Yb^{3+} ions are excited by NIR light typically around 980 nm. After a first energy transfer the first excited states of Er^{3+} will be populated either directly ($^4I_{11/2}$) or after a non-radiative relaxation ($^4I_{13/2}$). Subsequently, two more energy transfer processes are possible, and they would populate the higher states $^4F_{9/2}$, that can be relaxed to the ground state emitting a red photon (around 650 nm), and $^4F_{7/2}$ that is normally non-radiatively depopulated to the states $^2H_{11/2}$ and $^4S_{3/2}$ which account for green emissions (around 520 and 550 nm) when they are relaxed to the ground state.

Though upconversion in lanthanide ions often takes place through energy transfer, this is not the only possibility. There are more processes that can account for upconversion, yet some of them have very low probabilities of occurring compared to energy transfer and thus are often neglected.[20] This is mostly the case for mechanisms that do not use real intermediate states to reach higher energy states. Examples of this are cooperative sensitization, in which two sensitizers are simultaneously relaxed so an activator can take the added energy in one single step, or two-photon absorption excitation in which two excitation photons are absorbed simultaneously populating a higher energy state. It must be noted that although the probability of these processes is several orders of magnitude lower than energy-transfer-driven upconversion,[20] they can be still observed and they are in fact the main source of upconversion in other nanoscale systems not based on lanthanides, such as QDs.[58]

Following these ideas, it is clear that to enhance the probability of upconversion processes the existence of intermediate states is an advantage, even more so if they are characterized by adequately long lifetimes. There are two additional upconversion mechanisms that make use of the intermediate states and thus their efficiencies can become as high as energy transfer. These two mechanisms are known as *excited state absorption* (ESA), and *photon avalanche* (PA), and are schematically represented in Figure 5.4.

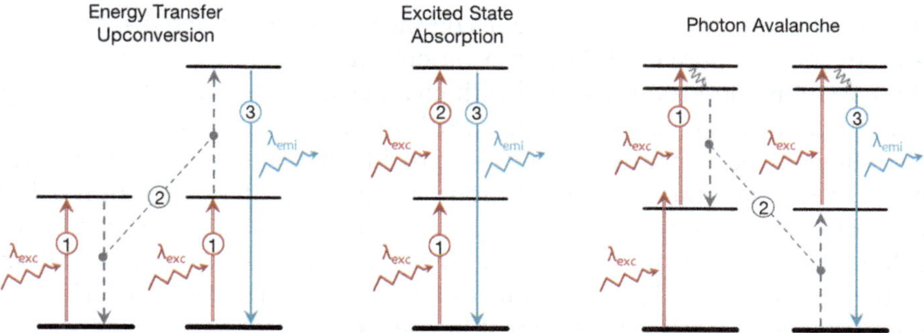

Figure 5.4 Upconversion processes in lanthanide-doped materials that use intermediate states to reach the higher levels.

The first one, ESA, involves the successive absorption of two or more excitation photons to emit only one of higher energy. This process is, for instance, an additional path to energy transfer to populate the visible emitting levels of Er^{3+}, as is shown in Figure 5.3B. Nevertheless, in the particular case of Er^{3+}/Yb^{3+} systems, although ESA mechanisms have been demonstrated and are relevant in some cases, they are generally less probable than energy transfer upconversion.

The second alternative upconversion mechanism, PA, is a more complicated process.[59,60] As shown in Figure 5.4, where the simplest situation is schematized, it involves both excited-state absorption (labelled as ①) and energy transfer (labelled as ②) mechanisms in such a way that a self-feeding loop of the intermediate state is created. Due to the high population that the intermediate state reaches in this way, it will be ready to provide electrons for upconversion, especially if its lifetime is particularly long, and thus the process can become very efficient. This mechanism has been extensively demonstrated for several lanthanide ions such as Pr^{3+}, Nd^{3+}, Sm^{3+}, Er^{3+} and Tm^{3+} in bulk crystals[59] and in some cases its efficiency is high enough to even be applied for laser generation.[61] Nevertheless, its study in nanoscale materials is not so common and therefore we will not elaborate on it further in this chapter.

5.3 Sensing Temperature with Lanthanides

The concept of lanthanide-based nanoparticle luminescence thermometry was first reported by Wang *et al.* early in the last decade.[62] This initial study focused on nanothermometry using several luminescent nanoparticles known at the time including CdTe semiconductor QDs, Mn^{2+}-doped ZnS and Mn^{2+}/Eu^{3+} co-doped ZnS, as well as BaFBr:Eu^{2+} in MCM-41 (Mobil Composition of Matter No. 41). In fact, the latter system co-doped with Eu^{2+} was shown to demonstrate a linear and reversible dependence on the 388 nm emission peak intensity ($4f^65d^1 \rightarrow 4f^7$ ($^8S_{7/2}$) transition) in the temperature range of 30–150 °C. The exceptional reversibility of these thermometric nanoparticles was considered to be of great importance as it emphasized their potential for use in biological applications, where a system of high stability and repeatability would be essential to monitor varying parameters *in vivo*. The linear thermometer relies on the variation of intensity; however, a more robust approach relies on a ratiometric technique, which removes the possibility of error associated with changes in the surroundings during measurements. In contrast, the Mn^{2+}/Eu^{3+} co-doped ZnS system, which was also shown to be highly interesting as it provides a dual-temperature-sensing probe, relies on a ratiometric technique (Figure 5.5). Upon excitation using light of a wavelength corresponding to 394 nm, the main emission is attributed to the tri-positive europium ions at 612 and 588 nm corresponding to the $^5D_0 \rightarrow ^7F_2$ and $^5D_0 \rightarrow ^7F_1$ transitions, respectively. The ratio of these intensities yields a linear relationship over a range of 30–140 °C. On the other hand, excitation at 360 nm results in prominent emission from the Mn^{2+} ions at 595 nm corresponding to the $^4T_1 \rightarrow ^6A_1$ transition. In providing

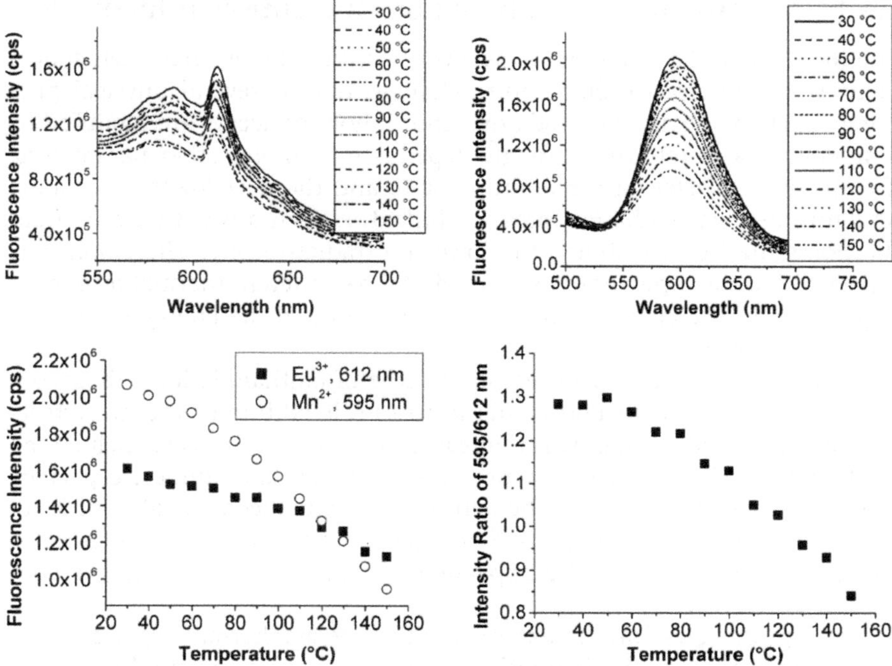

Figure 5.5 Top: Luminescence spectra of $ZnS:Mn^{2+}/Eu^{3+}$ nanoparticles obtained over the temperature range of 30–150 °C. Top left: The peak at 612 nm is ascribed to the $^5D_0 \rightarrow {}^7F_2$ transition of Eu^{3+} following 394 nm excitation. Top right: The peak at 595 nm ascribed to the $^4T_1 \rightarrow {}^6A_1$ transition of Mn^{2+} following 360 nm excitation. Bottom left: Variation in luminescence peak intensity as a function of temperature [squares: 612 nm emission (394 nm excitation), circles: 595 nm emission (360 nm excitation)]. Bottom right: Peak intensity ratio for the two peaks at 595 and 612 nm as a function of temperature.
Reproduced with permission from ref. 62. © American Chemical Society 2002.

a ratio of this emission peak to that obtained from the excitation of Eu^{3+} at 394 nm, a second thermometric relationship can be established over a similar range of temperatures.

The interest in nanothermometry and the temperature-sensitive emissions of the lanthanides continued to grow with works extended to other ions and techniques. Considering the characteristics of the lanthanide ions described, it is clear that they can be used as optical probes in the UV, visible or NIR ranges, and either through direct excitation or upconversion. Most commonly, the temperature-sensitive characteristics of the luminescence produced by different materials are the intensity or intensities ratios, the bandwidth, the spectral position of the band, the polarization of light and the luminescence lifetime of the emission.[46] Nevertheless, due to the shielding of valence electrons in lanthanides, not all of these options are feasible or sensitive enough in lanthanide-doped materials.

5.3.1 Nanothermometers based on One Emission Intensity

The intensity of the different emissions of lanthanide ions has been shown to be very sensitive to temperature changes due to several physical phenomena. On one hand, non-radiative (multiphonon) decays are modified by temperature at a rate that strongly depends on, among other parameters, the energy gap between the emitting state and the next lower lying state. Radiative and non-radiative rates are linked in such a way that the larger the non-radiative probability, the lower the radiative one. Therefore, it is expected that the different emissions will follow different thermal responses depending on the energy state generating them and the energy spacing to the next energy state.

On the other hand, depending on the type of lanthanide ions, their concentration as dopants in a host and the host itself, additional thermal effects can be found. The mechanisms that can be involved in this case include the modification of energy transfer processes between lanthanides, especially when they are non-resonant, the activation or enhancement of additional energy transfer mechanisms to other quenching centres, to the host or charge-transfer states, or the appearance of phonon-assisted Auger conversion processes.[50]

Such a profusion of mechanisms can influence the thermal dependency of each emission intensity differently, and in some cases can produce a very complicated landscape and account for difficulties in predicting and understanding the system. Moreover, fluctuations in the concentration of nanoparticles or the excitation power can also account for non-controlled intensity changes. This last case is particularly relevant in upconversion systems, since in this case, the emitted light will non-linearly depend on the excitation power. This can be a problem for applications where the path that the excitation beam must travel is not fully controlled and thus the light can be scattered or absorbed to unknown extents, as is the case of *in vivo* bio-applications. In these circumstances, intensities can change in unexpected ways and thus lead to erroneous thermal readings.

It is possible to find some examples of this technique, such as the work of Peng *et al.* using Y_2O_3 nanoparticles doped with europium ions.[63] They presented the interesting phenomenon of a decreasing emission intensity associated with the $^5D_0 \rightarrow {}^7F_2$ transition with increasing temperature in the 10–300 K range when the nanoparticles were excited at 580 nm. Instead, when the same system was pumped at 488 nm the emission intensity grew with increasing temperature, reaching a maximum at around 550 K, and then decreased for higher temperatures. The reason offered for these differences is the modification that the excitation wavelength can trigger in the dynamics, population density and thermally activated electron distribution of the states. In any case, this is a proof of the complicated behaviour that can be found related to the thermal dependence of emission intensities of lanthanide ions, that in practical terms make their use for nanothermometry very restricted. Nevertheless, for applications in which the excitation can be standardized and the concentration of nanoparticles is highly controlled, thermometry based on the measurement of the luminescence intensity of lanthanide ions becomes an option to consider.

5.3.2 The Ratiometric Technique based on more than One Emission Intensity

One of the most common techniques to measure temperature using lanthanide ions is based on changes of the intensity of the luminescence, but instead of using one single emission line as previously explained, using a ratio of intensities, *i.e.*, using a second emission band as a reference. This option can solve problems associated with fluctuations in the concentration of nanoparticles or excitation power density reaching the sample. Yet, the complicated dynamics of lanthanide ions often involve more than one temperature-dependent phenomenon such as non-resonant energy transfer or multiphonon relaxation, as was mentioned in the previous section. Since these mechanisms would affect the various emitting levels differently, the overall behaviour can be difficult to control and predict.

Intensity ratios can then be helpful following direct excitation to high-energy states, since the dependence on power density will generally be removed. Instead, in the case of a system excited through upconversion (typically excited in the NIR at 980 nm) the situation can become more complicated. In this case, the different emission bands will be excited following different routes, each one with its particular dependence on excitation power that can be quadratic, cubic, *etc.* depending on the number of upconversion steps required to populate each state. Therefore, the intensity ratio defined using two emission bands can still hold a dependence on excitation power, and therefore, the thermometer can lose accuracy in applications that involve non-homogeneous environments, such as thermal bio-imaging.

Nevertheless, some nanostructure designs have been proposed to solve this problem while exciting through upconversion by carefully controlling the excitation path of the lanthanide ions. This is the case of the work of Zheng *et al.*[64] where they developed an upconversion self-referencing temperature sensor based on a core–shell system. In their work, they started with an $NaGdF_4$:Tm^{3+}/Yb^{3+} core to which they introduced an $NaGdF_4$ shell co-doped with Eu^{3+} and Tb^{3+}. Using 980 nm excitation, the authors were able to populate the UV-lying levels of Gd^{3+} in the core *via* successive multi-photon upconversion from the Tm^{3+}/Yb^{3+} combination followed by energy transfer to Gd^{3+}. Subsequently, energy was transferred from Gd^{3+} in the core to Gd^{3+} in the shell, and then to Tb^{3+} and Eu^{3+} resulting in red and green emissions at 615 and 545 nm, respectively, that were used to define the thermally dependent intensity ratio. In this case, the excitation of Tb^{3+} and Eu^{3+} is reached through upconversion using NIR excitation, but due to the very simple energy scheme of Gd^{3+} ions (see Figure 5.1), the energy transfer mechanism that populates the ions used as probes is greatly simplified compared to the case in which the donor is Tm^{3+} instead of Gd^{3+}.

An additional alternative to define a highly reliable intensity ratio that can be excited through upconversion mechanisms is the so-called *Luminescence Intensity Ratio* (LIR) technique, one of the most widespread strategies for thermometry involving lanthanide ions. The LIR technique is based on the intensity ratio between two different energy levels that are thermally coupled. This means that both levels are separated by an energy gap small enough to allow the promotion of electrons to the upper state using thermal energy. Since both levels are closely spaced, the non-radiative relaxation from the upper level to the lower one is very likely to be high. Therefore, both states are linked and share the electronic population in a way that the ratio of intensities between their emissions will be independent of the excitation source and fluctuations in the particle concentration, constituting thus a reliable system to monitor temperature.

The thermal population of the upper levels is described by Boltzmann's distribution; hence the equation used in the LIR technique to thermally calibrate a system is:[65]

$$\mathrm{LIR} = \frac{I_\uparrow}{I_\downarrow} = B \exp\left(-\frac{\Delta E}{k_\mathrm{B} T}\right) \tag{5.1}$$

where I_i is the intensity of the transition coming from the higher energy ($i = \uparrow$) and the lower energy ($i = \downarrow$) states, which are separated by an energy gap ΔE; k_B is Boltzmann's constant; T is the temperature, and B is a constant that depends on the experimental system and on spectroscopic parameters of the material:

$$B = \frac{c_\uparrow(\nu) A_\uparrow g_\uparrow h\nu_\uparrow}{c_\downarrow(\nu) A_\downarrow g_\downarrow h\nu_\downarrow} \tag{5.2}$$

where h is Plank's constant, A_i is the spontaneous emission rate of the level, g_i its degeneracy and $c_i(\nu)$ is the response of the detection system at the emission frequency, ν_i. It could appear that the dependence on the detection system can make the method less exportable. However, since both energy states involved are close in energy, the selected intensities are usually close as well, and therefore it is often acceptable to assume $c_\uparrow(\nu)\nu_\uparrow/c_\downarrow(\nu)\nu_\downarrow \sim 1$.[66] In this way, it appears that the main parameters that can modify B are the spontaneous emission rate, which generally varies with the host, and the degeneracy of the level, which depends on the symmetry of the optical centers.[67]

In practical terms, there are certain limitations on the characteristics of the two levels involved in LIR:[68,69]

- An energy gap between the states smaller than $\sim 2000\,\mathrm{cm}^{-1}$ is required to guarantee thermal coupling and a high population in the upper level.

- Since the selected emission lines should be easily resolved, the energy gap should also be larger than $\sim 200\,\mathrm{cm}^{-1}$.
- The luminescence intensity from both states involved should be high to ensure easy detection of the optical signal.

A key example of a lanthanide ion with appropriate states for thermometry based on LIR is Er^{3+}, whose green emitting states, $^{2}H_{11/2}$ and $^{4}S_{3/2}$, are separated by a gap that generally ranges from around 350 to $900\,\mathrm{cm}^{-1}$, depending on the host.[66] This is likely the most widespread ion for nano-thermometry with lanthanides, especially when the probe is designed for bio-applications, since the thermally coupled states can be conveniently excited through upconversion in the NIR. However, this is not the only lanthanide ion suitable for LIR, and probes based on Pr^{3+}, Nd^{3+}, Sm^{3+}, Eu^{3+}, Gd^{3+}, Dy^{3+} and Yb^{3+} have also been demonstrated for materials at the macroscale[68,69] and in some cases also at the nanoscale.[62,70–72] It must be noted that in some cases, it is even possible to define an intensity ratio associated with only one emitting level with two well-separated Stark sub-levels,[70,71] in this case the value of ΔE will generally be smaller, as it is the separation between Stark sub-levels (Figure 5.1).

The sensitivity, S, of a thermal probe is a measure of the extent of change of a thermally sensitive parameter following a change in temperature. This can be written then, using eqn (5.1) as:

$$S = \frac{\mathrm{d(LIR)}}{\mathrm{d}T} = B\exp\left(-\frac{\Delta E}{k_{\mathrm{B}}T}\right) \times \frac{\Delta E}{k_{\mathrm{B}}T^2} = \mathrm{LIR} \times \frac{\Delta E}{k_{\mathrm{B}}T^2} \tag{5.3}$$

The sensitivity is indicative of the performance of the sensor. Nevertheless, it does not include some variables that can be relevant in practical terms, such as the signal-to-noise ratio of the emission and the emission quantum yields of the materials, which entail experimental limitations. To account for these additional factors that can affect the applicability of the sensor, the thermal resolution $(\Delta T_{\mathrm{min}})$ is defined as the minimum temperature that can be effectively resolved. It can be statistically described at a certain temperature as the standard deviation of several measurements (σ), *i.e.*, the ratio of uncertainty, weighted by the thermal sensitivity at this temperature

$$\Delta T_{\mathrm{min}} = \frac{\sigma}{S} \tag{5.4}$$

As shown in eqn (5.3), the sensitivity depends on the value of LIR at each temperature, and therefore, both B and ΔE are involved in it. Due to the properties of lanthanide ions, with the valence electrons shielded from the environment, the effect of the crystal field is weak. In practical terms, this would mean that large variations in these two parameters, B and ΔE, are not expected *a priori* due to concentration changes of the dopant or an increased number of defects in the host lattice. However, it has been demonstrated that the sensitivity can indeed change within a certain range to reflect these modifications.[73] When this general statement is extended to the nanoscale, it implies that the synthesis process selected for the nanostructure could affect the sensitivity, and that the sensitivity of the nanoparticle can be different to the sensitivity of the bulk counterpart. As an example to illustrate this, we can consider the work of Wang and colleagues[73] that investigated the temperature dependence of the $^2H_{11/2} \rightarrow {}^4I_{15/2}$ and $^4S_{3/2} \rightarrow {}^4I_{15/2}$ transition emission intensities in $ZnO:Er^{3+}$ nanocrystals annealed at different temperatures up to 700 °C. The sensitivity of the sensor increased with the annealing temperature in an effect that can be attributed to variations in ΔE mostly linked to symmetry changes in the optical centres of Er^{3+} from a highly symmetric cubic centre at room temperature to a pseudo-octahedral symmetry after annealing. This finally results in an improved thermal sensitivity of 6.2×10^{-3} K^{-1} at 150 °C, providing a nanoscale sensor with high accuracy.

The effect that modifications of the surface of the nanostructure can have on its properties is also remarkable. The surface-area-to-volume ratio of the material is larger in the case of nanostructures than in macro- or micro-structures and thus the properties of the surface become strongly relevant in this case. Particularly, in the case of lanthanide-doped nanoparticles the effect of molecules in their close environment or adsorbed on their surface can potentially quench the emissions from the lanthanides. A well-known example of this is the reduction of upconversion intensity from lanthanide-doped nanoparticles when they are dispersed in water, due to the high vibrational energies of the solvent molecules whose strong NIR absorption greatly affects the population dynamics of lanthanides in their proximity.[74,75] A common strategy to reduce this quenching effect is the addition of a coating to the nanoparticles that can protect their surface. The presence of coatings can also have an effect on the final sensitivity of the sensor, although in general, the effect on its thermal resolution is more important since they can greatly enhance the emission intensity of lanthanides, especially when they are excited following upconversion routes.

An interesting example of this is the work of Sedlmeier *et al.*, who undertook a comparative study of the thermal response of several lanthanide ions (Tm^{3+}, Ho^{3+} and Er^{3+}) in an $NaYF_4$ host, all co-doped with Yb^{3+} as the sensitizer in order to improve the upconversion efficiency.[76] In addition to evaluation of the temperature sensitivity of the dopant ion, they also studied the impact of the nanoparticle capping ligand as well as the effect of a protective passive shell. Generally, the best results were observed for the Er^{3+}/Yb^{3+} dopant pair based on the ratiometric variation of the green emission bands (as discussed above). In fact average temperature resolutions of 2.1, 11.4 and >20.0 °C were observed for Er^{3+}/Yb^{3+}, Ho^{3+}/Yb^{3+} and Tm^{3+}/Yb^{3+}, respectively, in citrate-capped $NaYF_4$ nanoparticles. In comparing the capping ligand, the thermal resolution was higher in the case of citrate capping ligands (2.1 °C) *versus* EDTA ligands (3.4 °C). The authors discussed that ligand-dependent enhancement of the resolution was related to the small size of the nanoparticles and the higher degree of stability and dispersibility imparted by the citrate molecules. As a result, spectra with higher signal-to-noise ratios were obtained, providing an increase in thermal resolution. Lastly, the addition of a passive shell of $NaYF_4$ resulted in an enhanced upconversion efficiency and once again an improvement in signal-to-noise resulting in resolutions of $\sim 0.5\,°C$ in contrast to $>2\,°C$ obtained with the other systems.

To clearly understand the degree to which the sensitivity of the sensor can be modified without changing the host material, but through the application of different synthetic routes or through the use of shells, we have summarized several calibrations performed in different works on $NaYF_4{:}Er^{3+}/Yb^{3+}$, always using 980 nm as the excitation wavelength and exploiting the thermally coupled states, $^2H_{11/2}$ and $^4S_{3/2}$, of Er^{3+} as a probe (Table 5.1). This material has been widely studied since $NaYF_4$ is considered to be one of the hosts that account for higher upconversion efficiencies, and therefore, it has been used for several applications as shown in the next section.[77] From Table 5.1, it can be observed that materials synthesized by applying different techniques, appeared to possess different particle sizes and morphologies such as nanoparticles, nanorods or microprisms, as well as different crystal phases namely cubic (α) and hexagonal (β). All these differences account for some noticeable changes regarding the calculated ΔE between the levels, and on the spectroscopic and experimental parameters that define B, even in the few cases in which the experimental factor can be totally discarded as the samples were calibrated with the same system. Nevertheless, the sensitivity is only changed within a range of $1\times10^{-3}\,K^{-1}$, which, as will be discussed later, is a small variation. The thermal resolution can instead account for greater differences within the same group of samples, mostly due to the strong influence of particle phase and surface-area-to-volume ratio on the emission intensity. Unfortunately, this parameter can hardly be compared between different works, since certain characteristics of the experimental setup such as excitation power density or detection response strongly affect it and thus, it cannot be reasonably standardized.

The effect of surface modifications such as a silica coating or a core–shell structure with a Er^{3+}/Yb^{3+}-doped core and undoped $NaYF_4$ shell are also considered in the table. These two surface modifications are common in the field of lanthanide-doped nanoparticles, since they have been shown to increase the emission intensity by removing possible adsorbed quenchers on the surface or simply allowing the particles to be dispersed in water, as is the case of silica coatings.[81–83] Nevertheless, regarding the sensitivity of the thermal sensor, the effect is negligible in the case of the undoped $NaYF_4$ shell, and less than $2\times10^{-3}\,K^{-1}$ for the silica shell.[76,80] Instead, the presence of the shells would greatly change the thermal resolution, as happened in the previously described work of Sedlmeier *et al.*[76] Unfortunately, this comparison cannot be extended to additional works since experimental differences between authors are unavoidable.

Regarding the effect that the size and shape of the nanoparticles can have on the sensitivity of the sensor, it is relevant to consider the work of Li and

Table 5.1 Optical thermometers based on the luminescence intensity ratio technique using $NaYF_4$:Er^{3+}/Yb^{3+} after excitation at 980 nm and using the emissions from $^2H_{11/2}$ and $^4S_{3/2}$ as thermally coupled levels. The sensitivity has been calculated for room temperature (298 K).

Morphology	Synthesis	Concentration	Size/nm	Phase	ΔE/cm^{-1}	B	S/×10^{-3} K^{-1}	Ref.
Bulk	Purchased		600–1200	β	901.6	11.7	2.09	78
Nanowires	Hydrothermal		4800×160	β	908.9	14.0	2.44	78
Nanorods	Solvothermal	$[Er^{3+}]=2\%$, $[Yb^{3+}]=18\%$	1100×140	β	926.8	12.1	1.96	78
Hexagonal nanoplates	Thermal decomposition		100×62	β	964.2	18.0	2.53	78
Nanoparticles	Thermal decomposition		18	α	709.3	5.7	2.06	77
Microprisms	Hydrothermal		600×2000	β	746.6	8.1	2.56	79
Microparticles	Co-precipitation		95–110	α + β	725.3	8.2	2.80	76
Nanoparticles	Thermal decomposition	$[Er^{3+}]=2\%$, $[Yb^{3+}]=20\%$	26–30	β	776.2	11.1	3.18	76
Core/undoped $NaYF_4$ shell	Thermal decomposition		32–34	β	715.6	8.9	3.17	76
Nanorods (NR)	Thermolysis	$[Er^{3+}]=3\%$, $[Yb^{3+}]=23\%$	49×23	β	654.1	6.5	2.84	80
NR@SiO$_2$					767.9	14.4	4.24	80

co-workers[78] who investigated morphological effects on the thermal sensitivity of $NaYF_4$ nanoparticles co-doped with Er and Yb to achieve a more profound understanding of the factors that can impact sensitivity and to ultimately develop a thermometer with a high thermal response efficiency. The authors investigated bulk powders (0.8–1.2 µm), nanowires (4.8 µm× 160.8 nm), nanorods (1.1 µm×140.5 nm) and nanoplates (48.2×62.6 nm) and their respective sensitivities, following 980 nm excitation. Owing to the fact that the particle size differed among the various morphologies, the authors compared the impact of the surface-area-to-volume ratio on the thermal sensitivity and found that it changed as follows: bulk material < nanowires < nanorods and nanoplates. The highest sensitivity was observed at 393 K for all morphologies with a value of 0.45×10^{-3} K^{-1} for the nanoplates. Further work related to the size effect of the nanoparticles was carried out by Alencar *et al.* where nanoparticles of sizes ranging from 26–60 nm were investigated.[84] In this case, their work was based on 980 nm excitation of Er^{3+}-doped $BaTiO_3$ and showed that the particle size played an important role not only due to the lattice vibrational modes, but also due to the numbers of carbonate and hydroxyl ions adsorbed on the surface of the nanocrystals. Smaller particles were found to possess lower thermal sensitivity in their work; however, particles of the same size doped with a four-fold increase in Er^{3+} ions did not show any significant difference in their sensitivity.

Up to this point, the effect of the morphological characteristics of the lanthanide-based nanostructures on sensitivity has been discussed. However, to account for larger changes in sensitivity, it is possible to use different material hosts or different lanthanide ions so both B and ΔE will change to a larger extent. Table 5.2 summarizes the properties of a selected group of sensors proposed in the literature and their sensitivities are plotted in Figure 5.6 to show the dependence they have on temperature. Some of them are based on the intensity ratio of the green emitting bands of Er^{3+} ($^2H_{11/2}$ and $^4S_{3/2}$), but there are also sensors based on alternative ions, which are relevant not only from the point of view of sensitivity, but also because different ions pave the way for the use of different emission and excitation wavelengths and therefore, for different applications.

Seeking to develop nanothermometers that can be excited and subsequently emit in the NIR regions where biological tissues are mostly transparent, other ions such as Tm^{3+} and Nd^{3+} have been studied. In both cases the intensity ratio was defined between two Stark components. For Tm^{3+}-based nanothermometers, Dong *et al.* studied the influence of the temperature on the upconverted luminescence ($\lambda_{exc} = 920$ nm) of the NIR-emitting transition at 800 nm ($^3H_4 \rightarrow {}^3H_6$ transition) in CaF_2:Tm^{3+}/Yb^{3+} nanoparticles.[71] In the case of Nd^{3+}-doped nanoparticles, the $^4F_{3/2} \rightarrow {}^4I_{9/2}$ transition that emits in the NIR range (generally, 860–900 nm), is the one exploited for nanothermometry based on intensity ratio. This is the case, for instance, in the work of Wawrzynczyk *et al.*[90] who used Nd^{3+} doped in $NaYF_4$. By ratiometrically calculating the changes in the two Stark level emission intensities at 863 and 870 nm, they were able to develop a

Table 5.2 Optical thermometers based on micro- or nanoparticles using different hosts and ions with thermally coupled bands for sensing.

Material	Dopant	Size/nm	Wavelength		Levels	B	$\Delta E/\text{cm}^{-1}$	Ref.
			Excitation/nm	Emission/nm				
PbF_2	Er^{3+}/Yb^{3+}	250	980	522–550		6.3	862.5	85
Gd_2O_3	Er^{3+}/Yb^{3+}	17–50	976	523–548		4.0	515.0	86
$NaY(WO_4)_2$	Er^{3+}/Yb^{3+}	3000	980	530–550		30.4	734.1	87
$LiNbO_3$	Er^{3+}/Yb^{3+}	100	980	525–550		33.1	862.5	66
CaF_2	Er^{3+}/Yb^{3+}	11	920	522–538	$^2H_{11/2}$ and $^4S_{3/2}$	350.7	1214.4	71
$GdVO_4$	Er^{3+}/Yb^{3+}	4	980	525–550		2.8	1047.5	88
$BaTiO_3$	Er^{3+}/Yb^{3+}	58	980	526–547		16.4	786.6	
		60				10.0	828	84
		26				6.0	648.6	
YbAG	Er^{3+}/Mo^{3+}	40–80	976	526–547		8.82	621.5	89
ZnO	Er^{3+}	80	978	535–555		10.1	607.3	83
$NaYF_4$	Nd^{3+} (10%)	25	830	863–870	Starks ($^4F_{3/2} \rightarrow {}^4F_{9/2}$)	1.3	87.6	90
	Nd^{3+} (15%)	25	830	863–870		1.2	86.2	90
LaF_3	Nd^{3+}	15	808	866–887		1.0	143.2	70
CaF_2	Tm^{3+}/Yb^{3+}	11	920	790–800	Starks ($^3H_4 \rightarrow {}^3H_6$)	0.5	143.8	71
$NaLuF_4$	$Gd^{3+}/Tm^{3+}/Yb^{3+}$	3000×1200	980	307–312	^6P_J	2.0	460.2	72
	$Gd^{3+}/Tm^{3+}/Yb^{3+}$		980	277–280	^6I_J	2.5	283.6	
KLuW	$Ho^{3+}/Tm^{3+}/Yb^{3+}$	100	980	648–641	$^1G_4, {}^5F_3$	1.7	146.0	91
	$Ho^{3+}/Tm^{3+}/Yb^{3+}$		980	539–549	$^5S_2, {}^5F_4$	1.7	121.3	

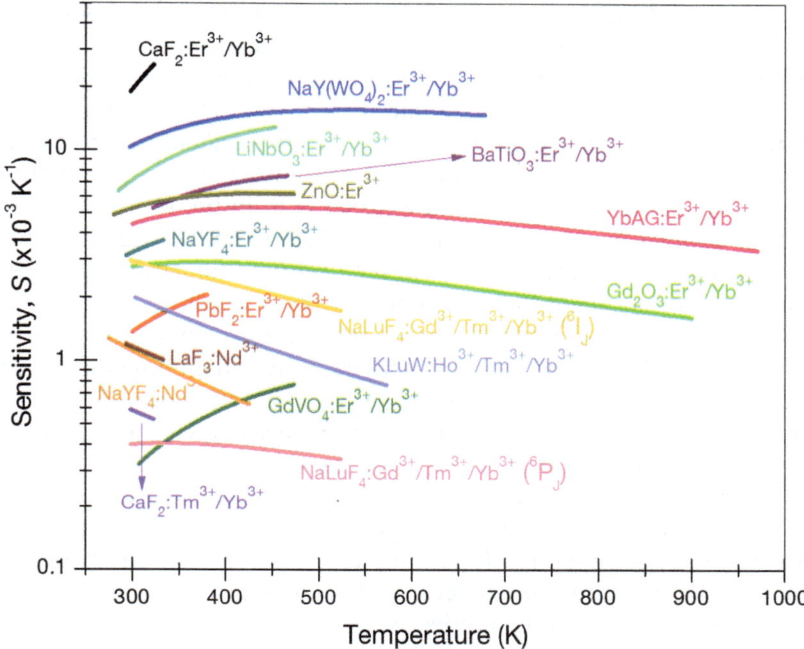

Figure 5.6 Calculated sensitivities of the different materials proposed for LIR-based nanothermometry whose characteristics are summarized in Table 5.2. The range in which each curve is displayed corresponds to the range of temperatures in which the material has been calibrated.

temperature correlation over the range of 0–150 °C with a thermal sensitivity of 0.1% K^{-1}. The excellent sensitivity combined with the Nd^{3+} NIR emissions (extending all the way to the IR region of the spectrum) render this ion highly suitable for biological applications where NIR excitation and emission are needed for tissue penetration. It is important to highlight that in the case of Nd^{3+}, the emissions observed are not upconverted but rather occur through direct excitation where the source wavelength was 830 nm and the observed emissions ranged from 840 to 950 nm, meaning at lower energies relative to the source.

Rocha *et al.* carried out an extensive study where they investigated the development of an Nd^{3+} nanothermometer,[70,92] this time using an LaF$_3$ host. Opting for a core–shell architecture (LaF$_3$:Nd^{3+}@LaF$_3$), a systematic investigation of the luminescence was carried out under NIR excitation at a wavelength of 808 nm. They showed that the quantum yield for the luminescence of Nd^{3+} was extremely high which is advantageous in terms of thermal resolution. It reached nearly 80% at 1 wt% doping levels; however it was only at a 15 wt% dopant concentration that the maximum luminescence brightness was achieved. The thermometer was established by monitoring the 863 and 885 nm emissions (from two different Stark levels) to define an intensity ratio, and they achieved a thermal resolution of 2 °C in the temperature range from 10 to 60 °C.

Further extending the use of different lanthanide ions, tri-doping was investigated as a means of achieving higher thermal sensitivity in the work of Savchuk *et al.*[91] where they evaluated KLuW nanoparticles co-doped with Ho^{3+}, Tm^{3+} and Yb^{3+}. The 100 nm sized particles could be excited with a 980 nm wavelength corresponding to the optimum excitation into the Yb^{3+} energy levels, or 808 nm for excitation of Tm^{3+} ions in the host. Excitation at 980 nm revealed blue, green and red emissions from both Tm^{3+} and Ho^{3+}. Thermally coupled levels such as the $^1G_4 \rightarrow {}^3F_4$ of Tm^{3+} (648 nm) and the $^5F_3 \rightarrow {}^5I_7$ of Ho^{3+} (661 nm) were monitored and the intensities were used to develop a ratio that could be correlated to the temperature. The transition intensity ratios clearly decreased over the temperature range, which spanned 300–600 K. A similar result was observed, over the same temperature range, when considering the thermally sensitive $^5F_4$, $^5S_2 \rightarrow {}^5I_8$ transitions (539 and 549 nm) of Ho^{3+} in developing a temperature-sensitive correlation. The temperature sensitivity reported was quite high with values of 1.7×10^{-3} K^{-1} and 1.9×10^{-3} K^{-1} upon considering a Tm^{3+}:Ho^{3+} ratio and thermally coupled transitions of Ho^{3+} alone, respectively. Excitation using an 808 nm source results in several emissions from Tm^{3+} and Ho^{3+} and once again a temperature-sensitive ratiometric correlation could be established, although this time the emitting levels applied were not thermally coupled; however the authors reported a three-fold increase in the maximum sensitivity reported at 808 nm relative to the ratios defined using 980 nm excitation.

An alternative proposed technique to improve the sensitivity and thermal resolution of the thermometers is co-doping the lanthanide ions with transition metal ions, which has been shown to improve the upconversion luminescence by approximately four-fold, ultimately resulting in a better thermometer. Dong and co-workers studied the effect of molybdenum doping on YbAG:Er^{3+} and the development of a nanothermometer over a very wide range from 300 to 900 K.[89] The upconversion efficiency observed in YbAG following Mo doping was on a par with that reported for $NaYF_4$, amounting to a value of 0.18–0.2%. Enhancement attributed to Mo doping was related to the formation of Yb^{3+}–MoO_4^{2-} dimers, in which, following 976 nm excitation, the ground state absorption promotes electrons to the $|{}^2F_{5/2}, {}^1A_1\rangle$ states followed by excited state absorption to the $^1T_1$ state of Mo. Subsequently, energy transfer to the luminescent lanthanide ion (Er^{3+}) occurred and upconversion emission was observed (Figure 5.7). Thermal sensitivity reached a maximum value of 4.8×10^{-3} K^{-1} at a maximum temperature of 467 K, while at more useful biological temperatures (300–350 K), the value varied from 3.8×10^{-3} K^{-1} to 4.2×10^{-3} K^{-1}.

It is clear that the gamut of options for nanothermometry using the LIR technique is broad. Nevertheless, regarding sensitivity this technique has an intrinsic limitation given by the Boltzmann distribution that rules it and the spectroscopic parameters involved. According to eqn (5.1), the main parameters that could be used to optimize the sensitivity are the energy gap, ΔE, and B; while B, as has been previously discussed, mainly depends on the spontaneous emission rates and the degeneracy of the states involved.

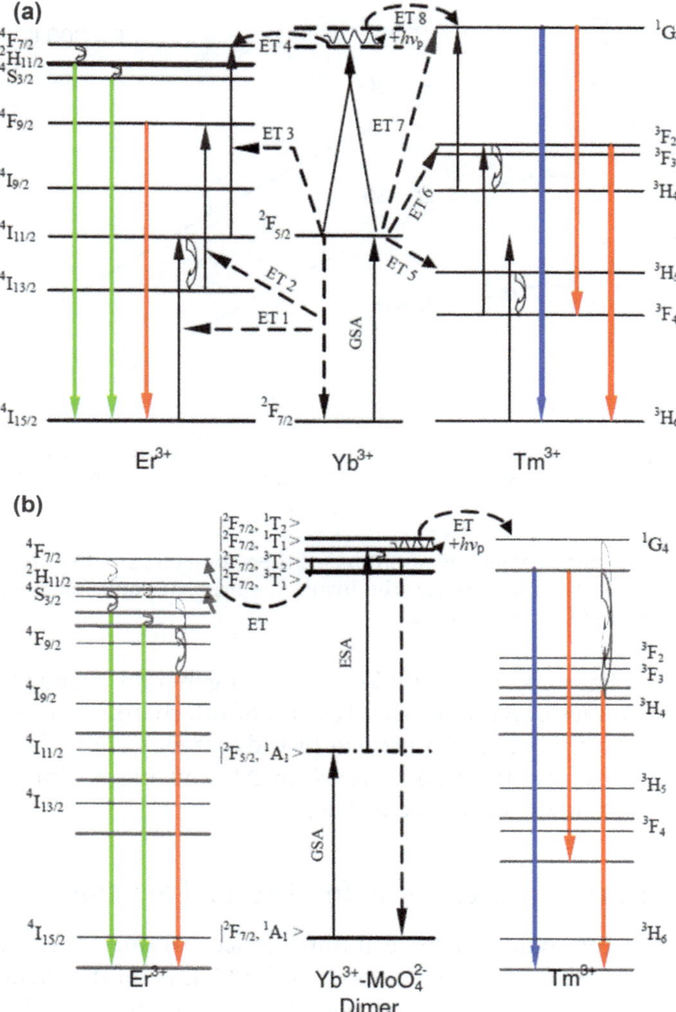

Figure 5.7 Upconversion sensitization mechanisms in (a) $Er^{3+}(Tm^{3+})$–Yb^{3+} co-doped systems and (b) $Er^{3+}(Tm^{3+})$–Yb^{3+}-Mo^{2+} co-doped systems. Reproduced with permission from ref. 89. © Elsevier 2014.

This last parameter, for lanthanides in most of the applied hosts is fixed by the relationship $J + 1/2$ (ref. 67) which dictates that the spontaneous emission rates are mainly responsible for the changes in B shown in Table 5.2. Additionally, it has been previously discussed that for a practical thermometer the energy gap must be in the range 200–2000 cm^{-1}.

In Figure 5.8, a calculation of the sensitivity at 300 K for the different allowed values of ΔE and for three representative values of B is shown. This figure demonstrates that there is an optimum value of ΔE at around 200 cm^{-1}. However, an energy gap of this width is normally related to

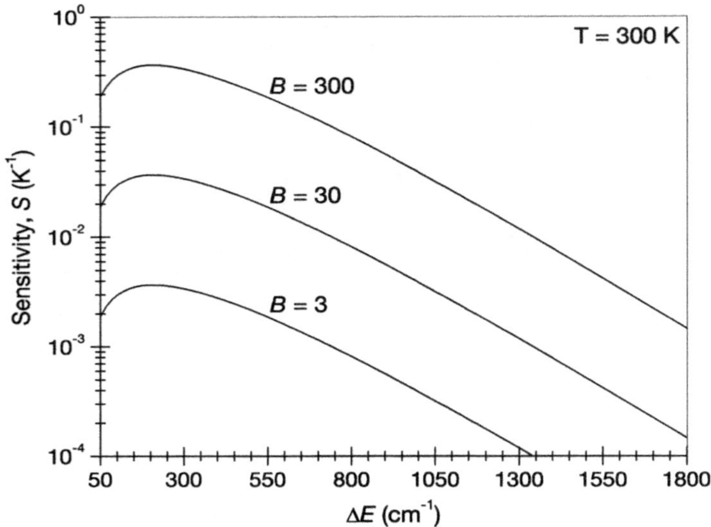

Figure 5.8 Calculated sensitivity of LIR-based thermometers at 300 K depending on the energy gap between the involved states, ΔE, and for three different values of the constant B.

thermometers based on Stark sub-levels, which generally account for detrimental small values of B (Table 5.2). The use of LIR thermometers based on two different energy bands can instead provide larger values of B, however, this subsequently results in an increase in ΔE, causing an inherent limitation in the sensitivity of the technique.

5.3.3 Luminescence Lifetime for Thermal Sensing

An alternative thermally sensitive luminescence parameter that generally does not require an exogenous reference is the lifetime of the excited states. The lifetimes of lanthanide-doped nanoparticles, defined as the time in which the initial emission intensity, I, drops to a value I/e, are normally in the range of milliseconds to microseconds. The experimental lifetime, τ_{exp}, or more accurately the inverse of τ_{exp}, can be generally defined as the sum of the different terms that contribute to it:

$$\frac{1}{\tau_{\text{exp}}} = \frac{1}{\tau_{\text{rad}}} + \frac{1}{\tau_{\text{NR}}} + \frac{1}{\tau_{\text{Q}}} \tag{5.5}$$

where τ_{rad} is the radiative contribution, τ_{NR} the non-radiative or multiphonon contribution and τ_{Q} is a term linked to the contribution of other quenching mechanisms, such as energy transfer to other lanthanide ions or additional quenchers.[50]

As previously discussed, multiphonon relaxation is expected to be enhanced at higher temperatures, thus shortening lifetimes. The third term, τ_{Q}, an additional depopulation pathway will generally contribute to shortening the lifetimes as well. As a matter of importance, temperature will have a strong effect on the lifetimes, especially if the quenching is related to non-resonant energy transfer processes that are dependent on the presence of phonons to occur. However, although the general trend is a reduction of the lifetime with temperature, energy transfer processes can be eventually activated at higher temperatures, and in these circumstances an acceptor ion could show a smooth increase in characteristic lifetime.[93] Nevertheless, the most common behaviour, as shown in Figure 5.9, is a sharp fall in lifetime within a certain temperature range that will depend on all the processes described above and thus, will change when the dopants or the host are changed.

These characteristics are common for every lanthanide-doped crystal, regardless of the size. However, a further effect related to quenching must be considered when moving to the nanoscale: the contribution of additional chemical species adsorbed on the surface of the particles. As was explained in the previous section, molecules such as OH^- can not only effectively reduce the luminescence intensity of the material, but also the characteristic lifetimes of the states. This quenching is typically more effective the higher the concentration of lanthanide ions in the host, since energy migration of the excitation energy and a larger number of dopants close to the surface will amplify the effect.[94] For these reasons, the behaviour of lifetimes with temperature is not necessarily the same in macro- and nanoscale crystals.

Indeed, Peng *et al.* published one of the first papers on luminescence lifetime based nanothermometry with lanthanide-doped nanoparticles.[63] The article investigates the effect of temperature on the different terms affecting the lifetimes [eqn (5.5)] of Y_2O_3:Eu^{3+} nanoparticles within the frame of analysing the lifetime features for nanothermometry, exciting either at

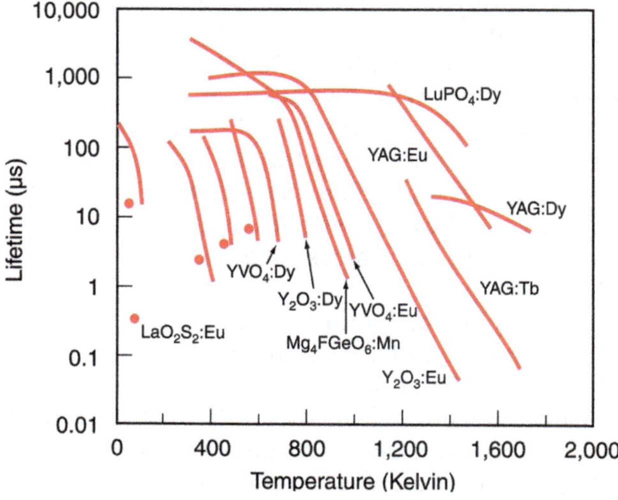

Figure 5.9 Lifetime dependence on temperature of several lanthanide-doped phosphors.
Reproduced with permission from ref. 93. © Macmillan Publishers 2014.

355, 488 or 580 nm. It also demonstrates that the lifetimes are shorter in the nanoparticles than in macroscale polycrystals. Such lifetime reduction in this case is related both to an increased radiative transition rate, which has almost no temperature dependence, and an increased quenching rate in the nanoparticles. Instead the non-radiative contribution through multiphonon relaxation is ruled out because of the large bandgap existing between the energy states.

Generally, the *quantum size effect*, defined as the consequence of the reduction of quantum allowed states in a small particle, results in an increased bandgap and an increased radiative transition rate in nanoparticles. However, this effect does not appear to be the cause in the system studied in this work. Actually, in lanthanide ions, the diameter of the electronic wave function is quite small (~ 0.1 nm) when compared to the particle size of Y_2O_3:Eu^{3+} nanoparticles. In fact, it is generally smaller than most of the lanthanide-doped nanoparticles reviewed in this present chapter with size ranges between 10–100 nm. An increase in the radiative transition rate was also observed in previous work by Igarashi *et al.*, who reported that the lattice distortion measured for nanoparticles was seven times larger than for microparticles.[95] The final outcome from such lattice distortions is a crystal field modification that will likely have an effect decreasing the lattice symmetry and consequently strengthen the oscillator intensity. An additional factor described by Peng *et al.* that can affect the electronic properties and lifetimes of the system, this time through the quenching term, is the increase of the surface-area-to-volume ratio, which is obviously much more important in nanoparticles compared to bulk crystal systems. The subsequent greater influence of surface defects may be assigned in this case as the cause of the increase of the quenching term in the Y_2O_3:Eu^{3+} nanosystem. Altogether, there are two main effects reducing the lifetimes of the nanoparticles compared to macro-sized polycrystals: lattice distortion affecting the radiative transition rate and a larger quenching rate due to a higher surface-area-to-volume ratio.

In the same year, Allison *et al.* measured the lifetimes of cerium (Ce^{3+})-doped YAG ($Y_3Al_5O_{12}$) nanoparticles and observed that they were roughly half as fast as those of micro particles of the same material.[55] On the nanothermometry side, the lifetimes measured by exciting with UV light (337.1 nm) and detecting the broad visible emission peak centred at 540 nm, showed a 33% variation in their values from 7 to 77 °C (nominal lifetimes in the range of 20 ns).

The values reported for the luminescence lifetimes of these Ce^{3+}-doped nanoparticles are rather short for lanthanide-doped materials. Normally short lifetimes are considered detrimental for applications requiring excitation through upconversion, since they will contribute to reducing the emitted intensity. However, if upconversion is not needed (for example in a non-biological application) there is an additional technical advantage for thermometry, since shorter lifetimes allow for increasing the measurement rate, thereby making the technique faster. Thus, the short decay times characteristic of some nanosystems, make the luminescence signal more convenient for standardized measurements in comparison with phosphors with millisecond decay times (when roughly the same number of photons are emitted and detected in both cases). On the technical downside of luminescence lifetime nanothermometry when compared to other techniques, a pulsed excitation laser source is required, and thus it is normally more expensive.

Some years later, Guo *et al.* reported the first use of Er^{3+}/Yb^{3+} co-doped yttria (Y_2O_3) UCNPs as luminescence lifetime based thermal probes, *i.e.*, using the luminescence lifetime of Er^{3+} instead of the spectral shape of its emission bands.[96] This work analyzed the upconverted emission lifetime of the $^2H_{11/2}$ and $^4S_{3/2}$ states over a broad temperature range spanning 300 to 1450 K. The authors reported very good thermal sensitivity since the measured lifetimes varied from 300 μs to 50 ns in the temperature range studied (~ 260 ns K^{-1} on average, following an exponential decrease with increased temperature). It was also noticeable that the presence of combustion gases around the nanoparticles did not trigger any change in the lifetimes, which supports the reliability of the technique for creating reference-free thermometers.

In recent years, more studies have been published on luminescence lifetime nanothermometry using lanthanide-doped nanoparticles. Nikolić *et al.* have published an interesting comparative study on the temperature-sensing possibilities of Eu^{3+}-doped TiO_2 nanoparticles.[97] They demonstrate similar relative sensitivities using both LIR and lifetime nanothermometry. On the other hand, as shown in Figure 5.10, the absolute sensitivity values for LIR

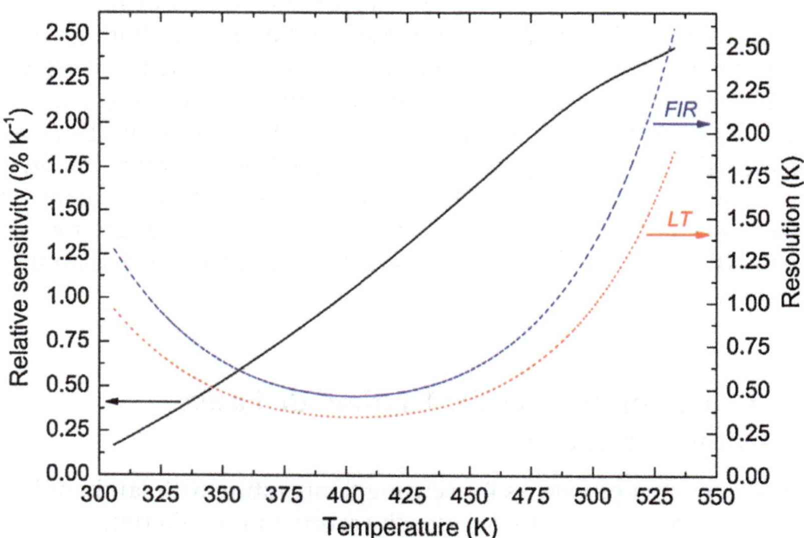

Figure 5.10　Relative sensor sensitivity (black line) and the resolutions of temperature sensing obtained from LIR (dashed blue line) and lifetime (dotted red line) measurements as a function of temperature.
Reproduced with permission from ref. 97. © American Chemical Society 2013.

measurements are between 3.3×10^{-3} and 17×10^{-3} K^{-1}, that is, seven times larger than the corresponding lifetime sensitivity values (for a further explanation of the concept of *sensitivity* (*S*) and *thermal resolution*, see previous section of this chapter). Lifetime measurements provide better resolution than LIR measurements, with values lower than 1 K over temperatures ranging from room temperature to 510 K, while the maximal resolution for lifetime measurements is 0.33 K and it is estimated at about 400 K.

Finally, Savchuk *et al.* have reported on the influence of "host engineering" on the sensitivity of luminescence lifetime nanothermometry achieved with Er/Yb co-doped nanoparticles. This work established the difference in the $^4S_{3/2}$ energy level lifetimes values of Er^{3+} emission at 545 nm, under pumping at 980 nm, comparing two similar fluoride-based nanoparticles NaYF$_4$:Er^{3+}/Yb^{3+} and NaY$_2$F$_5$O:Er^{3+}/Yb^{3+}.[98,99] The latter ones showed a larger thermal sensitivity compared to the former, due to the marked temperature dependence on their luminescence lifetime. The values reported for the lifetime thermal coefficient (*i.e.*, the sensitivity) were 5.4×10^{-3} and 15×10^{-3} K^{-1} (Figure 5.11).

The authors tentatively attributed the more pronounced temperature dependence of the luminescence lifetime in the NaY$_2$F$_5$O:Er^{3+}/Yb^{3+} nanoparticles to the fact that non-radiative relaxation and multiphonon phenomena, responsible for the shortening of the luminescence lifetime decays, are more significant in oxide materials than in fluoride materials.[100] Interestingly, the authors also report on the luminescence lifetime of the Er^{3+} red emission around 660 nm ($^4F_{9/2} \rightarrow {}^4I_{15/2}$) that shows a sensitivity value of 7×10^{-3} K^{-1}, in accordance with previous works dealing with the temperature dependence of the luminescence lifetimes of both green and red emission of Er^{3+} ions.[101] The reduced thermal dependence observed for the red emission was explained in terms of the fact that the number of phonons involved in the non-radiative de-excitations from the $^4F_{9/2}$ state is larger than that of the $^4S_{3/2}$ state. This is caused by the larger energy separation that exists between the $^4F_{9/2}$ state and the next lower energy state compared to the $^4S_{3/2}$ state. Due to that, the luminescence lifetime of the green emission of the Er^{3+} ions has a larger thermal sensitivity than the red emission.

5.3.4 Other Possibilities for Lanthanide-based Nanothermometry

The most exploited properties for sensing temperature with lanthanide ions are undoubtedly the intensity or intensity ratios and the lifetimes. However, these are not the only possibilities, and depending on the application, other options can be considered. In 1976, Kusama *et al.* published a very complete discussion on the best option for measuring the temperature of colour TV screens using Eu^{3+}-doped Y$_2$O$_2$S powder.[102] In their, work they concluded that the available options for LIR in this case do not offer enough sensitivity.

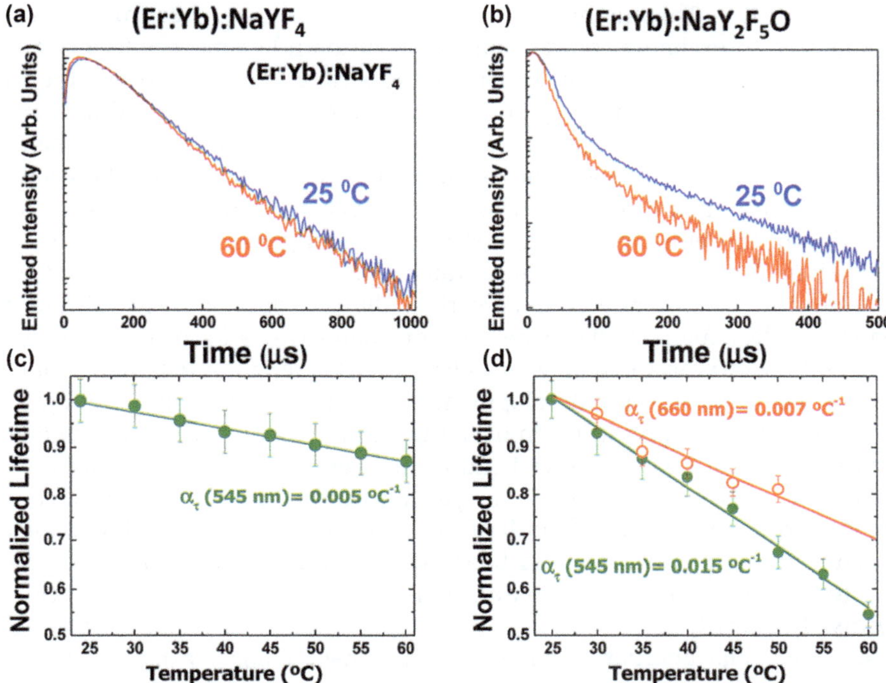

Figure 5.11 Luminescence decay curves of the 545 nm emission line of (a) $NaYF_4$:Er^{3+}/Yb^{3+} and (b) NaY_2F_5O:Er^{3+}/Yb^{3+} nanoparticles at 25 and 60 °C; and calculated lifetime values as a function of temperature for (c) $NaYF_4$:Er^{3+}/Yb^{3+} and (d) NaY_2F_5O:Er^{3+}/Yb^{3+} nanoparticles. The temperature dependence of the normalized lifetime of the red emission generated by the Er^{3+} ions in NaY_2F_5O:Er^{3+}/Yb^{3+} has also been included in (d) as open circles. In all cases the dots are experimental data and solid lines are the best linear fits.
Reproduced with permission from ref. 99. © AIP Publishing 2013.

The temperature dependence of the intensity of alternative emission bands is difficult to predict, and thus its determination must be empirical. Still, several intensity ratios are defined between Eu^{3+} emitting bands showing good thermal sensitivity. The relationships obtained though, are not the preferred linear behaviour, thus, the spectra must be recorded over a wide range of wavelengths, complicating the experiment, and moreover, the relationships detect important fluctuations in the defined ratios depending on small differences in Eu^{3+} concentration. For these reasons, the preferred parameter to define the thermometer was found to be the line-shift that can be observed under certain circumstances in lanthanide-doped materials. This phenomenon is attributed to the strain that temperature can induce on the environment of the ions, modifying the amplitude of lattice vibrations, and thus shifting the electronic energy states through electron–phonon interaction.

The shift of emission bands in lanthanides is an effect rarely exploited for thermometry. However, it can reach, according to Kusama *et al.*, a value of

$0.033\,\text{cm}^{-1}\,\text{K}^{-1}$. This method is highly reliable, since it is free from fluctuations due to excitation power density or lanthanide concentration, yet it requires good spectral resolution in the experimental setup to be able to provide competitive sensitivity and thermal resolution. The line-shift strategy has been observed at the nanoscale using the 863 nm emission band $(^4F_{3/2}\,(R_1)\rightarrow{}^4I_{9/2}\,(Z_1)$, where the symbols in brackets label the Stark sublevels, as described in Figure 5.1) of Nd^{3+}-doped LaF_3 nanoparticles.[92] The red-shift of this emission is demonstrated to change linearly and is defined by a 7×10^{-3} nm per °C modification rate in the range between 10 and 60 °C $(0.1\,\text{cm}^{-1}\,\text{K}^{-1})$. This rate is smaller than the one obtained with different nanoparticles, such as QDs, where 0.1 nm per °C rates have been demonstrated.[103] However, the bandwidth plays a role in terms of the applicability of this technique and the development of a measuring system based on it. The contrast of the line-shift can be defined as the modification rate divided by the bandwidth, so narrower bandwidths would account for higher contrast. The emission band of QDs is typically on the order of several tens of nanometres, while the bandwidth of the transition of Nd^{3+} mentioned ions is 3 nm. Consequently, rate and bandwidth compensate each other to give rise to two nanothermometers based on line-shift with comparable quality in terms of contrast.

An additional alternative for a thermal sensor using lanthanide-doped materials is the measurement of the bandwidth, an effect that has been demonstrated in $Y_2O_3{:}Eu^{3+}$ nanocrystals under 488 nm illumination.[63] Generally speaking, apart from the natural broadening of the emission lines given by Heisenberg's uncertainty principle, there are two main contributions to the bandwidth of the ions; one related to the intrinsic vibrations of the lattice, that can be labelled as a type of *homogeneous broadening*, and one related to the presence of different optical centres and defects, known as *inhomogeneous broadening*.[104] While the latter normally shows little dependence with temperature, the former can be greatly affected by it, since it is ruled by the characteristics of the lattice phonons. Nevertheless, as it happens with the line-shift strategy, the use of the bandwidth as a thermometer requires detecting systems with good spectral resolution.

For instance, in the above-mentioned work with $Y_2O_3{:}Eu^{3+}$, the effect is analyzed for the $^5D_0\rightarrow{}^7F_2$ transition in the range between 10 and 670 K, and two ranges of temperatures where the dependence of the bandwidth with temperature is different are defined.[63] Below 70 K the bandwidth remains constant within the resolution of the measurements $(2\,\text{cm}^{-1})$, while above this temperature the emission line is broadened following an almost linear function with an $\sim0.078\,\text{cm}^{-1}\,\text{K}^{-1}$ rate. In a different study, the bandwidth of several emission peaks of Tm^{3+}-doped $NaYbF_4$ microparticles coated with SiO_2 was analyzed.[105] In this case the study was performed from 100 to 700 K, and it was determined that the emissions corresponding to $^3H_4\rightarrow{}^3H_6$ and $^1D_2\rightarrow{}^3F_4$ transitions hold a linear dependence with temperature over the whole range. On the other hand, $^1G_4\rightarrow{}^3H_6$ and $^3F_2\rightarrow{}^3H_6$ transitions show more complicated dependencies that are therefore less relevant for thermometry.

5.4 Applications

The promise of exceptional sensitivity prompted the first instance of the integration of a lanthanide nanothermometer in 2009. Aigouy and co-workers[85] studied the temperature-dependent luminescence of 250 nm sized PbF_2 nanoparticles, which were co-doped with Er^{3+} and Yb^{3+}. The application developed relied on fixing the nanoparticle to an atomic force microscope (AFM) tip and the authors were able to correlate the changes induced in the electric current flow to the luminescence of the nano-particles, and derive the temperature from this relationship. Aigouy studied a 20 μm long and 1 μm wide nickel strip previously prepared *via* lithography and probed it with an AFM tip that had a nanoparticle glued to its tip. The tip was probed with 975 nm light to excite the lanthanide ions in the nanoparticle and the emission was directed to a photomultiplier tube for analysis (Figure 5.12).

Since the stripe was attached to nickel pads with electrical contacts and powered by an offset alternating current, both the temperature and lumi-nescence could be modulated and detected. Prior to the determination of the temperature, the authors proposed a normalization procedure based on room-temperature luminescence and thermally modulated temperature

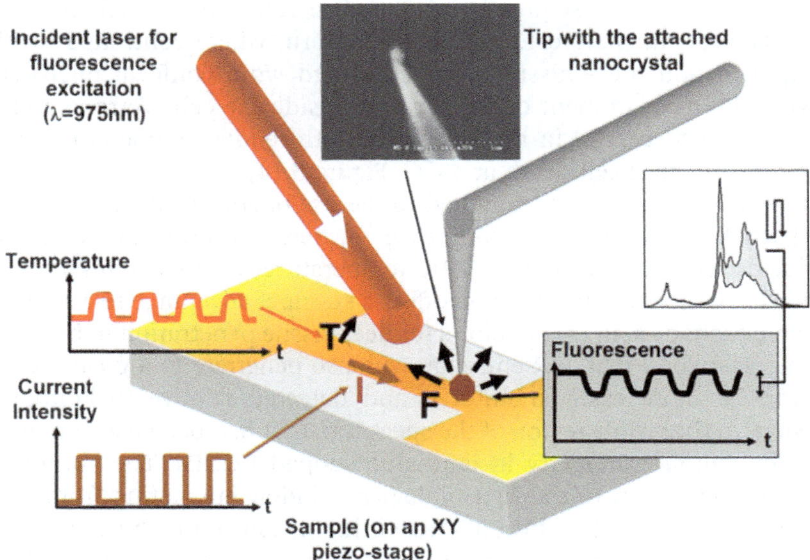

Figure 5.12 Diagram of the experimental setup involving a lanthanide-doped nano-particle glued to an AFM tip. The device is powered by a square electrical current (brown curve), which results in periodic Joule heating (red curve) and quenching of the luminescence of the particle (black curve). Inset: scanning electron microscopy image of the nanoparticle glued to the AFM tip.
Reproduced with permission from ref. 85. © The Council of Scientific & Industrial Research 2011.

luminescence in order to eliminate near-field optical effects to avoid masking of the thermal effects. The spatial resolution reported by the authors using their modified AFM tip was comparable to the size of the nanoparticle, meaning less than 500 nm while the thermal sensitivity was <5 K. The integration of temperature-sensitive nanoparticles with AFM tips is quite important as it opens up avenues towards integration in novel applications, not only in biology and nanomedicine but also in applications such as the characterization of electronic devices.

The first demonstration of a biological application of lanthanide nanothermometers was reported by Vetrone *et al.*[77] In their work, they demonstrated the possibility of acquiring the temperature of a single living cell, namely a HeLa cancer cell. The authors started with a water-dispersible $NaYF_4$:Er^{3+}/Yb^{3+} nanoparticle system, which is a critical aspect of biological applications. In order to demonstrate the potential for thermal imaging in solution, the authors set up a pump–probe experiment where they focused a 980 nm beam on the colloidal dispersion allowing for absorption of that wavelength by the water creating a heating effect. Subsequently, they probed with a 488 nm line, not absorbed by the water but used to excite the Er^{3+} ion, and the luminescence was collected, resulting in a temperature distribution along the pumped area monitoring the luminescence generated by the $^2H_{11/2} \rightarrow {}^4I_{15/2}$ and $^4S_{3/2} \rightarrow {}^4I_{15/2}$ transitions of the Er^{3+} ion (Figure 5.13). The nanoparticles were subsequently incubated in HeLa cancer cells for 1.5 h. The cells were transferred to a heating platform where controlled heating was applied and the emission was monitored *via* a confocal microscope allowing the measurement of temperature inside the cell. Vetrone and co-workers were successful in measuring the intracellular temperature in the range of 25 °C until cell death at 45 °C (Figure 5.14).

This concept was further extended to the use of NIR-to-NIR upconverting nanoparticles in the work of Dong *et al.*[71] where NIR optically excited lanthanide-doped UCNPs showed NIR temperature-sensitive emissions. As previously outlined, NIR excitation offers a succinct advantage relative to UV or visible excitation, in that it allows for deep tissue penetration in biological systems. It follows that NIR emission will also penetrate to a greater extent and can be more easily monitored and detected relative to lanthanide emission, in the visible region of the spectrum, that had been used up to that point for nanothermometry in lanthanide-doped UCNPs. The authors reported NIR emission in the first biological window emanating from Tm^{3+} and Yb^{3+} co-doped CaF_2. This CaF_2 host is highly biocompatible and as such can be used in biological applications. Incubation of the nanoparticles in HeLa cancer cells at 37 °C for 2 h was followed by 920 nm excitation. This wavelength was purposely used to avoid heating of the intracellular water observed upon excitation with a 980 nm beam. The results showed a very high multicontrast image for the Tm^{3+}-doped cells due to its very intense NIR emission. This was the result of luminescence efficiencies two orders of magnitude greater in the case of Tm^{3+}-doped nanoparticles relative to their Er^{3+}-doped counterparts. Phantom tissues of varying thickness were

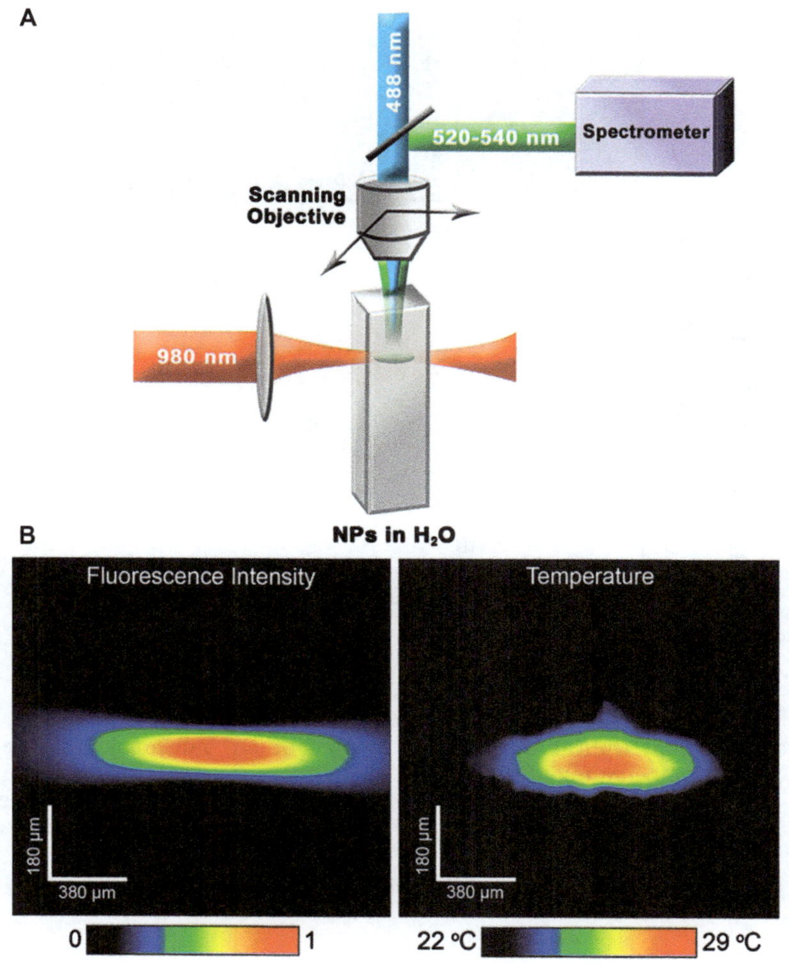

Figure 5.13 (A) Schematic representation of the pump–probe setup used to obtain the temperature profile resulting from the heating of a colloidal solution (*via* a 980 nm diode laser pump beam) of $NaYF_4$:Er^{3+}/Yb^{3+} nanoparticles in water and subsequent scanning with a 488 nm Ar^+ laser (probe beam). (B) Left: Confocal image (pump absorbed profile) of the upconverted luminescence following 980 nm excitation. Right: Thermal image of the area created by the 980 nm pump beam. Reproduced with permission from ref. 77. © American Chemical Society 2010.

subsequently placed in the confocal microscope between the colloidal suspension and the focusing/collecting objective. The authors demonstrated that temperature-sensitive NIR emission from the Tm^{3+}/Yb^{3+} system showed tissue penetration of 2 mm whereas the penetration depth of emission from the same system, which was instead doped with Er^{3+}/Yb^{3+} was four times less. These results were highly significant as they showed

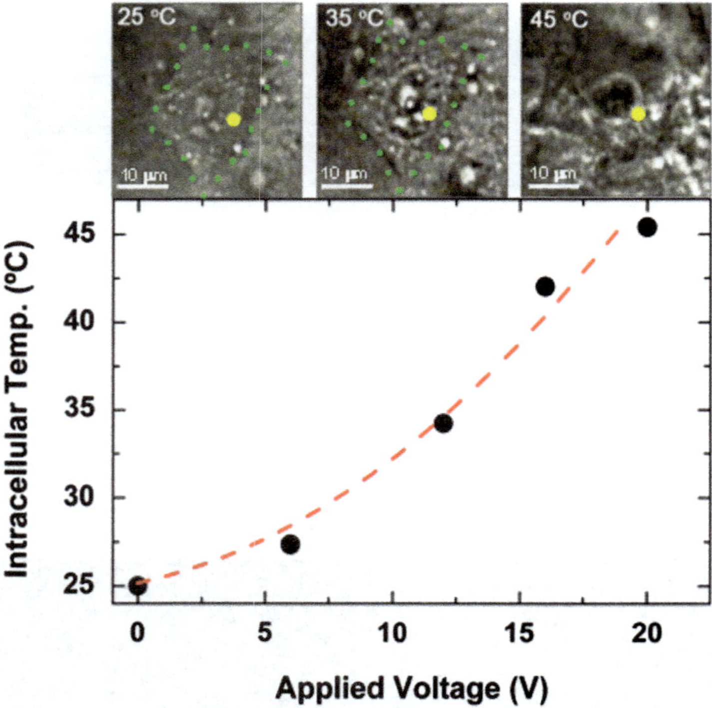

Figure 5.14 (Top) Optical transmission images of an individual HeLa cell at temperatures of 25, 35 and 45 °C. From the images, cell death was noted at 45 °C. (Bottom) Temperature of the HeLa cell, determined by the LIR method in $NaYF_4$:Er^{3+}/Yb^{3+} co-doped NPs, as a function of the applied voltage.
Reproduced with permission from ref. 77. © American Chemical Society 2010.

tissue penetration depths on a par with NIR CdTe QDs. The results obtained from the Tm^{3+} system were based on a ratiometric calculation of the variation of several emission lines at 800 nm. As these lines are ascribed to the various Stark levels of the $^3H_4$ transition of Tm^{3+}, their thermal dependence can be calculated in a ratio in order to achieve a means of thermal calibration. This differs to the Er^{3+} ion where the ratio calculated is based on two different transitions specifically the $^2H_{11/2} \rightarrow {}^4I_{15/2}$ and $^4S_{3/2} \rightarrow {}^4I_{15/2}$.

Rocha *et al.*[92] opted for a neodymium-based core–shell architecture (LaF_3:Nd^{3+}@LaF_3) and NIR excitation at a wavelength of 808 nm. Characteristic Nd^{3+} emission bands centred at 885 and 1060 nm were observed with characteristic high emission quantum yields that provided a thermal resolution of 2 °C in the temperature range from 10 to 60 °C by monitoring the 863 and 885 nm emission intensities of the ions as a thermal probe. The authors carried out phantom tissue studies using the nanoparticles and determined that penetration lengths of 2 mm can be achieved using direct excitation into the 808 nm energy level of Nd^{3+}. A combination of these Nd^{3+}

nanothermometers with gold nanorods was used to carry out sub-tissue hyperthermia where 808 nm excitation concomitantly initiates two processes: gold nanorod plasmonic excitation and subsequent heat generation, and a change in the luminescence profile of the Nd^{3+}-doped nanoparticles for thermometry (Figure 5.15).

Much of the work reported recently has focused solely on the development of nanothermometers relying on external heating agents such as gold nanoparticles, for example, as was previously seen.[92] More recently however, Rocha *et al.* studied the use of lanthanide-doped nanoparticles for both simultaneous heating and temperature sensing.[70] Relying on the efficient light-to-heat conversion of Nd^{3+}, NIR excitation can be used to excite the ion and convert the energy into heat through non-radiative depopulation processes. Non-radiative de-excitation of Nd^{3+} was shown to partially convert some of the 808 nm excitation light, resulting in the elevation of the temperature of a colloidal dispersion. The highest concentration of Nd^{3+} studied was 25 mol% and was found to be the most efficient, generating a temperature increase of more than 30 °C and reaching temperatures of nearly 58 °C, rendering it ideal for biological systems. Further demonstration of this system was shown in chicken breast tissue where heating on-target and off-target generated vastly different thermal images and resulted in temperature differentials of >15 °C at the surface of the tissue under

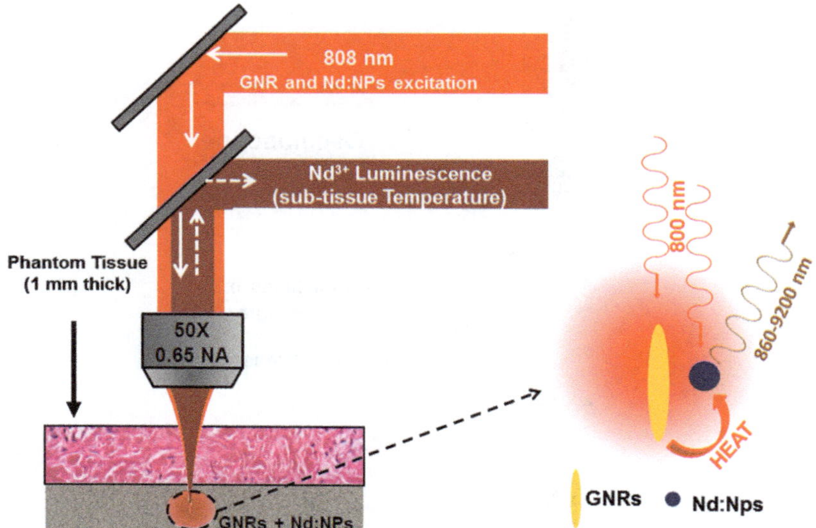

Figure 5.15 A cartoon of the experimental setup used for single-beam sub-tissue-controlled hyperthermia. The 808 nm laser beam is focused onto a plasmonic GNR dispersion, which also contains LaF_3:Nd^{3+} nanoparticles. The solution was placed under a 1 mm thick phantom tissue. Emission from the LaF_3:Nd^{3+} nanoparticles was collected by using the same objective, and the sub-tissue temperature was extracted from its spectral analysis.
Reproduced with permission from ref. 92. © AIP Publishing 2009.

analysis. In addition, the nanoparticles could be used to obtain temperature correlations concomitantly for both the injection site (at depths of 2 mm) and at the surface. The heating efficiency of the Nd^{3+}-doped nanoparticles was demonstrated by the fact that an elevation of the temperature by 10–15 °C could be achieved using very modest power densities of 3.7 W cm^2. The major advantage of this system is the ability to use a single excitation wavelength for both luminescence and luminescence-to-heat conversion, which simplifies the experimental approach and overrides the need for multiple excitation sources. Furthermore, it is advantageous in that it does not need the introduction of a heat source (gold nanoparticles for example) resulting in the development of more contained system without the need for additional chemical conjugation and purification procedures.

Only the most recent published reports on luminescence lifetime nanothermometry, by Savchuk *et al.*, present some applications.[98,99] $NaY_2F_5O:Er^{3+}/Yb^{3+}$ nanoparticles were tested in an *ex vivo* experiment of sub-tissue lifetime thermal sensing, being dispersed in water and injected into a fresh chicken breast at a depth of 1 mm (Figure 5.16). The lifetime-based nanothermometers monitored the temperature increment inside the chicken breast induced by a slightly focused pumping from a 1090 nm

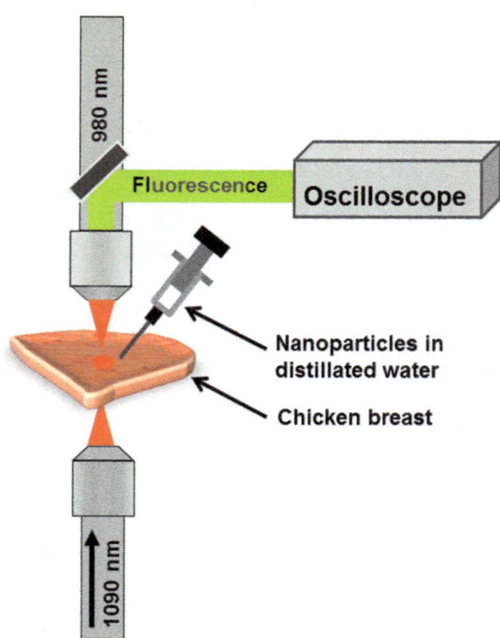

Figure 5.16 A representation of the experimental setup used to achieve *ex vivo* sub-tissue temperature determination in chicken breast using a heating laser (1090 nm) and a pumping laser for the luminescent nanoparticles (980 nm).
Reproduced with permission from ref. 98. © AIP Publishing 1997.

continuous laser. The lifetime values ranged from 160 µs when the heating laser was off, to 120 µs for a heating laser power close to 1.5 W. The data from the calibration curves—previously described in Section 5.3.3—together with the experimental data, allowed the authors of the work to establish the actual sub-tissue temperature, leading them to measure a temperature increment of 10 °C with respect to the non-heated tissue. This value shows a quite reasonable agreement with the magnitude of the 1090 nm laser-induced heating observed in other reported *in vivo* experiments using NIR thermometry.[106]

5.5 Conclusions and Perspectives

The past decade has witnessed tremendous advances in the synthesis and characterization of luminescent nanomaterials such as lanthanide-doped UCNPs. Scientists have actively targeted these lanthanide nanomaterials for biological and biomedical applications as they possess several interesting features, which have propelled them to the forefront of research. Their ability to undergo upconversion following excitation with NIR radiation is of the utmost importance in applications related to biology. Moreover, the progress achieved thus far regarding the synthesis and surface modification of these nanoparticles has been translated into standardized protocols allowing accessibility to the general scientific community. Multiple synthetic routes and well-understood surface chemistry not only allow for water dispersibility, but also for the development of a multimodal platform where the higher energy emissions can be used to trigger other light-activated modalities. Lanthanide-doped materials offer a multitude of other benefits that culminate in significant advantages ranging from their high resistance to environmental changes (no photobleaching or photoblinking) to luminescence that is unperturbed by the surroundings (due to the crystal field) and is minimally affected by size changes.

Applications of these lanthanide-doped nanoparticles have extended beyond active targeting and imaging to therapeutics such as drug delivery and photodynamic therapy. As their luminescent properties are becoming better understood, it is clear that additional modalities can be used to harvest additional information. This has been true in many cases but most notably for nanothermometry. The narrow emission bands of the Ln^{3+} ions are not only useful for multiplexing applications, but also important in nanothermometry as they can simplify the detection system and reduce analysis times. Building on their luminescent upconverting capacities and exploiting the fact that some lanthanide-doped UCNPs (as in the case of Tm^{3+}/Yb^{3+} co-doping) can undergo NIR-to-NIR upconversion, the resulting UV-visible luminescence has multiple functionalities, while the NIR emission can be used for nanothermometry affording precise temperature measurements. Furthermore, the ability to invoke a self-referencing LIR method results in the achievement of high thermal sensitivity and minimized errors in measurements associated with signal intensities and other environmental factors. This is the same for the lifetime nanothermometry method, which in itself is self-referenced.

The future of these nanoparticles certainly appears to be promising. It is in this regard that they continue to garner significant interest, and improvements to materials and their applications are currently being realized. This is the case for the efforts focused on enhancing the upconversion efficiency to allow for more intense emissions. With that said, the applicability of these nanoparticles, *in vivo*, has been successfully demonstrated on numerous occasions. Enhanced upconversion efficiency could be achieved *via* the investigation of novel hosts, or architectures (such as core–shell systems), as well as by using plasmonic enhancement of the luminescence, coupling UCNPs with plasmonic materials such as gold or silver. Also, new hybrid systems must be investigated where the localized measurement of temperature can be coupled with a nanoheater (*i.e.*, with gold nanorods). This would allow for the development of a heater/sensor system.

With respect to temperature measurements, while nanothermometry in the visible range (either directly excited or *via* upconversion) has been well documented, investigation of thermometry in the NIR range has just begun. Indeed, more work is required to define new ions or ion combinations capable of NIR emission that can provide higher sensitivities. Only then, when both the excitation and emission wavelengths lie in one of three biological windows, can these materials truly be considered for real-world applications.

The concept of nanothermometry is also currently being pushed beyond the unimodal functionality of temperature measurements using luminescent ions. Instead, new bodies of work are focusing on the development of lanthanide-based heater/thermometer nanoplatforms for simultaneous heat generation and temperature sensing. For example, the combination of plasmonic nanoparticles with UCNP counterparts fulfils this goal by exploiting the heating capacity, following irradiation of noble metal nanostructures, and the temperature sensitivity of the lanthanide ions in an inorganic host matrix. This allows for the imaging of the temperature profile of the nanoheater and its surrounding environment, and limits the irradiation power required to the minimum, thereby preventing excessive heating and avoiding damage to healthy tissues in biological systems.[107] This concept was also extended to the use of fluorescent molecules and even QDs[108–110] to measure temperatures in the range of 300–700 K with varying degrees of success and temperature accuracy, which can reach uncertainties of $\pm$ 20 K.[110]

The use of lanthanide-doped nanostructures in these platforms has been reported in the works of Carlson *et al.*[42,111,112] who used Er^{3+}-doped $Al_{0.94}Ga_{0.06}N$ in combination with gold nanoparticles, or gold nanodots for concomitant heating and temperature sensing. Early reports demonstrated the challenges that lie ahead due to insufficient optical resolution. In order to address this, more recent works have focused on the preparation of nanocomposites of $(Gd, Yb, Er)O_2$ for example with a decorated surface of gold nanoparticles. This approach reduces the thermometer to a similar scale in comparison to the heater, broadens the temperature-sensing range and forms a single platform for heating/sensing.[107] Alternatively, the combination of magnetic nanoparticles with UCNPs can also lead to interesting heater/sensor platforms. For example the work reported by Dong and Zink[80] evaluated the integration of iron oxide nanoparticles with $NaYF_4:Er^{3+}/Yb^{3+}$ nanoparticles in a silica nanoparticle effectively trapping both species and using magnetic field induction to induce a temperature rise and the optical properties for temperature determination over the range of 298–373 K.

Some of the other approaches involve the development of self-monitored photothermal agents without the requirements for a metal nanoheater, as was shown by Rocha *et al.*[70] At sufficiently highly Nd^{3+} concentrations, IR excitation using 808 nm will result in a heating effect, which can be monitored *via* the temperature-sensitive emission of the ion over the biologically relevant range of 298–333 K. This work was further extended by the same group in order to achieve an intratumoural thermal reading during the process of photothermal therapy and demonstrated clear differentiation between the tumour temperature, the surface temperature and the difference relative to control regions of the biological model.[113] In fact, the authors demonstrated a reduction in tumour volume over time following photothermal therapy using Nd^{3+}-doped LaF_3 combination nanoheaters/thermometers. In essence, this new direction aims to develop novel diagnostic and therapeutic techniques that can be used to harvest a multitude of information without the need for additional probes and vehicles and using minimally invasive approaches.

With respect to a longer term vision, for upconversion or NIR nanothermometry to be clinically viable, the complete toxicity profile must be established. While the chemical toxicity of the lanthanide ions is generally low, it is of the utmost importance to investigate the cytotoxicity of the nanoparticle constructs and the associated impact of surface modifications. Long-term studies will be required in order to clearly shed light on this matter while establishing a true understanding of the clearance mechanisms in the body.

Lanthanide-doped nanoparticles offer truly exciting and breath taking opportunities. These nanoparticles have graduated from a mere academic curiosity to the point where theranostic and nanomedicine applications are within the realm of possibility in the years to come.

References

1. B. Busche, R. Wiacek, J. Davidson, V. Koonsiripaiboon, W. Yantasee, R. S. Addleman and G. E. Fryxell, *Inorg. Chem. Commun.*, 2009, **12**, 312–315.
2. G. E. Fryxell, R. S. Addleman, S. V. Mattigod, Y. Lin, T. S. Zemanian, H. Wu, J. C. Birnbaum, J. Liu and X. Feng, in *Encyclopedia of Nanoscience and Nanotechnology*, ed. C. C. J. A. Schwarz and K. Putyera, Marcel Dekker, New York, NY, 2004, pp. 1135–1145.

3. S. Kathirvelu, L. D'Souza and B. Dhurai, *Indian J. Fibre Text. Res.*, 2009, **34**, 267–273.

4. U. Sutter, J. Loeffler, M. Bidmon, H. Valadon and R. Ebner, in *Nanoroad SME*, Steinbeis-Europa-Zentrum, Karlsruhe, 2006, pp. 1–105.

5. P. G. Tratnyek and R. L. Johnson, *Nano Today*, 2006, **1**, 44–48.

6. N. Vigneshwaran, P. V. Varadarajan and R. H. Balasubramanya, in *Nanotechnologies for the Life Sciences*, Wiley-VCH Verlag GmbH & Co. KGaA, 2007.

7. W. Yantasee, G. E. Fryxell, G. A. Porter, K. Pattamakomsan, V. Sukwarotwat, W. Chouyyok, V. Koonsiripaiboon, J. Xu and K. N. Raymond, *Nanomed. Nanotechnol. Biol. Med.*, 2010, **6**, 1–8.

8. D. K. Chatterjee, A. J. Rufaihah and Y. Zhang, *Biomaterials*, 2008, **29**, 937–943.

9. X. Gao, Y. Cui, R. M. Levenson, L. W. K. Chung and S. Nie, *Nat. Biotechnol.*, 2004, **22**, 969–976.

10. S. F. Lim, R. Riehn, W. S. Ryu, N. Khanarian, C.-k. Tung, D. Tank and R. H. Austin, *Nano Lett.*, 2005, **6**, 169–174.

11. X. Michalet, F. F. Pinaud, L. A. Bentolila, J. M. Tsay, S. Doose, J. J. Li, G. Sundaresan, A. M. Wu, S. S. Gambhir and S. Weiss, *Science*, 2005, **307**, 538–544.

12. M. Stroh, J. P. Zimmer, D. G. Duda, T. S. Levchenko, K. S. Cohen, E. B. Brown, D. T. Scadden, V. P. Torchilin, M. G. Bawendi, D. Fukumura and R. K. Jain, *Nat. Med.*, 2005, **11**, 678–682.

13. L. Wang and Y. Li, *Chem. Commun.*, 2006, 2557–2559.

14. K. C. Weng, C. O. Noble, B. Papahadjopoulos-Sternberg, F. F. Chen, D. C. Drummond, D. B. Kirpotin, D. Wang, Y. K. Hom, B. Hann and J. W. Park, *Nano Lett.*, 2008, **8**, 2851–2857.

15. M. Yu, F. Li, Z. Chen, H. Hu, C. Zhan, H. Yang and C. Huang, *Anal. Chem.*, 2009, **81**, 930–935.

16. F. Vetrone, R. Naccache, C. G. Morgan and J. A. Capobianco, *Nanoscale*, 2010, **2**, 1185–1189.

17. F. Auzel, *C. R. Acad. Sci. Paris B*, 1966, **262**, 1016–1019.

18. F. Auzel, *C. R. Acad. Sci. Paris B*, 1966, **263**, 819–821.

19. F. Auzel, *Chem. Rev.*, 2003, **104**, 139–174.

20. F. Auzel, in *Spectroscopic Properties of Rare Earths in Optical Materials*, ed. G. Liu and B. Jacquier, Springer/Tsinghua University Press, Beijing & Heidelberg, 2005.

21. R. Scheps, *Prog. Quantum Electron.*, 1996, **20**, 271–358.

22. K. König, *J. Microsc.*, 2000, **200**, 83–104.

23. G. Chen, J. Shen, T. Y. Ohulchanskyy, N. J. Patel, A. Kutikov, Z. Li, J. Song, R. K. Pandey, H. Ågren, P. N. Prasad and G. Han, *ACS Nano*, 2012, **6**, 8280–8287.

24. M. L. Cunningham, J. S. Johnson, S. M. Giovanazzi and M. J. Peak, *Photochem. Photobiol.*, 1985, **42**, 125–128.

25. R. M. Tyrrell and S. M. Keyse, *J. Photochem. Photobiol., B*, 1990, **4**, 349–361.

26. M. Bottrill and M. Green, *Chem. Commun.*, 2011, **47**, 7039–7050.

27. A. M. Derfus, W. C. W. Chan and S. N. Bhatia, *Nano Lett.*, 2003, **4**, 11–18.

28. D. K. Chatterjee and Z. Yong, *Nanomedicine*, 2008, **3**, 73–82.

29. G. Chen, T. Y. Ohulchanskyy, W. C. Law, H. Agren and P. N. Prasad, *Nanoscale*, 2011, **3**, 2003–2008.

30. E. N. M. Cheung, R. D. A. Alvares, W. Oakden, R. Chaudhary, M. L. Hill, J. Pichaandi, G. C. H. Mo, C. Yip, P. M. Macdonald, G. J. Stanisz, F. C. J. M. van Veggel and R. S. Prosser, *Chem. Mater.*, 2010, **22**, 4728–4739.

31. Y. Dai, X. Kang, D. Yang, X. Li, X. Zhang, C. Li, Z. Hou, Z. Cheng, P. a. Ma and J. Lin, *Adv. Healthcare Mater.*, 2013, **2**, 562–567.

32. G. K. Das, B. C. Heng, S.-C. Ng, T. White, J. S. C. Loo, L. D'Silva, P. Padmanabhan, K. K. Bhakoo, S. T. Selvan and T. T. Y. Tan, *Langmuir*, 2010, **26**, 8959–8965.

33. W. Feng, K. C. Dev, L. Zhengquan, Z. Yong, F. Xianping and W. Minquan, *Nanotechnology*, 2006, **17**, 5786–5791.

34. J. Y. Park, M. J. Baek, E. S. Choi, S. Woo, J. H. Kim, T. J. Kim, J. C. Jung, K. S. Chae, Y. Chang and G. H. Lee, *ACS Nano*, 2009, **3**, 3663–3669.

35. J. Ryu, H.-Y. Park, K. Kim, H. Kim, J. H. Yoo, M. Kang, K. Im, R. Grailhe and R. Song, *J. Phys. Chem. C*, 2010, **114**, 21077–21082.

36. C. Wang, L. Cheng and Z. Liu, *Biomaterials*, 2011, **32**, 1110–1120.

37. F. Wang, D. Banerjee, Y. Liu, X. Chen and X. Liu, *Analyst*, 2010, **135**, 1839–1854.

38. D. J. Bizzak and M. K. Chyu, *Rev. Sci. Instrum.*, 1994, **65**, 102–107.

39. D. J. Bizzak and M. K. Chyu, *Int. J. Heat Mass Transfer*, 1995, **38**, 267–274.

40. Z. P. Cai and H. Y. Xu, *Sens. Actuators, A*, 2003, **108**, 187–192.

41. B. S. Cao, Y. Y. He, Z. Q. Feng, Y. S. Li and B. Dong, *Sens. Actuators, B*, 2011, **159**, 8–11.

42. M. T. Carlson, A. Khan and H. H. Richardson, *Nano Lett.*, 2011, **11**, 1061–1069.

43. P. Haro-González, I. R. Martín, L. L. Martín, S. F. León-Luis, C. Pérez-Rodríguez and V. Lavín, *Opt. Mater.*, 2011, **33**, 742–745.

44. A. L. Heyes, S. Seefeldt and J. P. Feist, *Opt. Laser Technol.*, 2006, **38**, 257–265.

45. D. Li, Y. Wang, X. Zhang, K. Yang, L. Liu and Y. Song, *Opt. Commun.*, 2012, **285**, 1925–1928.

46. D. Jaque and F. Vetrone, *Nanoscale*, 2012, **4**, 4301–4326.

47. N. G. Connelly, T. Damhus, R. M. Hartshorn and A. T. Hutton, *Nomenclature of Inorganic Chemistry: IUPAC Recommendations 2005*, Royal Society of Chemistry Publishing/International Union of Pure and Applied Chemistry (IUPAC), Norfolk, UK, 2005.

48. B. G. Wybourne, *Spectroscopic Properties of Rare Earths*, Interscience Publishers, New York, NY, 1965.

49. A. J. Freeman and R. E. Watson, *Phys. Rev.*, 1962, **127**, 2058–2075.

50. B. Henderson and G. F. Imbusch, *Optical Spectroscopy of Inorganic Solids*, Clarendon Press, Oxford, UK, 1989.

51. G. H. Dieke, *Spectra and Energy Levels of Rare Earth Ions in Crystals*, Interscience Publishers, New York, NY, 1968.
52. W. T. Carnall, G. L. Goodman, K. Rajnak and R. S. Rana, *J. Chem. Phys.*, 1989, **90**, 3443–3457.
53. O. Kazuyoshi, W. Shinta, S. Yuki, T. Hiroaki, I. Takugo, G. B. Mikhail and T. Isao, *Jpn. J. Appl. Phys.*, 2004, **43**, L611.
54. R. T. Wegh, A. Meijerink, R.-J. Lamminmäki and H. Jorma, *J. Lumin.*, 2000, **87–89**, 1002–1004.
55. S. W. Allison, G. T. Gillies, A. J. Rondinone and M. R. Cates, *Nanotechnology*, 2003, **14**, 859–863.
56. R. K. Watts and H. J. Richter, *Phys. Rev. B: Solid State*, 1972, **6**, 1584–1589.
57. T. Miyakawa and D. L. Dexter, *Phys. Rev. B: Solid State*, 1970, **1**, 2961–2969.
58. L. Martinez Maestro, E. Martìn Rodriguez, F. Vetrone, R. Naccache, H. Loro Ramirez, D. Jaque, J. A. Capobianco and J. G. Solé, *Opt. Express*, 2010, **18**, 23544–23553.
59. M. F. Joubert, *Opt. Mater.*, 1999, **11**, 181–203.
60. M. F. Joubert, S. Guy and B. Jacquier, *Phys. Rev. B: Condens. Matter Mater. Phys.*, 1993, **48**, 10031–10037.
61. W. P. Risk, T. R. Gosnell and A. V. Nurmikko, *Compact Blue-Green Lasers*, Cambridge University Press, Cambridge, UK, 2003.
62. S. Wang, S. Westcott and W. Chen, *J. Phys. Chem. B*, 2002, **106**, 11203–11209.
63. H. Peng, H. Song, B. Chen, J. Wang, S. Lu, X. Kong and J. Zhang, *J. Chem. Phys.*, 2003, **118**, 3277–3282.
64. S. Zheng, W. Chen, D. Tan, J. Zhou, Q. Guo, W. Jiang, C. Xu, X. Liu and J. Qiu, *Nanoscale*, 2014, **6**, 5675–5679.
65. O. Svelto, *Principles of Lasers*, Plenum Publishing Corporation, New York, NY, 3rd edn, 1989.
66. M. Quintanilla, E. Cantelar, F. Cussó, M. Villegas and A. C. Caballero, *Appl. Phys. Express*, 2011, **4**, 022601.
67. S. Hüfner, *Optical Spectra of Transparent Rare Earth Compounds*, Academic Press, New York, NY, 1978.
68. V. K. Rai, *Appl. Phys. B*, 2007, **88**, 297–303.
69. S. A. Wade, S. F. Collins and G. W. Baxter, *J. Appl. Phys.*, 2003, **94**, 4743–4756.
70. U. Rocha, K. Upendra Kumar, C. Jacinto, J. Ramiro, A. J. Caamaño, J. García Solé and D. Jaque, *Appl. Phys. Lett.*, 2014, **104**, 053703.
71. N.-N. Dong, M. Pedroni, F. Piccinelli, G. Conti, A. Sbarbati, J. E. Ramírez-Hernández, L. M. Maestro, M. C. Iglesias-de la Cruz, F. Sanz-Rodríguez, A. Juarranz, F. Chen, F. Vetrone, J. A. Capobianco, J. García Solé, M. Bettinelli, D. Jaque and A. Speghini, *ACS Nano*, 2011, **5**, 8665–8671.
72. K. Zheng, Z. Liu, C. Lv and W. Qin, *J. Mater. Chem. C*, 2013, **1**, 5502–5507.

73. X. Wang, X. Kong, Y. Yu, Y. Sun and H. Zhang, *J. Mater. Chem. C*, 2007, **111**, 15119–15124.

74. N. Nuñez, M. Quintanilla, E. Cantelar, F. Cussó and M. Ocaña, *J. Nanopart. Res.*, 2010, **12**, 2553–2565.

75. H. Sun, L. Zhang, L. Wen, M. Liao, J. Zhang, L. Hu, S. Dai and Z. Jiang, *Appl. Phys. B: Lasers Opt.*, 2005, **80**, 881–888.

76. A. Sedlmeier, D. E. Achatz, L. H. Fischer, H. H. Gorris and O. S. Wolfbeis, *Nanoscale*, 2012, **4**, 7090–7096.

77. F. Vetrone, R. Naccache, A. Zamarrón, A. Juarranz de la Fuente, F. Sanz-Rodríguez, L. Martinez Maestro, E. Martín Rodriguez, D. Jaque, J. García Solé and J. A. Capobianco, *ACS Nano*, 2010, **4**, 3254–3258.

78. D. Li, Q. Shao, Y. Dong and J. Jiang, *Mater. Lett.*, 2013, **110**, 233–236.

79. S. Zhou, K. Deng, X. Wei, G. Jiang, C. Duan, Y. Chen and M. Yin, *Opt. Commun.*, 2013, **291**, 138–142.

80. J. Dong and J. I. Zink, *ACS Nano*, 2014, **8**, 5199–5207.

81. J.-C. Boyer and F. C. J. M. van Veggel, *Nanoscale*, 2010, **2**, 1417–1419.

82. G. Chen, T. Y. Ohulchanskyy, S. Liu, W.-C. Law, F. Wu, M. T. Swihart, H. Ågren and P. N. Prasad, *ACS Nano*, 2012, **6**, 2969–2977.

83. V. Mahalingam, F. Vetrone, R. Naccache, A. Speghini and J. A. Capobianco, *Adv. Mater.*, 2009, **21**, 4025–4028.

84. M. A. R. C. Alencar, G. S. Maciel, C. B. de Araújo and A. Patra, *Appl. Phys. Lett.*, 2004, **84**, 4753–4755.

85. L. Aigouy, E. Saïdi, L. c. Lalouat, J. Labéguerie-Egéa, M. Mortier, P. Löw and C. Bergaud, *J. Appl. Phys.*, 2009, **106**, 074301.

86. S. K. Singh, K. Kumar and S. B. Rai, *Sens. Actuators, A*, 2009, **149**, 16–20.

87. H. Zheng, B. Chen, H. Yu, J. Zhang, J. Sun, X. Li, M. Sun, B. Tian, S. Fu, H. Zhong, B. Dong, R. Hua and H. Xia, *J. Colloid Interface Sci.*, 2014, **420**, 27–34.

88. T. V. Gavrilovic, D. J. Jovanovic, V. Lojpur and M. D. Dramicanin, *Sci. Rep.*, 2014, **4**, 4209.

89. B. Dong, B. Cao, Y. He, Z. Liu, Z. Li and Z. Feng, *Adv. Mater.*, 2012, **24**, 1987–1993.

90. D. Wawrzynczyk, A. Bednarkiewicz, M. Nyk, W. Strek and M. Samoc, *Nanoscale*, 2012, **4**, 6959–6961.

91. O. A. Savchuk, J. J. Carvajal, E. W. Barrera, M. C. Pujol, X. Mateos, R. Solé, J. Massons, M. Aguiló and F. Díaz, *Proc. SPIE*, 2013, **8594**, 859406.

92. U. Rocha, C. Jacinto da Silva, W. Ferreira Silva, I. Guedes, A. Benayas, L. M. Maestro, M. Acosta Elias, E. Bovero, F. C. J. M. van Veggel, J. A. García Solé and D. Jaque, *ACS Nano*, 2013, 7, 1188–1199.

93. S. W. Allison and G. T. Gillies, *Rev. Sci. Instrum.*, 1997, **68**, 2615–2650.

94. M. Quintanilla, N. O. Núñez, E. Cantelar, M. Ocaña and F. Cussó, *J. Appl. Phys.*, 2013, **113**, 174308.

95. T. Igarashi, M. Ihara, T. Kusunoki, K. Ohno, T. Isobe and M. Senna, *Appl. Phys. Lett.*, 2000, **76**, 1549–1551.

96. X. Guo, H. Zhao, H. Song, E. Zhang, C. Combs, N. Clemens, X. Chen, K. K. Li, Y. K. Zou and H. Jiang, *Sens. Transducers*, 2011, **13**, 124–130.

97. M. G. Nikolić, Ž. Antić, S. Ćulubrk, J. M. Nedeljković and M. D. Dramićanin, *Sens. Actuators, B*, 2014, **201**, 46–50.

98. O. A. Savchuk, J. J. Carvajal, M. C. Pujol, J. Massons, P. Haro-Gonzalez, D. Jaque, M. Aguiló and F. Díaz, *Biophotonics: Photonic Solutions for Better Health Care IV*, 2014.

99. O. A. Savchuk, P. Haro-Gonzalez, J. J. Carvajal, D. Jaque, J. Massons, M. Aguilo and F. Diaz, *Nanoscale*, 2014.

100. C. Brecher, L. A. Riseberg and M. J. Weber, *Phys. Rev. B: Condens. Matter Mater. Phys.*, 1978, **18**, 5799–5811.

101. S. K. Ghoshal, M. R. Sahar, M. S. Rohani and S. Sharma, *Indian J. Pure Appl. Phys.*, 2011, **49**, 509–515.

102. H. Kusama, O. J. Sovers and T. Yoshioka, *Jpn. J. Appl. Phys.*, 1976, **15**, 2349–2358.

103. L. Martinez Maestro, E. Martín Rodríguez, F. Sanz Rodríguez, M. C. Iglesias-de la Cruz, A. Juarranz, R. Naccache, F. Vetrone, D. Jaque, J. A. Capobianco and J. G. Solé, *Nano Lett.*, 2010, **10**, 5109–5115.

104. J. García Solé, L. E. Bausá and D. Jaque, *An Introduction to the Optical Spectroscopy of Inorganic Solids*, John Wiley & Sons, Ltd, West Sussex, England, 2005.

105. X. Wang, J. Zheng, Y. Xuan and X. Yan, *Opt. Express*, 2013, **21**, 21596–21606.

106. L. Martinez Maestro, P. Haro-González, B. del Rosal, J. Ramiro, A. J. Caamaño, E. Carrasco, A. Juarranz, F. Sanz-Rodriguez, J. García Solé and D. Jaque, *Nanoscale*, 2013, **5**, 7882–7889.

107. M. L. Debasu, D. Ananias, I. Pastoriza-Santos, L. M. Liz-Marzán, J. Rocha and L. D. Carlos, *Adv. Mater.*, 2013, **25**, 4868–4874.

108. P. M. Bendix, S. N. S. Reihani and L. B. Oddershede, *ACS Nano*, 2010, **4**, 2256–2262.

109. J. Lee, A. O. Govorov and N. A. Kotov, *Angew. Chem., Int. Ed.*, 2005, **44**, 7439–7442.

110. K. Setoura, D. Werner and S. Hashimoto, *J. Phys. Chem. C*, 2012, **116**, 15458–15466.

111. M. T. Carlson, A. J. Green, A. Khan and H. H. Richardson, *J. Phys. Chem. C*, 2012, **116**, 8798–8803.

112. M. T. Carlson, A. J. Green and H. H. Richardson, *Nano Lett.*, 2012, **12**, 1534–1537.

113. E. Carrasco, B. del Rosal, F. Sanz-Rodríguez, Á. J. de la Fuente, P. H. Gonzalez, U. Rocha, K. U. Kumar, C. Jacinto, J. G. Solé and D. Jaque, *Adv. Funct. Mater.*, 2014, **25**, 615–626.

Organic Dye Thermometry

GUOQIANG YANG,*[a] XUAN LIU,[a] JIAO FENG,[a] SHAYU LI[a]
AND YI LI[b]

[a] Beijing National Laboratory for Molecular Sciences, Key laboratory of
Photochemistry, Institute of Chemistry, Chinese Academy of Sciences,
Beijing 100190, China; [b] Key Laboratory of Photochemical Conversion and
Optoelectronic Materials, Technical Institute of Physics and Chemistry,
Chinese Academy of Sciences, Beijing 100190, China
*Email: gqyang@iccas.ac.cn

6.1 Introduction

The intrinsic limitations of mechanical and electrical thermometers
have encouraged the development of non-contact optical thermometers on the
micro- and nanoscales. Among the available methods, luminescence-based
temperature sensors have received much attention in recent years. Several
kinds of luminescent materials based on quantum dots, lanthanide phosphors,[1] organic dyes, polymer[2] or metal–ligand complexes[3] have been reported
for temperature detection. Because of their fast response, high spatial and
temporal resolution and biocompatibility, organic dye based luminescence
thermometers are among the earliest and most used systems.[4–6]

According to previous research,[7–9] the luminescence of most organic
compounds changes with temperature (T) *via* various mechanisms. There
are usually more electrons of the molecules that are excited to non-emissive
states at higher T, while the luminescence intensity decreased. However,
only a few organic dyes can be used in thermometry for the following reasons: (1) their luminescence spectra should undergo a large change to provide good sensitivity; (2) the photostability of the organic dyes should be

RSC Nanoscience & Nanotechnology No. 38
Thermometry at the Nanoscale: Techniques and Selected Applications
Edited by Luís Dias Carlos and Fernando Palacio
© The Royal Society of Chemistry 2016
Published by the Royal Society of Chemistry, www.rsc.org

good enough to ensure precision; (3) the probes should not be toxic, especially if an *in vivo* use is intended; and (4) luminescence quantum yields should be high enough at different temperatures.[10–12]

Commonly used organic dyes eligible for luminescence thermometry are shown in Chart 6.1. Rhodamine family and Rhodamine family/reference dyes are the most frequently used organic dyes in water. Fluorescein is used as a thermometer and pH probe simultaneously. Pyranine, 7-nitrobenz-2-oxa-1,3-diazol-4-yl (NBD) and 6-dodecanoyl-2-dimethylamino-naphthalene (Laurdan) have also been reported as molecular probes for temperature in aqueous environments. Triarylboron compounds and bispyren dyes can also be applied to measure temperature in organic solvents. Furthermore, Rhodamine family, triarylboron compounds, acridine yellow, C_{70}, 2,5-dihexyloxy-4-bromobenzaldehyde (Br6A), perylene/*N*-allyl-*N*-methylaniline and bis(benzoxazolyl)stilbene (BBS) incorporated in various polymers can be applied as solid thermometers. In this chapter we discuss the temperature range, sensitivity, spatial/temporal resolution, accuracy and the concept of temperature detection of these dyes based on the systems they are used in.

6.2 Temperature-sensitive Organic Dyes

6.2.1 Thermal Quenching Dyes

6.2.1.1 *Rhodamine Family*

The fluorescence of Rhodamine family dyes has long been known to be temperature sensitive. Also, their quantum yields are usually pretty high. For example, the fluorescence quantum yield of Rhodamine B (RhB) is 0.31 in water, and decreases significantly with an increase in temperature. The temperature dependency of RhB fluorescence intensity is about 2.3% K^{-1} in an aqueous environment. These features make them very popular as fluorescent probes for temperature determination.

Much work has focused on the application of RhB in microfluidic systems. Ross *et al.* reported the fluorescence intensity based measurement of water temperatures with a standard fluorescence microscope and a CCD camera. With the equipment employed, fluid temperatures ranging from room temperature to 90 °C were measured with a precision ranging from 0.03 to 3.5 °C. The accuracy of the temperature measurement was about 0.8 °C. The spatial and temporal resolutions achieved were 1 μm and 33 ms, respectively.[13]

Such systems can be used to monitor temperature instantaneously. Gaitan *et al.* reported a microchannel fabricated in poly(dimethylsiloxane) (PDMS) combined with a microwave transmission line. A selected region of the microchannel system could be heated by the equipment. The temperature was obtained by measuring the fluorescence intensity of RhB in a sodium chloride aqueous solution. The precision was 0.5 °C. The spatial resolutions achieved were better than 7 μm (the depth of the microchannel).[14]

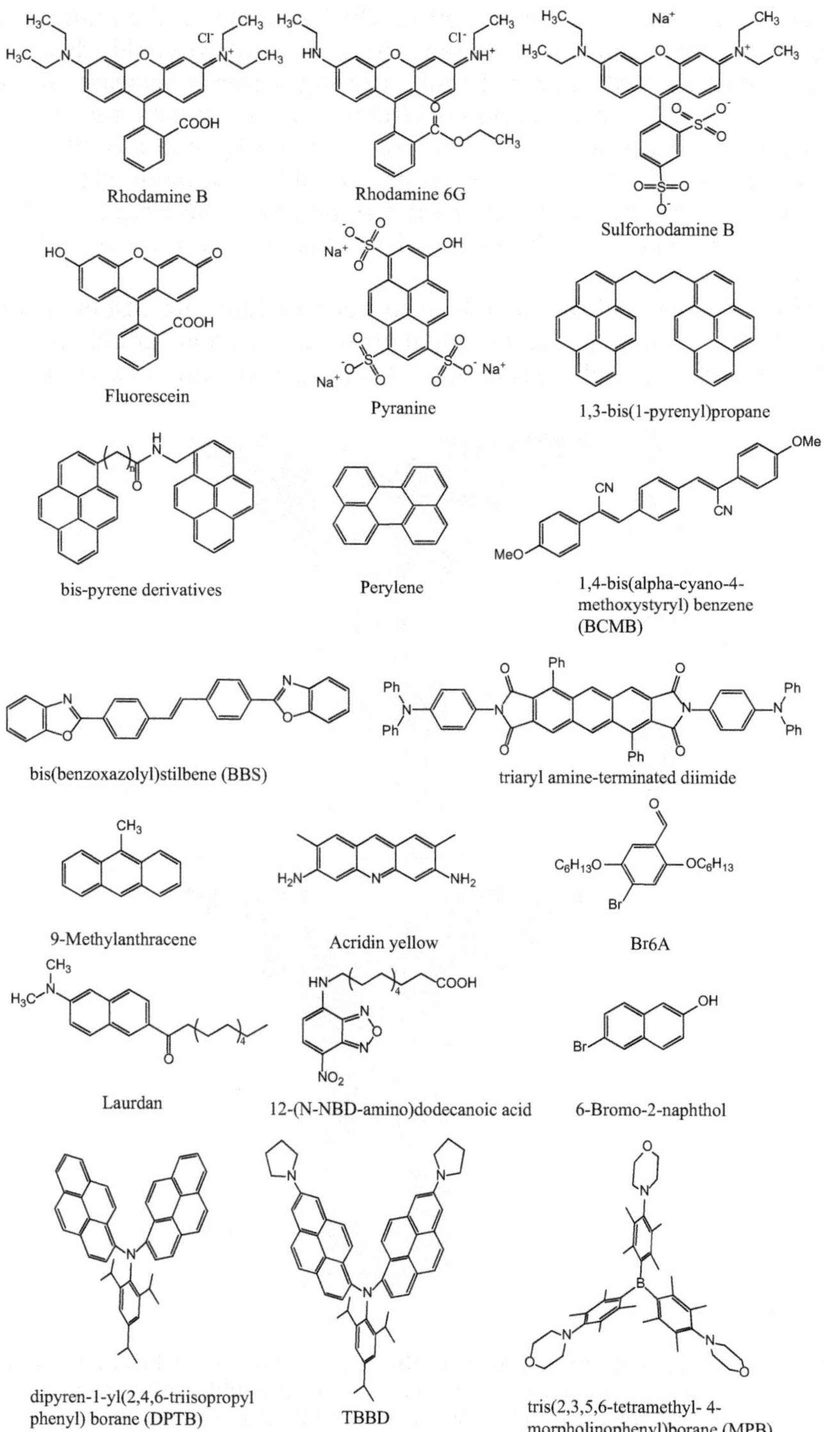

Chart 6.1 Structure of thermosensitive organic dyes.

However, there are a few interferences which could affect the fluorescence intensity of RhB in aqueous environments: (1) the RhB dye could adsorb and diffuse into the PDMS material walls, causing a steady increase in local concentration with time; (2) photobleaching is a common problem with luminescence probes; and (3) the fluorescence intensity of RhB is influenced by fluctuations in the excitation light. To obtain a more precise and accurate readout, improvements to the system were investigated.[15] Such improvements could also be extended to other luminescent organic dye thermometers.

Ren *et al.* discussed the notable absorption of RhB into PDMS channel walls. They found that the overall temperature readout could decrease by 10 °C for the second measurement 5 min later (Figure 6.1). By using a

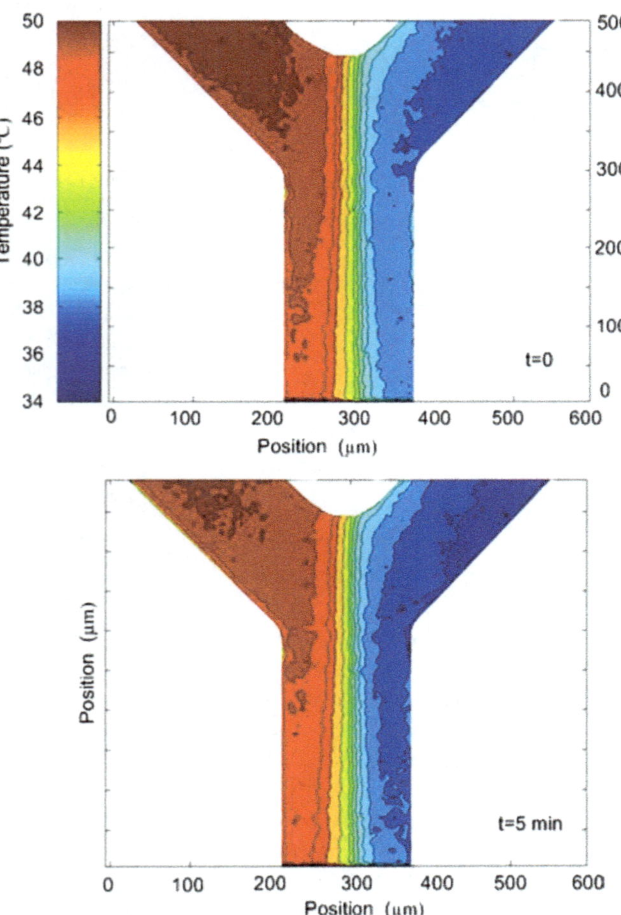

Figure 6.1 Temperature contour plots of the intersection obtained from two images taken 5 min apart while applying the photobleaching technique to prepare the area for fluorescence-based thermometry.
Adapted from ref. 16 with permission from the Royal Society of Chemistry.

high-intensity light source before taking images for analysis, photobleaching effects could remove the fluorescence of the absorbed RhB dye. The temperatures measured (in water) were 95, 70 and 55 °C. The sensitivity, spatial/temporal resolutions and accuracy achieved were expected to be similar to experiments mentioned earlier.[16]

Meanwhile, the introduction of a reference dye could improve the precision of the detection system, because the temperature intensity ratio would not be affected by excitation light fluctuations. A technique described by Sakakibara and Adrian could measure the instantaneous 3D temperature distribution in water using RhB and the nearly temperature-independent Rhodamine 110. The ratio of fluorescence intensities of these two dyes was calibrated against the temperature. The precision of the measurement was 1.4 °C. It was observed that the maximum sensitivity was approximately 1.6–1.7% K^{-1} in the temperature range of 15 to 40 °C, and the accuracy was 1.3 °C.[17]

Besides the use of a reference dye, a technique for measuring temperature by means of two-colour planar laser-induced fluorescence (PLIF) of a single dye can also provide absolute temperature. Guillard *et al.* reported such a technique applied to RhB in water. A heated turbulent jet injected into a co-flow at ambient temperature was presented as a demonstration. The temperature was determined with a precision of 0.25 °C. The temperature range displayed was 20–80 °C with an achieved sensitivity of 0.8% K^{-1}. The spatial resolution was 0.27×0.27 mm, which could be improved by using a better CCD.[18]

Both the two-colour laser-induced fluorescence (LIF) technique and an additional reference dye can be used in the same system. RhB and temperature-insensitive Sulforhodamine 101 (Sr101) were used to determine the temperature of ethanol and water by the PLIF method. The temperature sensitivities of the RhB–Sr101 dye combination in ethanol and water using calibration of the two-colour LIF system were 1.5% K^{-1} and 2.7% K^{-1}, and the precisions were 0.48–0.59 °C and 0.41–0.49 °C, respectively. The temperature accuracy in the microchannels was within 0.4 and 0.3 °C in the two solvents. The spatial resolution was 22.2×22.2 μm.[19] Besides Sr101, fluorescein-27 was also reported to be used in such a system with RhB.[20]

Absolute temperature can be provided by the fluorescence lifetime measurement in some systems. Granick *et al.* reported using a Rhodamine 6G-labelled DNA oligomer as a temperature indicator in de-ionized water. The fluorescence of the dye can be quenched by the conformational change of the designed sequence of DNA bases, which was faster at higher temperatures, resulting in shorter fluorescence lifetimes. The temperature range of the indicator was 15–35 °C with a sensitivity of about 2% K^{-1}. The temporal resolution was on the time scale of nanoseconds, while the spatial resolution was 40 nm based on the control experiment.[21]

Sulforhodamine B (SrB) is another fluorescent dye of the Rhodamine family that shows good temperature sensitivity. Methods using one or two emission bands were examined for SrB (acid form) in a PCR solution. The temperature precision was 0.3 °C, for two-colour procedures, and 0.6 °C, using a single colour. The temperature range measured was 45–95 °C with a sensitivity of 5.7% K^{-1}. The average accuracy was 0.5 °C. The sample volume was 25 μL.[22]

RhB dye could be applied to *in vivo* imaging because of its water solubility.[23] Wood *et al.* reported that RhB imaged by laser scanning confocal microscopy (LSCM) was used to measure the temperature of rat tendon samples. 2D and 3D images of temperature change and distributions in *in vivo* samples were obtained by the non-contact method. The temperature range measured was 25–40 °C and the sensitivity of the RhB dye for *in vivo* imaging was approximately 3% K^{-1} with a precision of 1.5 °C. More calibration was needed to provide the accuracy of the system. The spatial resolution was a few micrometres. Further experiments were described using brain tissue obtained from four-week-old rats. The maximum spatial resolution was 0.6 μm due to the camera.[24]

In addition to changing in liquids, the quantum yield of RhB also changes significantly in the solid state with temperature. Rhee *et al.* reported solid optical temperature sensors prepared by using RhB and silica gel (28–200 mesh)/sol–gel, with a mixture of 3-amino-propyl-trimethoxysilane (APTMS) and 3-glycidoxypropyl-trimethoxysilane (GPTMS) as support matrices. The temperature range measured was between 10 and 95 °C for the silica gel system (FOTS-SI) and 0–60 °C for the sol–gel system (FOTS-SO), respectively. The sensitivities were 0.685% K^{-1} and 0.627% K^{-1} for the two systems with precisions of less than 1 °C. After three month, the probes were still stable with only a 6.9% decrease in sensitivity for the FOTS-SI probe and no decrease for the FOTS-SO probe.[25]

Bergaud *et al.* used dried RhB as a probe for the temperature mapping of surface and resistively heated micro/nanowires. Thermal cycles and temperature calibrations were performed using a microscope heating stage. Dried RhB was applied in the temperature range between 25 and 70 °C with a sensitivity of 0.7% K^{-1}, which performed particularly well when high spatial resolution measurements were needed (Figure 6.2). The temperature precision was about 5–10 °C for a sub-micrometre spatial resolution. The temporal resolution achieved was 100 ms.[26] Similarly, to map the in-channel fluid temperature, an SU8/RhB thin layer was spin-coated onto a microheater. A temperature range between 30 and 80 °C was measured with a precision of 0.1 °C and the spatial resolution was on the microscale.[27]

Chu *et al.* reported an experiment to measure the imprint temperature of resist film with the fluorescence of RhB. To display high sensitivity and avoid precipitation problems, the maximal Rhodamine B concentration was 8×10^{-4} M. The temperatures measured were from room temperature to 250 °C with a sensitivity of 0.3% K^{-1}. The precision was about 8 °C and the spatial resolution was on the microscale.[28]

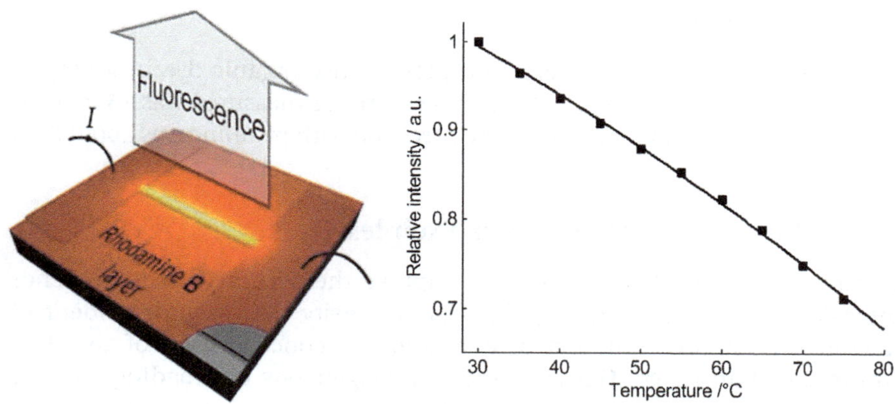

Figure 6.2 Temperature mapping of micro- and nanowires using a dried layer of RhB, where changes in intensity can be converted to temperature. Adapted with permission from ref. 26. © Wiley-VCH 2014.

6.2.1.2 Fluorescein

Fluorescein (Fl) is another highly luminescent probe with a fluorescence quantum yield of almost 100%. Like Rhodamines, the fluorescence of Fl decreases with increasing temperature. It also responds to changes in pH. Liu *et al.* reported a novel method to attach Fl (*via* its epoxy derivate) to a water-soluble polymer [poly(vinyl alcohol)], and the temperature/pH-sensitive qualities of its fluorescence were investigated.[29] The fluorescence quantum yield responded linearly to temperature in the range 0–60 °C with a sensitivity of 0.5% K^{-1}. In addition to water-soluble polymers, such a method could be applied to other water-soluble materials, such as chitosan[30] and epichlorohydrin.[31] Both of these responded linearly to temperature in the range 0–60 °C in water. They have good long-term stability and a fast equilibrium response. Their spatial and temporal resolutions should be as high as other single molecular probes. The pH- and temperature-sensitivities of their fluorescence can be advantageous for multifunctional applications.

Kim and Yoda reported that two fluorescent dyes, Fl and SrB, with inverted temperature sensitivity could be used as a thermometer. In aqueous solution, the Fl/SrB pair obtained temperatures ranging from 20 to 60 °C with a sensitivity of about 7% K^{-1}. The precision for this method was 1.1 °C, at a spatial resolution of 3.7 μm, and 0.3 °C, at a spatial resolution of 30 μm.[32]

Barilero *et al.* reported that Fl and the Texas Red fluorophores based on a nanothermal chemical reaction, either a DNA conformational transition or a protonation, which induced a variation of their emission intensity ratio as the temperature changes. The system can be used to detect temperatures between 5 and 35 °C with a precision better than 1 °C, provides sensitivity of up to 9% K^{-1}, and has a millisecond to microsecond response time. The spatial resolution was on the nanoscale.[33]

6.2.1.3 *Pyranine*

Pyranine is another temperature- and pH-sensitive organic dye in water, reported by Fradin and Wong.[34] The temperate range measured was 15 to 55 °C with a sensitivity of about 1% K^{-1}. The precision with pyranine was about 5 °C.

6.2.2 Intramolecular Excimer/Exciplex Dyes

Introduction of a reference dye can improve the accuracy of fluorescence intensity based thermometers, since the intensity ratio is independent of fluctuations in the excitation source, and the concentration of the dyes within a certain range. However, the inhomogeneous dye loading and the photostability difference between the two dyes could result in imprecision and inaccuracy. In order to obtain absolute temperatures, a single dye with a temperature-sensitive fluorescence intensity ratio is an ideal choice, but this kind of compound is rare. Single-dye dual-emission thermometers can be achieved *via* the intramolecular excimer/exciplex formation mechanism with temperature dependence. Baker *et al.* presented a ratiometric luminescent thermometer based on the temperature-dependent excited state equilibrium of 1,3-bis(1-pyrenyl)propane (BPP) in the ionic liquid (IL) 1-butyl-1-methyl-pyrrolidinium bis(trifluoromethylsulfonyl)imide ([C4mpy][Tf2N]).[35] The excimer-forming modality was used in this system, which operates in the range of 25 to 140 °C, with temperature precision better than 0.35 °C (Figure 6.3). The temperature sensitivity was about 0.9% K^{-1}. The temporal and spatial resolutions were estimated to be on the scale of micro- to nanoseconds and

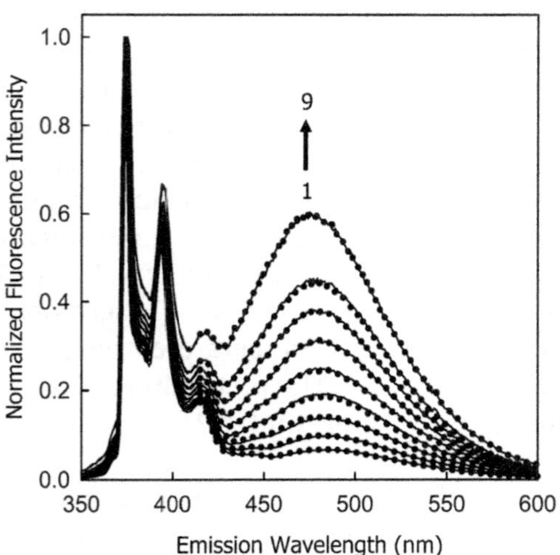

Figure 6.3 Emission spectra normalized to the intensity of the monomer band ($I_{376} = 1.00$). Spectra correspond to the following temperatures (°C): 30, 40, 50, 60, 70, 80, 90, 100 and 120.
Adapted from ref. 35 with permission from the Royal Society of Chemistry.

micro- to nanometres, respectively, similar to other single-fluorescence dyes. BPP was also used in solid polymers.[36] A similar intramolecular excimer formation mechanism for pyrene was achieved using *N*-(1-pyrenylmethyl)-1-pyrenebutanamide and *N*-(1-pyrenylmethyl)-1-pyreneacetamide in various organic solvents by Lou *et al.*,[37] and (1,1′-dipyrenyl)-methyl ether in synthetic phospholipid membranes by Georgescauld *et al.*[38]

In addition to bispyrene compounds, Albelda *et al.* reported a new temperature-sensitive fluorescent compound, which consists of a tri-podal polyamine with three naphthalene groups.[61] The mechanism of its thermoresponse fluorescence is the formation of an intramolecular excimer. The temperature range measured was 15–50 °C with an average relative sensitivity of 4.5% K^{-1} in aqueous solution.

A similar temperature-sensitive intramolecular exciplex formation of 1-(*N*-*p*-anisyl-*N*-methyl)-amino-3-anthryl-(9)-propane was observed in degassed methyl cyclohexane-isopentane solvent between −110 °C and −50 °C, reported by Pragst *et al.*[39,59]

6.2.3 Intermolecular Excimer/Exciplex Dyes

As well as the intramolecular excimer/exciplex, intermolecular excimer/exciplex formation can also be a mechanism for luminescent organic dye thermometers, although the accuracy and precision of such systems could be affected by inhomogeneous distribution of dyes and photobleaching.

6.2.3.1 *Perylene and* N-*Allyl-*N-*methylaniline*

Chandrasekharan and Kelly reported perylene and *N*-allyl-*N*-methylaniline form exciplexes in a polystyrene matrix that can be used as ratiometric fluorescent sensors for temperature (Figure 6.4).[40] It was inherently a

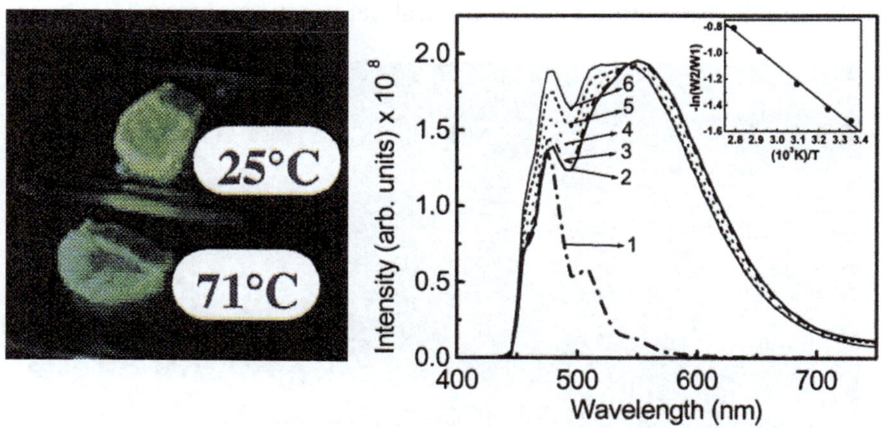

Figure 6.4   Fluorescence observed from a polymer/organic dye system prepared from perylene and *N*-allyl-*N*-methylaniline at different temperatures. Adapted with permission from ref. 40. © American Chemical Society 2014.

'two-colour' material, and calibration for excitation source fluctuations was unnecessary. The sensor was also independent of concentration changes in the fluorophores. The film showed a reversible change in luminescence in the temperature range of 25 to 85 °C, and the sensitivity was about 1% K^{-1}. With the spectrometer and scan rate employed, the precision was 2 °C, and the accuracy was 1 °C.

6.2.3.2   1,4-Bis(α-cyano-4-alkoxystyryl)benzenes

Blends of linear low-density polyethylene (LLDPE) and a series of highly emissive 1,4-bis(α-cyano-4-alkoxystyryl)benzene dyes were prepared using melt-processing techniques by Weder *et al.*[41–43] The dye had a strong tendency to form intermolecular excimers irreversibly. The emission characteristics of the blends strongly depended on the ratio of contributions from monomer and excimer emission of the dyes, with bathochromic shifts of up to 147 nm over the whole temperature range. The temperature range measured was 22–120 °C. The sensitivity was better than 5% K^{-1}. The response time was on the time scale of hours.

Further research reported a similar system composed of polyesters and excimer-forming 1,4-bis-(α-cyano-styryl)-benzenes. Blends of poly(ethylene terephthalate) (PET) or poly(ethylene terephthalate glycol) (PETG) and organic dyes were prepared by melt-processing. The temperature measured was from 90 to 120 °C (Figure 6.5).[41,44]

In addition, a series of such thermochromic polymer/dye blends based on similar organic dyes and glassy amorphous polymers was investigated. The irreversible property of fluorescence colour change, which responds to temperature, can be tailored by adjusting the polymer T_g by changing the composition of blends based on poly(alkyl methacrylate) co-polymers.[42,45] The temperature range measured was 60–115 °C. Such systems could be used as threshold temperature and mechanical deformation sensors.

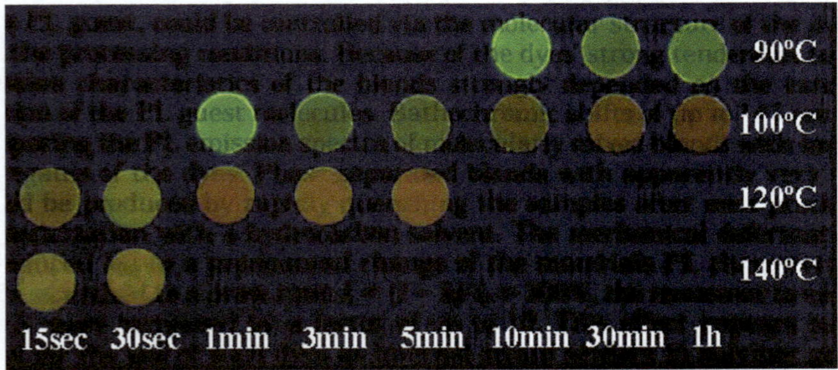

Figure 6.5   Pictures of initially quenched blend films of PET and 0.9% w/w C18-RG upon annealing for the time and at the temperature indicated.
Adapted with permission from ref. 44. © American Chemical Society 2014.

6.2.3.3 Bis(benzoxazolyl)stilbene

Pucci *et al.* reported that several novel polymeric film/organic dye sensors, which were also based on excimer fluorescence, were responsive to both mechanical and temperature changes, irreversibly. The thermometers were obtained through the dispersion of the food-grade dye bis(benzoxazolyl)-stilbene (BBS) into a biodegradable polyester [poly(1,4-butylene succinate), PBS].[46] Emission from BBS excimers emerged with dye concentrations higher than 0.05 wt% with bathochromic shifts about 50 nm. The optical behaviour of PBS/BBS blends was thermally affected in the range 50–80 °C, providing a sensitivity of 2% K^{-1}. The response time was over 5 h. The spatial resolution was better than 150 μm.

A similar threshold thermally sensitive polymer/dye system consisting of poly(lactic acid) (PLA, 85 wt%), PBS (15 wt%) and BBS (food-grade, <0.2 wt%) was prepared by the controlled incorporation method.[47] Films obtained from the blend with less than 0.2 wt% of dye showed fluorescence of BBS monomers (blue). Films containing 0.07 wt% BBS (the highest concentration allowing dye monomeric dispersion) were sensitive to thermovariations at temperatures higher than PLA's T_g (>60 to 70 °C). The temperature range measured was 60–150 °C with a sensitivity better than 10% K^{-1}. The maximum spatial resolution was 500 nm, while the response time was more than 10 min (Figure 6.6).

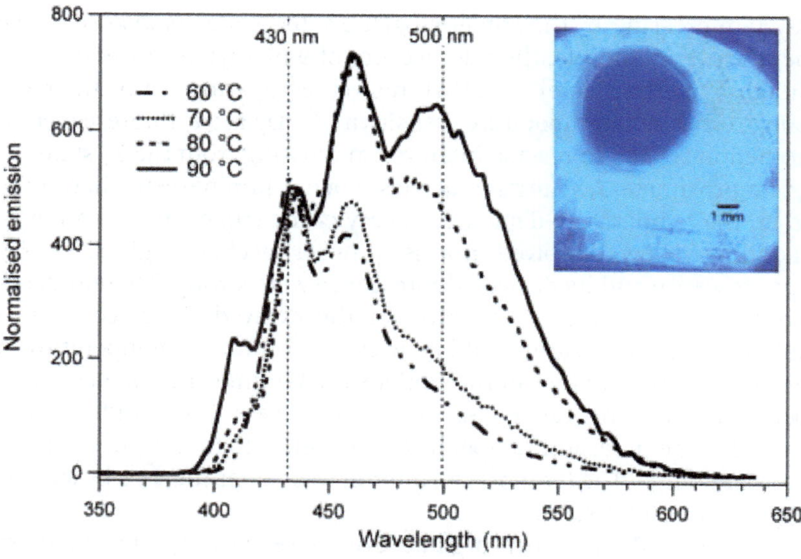

Figure 6.6 Fluorescence emission spectra of PLA/PBS/BBS films as a function of their annealing temperature (annealing time was 5 min).
Adapted from ref. 47 with permission from the Royal Society of Chemistry.

In addition to being used as threshold thermometers, both series of compounds could be used as time–temperature indicators (TTIs) at high temperature, as reported by Weder *et al.*[48] The approach used was to incorporate the excimer-forming dyes, BBS or a cyano-substituted oligo(*p*-phenylene vinylene) (C2-RY18) derivative, into ethylene/norbornene co-polymers with T_g values between 131 and 149 °C. These polymer/dye blends were melt-processed and quenched from homogeneous melts to below T_g to obtain monomeric dispersions. At various temperatures above T_g, self-assembly, with different speeds, of the dye molecules into excimers occurred with irreversible fluorescence colour changes. The TTI materials were useful in a temperature range from 130 to 200 °C. The temperature was measured from the response time of fluorescence variations, so the precision and accuracy could be high after further calibration.

6.2.3.4   Triaryl Amine-terminated Diimide

Meador *et al.* presented the synthesis and photophysics of triaryl amine-terminated diimide and its potential for use as a thermometer.[49] The system was composed of the organic dye and an LLDPE polymer. The mechanism was the breakdown of excimers at higher temperatures, although this does not fully explain the properties of the dye. The temperature range measured was 25–153 °C and the system was reversible. Further investigation is needed to provide the specific sensitivity and precision.

6.2.4   Intersystem Crossing Dyes

Intersystem crossing of the triplet/singlet excited states is also sensitive to temperature. For example, the fluorescence of 9-methylanthracene in a rigid poly(methyl methacrylate) (PMMA) matrix competing with intersystem crossing, exhibited temperature sensitivity.[50] Delayed fluorescence is a phenomenon related to reverse intersystem crossing from the T_1 state to the energetic proximate S_1. Harris *et al.* described a luminescent thermometer based on the temperature-dependent decay of the triplet state of an organic dye, acridine yellow, dissolved in a rigid mixed saccharide glass.[51] Both its phosphorescence and its delayed fluorescence were strong. Therefore, either the triplet-state lifetime or the ratio of the delayed fluorescence to the phosphorescence intensities could be used to monitor the temperature. The average relative sensitivities of the triplet-state lifetime and intensity ratio to temperature, over the range -50 to $+50$ °C were 2.0% and 4.5% K^{-1}, respectively. The precision temperature was better than 1 °C, over the same temperature range. The temporal resolution was better than 1 s based on the lifetime of the triplet state.

Other delayed fluorescence organic dyes have been applied in thermometry research. A new type of polymer thermometer based on the delayed fluorescence of C_{70} was investigated by Baleizao *et al.*, with polystyrene (PS), poly(*tert*-butyl methacrylate) (PtBMA), and poly(1-vinylnaphthalene) (P1VN) as matrices.[52] Temperature-dependent imaging of the C_{70} films was carried out using both intensity and lifetime methods. In the absence of oxygen, the fluorescence intensity of the C_{70} film increased markedly with temperature. The working range was from -80 to $+140\,^\circ$C for PtBMA with a sensitivity better than 0.5% K^{-1}. For the fluorescence lifetime method, the results showed a working temperature range of between 7 $^\circ$C and an estimated upper limit of 515 $^\circ$C with a sensitivity better than 0.5% K^{-1}. The temporal resolution was better than 1 ms.

In addition to temperature-sensitive delayed fluorescence of organic dyes, a metal-free organic phosphorescent material used as a temperature indicator, as reported by Kim *et al.*[53] The material was made by embedding a purely organic phosphor Br6A into an amorphous glassy polymer matrix (PMMA). The phosphorescence was strongly dependent on temperature. The *in situ* phosphorescence spectra decreased nearly linearly in the temperature range of 30–60 $^\circ$C with a sensitivity of about 2% K^{-1}. The spatial resolution was better than 100 µm and the temporal resolution was about 2 ms.

Phosphorescent lifetime can also be used as a temperature indicator. Schuh *et al.* reported a ternary complex of 6-bromo-2-naphthol and α-cyclodextrin used as a phosphorescent lifetime thermometer in aqueous solution.[60] The temperature range measured was 1.6–59.7 $^\circ$C. The sensitivity ranged from 11% K^{-1} at 1.6 $^\circ$C to 7.5% K^{-1} at 59.7 $^\circ$C. The accuracy was better than 0.1 $^\circ$C.

6.2.5 Twisted Intramolecular Charge-transfer Compounds

Twisted intramolecular charge transfer (TICT) compounds usually exhibit conformation changes in different excited states; local excited state (LE), intramolecular charge transfer (ICT), and/or TICT. These kinds of compounds could maintain their total luminescence intensity, or even exhibit an enhancement from lower to higher temperatures. If the excited states were emissive, a fluorescence colour change would be observed with a change in temperature.

6.2.5.1 7-Nitrobenz-2-oxa-1,3-diazol-4-yl

Thermal quenching fluorescence of TICT states can be used in intensity based temperature probes. For example, the change of fluorescence intensity observed for NBD was primarily due to heat-induced electronic changes leading to a TICT state, accompanied by a decrease in fluorescence lifetime.[54] Tromberg *et al.* reported two fluorescent membrane probes, NBD and Laurdan, for use as optical thermometers in living cells.[55] The maxima of the Laurdan fluorescence wavelength shifted as the membrane underwent a gel-to-liquid-crystalline phase transition induced by temperature variations. The results showed that NBD fluorescence lifetime recordings could provide a temperature precision of approximately 2 $^\circ$C, over a 15–67 $^\circ$C temperature range, and a sensitivity of about 1.5% K^{-1}. Laurdan's microenvironmental temperature precision was 0.1–1 $^\circ$C, over a temperature range of 33–46 $^\circ$C, with a sensitivity better than 5% K^{-1}. Both can be used as temperature indicators in living cells on the time scale of normal cellular processes.

6.2.5.2 Triarylboron

The thermal responsive process of TICT compounds can be concomitant with a luminescence colorimetric change that results from the shift of the thermal equilibrium between the LE state/ICT emission and TICT excited state emission. Therefore, with an LE/ICT state emission as a reference, and high total fluorescence intensity at high temperature, such TICT compounds are ideal dyes to be used as single-dye dual-emission luminescent thermometers. Recently, Yang *et al.* designed a thermosensitive luminescent pyrene-containing triarylboron molecule, dipyren-1-yl (2,4,6-triisopropylphenyl) borane (DPTB), combining the advantages of TICT compounds with two reverse luminescence intensity changes with high quantum yields.[56]

The design of the dye was based on the following considerations: (1) an electron-deficient boron atom with an empty p orbital was used as the electron acceptor, and aryl substituents are known as donors, so an ICT excited state should be present in the molecule; (2) the large steric hindrance in the compact Py–B–Py structure could result in a TICT excited state with a non-equivalent contribution of the two Py groups to the excited states; (3) any substituents that may enhance non-emissive decay were avoided; and (4) the sterically bulky substituent 2,4,6-triisopropylphenyl group was an effective stabilizer for the compound.

The dual fluorescence feature of DPTB was identified by double-exponential decays. The ratio of the shorter lifetime species to the longer lifetime species decreases with increasing emission wavelength, therefore, the shorter and longer lifetime species could be assigned to the LE excited state, and the TICT excited state of DPTB, respectively. The temperature strongly influenced the dynamic equilibrium between the LE and TICT excited states of DPTB. The luminescence colour was determined by the population of the two distinct excited state conformations (Figure 6.7). The lower energy TICT excited state is preferentially occupied with decreasing temperature, and as expected, a bathochromic shift of the luminescence is observed. Upon heating the system, the molecular motion that crossed the thermal barrier between the two excited states increased the population of the LE state, thus resulting in a hypsochromic luminescence.

This dye can be used as a luminescent colorimetric thermometer for *in situ* large-area or gradient temperature measurements in 2-methoxyethyl ether (MOE). The system can be applied over a temperature range of -50 to $+100\,^{\circ}\mathrm{C}$ with high stability and reversibility (Figure 6.8). The luminescence spectra intensity ratio can be correlated to the temperature values, with a sensitivity better than 1% K^{-1} and the accuracy was better than $1\,^{\circ}\mathrm{C}$. The luminescence colour shifted between green and blue, which can also be applied as a temperature indicator using a single CCD camera. In this case, the accuracy was calculated to be approximately $2\,^{\circ}\mathrm{C}$ and could be better

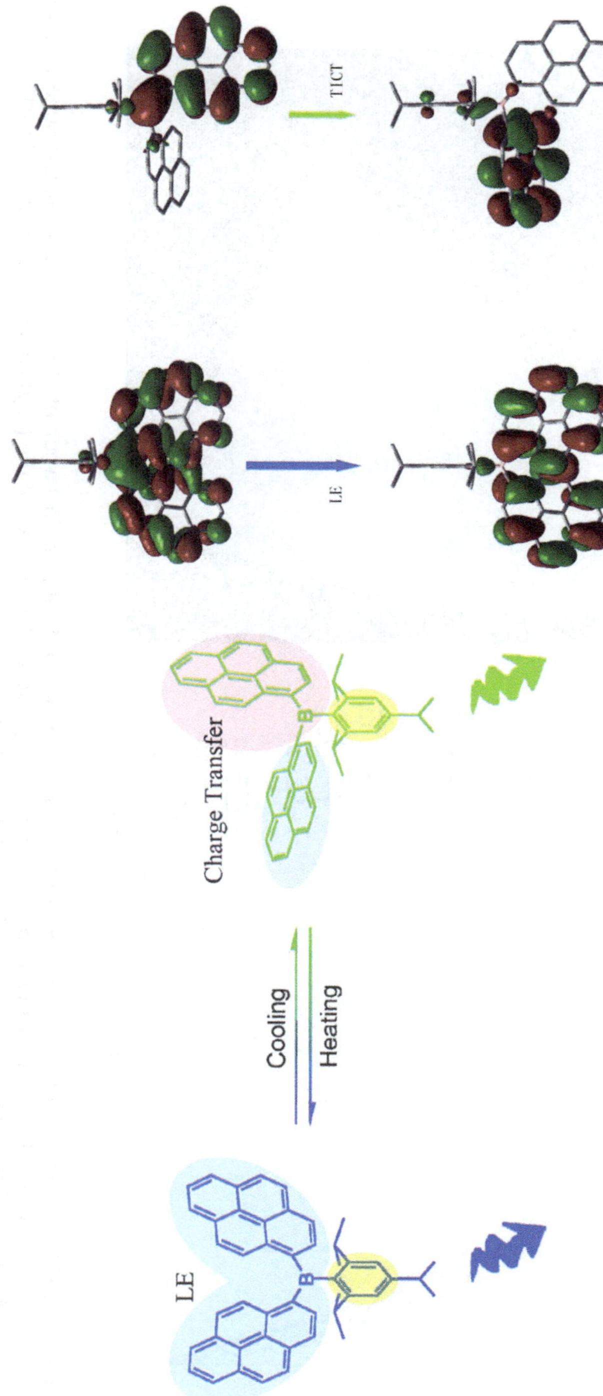

Figure 6.7 The triarylboron-based fluorescent probe DPTB. Contours are the frontier orbitals for DPTB. Adapted with permission from ref. 56. © Wiley-VCH 2014.

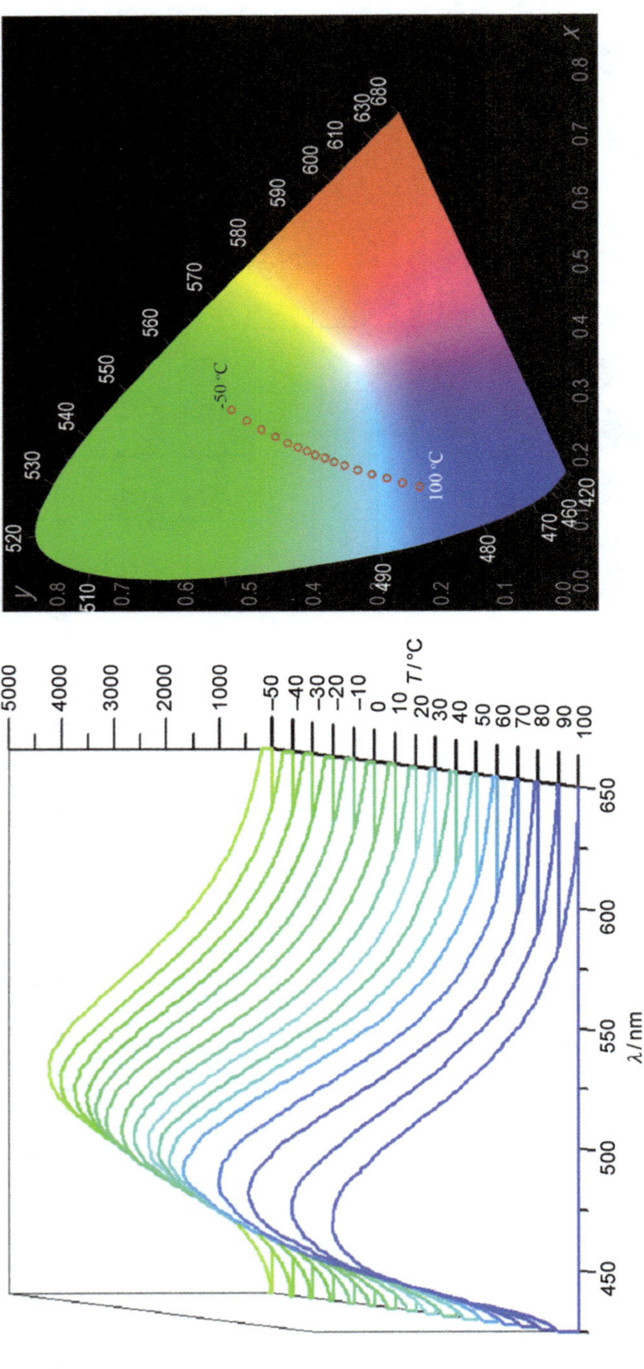

Figure 6.8 (Left) An increase of T causes the green TICT emission to disappear in favour of the blue local excited state emission. (Right) A CIE chromaticity diagram showing the temperature dependence of the (x, y) colour coordinates of DPTB. Adapted with permission from ref. 56. © Wiley-VCH 2014.

with a better CCD camera. The spatial resolution was the edge roughness of the DPTB/MOE system, estimated to be 30–40 μm with image capture times less than 10 ms.

In general, the performance of a luminescent material is also affected by microenvironments other than temperature, especially in liquids. Such a shortcoming could be overcome by encapsulating the probe solution in microparticles or nanoparticles made from organic polymers. Acting as barrier walls from liquids, microcapsules can protect the inside substances from external factors, such as oxidation, acidity, alkalinity, moisture and evaporation, which allows liquid materials to be used in various environments with high performance. Also, the limitation of the water insolubility of triarylboron dyes can be addressed. Yang *et al.* reported a luminescent microcapsule containing novel temperature-sensitive triarylboron dyes with self-referencing characteristics, which can be dispersed in different media (*e.g.*, organic solvents, water and polymers).[57]

DPTB only showed an intrinsic temperature-sensitive dual fluorescence in polar organic solvents like MOE, which makes it difficult to form microcapsules. To solve this problem, two similar triarylboron dyes, di-6-methoxypyren-1-yl-(2,4,6-triisopropylphenyl)borane (MPTB) and 1,1′-(6,6′-((2,4,6-triisopropylphenyl) boranediyl)bis(pyrene-6,1-diyl))dipyrrolidine (TBBD), were designed and synthesized. The introduction of an electron-donating group into the pyrene groups should render luminescence with stronger ICT characteristics, which would increase the polarity of the molecule and reduce the requirement for solvent polarity. This should be the easier way to fabricate microcapsules.

With the designed structure, the thermosensitive microcapsule was fabricated by encapsulating a 1,2,3,4-tetrahydronaphthalene-polystyrene (THN-PS) solution of TBBD into a cross-linked poly(urea-formaldehyde) microshell. This microcapsule could work under a wide range of temperatures (−30 to +140 °C) with a reversible fluorescence colour change from orange-red to green-yellow (Figure 6.9). The intensity ratio calibration gave sensitivities ranging from 1.1% K^{-1}, at −30 °C, to 5.9% K^{-1}, at 140 °C. The accuracy was better than 0.5 °C and the precision should be very high due to the elimination of errors due to inhomogeneous dye loading and photobleaching. The spatial resolution was better than 4 μm, according to the size of the microcapsules. The temporal resolution was on the scale of pico- or nanoseconds, from the time scale of the excitation and emission.

For colour-based procedures, the temperature-dependent spectra are transformed to Commission Internationale de l'Eclairage (CIE) 1931 co-ordinates (Figure 6.9). A fast-flowing system was simulated and measured as a demonstration. The hot and cold salt aqueous solutions containing 5% (w/w) microcapsules were injected into two end-ports of a T-shaped square tube with width 1 mm, respectively, and then out-flowed from the third port (Figure 6.10). The orange-red and green-yellow correspond to cold fluid and hot fluid, respectively. The colour-based approach is suitable for direct observation of a temperature gradient with the naked eye, or with a CCD camera.

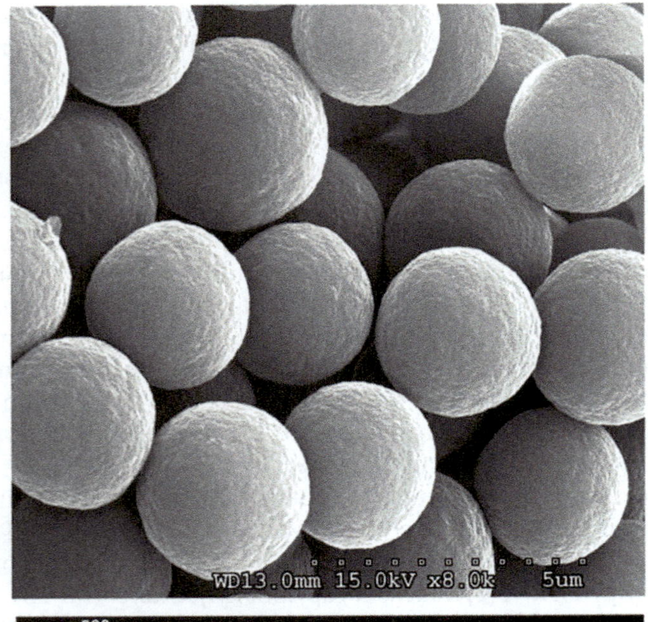
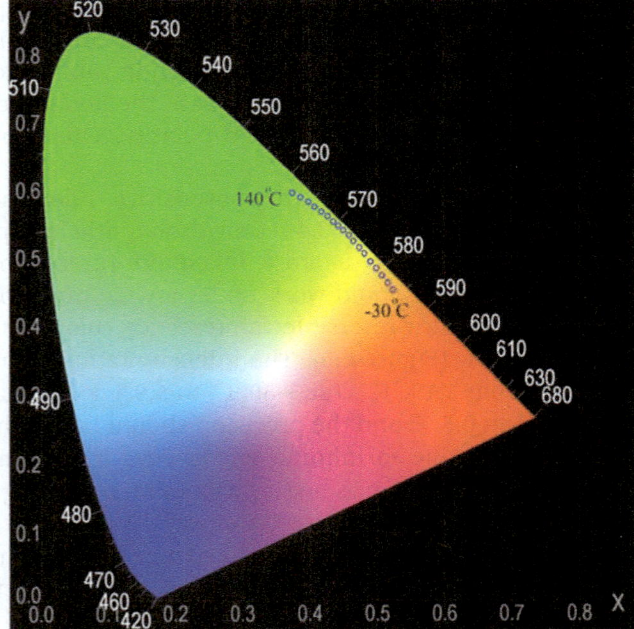

Figure 6.9 An SEM image of microcapsules and a CIE chromaticity diagram showing the temperature dependence of the (x, y) colour coordinates of TBBD-THN-PS microcapsules.
Adapted with permission from ref. 57. © Wiley-VCH 2014.

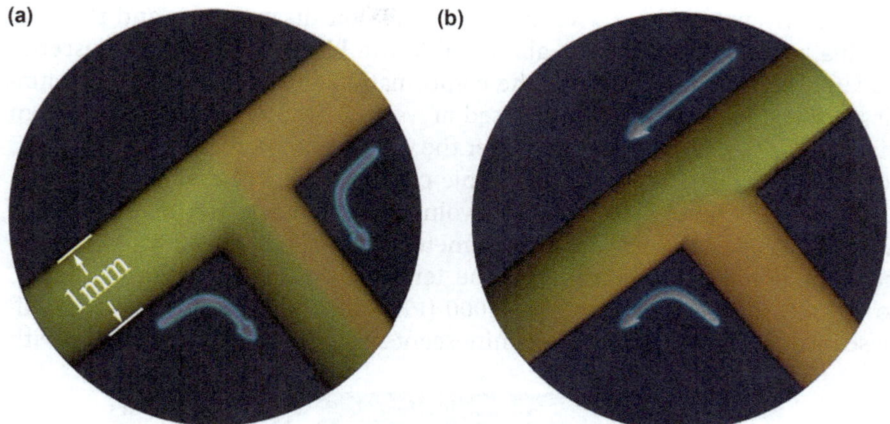

Figure 6.10   Fluorescence microscopy images of fluids with different temperatures. Fluid flow (a) in opposite directions and (b) in perpendicular directions. Adapted with permission from ref. 57. © Wiley-VCH 2014.

However, the triarylboron organic dyes mentioned in the previous section could only be used as temperature indicators in liquids, not in solid-state polymers because the high viscosity hinders the conformational change, due to the relatively large substituents (pyrene groups), which restrict the formation of their TICT excited states. A novel ICT triarylboron compound, tris(2,3,5,6-tetramethyl-4-morpholinophenyl)borane (MPB) was designed with smaller bulky substituents than previously reported compounds. Thus the exchange of its excited state conformations would require a smaller free volume, while retaining the compound's temperature-sensitive properties.[58] The fluorescence of MPB showed both temperature dependence and sensitivity to the free volume around it.

The population of the two excited states of MPB was affected by the temperature, and the luminescence colour was determined by the equilibrium of the two ICT/TICT distinct excited state conformations. This thermometer can be applied in most organic solvents with high stability and reversibility. For example, the temperature range was from -50 to $+100\,^{\circ}\mathrm{C}$ in MOE, similar to that of DPTB.

For the molecules of MPB, in addition to the temperature, the equilibrium between the two excited state conformations could also be affected by the free volume of its medium, because the conformational change between the two excited states requires extra space. The thermochromic properties of MPB in the high-viscosity liquid solvent, polyethylene glycol 200 (PEG 200) were evaluated. When the temperature ranged between 30 and $100\,^{\circ}\mathrm{C}$, the free volume of PEG 200 was larger than 10.5%, and it was large enough for the conformation to change freely between the two excited states. Thus, in this temperature range, the temperature effect on the population of the two excited states of MPB was similar to that in normal liquid solvents. When the temperature decreased from -20 to $-40\,^{\circ}\mathrm{C}$, the free volume of the PEG 200 changed from 5.3% to 3.3%, the TICT emission disappeared, and this was gradually accompanied by an abnormal 52 nm blue-shift of the fluorescence maximum. This indicated that the conformational change started to be hindered at $-20\,°C$ and was suppressed at $-40\,°C$. The fluorescence quantum yields remained higher than 0.19 over the whole temperature range of $-50\,°C$ to $+100\,°C$, and all the thermochromic processes were reversible.

The extraordinary property of free volume dependence of MPB expands its application as a fluorescent thermometer into solid-state polymers. It was perfectly suitable for monitoring the temperature in the range of $-20\,°C$ to $+40\,°C$ in polyethylene glycol 4000 (PEG 4000) (Figure 6.11). The luminescence intensity ratio or the luminescence colour could be correlated with

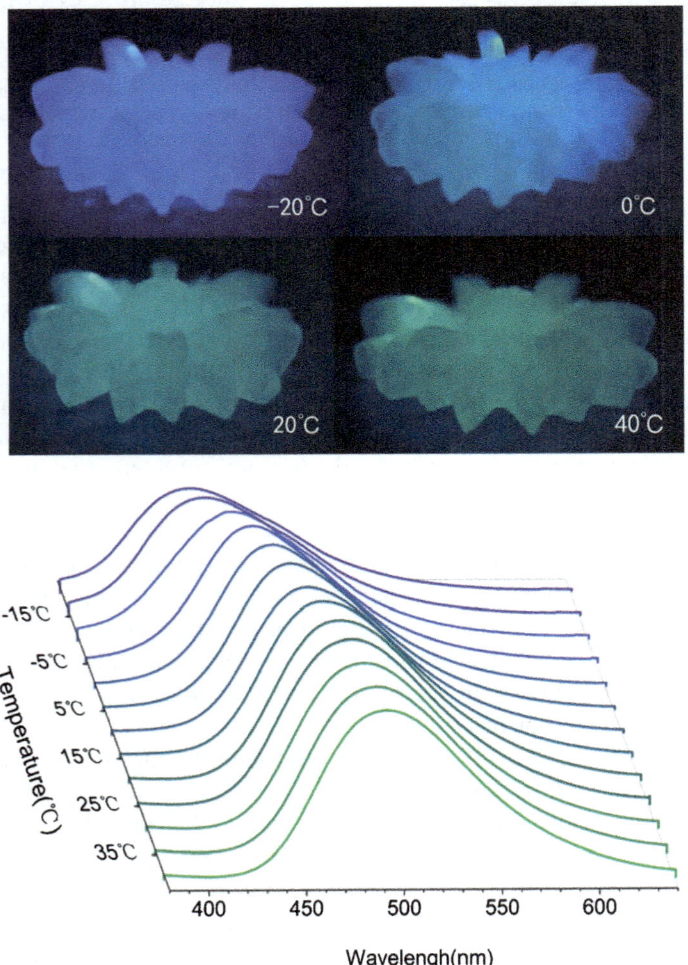

Figure 6.11 Fluorescence changes with temperature of an MPB-PEG 4000 polymer system.
Adapted from ref. 58 with permission from the Royal Society of Chemistry.

the temperature values. The slope of the curve gave temperature sensitivities ranging from 1.6% K^{-1} at $-20\,^{\circ}C$ to 14% K^{-1} at $10\,^{\circ}C$. The temporal and spatial resolutions were in the range of microseconds and micrometres, respectively. The precision and accuracy should be very high, due to the elimination of errors due to inhomogeneous dye loading and photobleaching, which are very rare in polymer/organic dye solid systems. Consequently, MPB is a rare candidate for a real-time and reversible temperature indicator with a concentration-independent feature.

6.3 Conclusions and Perspectives

In this chapter we have described advances in the organic dyes used as fluorescent thermometers for various systems at the micro- and nanoscales. The non-emissive conformation, excimer/exciplex, intersystem crossing, and TICT states of organic compounds emerge or disappear at higher temperatures, resulting in luminescence variations. Organic dyes show advances in higher spatial resolution, faster response and higher sensitivity. The most used organic dyes are members of the Rhodamine family, fluorescein, bis-pyren and triarylboron compounds in liquid solvents, and perylene/*N*-allyl-*N*-methylaniline, bis(benzoxazolyl)stilbene (BBS) and triarylboron in solid polymers.

For applications at the nanoscale, there are several disadvantages to using organic dyes as thermometers. The accuracy and precision could be affected by many factors: (1) fluctuations in the excitation source; (2) accumulation of dye molecules in certain areas in systems; (3) inhomogeneous dye concentrations; and (4) photobleaching. The interference of intensity fluctuations can be solved by the introduction of a reference dye, or measurement of the luminescence lifetime. The inhomogeneous dye concentration problem can be solved by using single-probe dual-emission dyes. The photobleaching effect can also be partially solved by using single-probe dual-emission dyes, because the photobleaching does not affect the luminescence intensity ratio significantly. However, the precision still decreases with long-term irradiation. So the ideal organic dye in luminescent thermometry should be self-referencing, reversible, photostable and should change with temperature significantly. Meanwhile, water solubility and biocompatibility are of interest in contexts with *in vivo* imaging, and will extend their application into the biosciences.

References

1. S. M. Borisov and O. S. Wolfbeis, *Anal. Chem.*, 2006, **78**, 5094.
2. K. Okabe, N. Inada, C. Gota, Y. Harada, T. Funatsu and S. Uchiyama, *Nat. Commun.*, 2012, **3**, 705.
3. X. D. Wang, R. J. Meier and O. S. Wolfbeis, *Adv. Funct. Mater.*, 2012, **22**, 4202.

4. A. Bousseksou, G. Molnár, L. Salmon and W. Nicolazzi, *Chem. Soc. Rev.*, 2011, **40**, 3313.
5. C. D. S. Brites, P. P. Lima, N. J. O. Silva, A. Millán, V. S. Amaral, F. Palacio and L. D. Carlos, *Nanoscale*, 2012, **4**, 4799.
6. F. P. Zamborini, L. Bao and R. Dasari, *Anal. Chem.*, 2012, **84**, 541.
7. F. L. Bai and L. A. Melton, *Appl. Spectrosc.*, 1997, **51**, 1276.
8. S. J. Hale and L. A. Melton, *Appl. Spectrosc.*, 1990, **44**, 101.
9. A. P. De Silva, H. Q. N. Gunaratne, K. R. Jayasekera, S. Ocallaghan and K. R. A. S. Sandanayake, *Chem. Lett.*, 1995, 123.
10. X.-d. Wang, O. S. Wolfbeis and R. J. Meier, *Chem. Soc. Rev.*, 2013, **42**, 7834.
11. S. Uchiyama, A. P. de Silva and K. Iwai, *J. Chem. Educ.*, 2006, **83**, 720.
12. J. F. Lou, T. M. Finegan, P. Mohsen, T. A. Hatton and P. E. Laibinis, *Rev. Anal. Chem.*, 1999, **18**, 235.
13. D. Ross, M. Gaitan and L. E. Locascio, *Anal. Chem.*, 2001, **73**, 4117.
14. J. J. Shah, S. G. Sundaresan, J. Geist, D. R. Reyes, J. C. Booth, M. V. Rao and M. Gaitan, *J. Micromech. Microeng.*, 2007, **17**, 2224.
15. R. Samy, T. Glawdel and C. L. Ren, *Anal. Chem.*, 2008, **80**, 369.
16. T. Glawdel, Z. Almutairi, S. Wang and C. Ren, *Lab Chip*, 2009, **9**, 171.
17. J. Sakakibara and R. J. Adrian, *Exp. Fluids*, 1999, **26**, 7.
18. M. Bruchhausen, F. Guillard and F. Lemoine, *Exp. Fluids*, 2005, **38**, 123.
19. V. K. Natrajan and K. T. Christensen, *Meas. Sci. Technol.*, 2009, **20**, 015401.
20. C. E. Estrada-Perez, Y. A. Hassan and S. C. Tan, *Rev. Sci. Instrum.*, 2011, **82**, 074901.
21. S. M. Jeon, J. Turner and S. Granick, *J. Am. Chem. Soc.*, 2003, **125**, 9908.
22. L. N. Sanford and C. T. Wittwer, *Anal. Biochem.*, 2014, **448**, 75.
23. J. F. Bermingham, Y. Y. Chen, R. L. McIntosh and A. W. Wood, *Bioelectromagnetics*, 2014, **35**, 181.
24. Y. Y. Chen and A. W. Wood, *Bioelectromagnetics*, 2009, **30**, 583.
25. H. D. Duong and J. I. Rhee, *Sens. Actuatuators, B*, 2007, **124**, 18.
26. P. Low, B. Kim, N. Takama and C. Bergaud, *Small*, 2008, **4**, 908.
27. W. Jung, Y. W. Kim, D. Yim and J. Y. Yoo, *Sens. Actuators, A*, 2011, **171**, 228.
28. F. H. Ko, L. Y. Weng, C. J. Ko and T. C. Chu, *Microelectron Eng.*, 2006, **83**, 864.
29. X. L. Guan, X. Y. Liu, Z. X. Su and P. Liu, *React. Funct. Polym.*, 2006, **66**, 1227.
30. X. Guan, X. Liu and Z. Su, *J. Appl. Poly. Sci.*, 2007, **104**, 3960.
31. X. L. Guan and Z. X. Su, *Polym. Adv. Technol.*, 2008, **19**, 385.
32. M. Kim and M. Yoda, *Exp. Fluids*, 2010, **49**, 257.
33. T. Barilero, T. Le Saux, C. Gosse and L. Jullien, *Anal. Chem.*, 2009, **81**, 7988.
34. F. H. C. Wong and C. Fradin, *J. Fluoresc.*, 2011, **21**, 299.
35. G. A. Baker, S. N. Baker and T. M. McCleskey, *Chem. Commun.*, 2003, 2932.

36. A. J. Bur, M. G. Vangel and S. C. Roth, *Polym. Eng. Sci.*, 2001, **41**, 1380.
37. J. F. Lou, T. A. Hatton and P. E. Laibinis, *Anal. Chem.*, 1997, **69**, 1262.
38. D. Georgescauld, J. P. Desmasez, R. Lapouyade, A. Babeau, H. Richard and M. Winnik, *Photochem. Photobiol.*, 1980, **31**, 539.
39. A. C. Spivey, A. J. Bissell and B. Stammen, *Synth. Commun.*, 1998, **28**, 623.
40. N. Chandrasekharan and L. A. Kelly, *J. Am. Chem. Soc.*, 2001, **123**, 9898.
41. B. R. Crenshaw and C. Weder, *Chem. Mater.*, 2003, **15**, 4717.
42. B. R. Crenshaw and C. Weder, *Adv. Mater.*, 2005, **17**, 1471.
43. C. Lowe and C. Weder, *Adv. Mater.*, 2002, **14**, 1625.
44. M. Kinami, B. R. Crenshaw and C. Weder, *Chem. Mater.*, 2006, **18**, 946.
45. B. R. Crenshaw, J. Kunzelman, C. E. Sing, C. Ander and C. Weder, *Macromol. Chem. Phys.*, 2007, **208**, 572.
46. A. Pucci, F. Di Cuia, F. Signori and G. Ruggeri, *J. Mater. Chem.*, 2007, **17**, 783.
47. A. Pucci, F. Signori, R. Bizzarri, S. Bronco, G. Ruggeri and F. Ciardelli, *J. Mater. Chem.*, 2010, **20**, 5843.
48. C. E. Sing, J. Kunzelman and C. Weder, *J. Mater. Chem.*, 2009, **19**, 104.
49. D. S. Tyson, A. D. Carbaugh, F. Ilhan, J. Santos-Perez and M. A. Meador, *Chem. Mater.*, 2008, **20**, 6595.
50. S. Schoof and H. Gusten, *Ber. Bunsen. Phys. Chem.*, 1989, **93**, 864.
51. J. C. Fister, D. Rank and J. M. Harris, *Anal. Chem.*, 1995, **67**, 4269.
52. C. Baleizao, S. Nagl, S. M. Borisov, M. Schaferling, O. S. Wolfbeis and M. N. Berberan-Santos, *Chem.–Eur. J.*, 2007, **13**, 3643.
53. D. Lee, O. Bolton, B. C. Kim, J. H. Youk, S. Takayama and J. Kim, *J. Am. Chem. Soc.*, 2013, **135**, 6325.
54. S. Feryforgues, J. P. Fayet and A. Lopez, *J. Photochem. Photobiol., A*, 1993, **70**, 229.
55. C. F. Chapman, Y. Liu, G. J. Sonek and B. J. Tromberg, *Photochem. Photobiol.*, 1995, **62**, 416.
56. J. Fen, K. J. Tian, D. H. Hu, S. Q. Wang, S. Y. Li, Y. Zeng, Y. Li and G. Q. Yang, *Angew. Chem., Int. Ed.*, 2011, **50**, 8072.
57. J. Feng, L. Xiong, S. Q. Wang, S. Y. Li, Y. Li and G. Q. Yang, *Adv. Funct. Mater.*, 2013, **23**, 340.
58. X. Liu, S. Li, J. Feng, Y. Li and G. Yang, *Chem. Commun.*, 2014, **50**, 2778.
59. F. Pragst, H.-J. Hamann, K. Teuchner, M. Naether, W. Becker and S. Daehne, *Chem. Phys. Lett.*, 1977, **48**, 36.
60. R. E. Brewster, M. J. Kidd and M. D. Schuh, *Chem. Commun.*, 2001, **12**, 1134.
61. M. T. Albelda, E. Garcia-Espana, L. Gil, J. C. Lima, C. Lodeiro, J. S. de Melo, M. J. Melo, A. J. Parola, F. Pina, C. Soriano and M. Teresa, *J. Phys. Chem. B*, 2003, **107**, 6573.

Polymeric Temperature Sensors

GERTJAN VANCOILLIE,[†] QILU ZHANG[†] AND
RICHARD HOOGENBOOM*

Supramolecular Chemistry Group, Department of Organic Chemistry,
Ghent University, Krijgslaan 281-S4, 9000 Ghent, Belgium
*Email: richard.hoogenboom@ugent.be

7.1 Introduction

The use of polymers is finding a permanent place in sensor development
as their chemical and physical properties can be tailored over a wide range
of characteristics.[1–3] Great progress has been made in the last 20 years
in polymeric sensors, especially for systems that make use of stimuli-
responsive polymers that respond sharply with a solution-phase transition to
environmental parameter changes such as temperature, pH value, or UV–vis
light, or respond to chemical changes.[3] The unique properties and easy
accessibility of polymeric sensors enable their development as important
alternative sensoric materials in areas such as biology, diagnostics, and
chemical analysis.[1,3–5]

For temperature-sensing purposes, the temperature-induced solution-
phase transition of a polymer can be translated into a sensory signal by
incorporated solvatochromic dyes[6,7] that specifically change their optical or
emissive properties in response to changes in the local environment
(Figure 7.1). Such a polymer sensor system is composed of two key parts:
(1) a thermoresponsive polymer that undergoes a phase transition due to the
environmental temperature, in particular, polymers with a lower critical

---

[†]Both authors contributed equally to this work.

RSC Nanoscience & Nanotechnology No. 38
Thermometry at the Nanoscale: Techniques and Selected Applications
Edited by Luís Dias Carlos and Fernando Palacio
© The Royal Society of Chemistry 2016
Published by the Royal Society of Chemistry, www.rsc.org

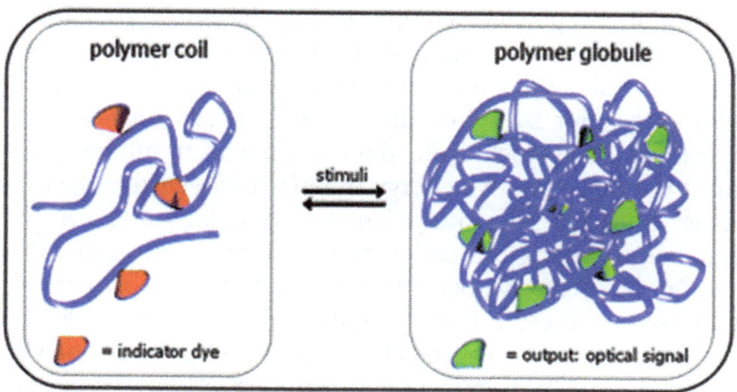

Figure 7.1 Schematic representation of polymeric sensors based on polymer phase transitions (coil-to-globule) and solvatochromic dyes. Reproduced from ref. 9.

solution temperature (LCST) are used, *i.e.*, the polymer is dissolved at lower temperatures and precipitates upon heating;[8] and (2) the chromophore can be a fluorescent or visible solvatochromic dye that generates an output signal that can be quantitatively detected. In aqueous solution, the chromophore changes its absorbance or emission behaviour upon variation of the microenvironment from exposure to water in the soluble polymer state to the less polar collapsed polymer globules. For visible solvatochromic dyes, the change of signal could be either variation of absorption intensity or maximum absorption wavelength. In terms of fluorescent dyes, more information can be provided because numerous parameters like fluorescence decay times, fluorescence intensity, quenching efficiency, energy transfer and fluorescence polarization can be determined.

The combination of thermoresponsive polymers and solvatochromic dyes has rendered the resulting thermometers beneficial properties originating from both components: (1) the solubility of the sensor materials can be easily tuned by varying the ratio of hydrophobic-to-hydrophilic monomers; (2) the polymer can provide structural stability and easy processability (*e.g.*, coating or formation of nanoparticles); and (3) embedding of the chromophore into the polymeric matrix can dramatically reduce photobleaching and chemical reactions of the dye. Due to their excellent properties, polymeric temperature sensors have been considered as very promising thermometers and have received significant interest over the last two decades. Numerous publications have appeared and have been summarized in recent reviews.[3,9]

The first reported examples of thermoresponsive polymers modified with (solvatochromic) chromophores as indicator dyes were designed as a new method to study polymer phase transitions. At the end of the 1980s, Kungwatchakun and Irie[10] studied azobenzene containing poly(*N*-isopropylacrylamide) (PNIPAM) co-polymers as photoresponsive systems, and Binkert *et al.*[11] reported a fluorescein-labelled PNIPAM to investigate the local mobility of the polymer chains during the phase transition by fluorescence spectroscopy. Fundamental studies on fluorophores as indicators for stimuli-responsive polymers were reported by the Winnik group[12,13] in 1990 and shortly after by Schild and Tirrell[14] based on PNIPAM in combination with the solvatochromic fluorescent chromophore pyrene. Both groups used the non-radiative energy transfer between donor and acceptor chain labels to explore the interpolymer interactions and the changes in chain dimensions during the coil-to-globule transition. Since 2003 a large number of fluorescent/solvatochromic dye containing polymers have been developed based on these initially reported concepts. These dye functional polymers are studied for two purposes: (1) the development of new polymeric sensor systems; and (2) to gain in-depth understanding of polymer chain conformations and/or phase transitions.[15,16]

7.2 Dye Incorporation and Positioning in the Polymer Chain

The synthesis of a polymeric thermometer starts with the design of a suitable polymerization method and dye-incorporation approach. Most of these thermometers have been synthesized using free radical polymerization due to its high functional group tolerance compared to other polymerization techniques. In recent years, controlled radical polymerization (CRP) methods were also frequently performed if well-defined polymer architecture, composition or chain length were desired for the target polymer. Among the reported CRP techniques, atom-transfer radical polymerization (ATRP)[17–19] or reversible addition-fragmentation chain-transfer polymerization (RAFT)[20–23] are mostly used for the construction of dye-functionalized polymer chains. Besides linear chains, other architectures, such as block,[24,25] comb-like[26–29] or star co-polymers[30] became available by using CRP methods. By adding bi-functional monomers or cross-linking agents to an emulsion polymerization, nano/micro gels/particles consisting of dye-functionalized polymers can be synthesized. The choice of polymerization technique is mostly determined by the level of compatibility with the incorporated dye and/or incorporation approach.

In general, a polymer can be functionalized on a few different positions along the backbone of the chain, namely the α-chain end, the ω-chain end and/or the side-chains. These positions can be functionalized either during the polymerization through the use of dye-functionalized monomers and/or initiators/chain-transfer agents/terminating agents, or after the polymerization through post-polymerization modifications. These synthetic methodologies can be organized in three main approaches: (1) the dye-monomer approach; (2) end-functionalization using a dye-bearing initiator, terminator or chain-transfer agent; and (3) post-polymerization modification allowing both side-chain and chain end modification (Figure 7.2). All of these methods have been extensively studied in various reviews.[31–37]

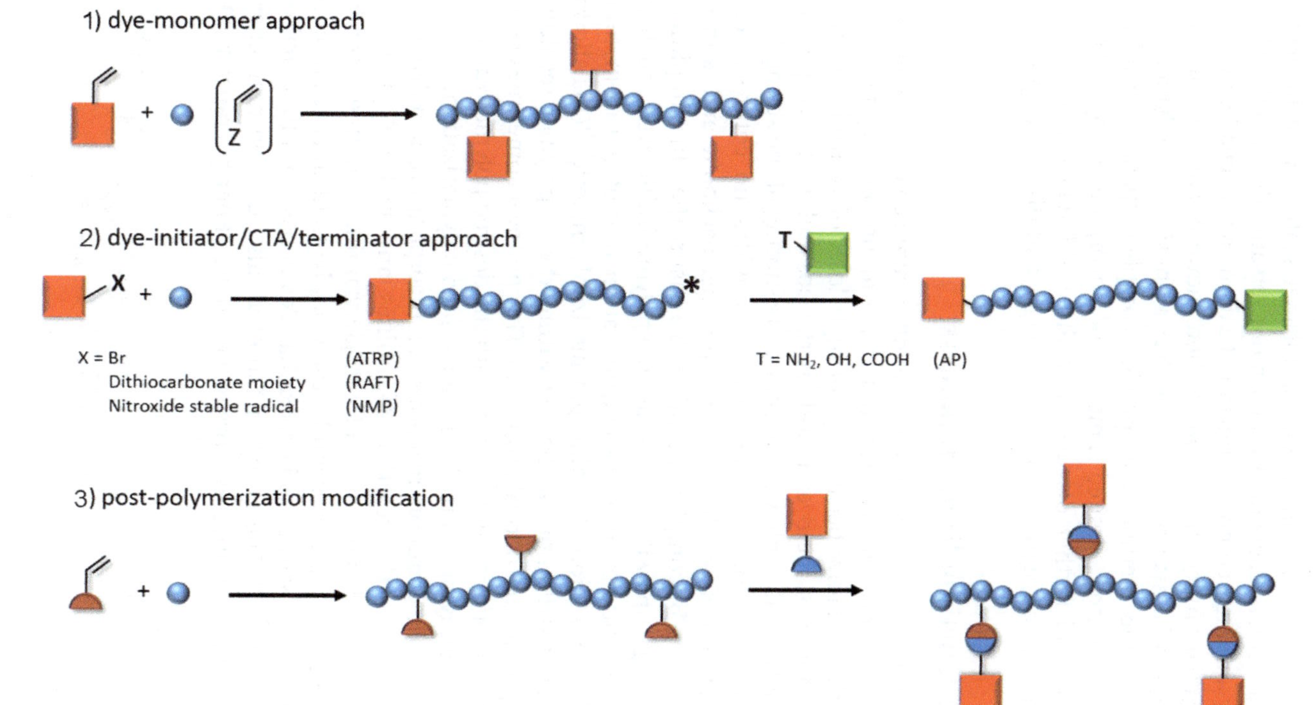

Figure 7.2   Overview of the different dye-incorporation approaches whereby the dyes are represented as squares and monomers as single spheres.

While the first method allows for easy control of the ratio between the co-monomers, *i.e.*, the dye content, it requires a polymerizable dye that is compatible with the chosen polymerization technique. The final method bypasses this problem by starting from a general polymer scaffold with incorporated reactive co-monomers allowing a much more versatile dye functionalization methodology although complete functionalization can be challenging.[9] Both of these side-chain functionalization methods however suffer from a lack of control of the spatial distribution of the dyes alongside the polymeric chain, whereas the second method does provide complete spatial control at the expense of a severely limited amount of incorporated dye. The rest of this chapter will further discuss these three different dye-incorporation approaches as well as highlight their advantages and disadvantages.

7.2.1 Dye-functionalized Monomers

The first method to prepare polymeric temperature sensors by co-polymerization of a solvatochromic dye-functionalized monomer is by far the most reported method mainly because of its high level of control over the incorporated amount of the dye, as well as the possibility of incorporating a large quantity of the dye, allowing accurate fine-tuning of the strength of the generated signal.[32] However, this method provides no control over the spatial distribution of the dye along the polymer backbone and is in most cases considered to be completely random. To ensure that this simplification is correct, most groups will try to design the polymerizable dye in such a way that there is no difference in co-polymerization rate of the co-monomers involved by incorporation of similar polymerizable moieties. This allows for a homogenous distribution of the dye within the co-polymer resulting in defined and predictable optical properties. This however sometimes requires complicated synthetic routes to end up with the desired polymerizable group without the certainty of a completely random co-polymerization.

One of the most common reactions for incorporating a polymerizable group into a dye is the reaction of amines and hydroxyl groups with (meth)acryloyl chloride resulting in (meth)acrylamides and (meth)acrylates, respectively (Figure 7.3). For example, the work by Uchiyama *et al.* on polymeric thermometers with 2,1,3-benzoxadiazoles is based on this method of dye incorporation.[38–47] The group of Liu *et al.* also use the monomer approach for the incorporation of Förster Resonance Energy Transfer (FRET)

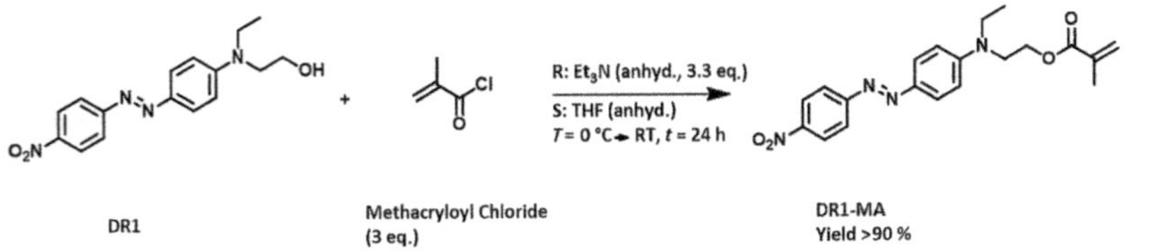

Figure 7.3   Reaction conditions for the synthesis of Disperse Red 1 methacrylate (DR1-MA) from Disperse Red 1 (DR1) reacted with methacryloyl chloride.[60]

donors and acceptors, including Rhodamine B acrylates, Rhodamine B acrylamides, spyropyran methacrylates and 2,1,3-benzoxadiazole acrylate into thermoresponsive polymers.[24,48–50] Other dyes functionalized by this method include tetraphenylethene (TPE),[15] 4-amino-1,8-naphthalimide,[51] pyrene,[42,59] *N*-carbazole,[30,52,53] 9-(4-*N*, *N*-dimethylaminophenyl) phenanthrene (DP),[54,55] coumarins[56,57] and Disperse Red 1 (DR1).[58–60]

Styrenic dyes can easily be synthesized through a reaction with 4-vinyl-benzyl chloride as was reported by Schubert *et al.* in 2012 for 4-hydroxy-1,3-thiazole blue-emitting solvatochromic dyes on a gram scale, using a one-step synthesis.[61] This was a similar method to that described by Koopmans and Ritter for a Reichardt's dye styrene monomer.[62]

More complex synthetic routes include the synthesis of the D–π–A pyran-based dyes.[63–66]

7.2.2 Dye-functionalized Initiators/Terminators/Transfer Agents

This dye-incorporation approach is not commonly used although it allows for unique control of the positioning of the dye. By careful design of the reagents used, one or two different dyes can be selectively incorporated at the α- and/or ω-chain end of the polymer chain.[31] The main drawback of this approach is that the amount of incorporated dye is limited to one or two per polymer chain and in some cases the dye-functionalization requires multiple reaction steps. When employing this approach, the identity of the functional group (X– or T– in Figure 7.2) is strongly dependent on the polymerization technique.

Perhaps the most straightforward CRP method for this approach is ATRP, requiring a secondary or tertiary alkylbromide group on the dye to successfully initiate the polymerization, leading to incorporation of the dye at the α-chain end. An example of this has been reported by Wang *et al.*, where they used a tertiary-bromide-bearing polyfluorene as an initiator in order to create a block co-polymer with *N*-isopropylacrylamide (NIPAAm).[67] Kakuchi *et al.* synthesized PNIPAAm of different lengths bearing a pyrene α-end-group *via* ATRP initiated by 1-pyrenyl-2-chloropropionate (Figure 7.4). Their investigation of the temperature-responsive properties of the polymer was,

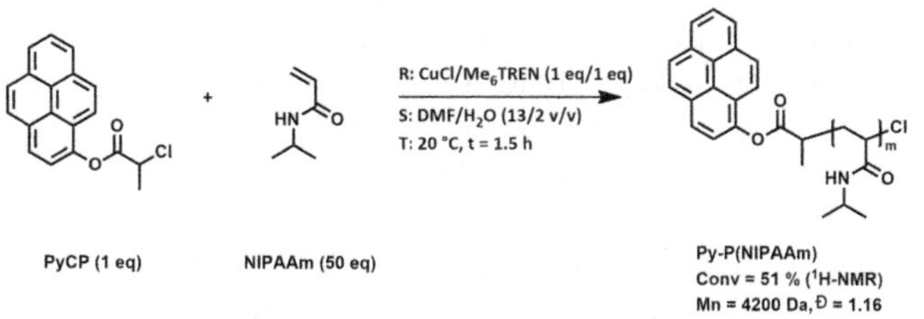

Figure 7.4 Polymerization conditions for a 1-pyrenyl-2-choropropionate initiated ATRP of NIPAAm.[68]

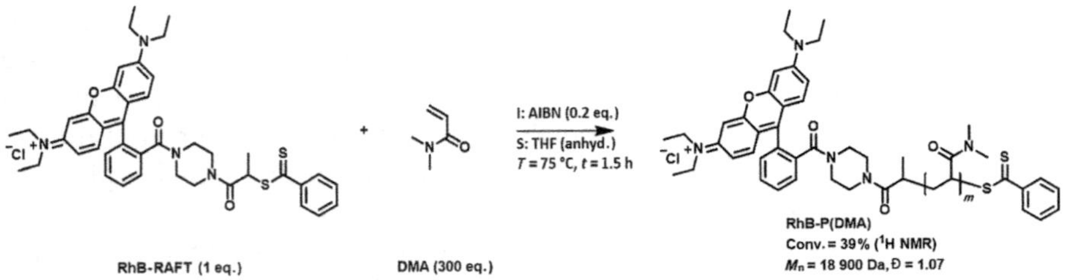

Figure 7.5 Synthesis conditions for the RAFT polymerization of DMA using an RhB-functionalised RAFT agent.[71]

however, restricted to the effect of the hydrophobic end-group in regard to the polymer length.[68] Another example was reported by Akiyama and Tamaoki for the ATRP of NIPAAm using an azobenzene derivative containing a 2-chloropropionyl group.[69]

RAFT polymerization can also be employed to introduce a dye at the α-chain end by functionalization of the leaving R group of the chain-transfer agent (CTA). Martinho *et al.* employed a tailor-made biotin-containing CTA in order to prepare water-soluble polymers with 95% biotin end-functionality.[70] They also synthesized CTAs bearing a malachite green, Rhodamine B (RhB) (Figure 7.5) or phenanthrene moiety to prepare polymers that bear these dyes at the α-end group.[71,72] A similar methodology was used by Liu and co-workers to synthesize an α-chain end pyrene-functionalized NIPAAm–mOEGMA co-polymer.[73]

Although the functionalization of the ω-chain end is mostly reported using the post-polymerization modification approach, some reports are available where a dye is used to terminate the polymerization, effectively functionalizing the ω-chain end. This one-step approach is mostly used in cationic or anionic polymerizations (AP).[31] Schubert *et al.* have used this approach to synthesize fluorescein-terminated poly(2-oxazoline)s.[74]

7.2.3 Post-polymerization Modification

This final dye-incorporation approach is based on the incorporation of reactive functional groups into the polymer either as side-chains or on the chain ends that are inert towards the polymerization conditions. Subsequently, these reactive moieties can be transformed into a wide variety of other functional groups by post-polymerization modification, including various dye molecules.[33] Since positioning of dyes at both the side-chains and chain ends is possible by this approach, most of the advantages and disadvantages that are discussed in the previous section are also valid for this approach. Nonetheless, the most commonly used reactive co-monomers/initiators/CTAs are much more easily available, either commercially or through synthesis, compared to their dye-functionalized analogues. The main advantage of this final method is that the reactive co-polymer can be used as a versatile platform for functionalization with a wide variety of dyes.

Multiple organic reactions have been used for this post-polymerization modification approach including click-reactions,[33–36,75] esterifications of activated esters,[37] DCC coupling[76] or even simple etherifications.[77–80]

Zhang *et al.* for example used the copper(I)-catalysed azide–alkyne cycloaddition (CuAAC) reaction for the attachment of azide-containing pyrene to an alkyne-containing poly(methoxy oligoethylene glycol methacrylate).[26]

Another common post-polymerization modification technique involves the reaction of incorporated activated esters, mostly bearing a succinimide or pentafluorophenyl (PFP) group with amine-containing dyes. These activated esters are available in a wide variety of monomers including *e.g.*, (meth)acrylates, styrenics and norbornenes, but this method can also be employed to incorporate the dye on a different position by using activated ester initiators and/or chain-transfer agents.[37] Hirai *et al.* were able to synthesize RhB-containing co-polymers by a very simple post-polymerization modification of the co-polymer NIPAAm with *N*-acryloxysuccinimide (NASI). The activated ester was subsequently used to label the thermoresponsive co-polymer with RhB functionalized with diethylenetriamine or an amine-containing anthracene.[81–83] Winnik *et al.* used the same poly(NIPAAm-*co*-NASI) platform to label the co-polymer through reaction with 4-(1-pyrenyl)-butylamine hydrochloride (Figure 7.6),[12,13,84] while Solovyov *et al.* used it for the incorporation of a porphyrin derivative.[85]

Zentel *et al.* used a pentafluoro-bearing CTA to synthesize a methoxy diethylene glycol acrylate (mDEGA) co-polymer that was modified afterwards with an Oregon Green derivative at the α-chain end (Figure 7.7). Since both chain ends are sensitive towards amines, Oregon Green was added in equimolar quantities in the presence of 1,8-bis(dimethylamino)naphthalene as proton sponge, resulting in a relatively low PFP conversion of 71.1%. The ω-chain was subsequently modified by aminolysis of the remaining dithiobenzoate moiety and disulphide formation between the formed thiol and Texas Red methanethiosulfonate.[86] A similar RAFT agent was used for the α- and ω-chain end modification of an mOEGMA polymer with

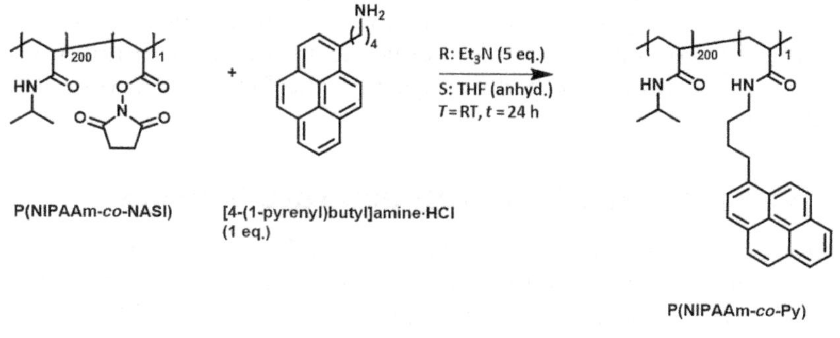

Figure 7.6　Post-polymerization modification conditions for the labelling of a poly(NIPAAm-*co*-NASI) co-polymer with an amine-functionalized pyrene dye.[13]

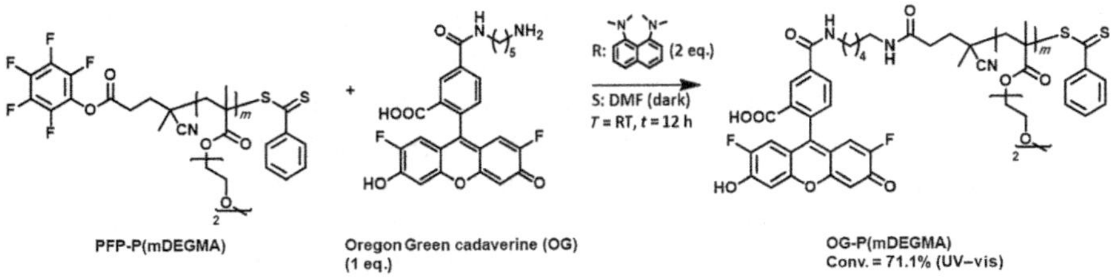

Figure 7.7 Post-polymerization modification conditions for the labelling of poly(mDEGMA) with Oregon Green at the α-chain end using a PFP-modified RAFT agent.[86]

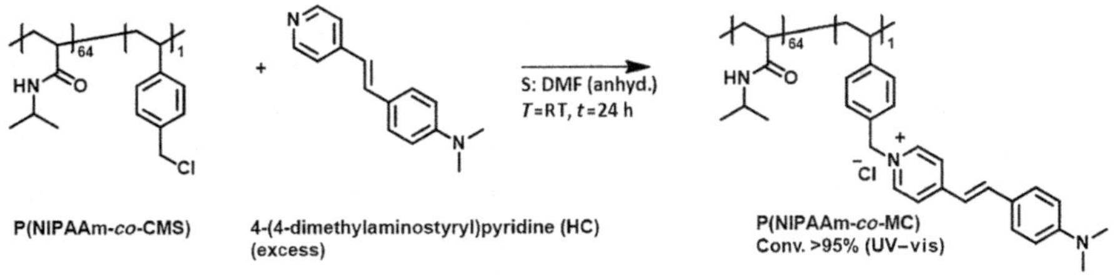

Figure 7.8 Post-polymerization modification conditions for the labelling of poly(NIPAAm-*co*-CMS) with 4-(4-dimethylaminostyryl)pyridine.[77]

azobenzene moieties.[87] Alternatively, post-polymerization modification of amino-functionalized polymers can be performed with dyes containing activated ester. For example, Pluronic F127 (PF127) was end-functionalized by post-modifying the hydroxyl groups to amines using 4-nitrophenyl chloroformate and diethylamine, making the PF127 reactive towards the NHS-activated ester form of Cy5.5.[88]

Solovyov also synthesized an active pre-polymer to attach porphyrin units to the side-chains by using allybromide or 4-bromobutyl acrylate that was co-polymerized with NIPAAm using FRP. subsequently, the co-polymer showed strong fluorescence enhancement upon phase transition resulting from the reduced intramolecular interactions inside the polymeric globule.[79,80] Hirai *et al.* exploited hemicyanine (HC) as a solvatochromic dye by attaching it to a co-polymer of NIPAAm and chloromethylstyrene *via* its pyridinium group (Figure 7.8).[77] A similar method was used for the functionalization of Rose Bengal to PNIPAAm.[78]

7.3   Polymers used for Temperature Sensors

Polymers used for temperature sensors can simply act as a physical support for small-molecule temperature probes to enhance their stability and processability;[89] however, increasing attention has been paid to complex thermometers based on the combination of solvatochromic dyes and thermoresponsive polymers that respond sharply to environmental temperature changes with a solubility phase transition (Scheme 7.1).[8,90] In this chapter,

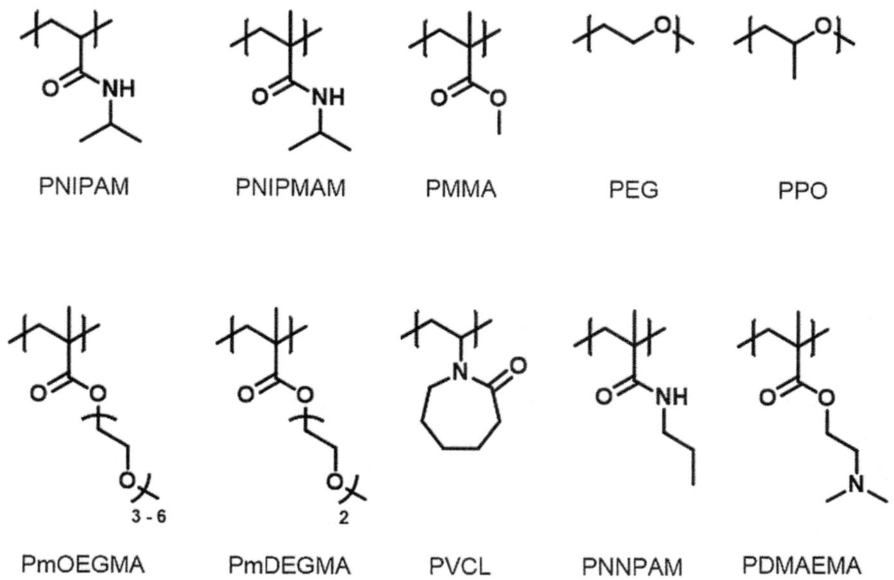

Scheme 7.1   Chemical structures of thermoresponsive polymers employed for the synthesis of polymeric thermosensors described in this chapter.

we mainly focus on the latter system where the thermosensitivity of the polymers is responsible for the temperature sensing. For the development of polymeric temperature sensors in which the polymer acts only as the immobilizing matrix of sensory probes, the reader is referred to a recent review[91] and another chapter in this book where specific temperature probes are discussed.

7.3.1   Classification of Polymers used for Temperature Sensors

The polymeric thermometers as discussed in this chapter are based on thermoresponsive polymers, *i.e.*, polymers that undergo a reversible phase transition from a molecularly dissolved hydrated state in aqueous solution (hydrophilic) to a dehydrated state (hydrophobic) in response to temperature changes (see Scheme 7.2). This sharp coil-to-globule transition has a strong influence on the microenvironment of the repeating units of the polymer as the (majority of) water molecules are expelled from the collapsed globules during the phase transition and, therefore, a hydrophilic-to-hydrophobic change occurs in the microenvironment of the polymer. By attaching a solvatochromic chromophore to the polymer chain, this microenvironmental polarity change can be translated into a colorimetric or fluorescent sensing signal.

There are two types of thermoresponsive polymers: when the phase separation occurs at elevated temperatures, this is referred to as *LCST behaviour* while the reversed-phase behaviour is known as *upper critical solution temperature (UCST) behaviour*. The phase transition is often accompanied by a

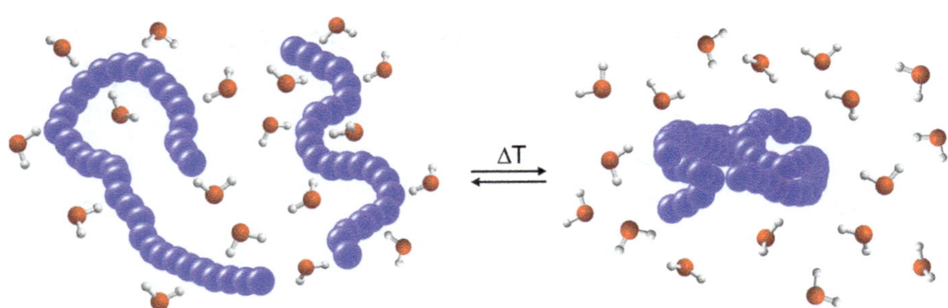

Scheme 7.2 Scheme showing a thermoresponsive polymer reaction in aqueous solution.

transition from a clear solution to a cloudy solution, and the temperature at which this transition occurs is called the *cloud point temperature* (T_{CP}). In general, the LCST phase transition is an entropic event driven by the release of hydrated water molecules upon heating, while the UCST transition is an enthalpic event driven by stronger interpolymer attraction at lower temperatures.

The most commonly studied and first-reported thermoresponsive polymer in aqueous solution is PNIPAM with an LCST of *ca.* 32 °C,[92,93] which is close to human body temperature. Its phase-transition temperature is relatively insensitive to changes in concentration and pH making it quite robust. Moreover, the extremely low toxicity of PNIPAM has made its use for biomedical applications possible. Hence, it is not surprising that PNIPAM is the most popular thermoresponsive polymer for developing polymeric temperature sensors.[10,13,39,40,44,55,85,94–96] In addition to PNIPAM, other analogous of polyacrylamides, *e.g.*, poly(*N-n*-propylacrylamide) (PNNPAM) and poly(*N*-isopropylmethacrylamide) (PNIPMAM), have also been reported as a basis for temperature-sensing co-polymers.[39,77,97,98] Both homo- and co-polyacrylamides exhibit very high glass-transition temperatures (T_g), which have been reported to lead to vitrification of the highly concentrated polymer phase during phase separation, potentially inducing hysteresis between heating and cooling.[99] This translates into a typical limitation of PNIPAM as a polymeric matrix for a thermometer, *i.e.*, differences in the output for the same temperature value, depending on whether the temperature is increasing or decreasing.

Recently, poly(oligoethylene glycol (meth)acrylate)s (POEG(M)A)s with low T_g values have been introduced as thermoresponsive alternatives to PNIPAM showing excellent reversibility of the phase transition without hysteresis (see Figure 7.9).[100–104] In addition, the phase-transition temperature can easily be tuned by co-polymerization of different commercially available OEG(M)A monomers. These polymers, furnished by 'stealthy' oligo ethylene glycol side-chains, also show superb biocompatibility making them very well suited for biomedical applications, similar to PNIPAM. Hence, POEG(M)As have frequently been employed as thermoresponsive polymeric matrices for thermometers in recent years.[52,56–58,60,105]

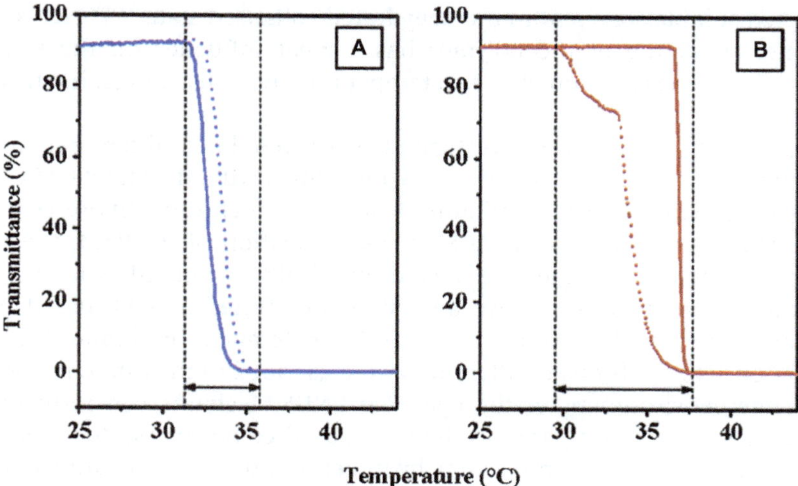

Figure 7.9   Transmittance as a function of temperature measured for aqueous solutions of (A) poly(mDEGMA-*co*-mOEGMA) and (B) PNIPAM. Solid lines are heating cycles, and dotted lines are cooling cycles.
Reproduced with permission from ref. 102. © American Chemical Society 2006.

Apart from PNIPAM and POEG(M)A, various other LCST polymers have also been reported as polymeric matrices for thermometers, such as poly [2-(*N,N*-dimethyl amino)ethyl methacrylate] (PDMAEMA),[95] poly(propylene oxide) (PPO)[88] and poly(*N*-vinylcaprolactam) (PVCL).[28]

However, the temperature-sensing regime of LCST-based sensors is often limited to a narrow temperature range (around 10 to 20 °C) due to the sharp entropic LCST phase transition. Instead of the entropy driven phase separation for LCST polymers, polymers with UCSTs are mostly driven by enthalpy, causing a broader transition range. As a result, temperature sensors based on UCST polymers are expected to give a broader temperature-sensing regime compared with LCST polymers.

However, polymers with UCST transitions have been rarely used for thermometers, possibly due to the fact that polymers that show UCST behaviour in aqueous solution are quite rare.[106] Nevertheless, based on the UCST transition of poly(methyl methacrylate) (PMMA) in ethanol/water solvent mixtures,[107–109] Hoogenboom *et al.* have reported temperature sensors by incorporating DR1 or pyrene into PMMA.[59] The sensors indeed had a much broader temperature-sensing regime compared to those based on LCST polymers.

7.3.2   Chemical Structures of Thermoresponsive Polymers for Thermometers

The chemical structure of the polymer used in thermometers acts as the chemical environment of the incorporated solvatochromic dyes, both in the hydrated soluble state and in the dehydrated collapsed state.[51,60] Hence, the chemical structure of the polymer has a great influence on the sensory properties of the thermometer, *e.g.*, temperature response range, intensity of fluorescence.

Laschewsky *et al.*[51] have investigated structure-related differences or solvatochromic shifts in sensing behaviour for PNIPAM, PmOEGMA and PmOEGA modified with naphthalimide as a solvatochromic fluorescent dye (see Figure 7.10). Due to the LCST phase transition of PNIPAM, the fluorescence intensity of the dye was dramatically increased, whereas the emission properties of the dye are rather unaffected as OEG-based polyacrylates and methacrylates undergo their temperature-induced phase transition. These observed differences were ascribed to the difference of the local microenvironment of the dye. The PNIPAM chains can form much denser aggregates, compared to the OEG-based co-polymers, due to strong intramolecular and intermolecular hydrogen bonds at temperatures above the LCST. In addition, the hydrophilic OEG side-chains of POEG(M)As hinder efficient dehydration of the collapsed polymer globules upon heating, resulting in a smaller change in the polarity of the microenvironment of the dye. Similar conclusions were made when employing a different solvatochromic fluorescent dye, namely 7-(diethylamino)-3-carboxy-coumarin (DEAC).[56] The dye is mostly insensitive to the polymer phase transition of dye-functionalized poly(mDEGMA$_{81}$-*co*-mOEGMA$_{19}$). In contrast, PNIPAM-based sensors exhibit a drastically increased emission after heating above the LCST.

The influence of polymer chemical structure on sensing behaviour has been reported by Hoogenboom *et al.*[60] The authors have reported that PmDEGMA decorated with DR1 reveals dual pH- and temperature-sensing behaviour with both an absorption intensity and peak shift output. When replacing the di(ethylene glycol) side-chains with more hydrophilic OEG side-chains, the temperature-induced absorption peak shift of the incorporated polarity sensitive dye was lost, ascribed to the less efficient dehydration of the more hydrophilic PmOEGMA.

Ionic components can be introduced into thermoresponsive polymeric materials *via* monomer or initiator in the expectation that electrostatic repulsion would prevent the formation of large aggregates in the dehydrated state, which can cause hysteresis between heating and cooling. Gota *et al.*[46] have investigated the influence of incorporating an ionic component into fluorescent polymeric thermometers. Fluorescent polymeric thermometers consisting of only *N*-alkylacrylamide and the fluorescent dye show rather low temperature resolution in their functional ranges due to the occurrence of interpolymer aggregation, whereas much a high temperature resolution ($<0.2\,^{\circ}\mathrm{C}$) is obtained by adding an ionic component during co-polymerization. Such an ionic fluorescent nanogel was then applied for intracellular thermometry in living cells.[38,47] With increasing temperature, the thermoresponsive ionic gel produces stronger fluorescence in the cytoplasm, allowing the monitoring of temperature

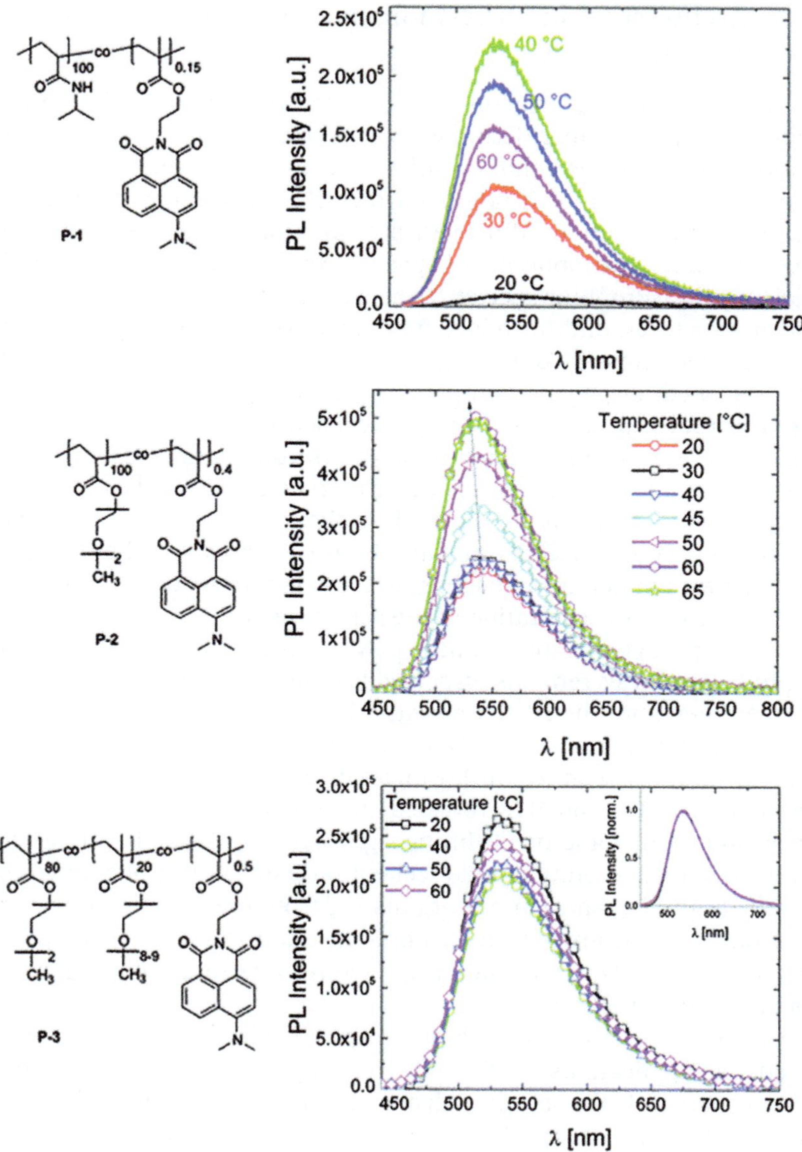

Figure 7.10 The fluorescence spectra of naphthalimide-functionalized PNIPAM, PmDEGA and poly(mDEGMA-*co*-mOEGMA) in PBS (0.1 g L^{-1}) at various temperatures.
Reproduced from ref. 51.

differences of less than 0.5 °C without any interference due to precipitation or interaction with cellular components. An even higher temperatures, resolution was achieved by the application of fluorescence lifetime imaging as the output signal.

7.3.3 Architectures of Thermoresponsive Polymers for Thermometers

The simplest polymer structure for use in thermometers is a thermo-responsive homo-polymer decorated with a solvatochromic dye. In real applications, however, the solution will be become turbid during sensing due to the aggregation of such homo-polymer chains, leading to lower transmittance of the solution. Moreover, precipitation of the co-polymer could happen when the hydrophobic collapsed polymer globules are not stabilized sufficiently by hydrophilic species. To avoid such drawbacks of simple homo-polymer structures, and introduce functionalities, more complex polymeric structures, like cross-linked particles, block and brush co-polymers, have been developed for polymer temperature sensor applications.

Dye-incorporated nano- or microparticle sensors[44,54,55,110,111] are usually prepared by (mini-)emulsion radical polymerization in the presence of cross-linkers and ionic species. The resulting particles, dispersed in aqueous solution, undergo temperature-induced swollen-to-shrunken transitions, as normally noticed by a decrease in particle size. Such systems have frequently been used for temperature imaging, as the phase-transition temperature is impervious to the concentration of the sensory materials. Moreover, instead of macroscopic aggregation of linear polymer chains, the shrunken state of cross-linked particles remains dispersed leading to a higher reversibility. For instance, Chen and Chen[111] have synthesized a cross-linked PNIPAM-based fluorescent nanothermometer known as Thermo-3HF (where 3HF refer to 3-hydroxyflavones) (Figure 7.11). Single-run and multiple-run reversibility experiments of Thermo-3HF revealed no hysteresis during the cycles of heating and cooling and no declining signals during multiple-run tests. The high temperature resolution and excellent reversibility was attributed to the highly hydrophilic, ionic surface originating from the anionic initiator used for the radical emulsion polymerization. The electrostatic repulsion prevents individual nanogels from incurring severe intermolecular aggregation through electrostatic repulsion.

Block co-polymers with one thermoresponsive block are also interesting for sensing applications.[58,67,88,112–114] The dye can be incorporated into different parts of the block polymer chain making the polymer very robust for different tasks. Han *et al.*[112] have reported the sensory amphiphilic co-polymer, poly(NIPAM-*b*-1-(4-vinylbenzyl)-2-naphthyl-benzimidazole) (poly(NIPAM-*b*-VBNBI)). The co-polymer can self-assemble into micelles in aqueous solution with the hydrophobic-dye-containing block as the core, and the thermoresponsive block (PNIPAM) as a corona. The system exhibited a reversible fluorescent response to changes in temperature around the LCST of PNIPAM due to a coil-to-globular induced microviscosity increase of the dye microenvironment. Another amphiphilic block co-polymer, poly(St-*co*-NBDAE-*co*-SPMA)-*b*-poly(NIPAM-*co*-RhBAM), bearing NBDAE and SPMA moieties in the hydrophobic polystyrene (PS) block, and RhBAM moieties in the thermoresponsive PNIPAM block was reported by Liu *et al.*[24] A sensitive

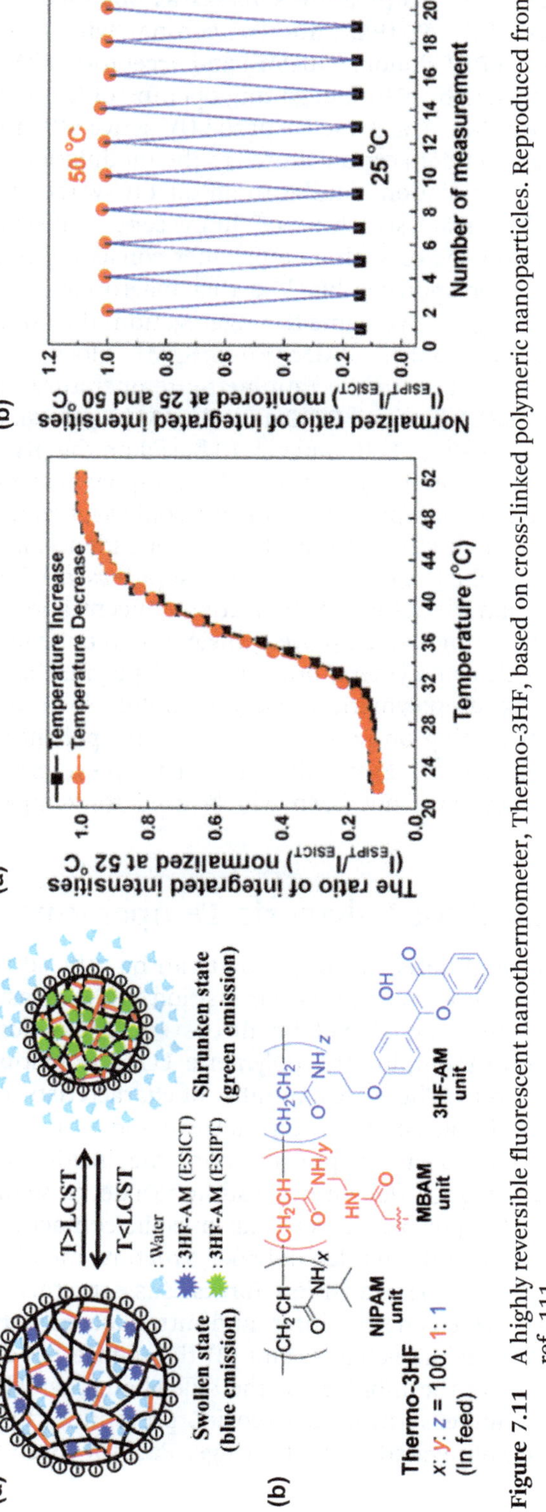

Figure 7.11 A highly reversible fluorescent nanothermometer, Thermo-3HF, based on cross-linked polymeric nanoparticles. Reproduced from ref. 111.

ratiometric fluorescent temperature sensor was obtained through thermo-induced collapse of the PNIPAM micellar corona owing to the closer proximity between the FRET donors (NBDAE) and acceptors (RhBAM). The FRET could be also adjusted by pH-induced ring-opening of RhBAM leading to pH sensing by the micelles. The presence of the UV-active SPMA moieties in the core of the micelle renders extra features to the modulation for multicolour emission by light irradiation, similar to an on–off switch. The principal of this reversible three-state switching of fluorescence emission is shown in Figure 7.12. A drawback of such a co-polymer consisting of a hydrophobic block and a thermoresponsive block is that macroscopic aggregation will occur during sensing. To overcome this aggregation, the double hydrophilic poly(NIPAM-*co*-FITC)-*b*-poly(mOEGMA-*co*-RhBAM) block co-polymer was synthesized.[25] In addition to the ratiometric temperature sensory manner originating from FRET between FITC and RhBAM at pH <6, the co-polymer is also capable of sensing pH in range 2–10 based on the pH responsiveness of the two dyes. Such double hydrophilic block co-polymers go from unimers to micelles during sensing preventing macroscopic aggregation. A drawback of such systems is their strong concentration dependence as this influences the micellization, which can even be fully suppressed when the concentration is lower than the critical micellization concentration.

In addition to the mostly employed linear homo-polymers, cross-linked particles and block co-polymers, other types of polymer architecture have also been reported as polymeric sensory scaffolds, such as a brush polymer,[27] a polymer coating on nanoparticles,[48] a star polymer,[115] and a dendronized co-polymer.[116] Even though such polymers bear unique functionalities, they have not been widely used for temperature sensors to date.

7.4 Dyes used for Polymeric Temperature Sensors

In this chapter, the different concepts that can be utilized to translate the temperature-induced polymeric phase transition into a measurable UV and/or fluorescent output signal will be discussed. The reported polymeric thermometers mostly employ the polymeric LCST transition as the temperature-sensing event. There are a number of characteristic changes during this LCST polymeric phase transition that can serve to induce the output signal (Scheme 7.3): (1) upon precipitation, the inside of the collapsed polymeric globule strongly dehydrates leading to a decrease in polarity; (2) at low temperatures the polymer tries to maximize its contact with the solvent, taking on a very open and extended polymer structure, mostly represented as a random coil. The LCST transition forces this structure to become increasingly dense, decreasing the inter- and intradistance between the polymer backbone and its side-chains; and (3) this dense polymeric structure also strongly hinders the mobility of the side-chains, which is sometimes referred to as an increase in local viscosity, preventing, for example, free rotation between conjugated aromatic rings. Each of these three concepts

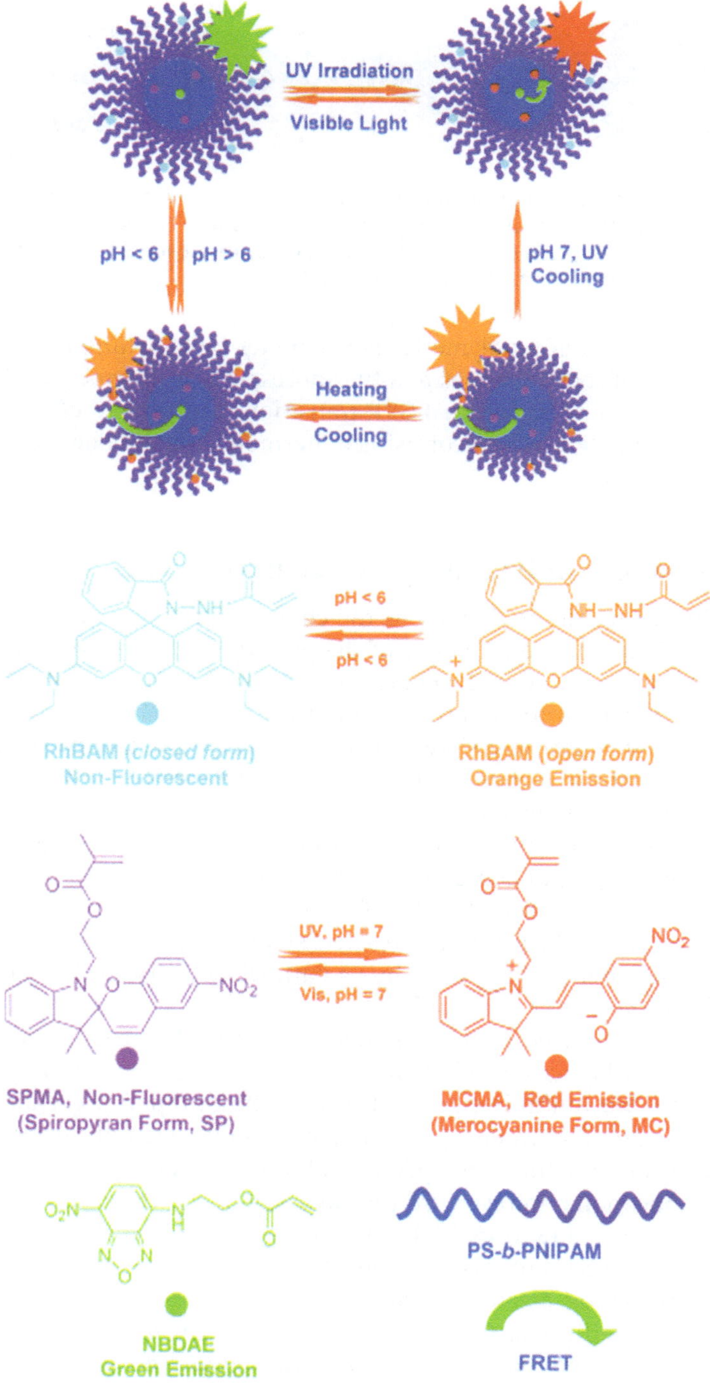

Figure 7.12 Schematic representation of a block co-polymer based reversible three-state switchable multicolour luminescent system from poly(St-*co*-NBDAE-*co*-SPMA)-*b*-poly(NIPAM-*co*-RhBAM).
Reproduced with permission from ref. 24. © 2010 WILEY-VCH Verlag GmbH & Co. KGaA, Weinheim.

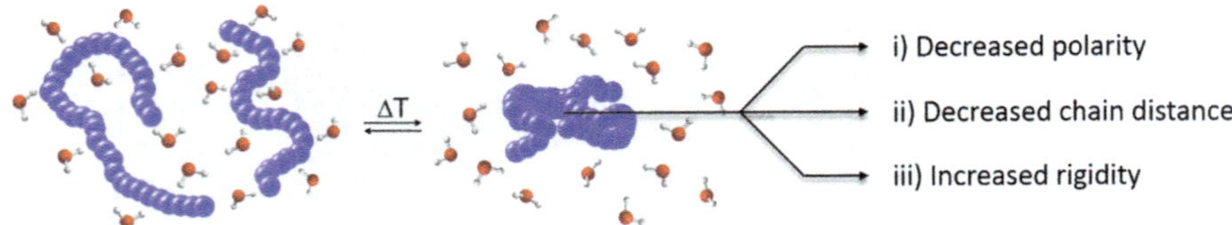

Scheme 7.3   Schematic representation of a polymer phase transition in aqueous solution for an LCST polymer if $\Delta T >0$ or a UCST polymer if $\Delta T <0$ and a description of the significant changes in the local environment of the polymer for an LCST transition.[104]

can form the basis of a polymeric thermometer by attaching a dye that responds to any of these changes to a thermoresponsive polymer. While these concepts are described for an LCST transition, the reverse effects can be employed for a polymeric sensor using a thermoresponsive polymer showing a UCST transition.

7.4.1   Changes in Polarity: Solvatochromic Dyes

Solvatochromism can be defined as the change in fluorescence/absorbance of a molecule under the influence of the polarity of the microenvironment. This effect was reviewed by Reichardt in 1994, and its origin was attributed to the uneven stabilization of the ground state (D) and the excited state (D^*) of a dye. As schematically shown in Figure 7.13, this unequal solvent (S) stabilization leads to a change in the energy difference (ΔE) and therefore a shift in the maximum absorption, excitation or emission wavelength. If with increasing solvent polarity ($P_{(S)}$) the ground state is more stabilized, this will increase the energy gap between the ground state and the first excited state leading to a hypsochromic (blue) shift or so-called *negative solvatochromism* (Figure 7.13, left). When the first excited state is stabilized more, this will lead to a smaller energy gap and thus a bathochromic (red) shift or *positive solvatochromism* (Figure 7.13, right).[7]

This polarity induced change in energy between the ground and excited states greatly influences the electronic structure of the dye and can therefore be monitored *via* multiple signals. These potential reporter events include both the position and intensity of the maximum wavelength in the absorbance, excitation and emission spectra and also the fluorescence lifetime.

Most publications however use the intensity of the fluorescent signal as the reporter event due to its higher sensitivity, as well as the possibility to readout the fluorescent signal from living matter. These fluorescence changes (quenching or enhancement) are however also dependent on the experimental conditions, such as the excitation light intensity and the solution turbidity, potentially leading to scattering and concentration of the fluorescent thermometer.[45] Recently, Neher *et al.* demonstrated that the polarity of the co-monomer side-chains can also greatly influence the extent of the fluorescence intensity change. They showed that by careful design of a

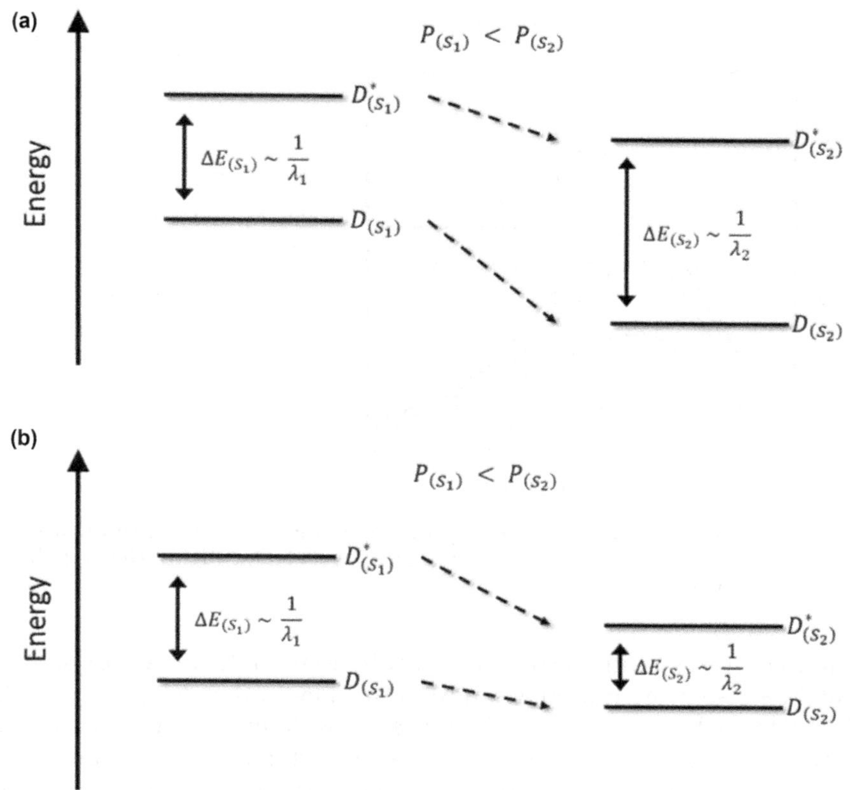

Figure 7.13 Schematic representation of the solvatochromic effect, in which the decreased polarity has a stabilizing effect on the molecule. (a) Hypsochromic (blue) shift or negative solvatochromism. (b) Bathochromic (red) shift or positive solvatochromism.

1,8-naphthalimide-decorated thermoresponsive co-polymer, the extent of fluorescence enhancement could be controlled by the identity of the co-monomer. Whereas the signal for a poly(NIPAAm) co-polymer showed an enhancement factor of 25, the fluorescence was only doubled for a poly(mDEGA) co-polymer and was even weaker for a poly(mOEGMA) co-polymer. The hydrophilic OEG side-chains in the latter two cases most likely prevent complete dehydration of the precipitated polymeric globule, causing the decrease in fluorescence enhancement.[51]

The concept of using the solvatochromic properties of a dye molecule to translate the polymeric phase transition was first reported by Iwai *et al.* They reported that by incorporating only 0.1 mol% of a benzofuran-based monomer into PNIPAAm *via* FRP, the resulting co-polymer would show a temperature-dependent fluorescence intensity enhancement when the co-polymer collapsed from coil-to-globule, which was shown to be similar to the effect that decreasing solvent polarity has on the dye molecule (Figure 7.14).[40] In their first publication in 2003 this concept was proven and

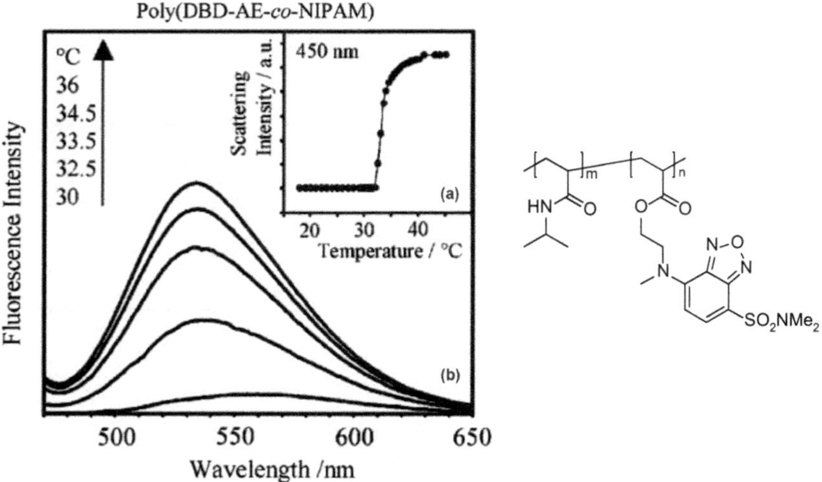

Figure 7.14   Left: (a) Scattering intensity and (b) fluorescence spectra (normalized at 30 °C) of poly(DBD-AE-*co*-NIPAM) (λ_{ex} 444 nm). Right: The co-polymer structure.[40]

optimized, starting with the dye monomer in terms of fluorescence quantum yield. The co-polymerization with PNIPAAm showed that 4-*N*-(2-acryloyloxyethyl)-*N*-methylamino-7-*N*,   *N*-dimethylaminosulfonyl-2,1,3-benzoxadiazole (DBD-AE) outperformed similar dyes with a fluorescence enhancement factor of 13.3. Control experiments revealed that both the incorporation of the dye as a co-monomer and the temperature-responsive behaviour of the resulting co-polymer are necessary requirements for the temperature-dependent fluorescence enhancement.[40]

The versatility of this benzofuran-based thermometer concept was highlighted by using different acrylamide monomers or combinations thereof to successfully change the temperature-sensing range while retaining the high sensitivities and reproducibilities observed for PNIPAAm.[39,40] This concept was then further extended to a dual-responsive polymeric AND logic gate using pH as a second stimulus. The co-polymerization of DBD-AE with *N,N*-dimethylaminopropylacrylamide (DMAPAM) and *N-t*-butylacrylamide yielded an LCST co-polymer that is cationically charged at pH 8 and lower ($pK_a \approx 9.8$). The partial ionization at pH 8 results in a shift of the LCST towards higher temperatures and a decrease of the fluorescence enhancement through the prevention of intermolecular aggregation. At pH 5, this aggregation is completely eliminated resulting in a small but stable fluorescence signal.[42] In 2005 they reported the emulsion co-polymerization of DBD-AE with different acrylamide-based monomers to create fluorescent microgel thermometers ($D = 100–250$ nm).[44,117] A year later Uchiyama *et al.* continued this work by replacing the ester linkage in the benzofuran with a more stable acrylamide linkage and co-polymerizing it with an extra ionic co-monomer. The added 3-sulfopropyl acrylate (SPA) prevented the aggregation

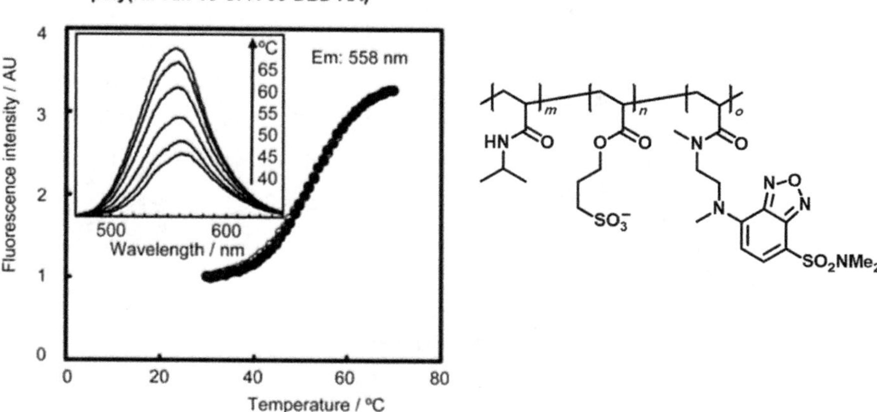

Figure 7.15   Left: Fluorescence response as a function of temperature for P(NIPAM-*co*-SPA-*co*-DBD-AA) (0.01 w/v%) in water ($\lambda_{ex} = 456$ nm). Right: The co-polymer structure.[46]

of the co-polymer upon heating, decreasing the fluorescence enhancement factor but widening the coil-to-globule transition over a broader temperature range. This improved both the active sensing temperature range to roughly 40 °C and the resolution of the fluorescence thermometer to below 0.2 °C (Figure 7.15).[46]

Other co-monomer combinations allowed the creation of an on–off pH switch[41] and a fluorogenic ion sensor[43] using the same reporter signal. Finally in 2008, they published their first fluorescent nanogel thermometer for *in vivo* use by cross-linking NIPAAm and DBD-AA in an emulsion polymerization. The use of a large amount of an ammonium persulfate initiator resulted in a highly hydrophilic surface decorated with charged sulphate groups, preventing precipitation upon temperature increase and non-specific interaction in the cytoplasm of the cell, while retaining the thermometry properties (Figure 7.15).[46,47]

Another polymeric thermometer utilizing solvatochromic dyes has been reported based on RhB derivatives. Hirai *et al.* showed that by decorating PNIPAAm with RhB, both the fluorescence and absorbance show a strong temperature-dependent increase. The biggest difference with the previously discussed benzofuran system is the additional fluorescence quenching upon further heating of the polymeric mixture, limiting the overall fluorescent signal to a 10 °C temperature window (Figure 7.16, left). This quenching at higher temperatures is ascribed to further polymeric coagulation leading to micrometre-size particles ($D \approx 300$ μm) that reduce the RhB excitation through scattering of the incident beam.[81,82] A similar fluorescence quenching mechanism involving the scattering of the incident beam has been reported for an *N*-carbazole-decorated thermoresponsive polymer.[30,52,53] The same group also reported a fluorescence thermometer using a hemicyanine dye as solvatochromic reporter. This dye showed an equilibrium shift from the

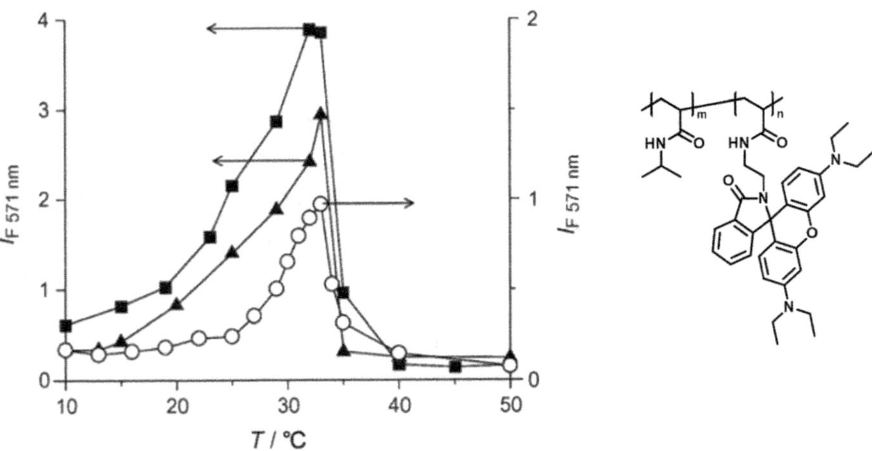

Figure 7.16   Left: Temperature-dependent   change   in   fluorescence   intensity ($\lambda_{ex} = 530$ nm, $\lambda_{em} = 571$ nm) of poly(NIPAAm-*co*-RhB-A) in water at different concentrations: squares represent 1.0 g L^{-1}, triangles represent 0.5 g L^{-1} and circles represent 0.2 g L^{-1}. Right: The co-polymer structure.[81]

non-fluorescent benzenoid form to the fluorescent quinoid form upon a decrease in polarity. The synthesized co-polymer with NIPAAm showed not only fluorescence enhancement but also a large increase in absorbance as a result of this equilibrium shift.[77]

A different type of fluorescent dye was reported by Kim *et al.* based on a donor–π–acceptor (D–π–A) conjugated fluorophore. Excitation of this dye leads to a large redistribution of the electron density or internal charge transfer (ICT), causing a non-radiative stabilization of the excited state.[118] This charge transfer is highly susceptible to changes in temperature, polarity and aggregation and is the origin of the solvatochromic effect in this type of D–π–A dye.[63] These multiple effects however complicate the predication and explanation of fluorescence (FL) intensity change upon phase transition of a labelled D–π–A co-polymer. In 2010 the synthesis of 4-(dicyanomethylene)-2-methyl-6-(*p*-(dimethylamino)styryl)-4*H*-pyran (DCM) and its co-polymerization with NIPAAm was reported. The resulting co-polymer showed remarkable fluorescence enhancement in aqueous solution upon heating, in addition to a FL blue-shift and absorbance increase due to the decrease in polarity.[65] Co-polymerization with spiropyran (SP) allowed for an extra UV–vis switch on the earlier reported fluorescence enhancement.[64] A simpler but also effective dye showing ICT and the corresponding solvatochromic behaviour in its fluorescence was reported by Yamamoto and co-workers. The 9-(4-*N*,*N*-dimethylaminophenyl)-phenanthrene (DMA-Phen or DP) methacrylate derivative was co-polymerized with NIPAAm and a cross-linker using FRP to yield a thermoresponsive hydrogel. Upon temperature-induced dehydration and collapse of the microgel, a large FL blue-shift and FL enhancement was observed (Figure 7.17).[54,55,117] Other dyes showing similar behaviour originating

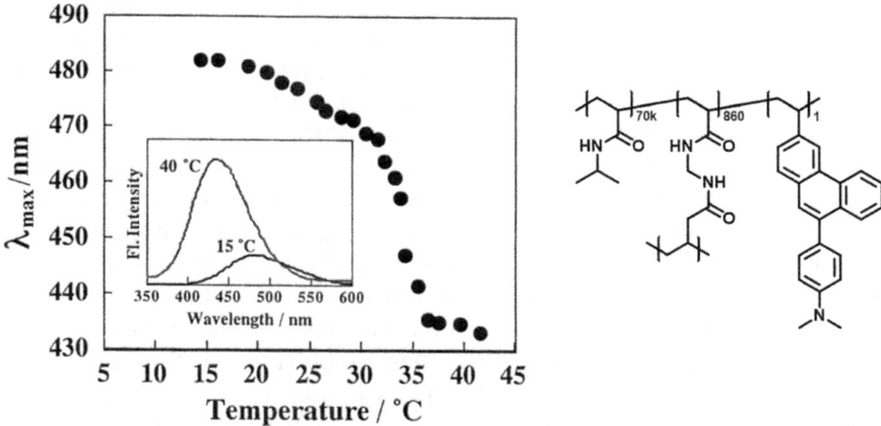

Figure 7.17 Left: Changes in fluorescence λ_{max} upon heating a DP-labelled PNIPAAm hydrogel and fluorescence spectra at 15 and 40 °C ($\lambda_{em} = 320$ nm).[55] Right: The polymeric hydrogel structure.

from ICT include aminocoumarin derivatives,[56] 1,8-naphthalimides,[51,110,119] amino-2,3-dimorpholinoquinoxaline[120] and (8-methoxy-4-methyl-2*H*-benzo[*g*]-chromen-2-one).[121] Interestingly enough, in 2009, Tian *et al.* published the synthesis of a similar dicyanomethylene-4*H*-pyran moiety and co-polymerized it with NIPAAm and the resulting co-polymer showed fluorescence quenching upon temperature increase in a water/ethanol mixture (5 : 1).[66] 3-Hydroxyflavones are another a class of dyes recognized for their ICT characterized by two distinct FL emissions at roughly 440 and 530 nm ($\lambda_{ex} = 355$ nm) corresponding to the tautomeric equilibrium within the molecule. This results in a FL red-shift upon polymeric LCST phase transition or a large increase in the ratio between the emission intensity at 530 nm to that at 440 nm.[111] 2-(2-Hydroxyphenyl)benzoxazole can also be coupled to thermoresponsive polymers in order to generate a ratiometric signal for a temperature readout.[122] Although both effects are equally applicable as reporter events, the presence of both FL quenching and enhancement for these similar dyes proves that their solvatochromic behaviour most likely results from a complex interplay of various effects.

Temperature-responsive co-polymers are also often used as tools for strengthening a sensor signal through fluorescence enhancement.[3] Liu *et al.* developed a RhB hydrazide based acrylamide monomer (RhBHA), which is non-fluorescent due to the presence of an internal spirolactam. Ring-opening under acidic conditions or in the presence of Hg^{2+} releases the fluorescence properties of the RhB.[123,124] The performance of the sensor was improved by co-polymerizing RhBHA with NIPAAm and using the fluorescence enhancement upon temperature-induced precipitation to boost the sensor signal.[58] Similar structures were designed by Liu *et al.* for the detection of Cu(ɪɪ) with temperature-controllable sensitivity. By emulsion co-polymerizing a NIPAAm with a Cu(ɪɪ)-binding unit like 1,10-phenanthroline, a thermoresponsive microgel was synthesized showing fluorescence quenching upon Cu(II) chelation. Upon temperature increase, the collapse of the particles led to both fluorescence enhancement and a more efficient analyte binding, leading to a more sensitive sensor.[125,126]

The cases described above all seem to prefer the change in fluorescence signal as the reporter event over the change in absorbance. While in some cases both responses are present, the preference for the FL signal is most likely due to its higher sensitivity as well as its potential applicability in biological media.[65,81,82] Nonetheless, an absorbance output signal might also be interesting as it facilitates a visual readout. One of the first groups to report a polymeric thermometer with an optically noticeable output signal was Koopmans and Ritter in 2007. Their co-polymer of NIPAAm and a (pyridinylidene)ethylidene-phenolate-type dye was able to change colour with variations in the pH, temperature and/or solvent polarity of the polymeric solution.[62] The azobenzene-type dyes are amongst the most used for polymeric thermometers with an absorbance output signal. These dyes are well known for their solvatochromic properties and for photo-isomerization between their *cis-* and *trans-*isomers, which has been used extensively for light-responsive polymeric materials.[87,127–132] Hoogenboom *et al.* have reported another application of a change in absorbance as the reporter event for a temperature increase. They used the absorbance intensity ratio of a DR1-containing co-polymer as the reporter event for either an LCST transition in a co-polymer with poly(mDEGMA) or a UCST transition in a co-polymer with methyl methacrylate (MMA).[59,60] The same dye was used by Zhang *et al.* in their thermoresponsive comb-like co-polymers of DR1-MA and the methacrylate derivative of first- or second-order ethylene glycol dendrimers. The dehydration of the thermoresponsive dendronized co-polymer upon temperature increase showed a characteristic bathochromic (red) shift indicative of a decrease in polarity of the microenvironment of the incorporated DR1.[27,116] Another possible absorbance output signal is the change in λ_{max} of the absorbance spectrum as reported by Hirai *et al.* for a NIPAAm-spiropyran-acrylate co-polymer (Figure 7.18). The decrease in polarity upon temperature increase caused a shift in the isomerization equilibrium of the dye from the zwitterionic to the quinodal form, resulting in a linear increase in λ_{max} as a function of temperature over quite a broad temperature range.[96] Spiropyran (SP) has also been reported to show a decrease in absorbance intensity upon a polymeric LCST phase transition.[133,134]

Finally, while the fluorescence intensity can be used as a reporter event for temperature sensors, it is also strongly dependent on experimental conditions. Therefore some publications have focussed on the use of fluorescence lifetime (calculated as the average decay time) as the output signal for sensing changes in temperature, pH or polarity depending on the type of dye. Since this reporter event is virtually independent of experimental conditions like dye concentration, beam intensity and photon path length, it is possible to make a universal calibration for such polymeric sensors depending on the identity of the incorporated dye.[45] Uchiyama used this

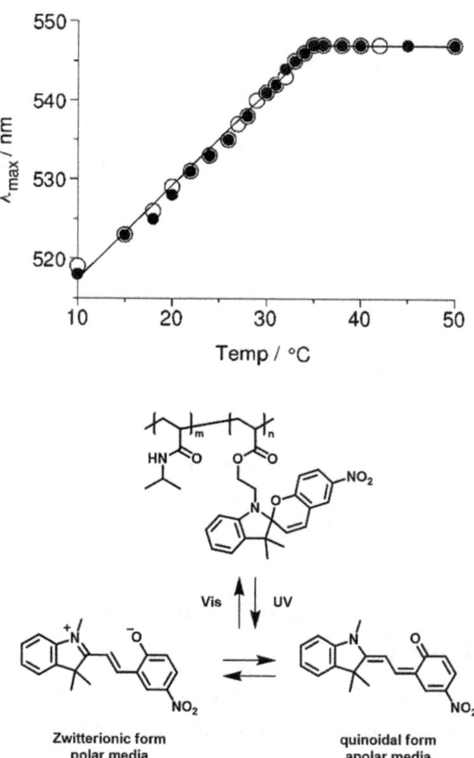

Figure 7.18 Top: Change in λ_{max} of poly(NIPAAm-*co*-SP) in water upon heating (open circles) and cooling (filled circles). Bottom: The co-polymer structure.[96]

reporter event for the measurement of intracellular temperature differences as seen in Figure 7.19 using a NIPAAm-benzofuran co-polymer. The co-polymer was excited at 275 nm and measured at 560 nm over a 100 ns time span. The resulting fluorescence decay curve could be fitted with a double exponential function of which the calculated average lifetime increased from 3.92 to 14.1 ns when the temperature increased from 25 to 40 °C. Further studies using benzofuran-based model compounds proved that the sharp increase in lifetime with increasing temperature results from both the decrease in local polarity around the dye, as well as the loss of hydrogen bonding in the dehydrated polymeric globule.[45] Eventually the fluorescence nanogel thermometer (FNT) formed was able to map the temperature differences inside a living cell using fluorescence lifetime imaging microscopy. The environmental independence of the FNT with its high-temperature resolution (0.18–0.58 °C) was able to differentiate both the nucleus and centrosome from the cytoplasm of a COS7 cell as well as visualize the heat production from the mitochondria (Figure 7.19).[38,45] It was shown by de Mello *et al.* that RhB also shows temperature-dependent fluorescence lifetimes making it interesting for similar applications.[135]

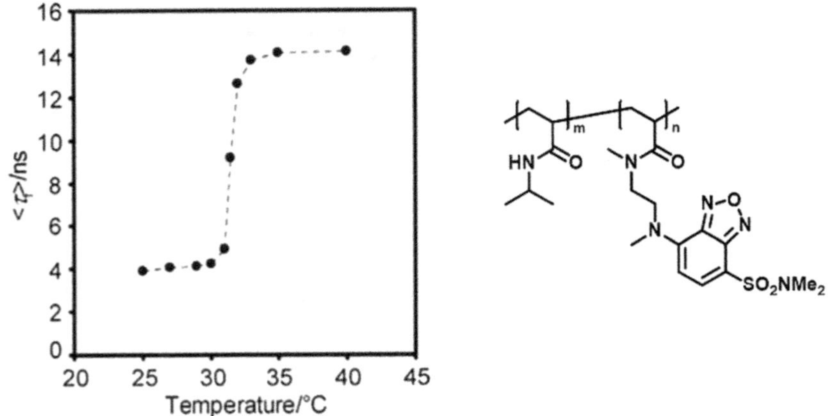

Figure 7.19   Left: Average fluorescence lifetime (τ_f) as a function of temperature for a poly(NIPAAm-*co*-DBD-AA) co-polymer in aqueous solution ($\lambda_{ex} = 275$ nm, $\lambda_{em} = 560$ nm). Right: The co-polymer structure.[45]

7.4.2 Changes in Interchain Distance: Förster Resonance Energy Transfer

FRET involves the non-radiative transfer of energy from an excited donor fluorophore to a second acceptor fluorophore that subsequently emits the transferred energy. This means that the FRET process results in the emission spectrum of the second fluorophore while using the excitation wavelength of the first fluorophore. The non-radiative energy transfer is strongly affected by the distance between the two moieties and is only effective at very short distances.[136,137] This distance dependence is most commonly utilized in DNA and protein folding research.[138–140] However, by incorporating two fluorophores into the same thermoresponsive polymer chain, the LCST transition could enhance the FRET process and change the fluorescence spectrum upon temperature increase as schematically illustrated in Figure 7.20. This also means that while the presence of the FRET process influences the total spectrum, careful selection of a single emission wavelength can lead to both fluorescence enhancement ($\lambda_{em,acceptor}$) and fluorescence quenching ($\lambda_{em,donor}$) with temperature increase, both in absolute values and in the ratio of the two.

Martinho *et al.* used this methodology to determine the critical micelle concentration (CMC) of their phenanthrene end-functionalized amphiphilic block co-polymer by adding anthracene into the mixture. By exciting the phenanthrene and measuring the anthracene emission, micellization would be detected as an increased anthracene emission due to the FRET process that takes place in the hydrophobic core of the micelle.[72] Winnik *et al.* labelled PNIPAAm with both naphthalene and pyrene and used the FRET process to investigate the mixing behaviour of polymer above and below the LCST, and proved that there is no interpolymer interaction below

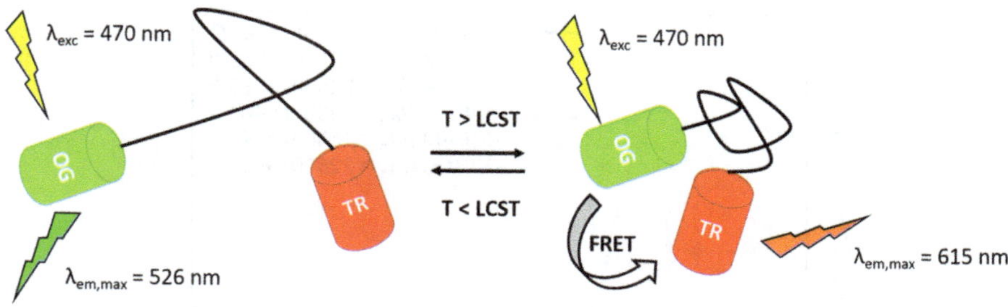

Figure 7.20 Schematic explanation of FRET temperature sensing based on the results of Zentel *et al.* using a telechelic, thermoresponsive polymer, end-modified with Oregon Green (OG) as a FRET donor and Texas Red (TR) as the FRET acceptor.[86]

the LCST.[12] The same FRET pair was used a few years later to investigate the solution properties in water of PVCL.[28]

Lyon *et al.* utilized the FRET process in the characterization of thermo-responsive core–shell microgels labelled with two cyanine-based dyes. The ratio between the emission of the Cy5 (donor) and Cy5.5 (acceptor) as a function of temperature provided valuable insight in the effect of the added shell on the temperature-induced core compression.[141] An example of a polymeric thermometer that utilizes this concept was reported by Choi *et al.* in 2009 based on the thermoresponsive PF127, which was end-functionalized with the near-infrared (NIR) fluorescent dye Cy5.5 on both chain ends. A near-linear decrease of the NIR fluorescence intensity was observed upon temperature increase due to an increase in FRET and shielding of the in-corporated dyes through micelle formation.[88] Finally, as described in Figure 7.20 a thermoresponsive poly(mDEGMA) was prepared *via* RAFT and subsequently asymmetrically end-modified with Oregon Green and Texas Red on the α- and ω-chain ends, respectively by post-polymerization modi-fication.[86] Another FRET example was employed by Chujo *et al.* during the synthesis of thermoswitchable luminescent gold nanoparticles (AuNPs). A tailor-made CTA functionalized with a boradiazaindacene (BODIPY) moi-ety on the leaving group was used to polymerize NIPAAm. The resulting end-functionalized polymer was subsequently *in situ* reduced with $HAuCl_4$ in order to create stabilized AuNPs. Upon temperature-induced phase transi-tion of the polymer, a FRET pair was created between the BODIPY end-groups and the AuNP itself, causing fluorescence quenching (Figure 7.21).[142] Besides AuNPs, conjugated polyelectrolytes like polythiophene can also be added as FRET acceptors to a solution of a labelled thermoresponsive polymer, as reported by Neher *et al.* for a PNIPAAm labelled with a coumarin derivative as a FRET donor.[57]

Another example was reported by Jo *et al.* in 2009, who presented a PNIPAAm synthesized using ATRP with a pyrene initiator. The resulting controlled polymer was further modified with C_{60} molecules at either the ω-chain end or in the side-chains. The decreasing distance between

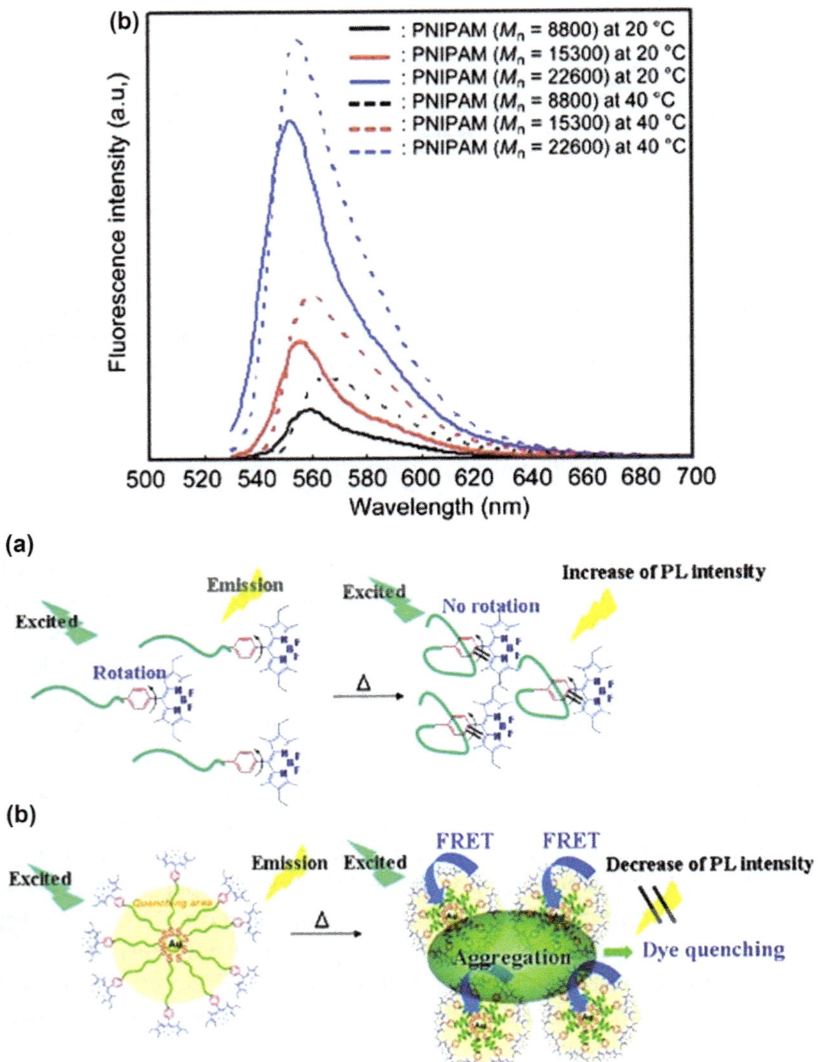

Figure 7.21 Top: Fluorescence intensity of BODIPY-PNIPAAm-stabilized AuNPs at 20 °C (solid line) and 40 °C (dashed line) in aqueous solution (λ_{ex} = 526 nm). Bottom: Proposed mechanism for (a) fluorescence enhancement in BODIPY-PNIPAAm and (b) fluorescence quenching through FRET of the corresponding AuNP.[142]

the pyrene donor and the C_{60} acceptor when the co-polymer is heated above its LCST provides fluorescence quenching through resonance energy transfer.[16]

Finally, the FRET process has also been used to strengthen the sensor signal provided by a different trigger, similarly to the previous translating method. Liu *et al.* used a RhB/fluorescein FRET pair to increase the sensitivity of the resulting pH sensor.[25] The group of Liu has reported multiple thermoresponsive microgel sensors based on a benzofuran/RhB FRET pair and the temperature-dependent volume phase transition (VPT), *i.e.*, swelling or deswelling, of a PNIPAAm particle. In 2011 they published the synthesis of light-responsive microgels with FRET as the reporter event using free radical emulsion polymerization of NIPAAm with the photo-cleavable 5-(2′-(dimethylamino)ethoxy)-2-nitrobenzyl acrylate (DMNA), 4-(2-acryloyloxyethylamino)-7-nitro-2,1,3-benzoxasiazole (NBAE) as a FRET donor, and RhB as the FRET acceptor. Under the influence of UV light, a nitrobenzyl moiety is cleaved off from DMNA leaving a carboxylic acid group on the chain causing swelling of the microgel. This change in VPT can be measured by following the ratio of the fluorescence emission of the FRET acceptor to the FRET donor which increases upon temperature increase through further deswelling of the particle.[49]

A similar methodology was used for the sensing of K^+ ions by covalently incorporating crown-ethers into the microgel.[50] By replacing RhB with SP as the FRET acceptor, the energy transfer can be switched on and off by irradiation with UV and visible light, respectively. The UV light ring-opens the non-fluorescent SP moiety into its fluorescent merocyanine form while irradiating with visible light shifts the equilibrium back to the SP form. The co-polymerization of these FRET dyes in the block co-polymer P[(NIPAAm-*co*-NBDAE)-*b*-(NIPAAm-*co*-SP)] allows for further enhancement of the FRET signal upon UV irradiation by temperature-induced collapse of the co-polymer.[24,48]

A final example that fits in this chapter of employing reduced interchain distance to induce an output signal, albeit not based on FRET, was reported by Ito and co-workers in 2009. They synthesized a thermo-responsive four-arm star-shaped polymer, initiated by a tetra-functional porphyrin moiety on which *N,N*-diethylacrylamide (DEA) was polymerized using RAFT polymerization. Below 27 °C, the hydrated polymer would sterically hinder the interaction between the porphyrin units, thereby reducing the amount of fluorescence quenching. When heated above its LCST, the reduced volume of the polymer would allow the porphyrin cores to approach each other which could be measured as fluorescence quenching.[115]

7.4.3 Changes in Rigidity/Viscosity

Another way of translating the polymeric phase transition into an output signal is based on the increased rigidity within the collapsed polymer globules, *i.e.*, the lower mobility of the chains. This method was first reported by Solovyov in 1999 for a porphyrin-labelled PNIPAAm. Below the cloud point, the highly hydrophobic porphyrin units showed a strong intermolecular interaction with each other leading to a large broadening of the Soret band *i.e.*, a decrease in the fluorescence intensity. Upon collapse of the co-polymer upon heating, the decreased mobility of the side-chains inside the globule, in combination with the lower polarity of the microenvironment, prevents the interaction, causing a very strong fluorescence enhancement of what are now isolated porphyrin units.[79,80] Recently, Yin *et al.* however reported a NIPAAm, hydroxyethylmethacrylate (HEMA) and methacrylic acid (MAA) tri-block co-polymer with tetra(4-carboxylatophenyl)porphyrin moieties attached to the middle block. In aqueous media, micelles were formed with a NIPAAm core and separate porphyrin units that only aggregated upon temperature increase, which is the reverse behaviour to the previously discussed example. Nevertheless, the temperature-dependent change in polarity and aggregation induced a large colour shift as a result of the change in absorption wavelength. Both the LCST and the colour shift could be controlled by the complexation of different metal ions by the porphyrin moieties.[76]

Another possibility of using this increased rigidity for temperature sensing has been proposed by Tang *et al.* for tetraphenylethene (TPE) dyes.[143] Upon precipitation of the TPE-functionalized PNIPAAm, the increased rigidity restricts the free rotation of the phenyl rings around the central double bond causing the non-emissive solvated TPE to become highly fluorescent (Figure 7.22).[15,144] A similar concept was developed using BODIPY sensing moieties.[145] It is known that the fluorescence intensity and lifetime of BODIPY dyes increase with an increase in viscosity of the microenvironment through restriction of the rotation of the *meso*-phenyl groups of the excited state, hence supressing the non-radiative stabilization. It was reported by Hirai *et al.* that a co-polymer of PNIPAAm with BODIPY side-chains showed a large fluorescence enhancement upon the polymer LCST phase transition and they proved that this was a result of the increase in microenvironmental viscosity and not due to the decreased polarity.[95,105,146]

Finally, a 2-naphthyl-benzimidazole has been shown to translate the increased viscosity upon LCST transition of the corresponding NIPAAm co-polymer.[112]

7.4.4 Excimer Formation/Deformation

Pyrene is one of the best-known fluorescent monomers to be used extensively as a so-called *polarity probe*, as its emission spectrum changes in response to the surrounding polarity. Furthermore, it is well known that in aqueous solution, pyrene will form dimers in the excited state, so called *excimers*. These excimers are characterized by a specific broad fluorescent emission band around 480 nm, which can be clearly distinguished from the sharp characteristic pyrene monomer emission peaks at 375–396 nm. The use of pyrene as a polarity probe allowed the investigation of the micellization behaviour of block co-polymers in aqueous solution by following the pyrene monomer emission peaks. The shape of this emission region is dependent on polarity and can be expressed as the ratio between peak I and peak III as a measure for the local polarity.[26,29,73,147–150] It was also used to determine T_g in polymeric thin films, whereby stronger excimer formation

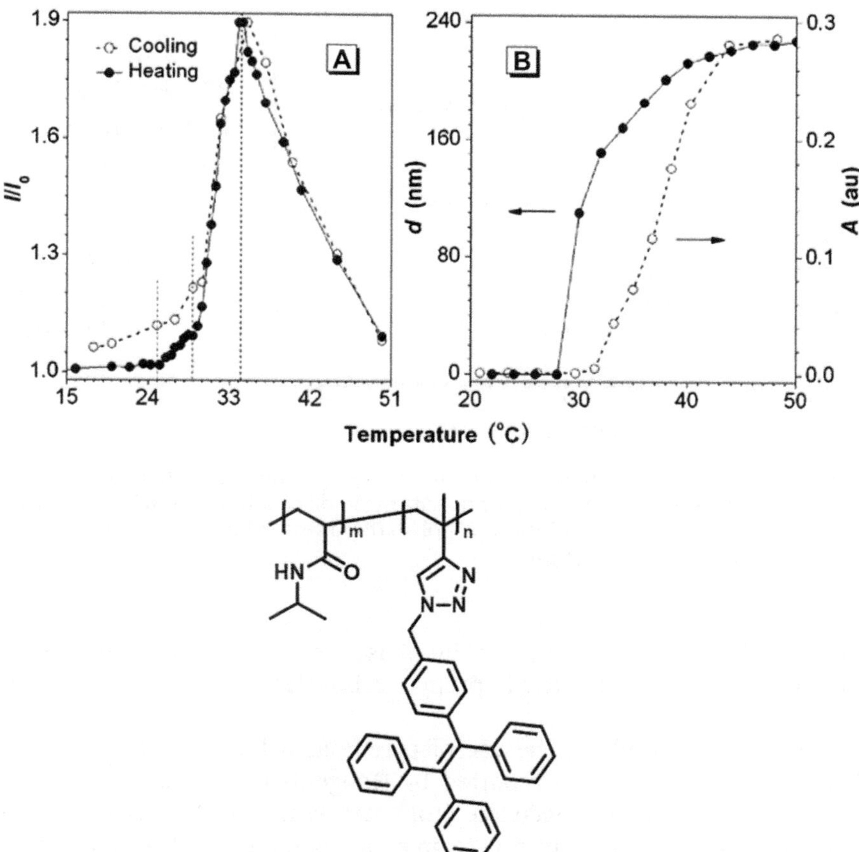

Figure 7.22 Top: Effect of temperature on (A) FL intensity at 468 nm ($\lambda_{ex} = 322$ nm, I_0 = FL intensity at 468 nm and 14 °C) and (B) particle size (d) and solution turbidity ($\lambda_{max} = 650$ nm) of poly(NIPAAm-*co*-TPE) (1 mg mL^{-1}). Bottom: The co-polymer structure.[15]

was observed above the T_g due to the higher mobility of pyrene.[151] This polarity and mobility sensitive excimer formation of pyrene also proved useful in the investigation of the temperature-induced phase transition of PNIPAM by Winnik and co-workers. The phase-separation temperature could be accurately determined using the ratio of excimer-to-pyrene mono-mer emission or the difference in quantum yields, as seen in Figure 7.23. It was theorized that below the LCST, the pyrene side-chains would form hydrophobic domains in between the otherwise water-soluble co-polymer, resulting in a relatively high contribution of excimer emission intensity. These hydrophobic domains are disrupted by the precipitation of the entire co-polymer, driven by the lower polarity of the collapsed polymer microenvironment, leading to a loss of excimer emission and an increase in isolated pyrene monomer emission within the polymer globule.[13]

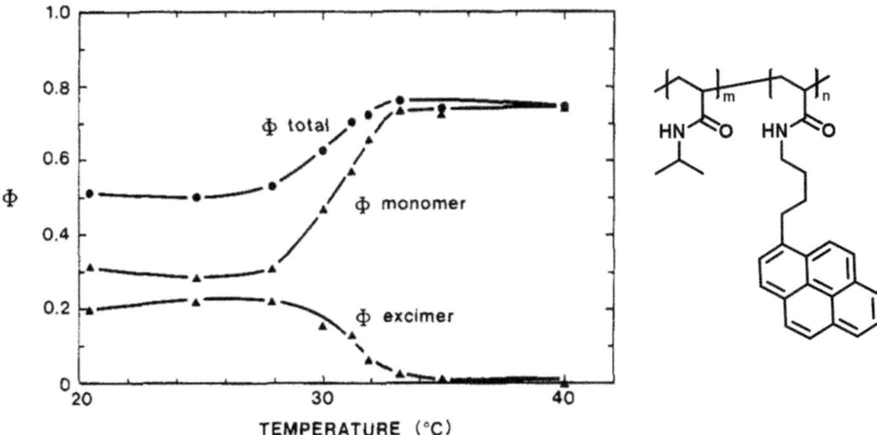

Figure 7.23　Left: Total fluorescence quantum yield (Φ) and fractional contribution of monomer and excimer quantum yields as a function of temperature for an aqueous solution of poly(NIPAAM-*co*-Py) (0.1 g L^{-1}). Right: The co-polymer structure.[13]

This allowed for extensive study of the phase separation of pyrene-labelled PNIPAM[84,152–159] and hydroxyl propyl cellulose[160,161] under different conditions.

A very recent example of the use of pyrene as a fluorescent probe in a polymeric thermometer was reported by Hoogenboom *et al.* The co-polymerization of a pyrene-methacrylate monomer with mDEGMA yielded a co-polymer with an LCST of around 17 °C in aqueous solution. This LCST phase transition could be followed using the emission intensity ratio between the excimer and pyrene monomer as a function of temperature, revealing a linear decrease in between 10 and 20 °C, which they regarded as the temperature-sensing regime of this polymeric thermometer.[58] By using a pyrene-labelled PMMA the same ratio could be used to translate the UCST behaviour of PMMA into an output signal by the increase in the excimer-to-pyrene monomer ratio.[59]

Excimer formation is not limited to pyrene and is also observed with other polycyclic aromatic hydrocarbons including naphthalene[162] and anthracene.[83] Another example of the use of excimers in polymeric thermometers was reported by Pan *et al.* for a PS-PNIPAM block co-polymer functionalized with *N*-carbazole ethylamine moieties attached to the second block. These co-polymers formed stable micelles in water characterized by a PS core with a fluorescent PNIPAM shell revealing excimer emission. Upon increasing the temperature, the mobility and polarity within the shell decreased, resulting in disruption of the excimer emission causing a sharp decrease in fluorescence intensity ($\lambda_{ex} = 295$ nm).[114] Similar results were obtained for a NIPAM co-polymer with Rose Bengal as the dye, which showed an increase in absorbance upon temperature increase.[78]

Liu *et al.* proposed a PNIPAM-based thermoresponsive di-block co-polymer sensor, prepared by ATRP initiated with a bromide derivative of a water-soluble conjugated polyfluorene (PF). The latter is highly fluorescent and forms a water-soluble micellar shell upon the LCST transition of PNIPAM. The increased rigidity of the PNIPAM core due to the temperature-induced coil-to-globule transition forces the PF blocks closer together, resulting in excimer emission as shown by the red-shift in $\lambda_{em,max}$ from roughly 435 to 470 nm.[67] A similar approach was reported a year later in which a hydrophobic PF was attached as a side-group to the PNIPAM. The hydrophobic PF dyes formed hydrophobic aggregates below the LCST transition leading to quenching of the fluorescence intensity. This quenching was diminished when the co-polymer was heated above its LCST, due to the restrictive rigidity and decreased polarity within the polymeric globule.[163]

7.5 Potential Applications of Polymeric Temperature Sensors

In this section, the potential applications of polymeric temperature sensors are highlighted based on selected examples. Of course, all reported polymeric thermometers can be used as sensors for the detection of temperature in general. In this section, we only highlight some of the novel creative strategies and concepts for more advanced applications of polymeric sensors. More specifically, recent studies have demonstrated the extension of this temperature-sensing concept towards intracellular temperature imaging, dual sensing, ion sensing and logic gates, as will be discussed in the following.

7.5.1 Intracellular Temperature Detection

The temperature and its distribution in a living cell are very important parameters that relate to various cellular events and strongly influence biochemical reactions inside living cells. Hence, intracellular temperature mapping within a living cell is keenly desired in various field of life science. However, detection of temperature with high resolution and accuracy in a living cell is still a challenge for conventional thermometers.[38] Recently, a series of high-resolution thermometers based on dye-incorporated polymers was developed, which have shown great promise for the detection and mapping of temperature.[38,47,110,164–166]

Uchiyama *et al.* have developed intracellular thermometry with a fluorescent nanogel thermometer based on the thermoresponsiveness of PNIPAM in combination with environment-sensitive fluorophores.[45,47,166] The sensor encompasses thermoresponsive PNIPAM, a fluorescent benzoxadiazole dye and negatively charged units to enrich the hydrophilicity of the thermometer and to prevent interpolymer aggregation within a cell. The nanosensor undergoes strong fluorescence enhancement due to the temperature-induced collapse of the gel inside the cytoplasm of a COS7 cell. The sensor system was then optimized by using fluorescence lifetime imaging microscopy, which improved the spatial and temperature resolutions of the thermometry down to 200 nm and 0.18–0.58 °C, respectively (Figure 7.24).[38] In addition to the high resolution, the polymeric sensor is highly biocompatible and negligible interactions occurred with cellular components, making this system very attractive for such applications.

Yin *et al.* have extended the intracellar thermosensor concept to a system that responds to both temperature and pH.[110] The sensory polymeric nanogel was prepared by combining a thermoresponsive PNIPAM hydrogel and a pH-responsive fluorescent benzo[*de*] isoquinoline dye by co-polymerization of NIPAM, the dye-functionalized monomer and a cross-linker. The resulting polymeric nanogel could sense temperature in living cells between 32 and 40 °C by fluorescence enhancement. In addition, due to the pH sensitivity of the dye, the fluorescence intensity decreased when increasing the pH from 4.0 and 10 at 25 °C providing the possiblity of pH sensing.

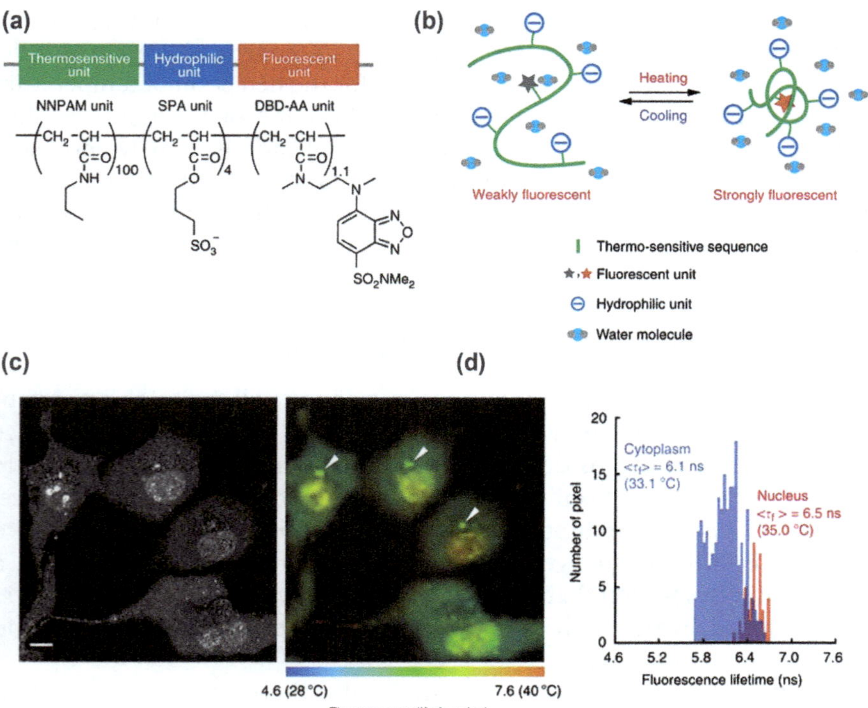

Figure 7.24   (a) Chemical structure and, (b) mechanism of the PNIPAM nanogel used for intracellular temperature imaging. (c) Confocal fluorescence image and fluorescence lifetime image of FPT in a COS7 cell. (d) Temperature of the cytoplasm and nucleus detected by fluorescence lifetime in COS7 cells.
Reproduced with permission from ref. 38. © Macmillan Publishers 2012.

As a drawback, this temperature and pH sensing system could only sense temperature or pH separately. The dual-sensing systems that sense temperture and pH simultaneously are highlighted in the next section.

7.5.2 Dual Temperature and pH Sensors

A dual sensor that simultaneously responds to both temperature and pH value was synthesized by Hoogenboom *et al.*, and is illustrated in Figure 7.25.[60] The sensor is based on a thermoresponsive PmDEGMA co-polymer bearing DR1 as a solvatochromic dye in the side-chain. This dual-sensitive co-polymer shows temperature responsiveness in the range from 10 to 20 °C due to the phase separation of the PmDEGMA; and the chromophore exhibits a colour change from 487 to 532 nm under acidic conditions due to protonation of the chromophore, while under basic conditions this shift is lost. By measuring a single absorption spectrum, the combination of the absorption maximum as well as the absorbance ratio of the DR1 dye provides information on both the temperature and the pH value of the solution. In a similar approach, DR1 was incorporated into a thermo-responsive dendronized polymer[116] and the authors demonstrated that the dual-sensing behaviour can be tuned by the generation of the dendronized polymers, which affect the dye localization from the interior onto the periphery of the collapsed polymer aggregates.

Dual temperature and pH sensors based on platinum(II)metallosupra-molecular tri-block co-polymers with hydrophilic alkynyl ligands have been reported by Yam and co-workers (Figure 7.26)[113] yielding a NIR-emitting dual sensor for pH and temperature in the range of pH 4–10 and 20–50 °C. Such dual-responsive behaviour was ascribed to the modulation of the self-assembly of the complex moieties by temperature-induced micellization and

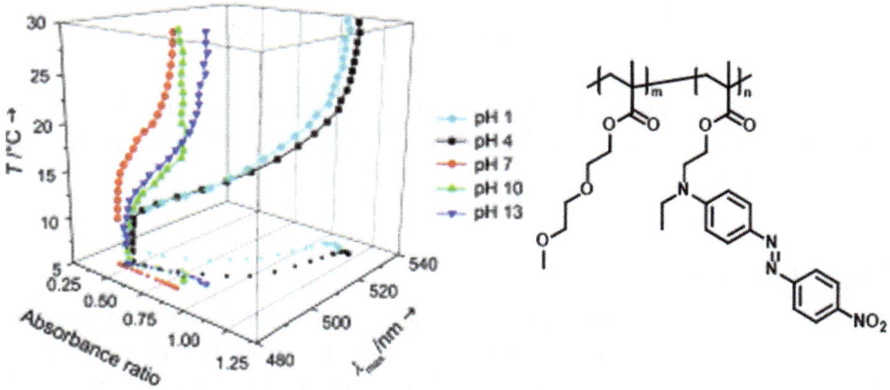

Figure 7.25 Dual sensor for temperature and pH.
Reproduced with permission from ref. 60. © 2009 WILEY-VCH Verlag GmbH & Co. KGaA, Weinheim.

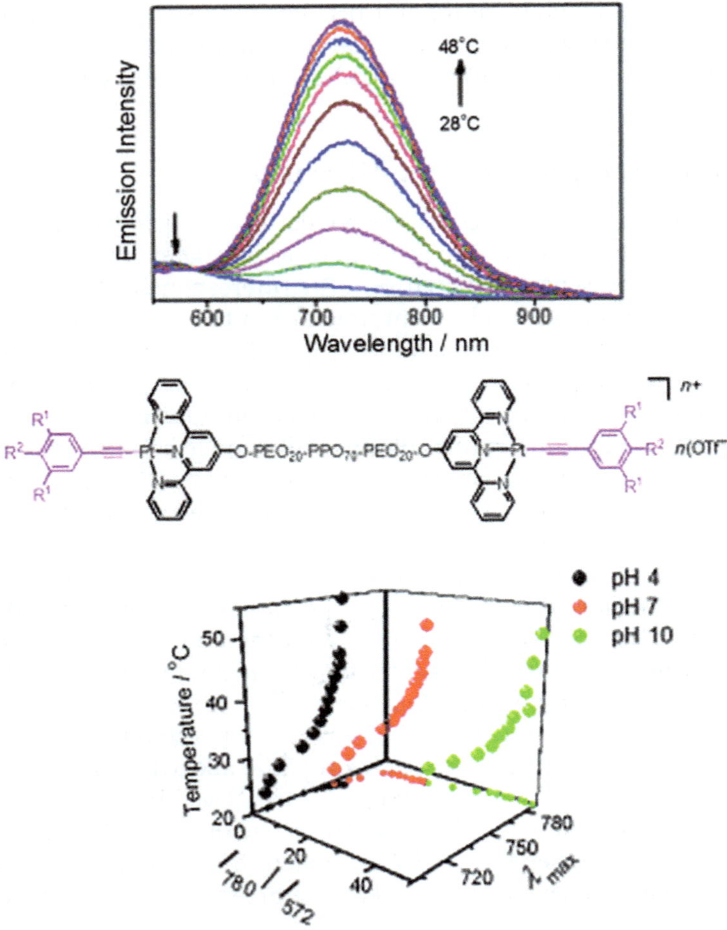

Figure 7.26 Dual temperature and pH sensor system based on metallosupramole-
cular tri-block co-polymers.
Reproduced with permission from ref. 113. © Wiley-VCH 2013.

the changes in the hydrophilicity as well as electrostatic interactions upon
protonation/deprotonation of the CH_2NMe_2 (R1 in Figure 7.26) groups on
the alkynyl ligand.

7.5.3 Ion Sensors

Thermoresponsive PNIPAM microgel-based ratiometric fluorescent sensors
for both K^+ ions and temperature were reported by Liu and co-workers
(Figure 7.27).[50] The system showed facile modulation of FRET efficiency due
to thermo-induced collapse and swelling of thermoresponsive microgels
functionalized with NBDAE as a FRET donor and RhBEA as the acceptor.
Crown-ether moieties were incorporated within the microgels to

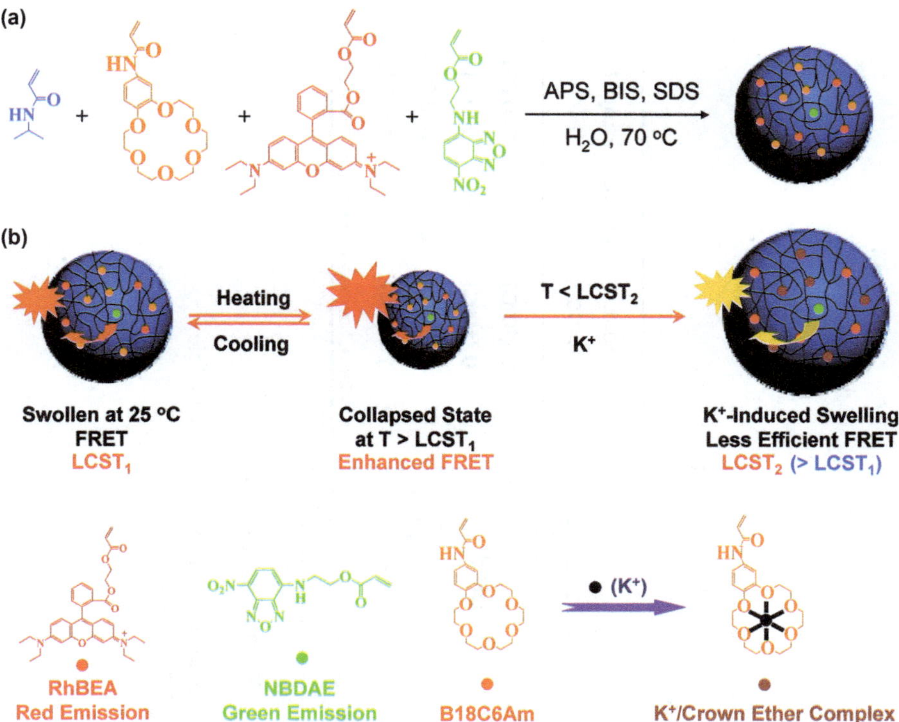

Figure 7.27 (a) Synthesis of thermoresponsive poly(NIPAM-B18C6Am-NBDAE-RhBEA) microgels *via* emulsion polymerization. (b) Schematic illustration of temperature- and K^+ ion-responsive poly(NIPAM-B18C6Am-NBDAE-RhBEA) microgels.
Reproduced with permission from ref. 50. © American Chemical Society 2010.

preferentially capture K^+ *via* the formation of $1:1$ molecular recognition complexes, resulting in the enhancement of microgel hydrophilicity and elevated phase-transition temperatures. Thus, the gradual addition of K^+ into microgel dispersions at intermediate temperatures, *i.e.*, in between the phase-transition temperatures of the microgels in the absence and presence of K^+ ions, respectively, can lead directly to the reswelling of initially collapsed microgels. This process can also be monitored by changes in fluorescence intensity ratios, *i.e.*, FRET efficiencies, enabling efficient K^+ detection.

Heavy metal ion pollution poses a huge threat to human beings and our environment. However, the detection of heavy metal ions based on small fluorescent molecules is suffering from poor solubility and biocompatibility. To address these drawbacks, polymeric ion sensors are being developed, providing better solubility and even high sensitivity.[58,125,167] A RhB hydrazide functionalized thermoresponsive block co-polymer, PEO-*b*-poly(NIPAM-*co*-RhBHA), was synthesized for Hg^{2+} sensing.[58] In its dissolved state, the RhBHA can be ring-opened by adding Hg^{2+} ions to yield a highly fluorescent acyclic species

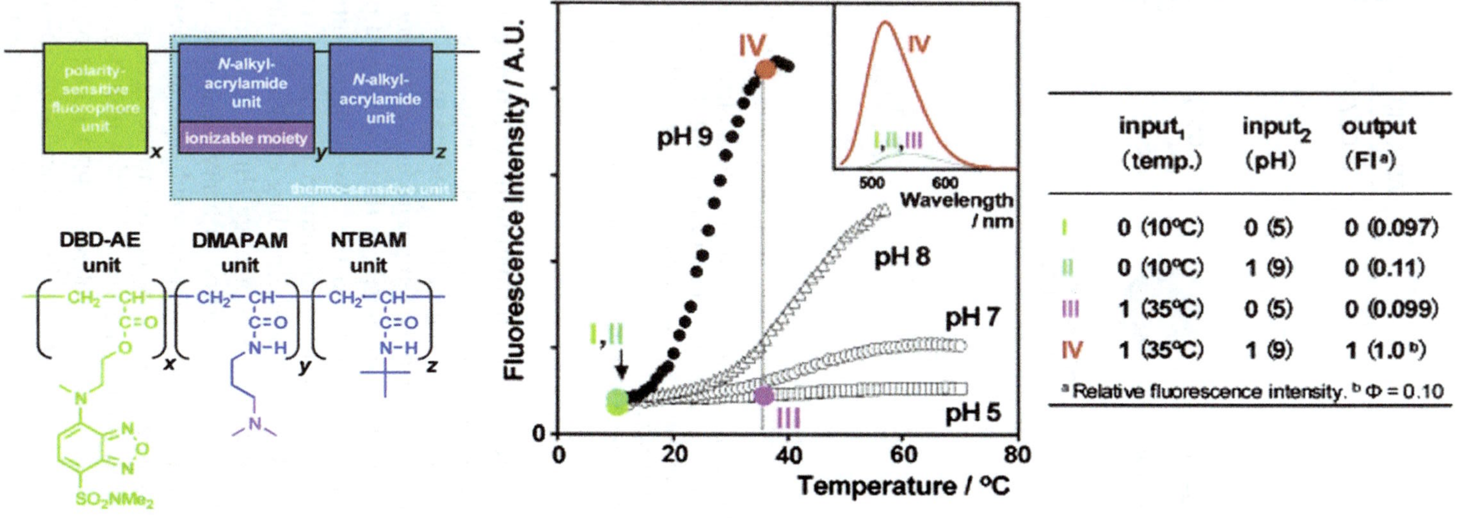

	input₁ (temp.)	input₂ (pH)	output (FI[a])
I	0 (10°C)	0 (5)	0 (0.097)
II	0 (10°C)	1 (9)	0 (0.11)
III	1 (35°C)	0 (5)	0 (0.099)
IV	1 (35°C)	1 (9)	1 (1.0[b])

[a] Relative fluorescence intensity. [b] Φ = 0.10

Figure 7.28 Logic gates based on P(DBD-*co*-DMAPAM-*co*-NTBAM), chemical structure and working principle. Reproduced with permission from ref. 42. © American Chemical Society 2004.

leading to ion sensing. Dramatically improved detection sensitivity of Hg^{2+} ions, however, could be achieved by thermo-induced micellization of the co-polymer leading to fluorescence enhancement of RhBHA acyclic residues within the hydrophobic core. Similarly, the detection of Cu^{2+} ions by phenanthroline could be improved *via* thermo-induced microgel collapse.[126] In this system, the fluorescence intensity of phenanthroline-incorporated microgels is quenched by complexation with Cu^{2+} ions, where the quenching efficiency by Cu^{2+} ions is considerably enhanced in the collapsed microgel state.

Yan *et al.* have reported a porphyrin-containing tri-block co-polymer for the dual sensing of metal ions and temperature.[76] The sensor exhibits full colour-tunable behaviour as a cationic detector and colorimeter.

7.5.4 Logic Gates

Logic gates that can (simultaneously) respond to multiple inputs with an output signal are extremely interesting for developing molecular memory systems. The fabrication of logic gates based on dye-incorporated polymers has attracted significant attention; especially temperature-sensitive polymers, due to the importance of temperature as an input signal.[42,66,168,169] Uchiyama *et al.* have developed a fluorescent polymeric AND logic gate with both temperature and pH as input signals (Figure 7.28).[42] The fluorescence intensity is enhanced by thermo- and/or pH-induced phase transitions, which serve as output signals for the logic gate. As shown in Figure 7.28, the fluorescence intensity (output) of the co-polymer is distinctly high (1) only when ($input_1$, $input_2$) is (1, 1), that is when the temperature is 35 °C and pH is 9. In contrast, the fluorescence intensity has a low level when ($input_1$, $input_2$) is (0, 0), (0, 1) or (1, 0).

7.6 Conclusion and Outlook

Polymeric temperature sensors have been attracting increasing attention during recent decades mainly due to the wide applicability and importance of the temperature stimulus, and the adaptability of their sensing behaviour towards different applications. Polymeric thermometers based on the combination of solvatochromic dyes and thermoresponsive polymers that undergo temperature-induced phase transitions are the most common in the current literature due to their uniquely controllable chemical and physical properties. This allows for a large variety of both output signals and sensor response regions with only limited changes to the polymeric design, such as the polymer architecture, identity of the (co-)monomers and the dye-incorporation approach. By careful consideration of the choice of dye, a wide variety of output signals is available including a shift in the absorbance/ excitation/emission wavelength and a change in the fluorescence intensity or lifetime. The latter response has allowed the development of polymeric thermometers that can be used for *in vivo* applications due to its relative independence from experimental conditions. Besides these new biomedical applications, which are made possible due to the improvements discussed in this chapter, other recently emerging application areas include temperature imaging, logic gate operation, dual sensors, and metal ion detection.

Based on the number and complexity of the currently developed sensors employing dye-labelled temperature-responsive polymers we foresee the development of much more versatile and sophisticated systems in the future. Based on the current limitations of these systems, it is expected that future research will focus on the long-term stability and reversibility of these polymeric temperature sensors, which are essential for continuous sensing applications. Another challenge is related to the investigation of polymeric thermometers with a broad sensing range. In addition, the development of dual- or multi-responsive sensors for specific applications will be an important research topic for the coming years. Finally, we are convinced that polymeric temperature sensors will find more applications in the future, not only in terms of proof-of-principle concepts but also for the engineering of commercial thermometer devices.

References

1. B. Adhikari and S. Majumdar, *Prog. Polym. Sci.*, 2004, **29**, 699–766.
2. C. Li and S. Liu, *Chem. Commun.*, 2012, **48**, 3262–3278.
3. J. Hu and S. Liu, *Macromolecules*, 2010, **43**, 8315–8330.
4. A. M. Horgan, J. D. Moore, J. E. Noble and G. J. Worsley, *Trends Biotechnol.*, 2010, **28**, 485–494.
5. U. Lange, N. V. Roznyatouskaya and V. M. Mirsky, *Anal. Chim. Acta*, 2008, **614**, 1–26.
6. M. I. J. Stich, L. H. Fischer and O. S. Wolfbeis, *Chem. Soc. Rev.*, 2010, **39**, 3102–3114.
7. C. Reichardt, *Chem. Rev.*, 1994, **94**, 2319–2358.
8. D. Roy, W. L. A. Brooks and B. S. Sumerlin, *Chem. Soc. Rev.*, 2013, **42**, 7214–7243.
9. C. Pietsch, U. S. Schubert and R. Hoogenboom, *Chem. Commun.*, 2011, **47**, 8750–8765.
10. D. Kungwatchakun and M. Irie, *Makromol. Chem. Rapid Commum.*, 1988, **9**, 243–246.
11. P. Vyskocil, J. Ricka and T. Binkert, *Helv. Phys. Acta*, 1989, **62**, 243–245.
12. F. M. Winnik, *Polymer*, 1990, **31**, 2125–2134.
13. F. M. Winnik, *Macromolecules*, 1990, **23**, 233–242.
14. H. G. Schild and D. A. Tirrell, *Macromolecules*, 1992, **25**, 4553–4558.
15. L. Tang, J. K. Jin, A. J. Qin, W. Z. Yuan, Y. Mao, J. Mei, J. Z. Sun and B. Z. Tang, *Chem. Commun.*, 2009, 4974–4976.
16. S. W. Hong, D. Y. Kim, J. U. Lee and W. H. Jo, *Macromolecules*, 2009, **42**, 2756–2761.
17. J. S. Wang and K. Matyjaszewski, *J. Am. Chem. Soc.*, 1995, **117**, 5614–5615.
18. K. Matyjaszewski and J. H. Xia, *Chem. Rev.*, 2001, **101**, 2921–2990.

19. K. Matyjaszewski and N. V. Tsarevsky, *J. Am. Chem. Soc.*, 2014, **136**, 6513–6533.
20. G. Moad, E. Rizzardo and S. H. Thang, *Aust. J. Chem.*, 2009, **62**, 1402–1472.
21. G. Moad, E. Rizzardo and S. H. Thang, *Aust. J. Chem.*, 2006, **59**, 669–692.
22. G. Moad, E. Rizzardo and S. H. Thang, *Aust. J. Chem.*, 2005, **58**, 379–410.
23. S. Perrier and P. Takolpuckdee, *J. Polym. Sci. Pol. Chem.*, 2005, **43**, 5347–5393.
24. C. Li, Y. Zhang, J. Hu, J. Cheng and S. Liu, *Angew. Chem., Int. Ed.*, 2010, **49**, 5120–5124.
25. J. M. Hu, X. Z. Zhang, D. Wang, X. L. Hu, T. Liu, G. Y. Zhang and S. Y. Liu, *J. Mater. Chem.*, 2011, **21**, 19030–19038.
26. X. W. Zhang, X. M. Lian, L. Liu, J. Zhang and H. Y. Zhao, *Macromolecules*, 2008, **41**, 7863–7869.
27. L. Liu, W. Li, K. Liu, J. Yan, G. Hu and A. Zhang, *Macromolecules*, 2011, **44**, 8614–8621.
28. A. Laukkanen, F. M. Winnik and H. Tenhu, *Macromolecules*, 2005, **38**, 2439–2448.
29. A. Laukkanen, L. Valtola, F. M. Winnik and H. Tenhu, *Polymer*, 2005, **46**, 7055–7065.
30. W. Zhang, W. Zhang, Z. Zhang, Z. Cheng, Y. Tu, Y. Qiu and X. Zhu, *J. Polym. Sci. Pol. Chem.*, 2010, **48**, 4268–4278.
31. M. Beija, M. T. Charreyre and J. M. G. Martinho, *Prog. Polym. Sci.*, 2011, **36**, 568–602.
32. A. M. Breul, M. D. Hager and U. S. Schubert, *Chem. Soc. Rev.*, 2013, **42**, 5366–5407.
33. M. A. Gauthier, M. I. Gibson and H. A. Klok, *Angew. Chem., Int. Ed.*, 2009, **48**, 48–58.
34. H. C. Kolb, M. G. Finn and K. B. Sharpless, *Angew. Chem., Int. Ed.*, 2001, **40**, 2004–2021.
35. U. Mansfeld, C. Pietsch, R. Hoogenboom, C. R. Becer and U. S. Schubert, *Polym. Chem.*, 2010, **1**, 1560–1598.
36. C. R. Becer, R. Hoogenboom and U. S. Schubert, *Angew. Chem. Int. Ed.*, 2009, **48**, 4900–4908.
37. P. Theato, *J. Polym. Sci. Pol. Chem.*, 2008, **46**, 6677–6687.
38. K. Okabe, N. Inada, C. Gota, Y. Harada, T. Funatsu and S. Uchiyama, *Nat. Commun.*, 2012, **3**, 705.
39. S. Uchiyama, Y. Matsumura, A. P. de Silva and K. Iwai, *Anal. Chem.*, 2004, **76**, 1793–1798.
40. S. Uchiyama, Y. Matsumura, A. P. de Silva and K. Iwai, *Anal. Chem.*, 2003, **75**, 5926–5935.
41. S. Uchiyama and Y. Makino, *Chem. Commun.*, 2009, 2646–2648.
42. S. Uchiyama, N. Kawai, A. P. de Silva and K. Iwai, *J. Am. Chem. Soc.*, 2004, **126**, 3032–3033.
43. M. Onoda, S. Uchiyama and T. Ohwada, *Macromolecules*, 2007, **40**, 9651–9657.

44. K. Iwai, Y. Matsumura, S. Uchiyama and A. P. de Silva, *J. Mater. Chem.*, 2005, **15**, 2796–2800.

45. C. Gota, S. Uchiyama, T. Yoshihara, S. Tobita and T. Ohwada, *J. Phys. Chem. B*, 2008, **112**, 2829–2836.

46. C. Gota, S. Uchiyama and T. Ohwada, *Analyst*, 2007, **132**, 121–126.

47. C. Gota, K. Okabe, T. Funatsu, Y. Harada and S. Uchiyama, *J. Am. Chem. Soc.*, 2009, **131**, 2766–2767.

48. T. Wu, G. Zou, J. Hu and S. Liu, *Chem. Mater.*, 2009, **21**, 3788–3798.

49. J. Yin, H. Hu, Y. Wu and S. Liu, *Polym. Chem.*, 2011, **2**, 363–371.

50. J. Yin, C. Li, D. Wang and S. Liu, *J. Phys. Chem. B*, 2010, **114**, 12213–12220.

51. S. Inal, J. D. Kolsch, L. Chiappisi, D. Janietz, M. Gradzielski, A. Laschewsky and D. Neher, *J. Phys. Chem. C*, 2013, **1**, 6603–6612.

52. B. H. Lessard, E. J. Y. Ling and M. Marić, *Macromolecules*, 2012, **45**, 1879–1891.

53. B. Lessard and M. Marić, *J. Polym. Sci. Pol. Chem.*, 2011, **49**, 5270–5283.

54. Y. Matsumura and K. Iwai, *Polymer*, 2005, **46**, 10027–10034.

55. K. Iwai, K. Hanasaki and M. Yamamoto, *J. Lumin.*, 2000, **87-9**, 1289–1291.

56. S. Inal, J. D. Kolsch, F. Sellrie, J. A. Schenk, E. Wischerhoff, A. Laschewsky and D. Neher, *J. Mater. Chem. B*, 2013, **1**, 6373–6381.

57. S. Inal, J. D. Kölsch, L. Chiappisi, M. Kraft, A. Gutacker, D. Janietz, U. Scherf, M. Gradzielski, A. Laschewsky and D. Neher, *Macromol. Chem. Phys.*, 2013, **214**, 435–445.

58. C. Pietsch, A. Vollrath, R. Hoogenboom and U. S. Schubert, *Sensors*, 2010, **10**, 7979–7990.

59. C. Pietsch, R. Hoogenboom and U. S. Schubert, *Polym. Chem.*, 2010, **1**, 1005–1008.

60. C. Pietsch, R. Hoogenboom and U. S. Schubert, *Angew. Chem., Int. Ed.*, 2009, **48**, 5653–5656.

61. A. M. Breul, C. Pietsch, R. Menzel, J. Schafer, A. Teichler, M. D. Hager, J. Popp, B. Dietzek, R. Beckert and U. S. Schubert, *Eur. Polym. J.*, 2012, **48**, 1339–1347.

62. C. Koopmans and H. Ritter, *J. Am. Chem. Soc.*, 2007, **129**, 3502–3503.

63. G. Kwak, H. Kim, I. K. Kang and S. H. Kim, *Macromolecules*, 2009, **42**, 1733–1738.

64. S. H. Kim, I. J. Hwang, S. Y. Gwon and Y. A. Son, *Dyes Pigm.*, 2010, **87**, 158–163.

65. S. H. Kim, I. J. Hwang, S. Y. Gwon and Y. A. Son, *Dyes Pigm.*, 2010, **87**, 84–88.

66. Z. Q. Guo, W. H. Zhu, Y. Y. Xiong and H. Tian, *Macromolecules*, 2009, **42**, 1448–1453.

67. W. Z. Wang, R. Wang, C. Zhang, S. Lu and T. X. Liu, *Polymer*, 2009, **50**, 1236–1245.

68. Q. Duan, Y. Miura, A. Narumi, X. Shen, S. Sato, T. Satoh and T. Kakuchi, *J. Polym. Sci. Pol. Chem.*, 2006, **44**, 1117–1124.

69. H. Akiyama and N. Tamaoki, *Macromolecules*, 2007, **40**, 5129–5132.

70. P. Relogio, M. Bathfield, Z. Haftek-Terreau, M. Beija, A. Favier, M. J. Giraud-Panis, F. D'Agosto, B. Mandrand, J. P. S. Farinha, M. T. Charreyre and J. M. G. Martinho, *Polym. Chem.*, 2013, **4**, 2968–2981.
71. M. Beija, C. A. M. Afonso, J. P. S. Farinha, M. T. Charreyre and J. M. G. Martinho, *Polymer*, 2011, **52**, 5933–5946.
72. T. J. V. Prazeres, M. Beija, M. T. Charreyre, J. P. S. Farinha and J. M. G. Martinho, *Polymer*, 2010, **51**, 355–367.
73. J. Y. Rao, J. Xu, S. Z. Luo and S. Y. Liu, *Langmuir*, 2007, **23**, 11857–11865.
74. K. Kempe, A. Vollrath, H. W. Schaefer, T. G. Poehlmann, C. Biskup, R. Hoogenboom, S. Hornig and U. S. Schubert, *Macromol. Rapid Commun.*, 2010, **31**, 1869–1873.
75. D. Fournier, R. Hoogenboom and U. S. Schubert, *Chem. Soc. Rev.*, 2007, **36**, 1369–1380.
76. Q. Yan, J. Yuan, Y. Kang, Z. Cai, L. Zhou and Y. Yin, *Chem. Commun.*, 2010, **46**, 2781–2783.
77. Y. Shiraishi, R. Miyarnoto and T. Hirai, *Langmuir*, 2008, **24**, 4273–4279.
78. M. Nowakowska, M. Kepczynski and M. Dabrowska, *Macromol. Chem. Phys.*, 2001, **202**, 1679–1688.
79. Y. S. Avlasevich, T. A. Chevtchouk, V. N. Knyukshto, O. G. Kulinkovich and K. N. Solovyov, *J. Porphyrins Phthalocyanines*, 2000, **4**, 579–586.
80. V. N. Knyukshto, Y. S. Avlasevich, O. G. Kulinkovich and K. N. Solovyov, *J. Fluoresc.*, 1999, **9**, 371–378.
81. Y. Shiraishi, R. Miyamoto and T. Hirai, *J. Photochem. Photobiol. A*, 2008, **200**, 432–437.
82. Y. Shiraishi, R. Miyamoto, X. Zhang and T. Hirai, *Org. Lett.*, 2007, **9**, 3921–3924.
83. Y. Shiraishi, R. Miyamoto and T. Hirai, *Tetrahedron Lett.*, 2007, **48**, 6660–6664.
84. M. F. Ottaviani, F. M. Winnik, S. H. Bossmann and N. J. Turro, *Helv. Chim. Acta*, 2001, **84**, 2476–2492.
85. Y. S. Avlasevich, V. N. Knyukshto, O. G. Kulinkovich and K. N. Solovyov, *J. Appl. Spectrosc.*, 2000, **67**, 663–669.
86. P. J. Roth, M. Haase, T. Basche, P. Theato and R. Zentel, *Macromolecules*, 2010, **43**, 895–902.
87. F. D. Jochum, L. zur Borg, P. J. Roth and P. Theato, *Macromolecules*, 2009, **42**, 7854–7862.
88. S. Y. Lee, S. Lee, I. C. Youn, D. K. Yi, Y. T. Lim, B. H. Chung, J. F. Leary, I. C. Kwon, K. Kim and K. Choi, *Chem.- Eur. J.*, 2009, **15**, 6103–6106.
89. A. Balamurugan, V. Kumar and M. Jayakannan, *Chem. Commun.*, 2014, **50**, 842–845.
90. R. X. Liu, M. Fraylich and B. R. Saunders, *Colloid Polym. Sci.*, 2009, **287**, 627–643.
91. X.-d. Wang, O. S. Wolfbeis and R. J. Meier, *Chem. Soc. Rev.*, 2013, **42**, 7834–7869.
92. H. G. Schild, *Prog. Polym. Sci.*, 1992, **17**, 163–249.
93. M. Heskins and J. E. Guillet, *J. Macromol. Sci. Part A*, 1968, **2**, 1441–1455.

94. N. Jordão, R. Gavara and A. J. Parola, *Macromolecules*, 2013, **46**, 9055–9063.
95. A. Nagai, K. Kokado, J. Miyake and Y. Cyujo, *J. Polym. Sci. Pol. Chem.*, 2010, **48**, 627–634.
96. Y. Shiraishi, R. Miyamoto and T. Hirai, *Org. Lett.*, 2009, **11**, 1571–1574.
97. Y. Zhao, L. Tremblay and Y. Zhao, *J. Polym. Sci. Pol. Chem.*, 2010, **48**, 4055–4066.
98. G. Liu, W. Zhou, J. Zhang and P. Zhao, *J. Polym. Sci. Pol. Chem.*, 2012, **50**, 2219–2226.
99. K. Van Durme, G. Van Assche and B. Van Mele, *Macromolecules*, 2004, **37**, 9596–9605.
100. J.-F. Lutz, *J. Polym. Sci. Pol. Chem.*, 2008, **46**, 3459–3470.
101. J.-F. Lutz, *Adv. Mater.*, 2011, **23**, 2237–2243.
102. J.-F. Lutz, O. Akdemir and A. Hoth, *J. Am. Chem. Soc.*, 2006, **128**, 13046–13047.
103. S. Han, M. Hagiwara and T. Ishizone, *Macromolecules*, 2003, **36**, 8312–8319.
104. G. Vancoillie, D. Frank and R. Hoogenboom, *Prog. Polym. Sci.*, 2014, **39**, 1074–1095.
105. R. París, I. Quijada-Garrido, O. García and M. Liras, *Macromolecules*, 2011, **44**, 80–86.
106. J. Seuring and S. Agarwal, *Macromol. Rapid Commun.*, 2012, **33**, 1898–1920.
107. Q. Zhang, P. Schattling, P. Theato and R. Hoogenboom, *Polym. Chem.*, 2012, **3**, 1418–1426.
108. R. Hoogenboom, C. R. Becer, C. Guerrero-Sanchez, S. Hoeppener and U. S. Schubert, *Aust. J. Chem.*, 2010, **63**, 1173–1178.
109. R. Hoogenboom, S. Rogers, A. Can, C. R. Becer, C. Guerrero-Sanchez, D. Wouters, S. Hoeppener and U. S. Schubert, *Chem. Commun.*, 2009, 5582–5584.
110. L. Yin, C. He, C. Huang, W. Zhu, X. Wang, Y. Xu and X. Qian, *Chem. Commun.*, 2012, **48**, 4486–4488.
111. C.-Y. Chen and C.-T. Chen, *Chem. Commun.*, 2011, **47**, 994–996.
112. B. Han, N. Zhou, W. Zhang, Z. Cheng, J. Zhu and X. Zhu, *J. Polym. Sci. Pol. Chem.*, 2013, **51**, 4459–4466.
113. C. Y.-S. Chung and V. W.-W. Yam, *Chem. – Eur. J.*, 2013, **19**, 13182–13192.
114. Q. Yan, J. Y. Yuan, W. Z. Yuan, M. Zhou, Y. W. Yin and C. Y. Pan, *Chem. Commun.*, 2008, 6188–6190.
115. S.-I. Yusa, T. Endo and M. Ito, *J. Polym. Sci. Pol. Chem.*, 2009, **47**, 6827–6838.
116. L. Liu, W. Li, J. Yan and A. Zhang, *J. Polym. Sci. Pol. Chem.*, 2014, **52**, 1706–1713.
117. Y. Matsumura and K. Iwai, *J. Colloid Interface Sci.*, 2006, **296**, 102–109.
118. J. F. Callan, A. P. de Silva and D. C. Magri, *Tetrahedron*, 2005, **61**, 8551–8588.
119. S. Saha and A. Samanta, *J. Phys. Chem. A*, 2002, **106**, 4763–4771.

120. Y. Matsumura and A. Katoh, *J. Lumin.*, 2008, **128**, 625–630.
121. S. Uchiyama, K. Takehira, T. Yoshihara, S. Tobita and T. Ohwada, *Org. Lett.*, 2006, **8**, 5869–5872.
122. C. C. Yang, Y. Tian, C. Y. Chen, A. K. Y. Jen and W. C. Chen, *Macromol. Rapid Commun.*, 2007, **28**, 894–899.
123. H. N. Kim, M. H. Lee, H. J. Kim, J. S. Kim and J. Yoon, *Chem. Soc. Rev.*, 2008, **37**, 1465–1472.
124. S. K. Ko, Y. K. Yang, J. Tae and I. Shin, *J. Am. Chem. Soc.*, 2006, **128**, 14150–14155.
125. J. Yin, X. F. Guan, D. Wang and S. Y. Liu, *Langmuir*, 2009, **25**, 11367–11374.
126. T. Liu, J. Hu, J. Yin, Y. Zhang, C. Li and S. Liu, *Chem. Mater.*, 2009, **21**, 3439–3446.
127. W. Yuan, G. Jiang, J. Wang, G. Wang, Y. Song and L. Jiang, *Macromolecules*, 2006, **39**, 1300–1303.
128. P. Ravi, S. L. Sin, L. H. Gan, Y. Y. Gan, K. C. Tam, X. L. Xia and X. Hu, *Polymer*, 2005, **46**, 137–146.
129. G. J. Mohr and O. S. Wolfbeis, *Anal. Chim. Acta*, 1994, **292**, 41–48.
130. C. Luo, F. Zuo, X. Ding, Z. Zheng, X. Cheng and Y. Peng, *J. Appl. Polym. Sci.*, 2008, **107**, 2118–2125.
131. F. D. Jochum and P. Theato, *Chem. Commun.*, 2010, **46**, 6717–6719.
132. H. Akiyama and N. Tamaoki, *J. Polym. Sci., Part A: Polym. Chem.*, 2004, **42**, 5200–5214.
133. K. Sumaru, M. Kameda, T. Kanamori and T. Shinbo, *Macromolecules*, 2004, **37**, 7854–7856.
134. M. Kameda, K. Sumaru, T. Kanamori and T. Shinbo, *Langmuir*, 2004, **20**, 9315–9319.
135. R. K. P. Benninger, Y. Koc, O. Hofmann, J. Requejo-Isidro, M. A. A. Neil, P. M. W. French and A. J. de Mello, *Anal. Chem.*, 2006, **78**, 2272–2278.
136. T. Forster, *Naturwissenschaften*, 1946, **33**, 166–175.
137. T. Förster, *Ann. Phys.*, 1948, **437**, 55–75.
138. A. A. Deniz, M. Dahan, J. R. Grunwell, T. J. Ha, A. E. Faulhaber, D. S. Chemla, S. Weiss and P. G. Schultz, *Proc. Natl. Acad. Sci. U. S. A.*, 1999, **96**, 3670–3675.
139. F. He, Y. L. Tang, S. Wang, Y. L. Li and D. B. Zhu, *J. Am. Chem. Soc.*, 2005, **127**, 12343–12346.
140. N. T. Chen, S. H. Cheng, C. P. Liu, J. S. Souris, C. T. Chen, C. Y. Mou and L. W. Lo, *Int. J. Mol. Sci.*, 2012, **13**, 16598–16623.
141. C. D. Jones, J. G. McGrath and L. A. Lyon, *J. Phys. Chem. B*, 2004, **108**, 12652–12657.
142. A. Nagai, R. Yoshii, T. Otsuka, K. Kokado and Y. Chujo, *Langmuir*, 2010, **26**, 15644–15649.
143. Y. N. Hong, J. W. Y. Lam and B. Z. Tang, *Chem. Commun.*, 2009, 4332–4353.
144. Q. Zeng, Z. Li, Y. Q. Dong, C. A. Di, A. J. Qin, Y. N. Hong, L. Ji, Z. C. Zhu, C. K. W. Jim, G. Yu, Q. Q. Li, Z. A. Li, Y. Q. Liu, J. G. Qin and B. Z. Tang, *Chem. Commun.*, 2007, 70–72.

145. M. A. H. Alamiry, A. C. Benniston, G. Copley, K. J. Elliott, A. Harriman, B. Stewart and Y. G. Zhi, *Chem. Mater.*, 2008, **20**, 4024–4032.

146. D. P. Wang, R. Miyamoto, Y. Shiraishi and T. Hirai, *Langmuir*, 2009, **25**, 13176–13182.

147. C. L. Zhao, M. A. Winnik, G. Riess and M. D. Croucher, *Langmuir*, 1990, **6**, 514–516.

148. G. Kwon, M. Naito, M. Yokoyama, T. Okano, Y. Sakurai and K. Kataoka, *Langmuir*, 1993, **9**, 945–949.

149. F. M. Winnik, *Chem. Rev.*, 1993, **93**, 587–614.

150. J. Y. Zhang, Y. T. Li, S. P. Armes and S. Y. Liu, *J. Phys. Chem. B*, 2007, **111**, 12111–12118.

151. S. Kim, C. B. Roth and J. M. Torkelson, *J. Polym. Sci., Part A: Polym. Chem.*, 2008, **46**, 2754–2764.

152. H. Ringsdorf, J. Venzmer and F. M. Winnik, *Macromolecules*, 1991, **24**, 1678–1686.

153. H. Ringsdorf, J. Simon and F. M. Winnik, *Macromolecules*, 1992, **25**, 7306–7312.

154. H. Ringsdorf, J. Simon and F. M. Winnik, *Macromolecules*, 1992, **25**, 5353–5361.

155. F. M. Winnik, *Macromolecules*, 1990, **23**, 1647–1649.

156. F. M. Winnik, H. Ringsdorf and J. Venzmer, *Langmuir*, 1991, **7**, 905–911.

157. F. M. Winnik, H. Ringsdorf and J. Venzmer, *Langmuir*, 1991, **7**, 912–917.

158. F. M. Winnik, M. F. Ottaviani, S. H. Bossman, W. S. Pan, M. Garciagaribay and N. J. Turro, *J. Phys. Chem.*, 1993, **97**, 12998–13005.

159. P. Kujawa, V. Aseyev, H. Tenhu and F. M. Winnik, *Macromolecules*, 2006, **39**, 7686–7693.

160. F. M. Winnik, N. Tamai, J. Yonezawa, Y. Nishimura and I. Yamazaki, *J. Phys. Chem.*, 1992, **96**, 1967–1972.

161. F. M. Winnik, *Langmuir*, 1990, **6**, 522–524.

162. K. Prochazka, D. Kiserow, C. Ramireddy, Z. Tuzar, P. Munk and S. E. Webber, *Macromolecules*, 1992, **25**, 454–460.

163. C. C. Yang, Y. Q. Tian, A. K. Y. Jen and W. C. Chen, *J. Polym. Sci. Pol. Chem.*, 2006, **44**, 5495–5504.

164. Y. Y. Li, H. Cheng, J. L. Zhu, L. Yuan, Y. Dai, S. X. Cheng, X. Z. Zhang and R. X. Zhuo, *Adv. Mater.*, 2009, **21**, 2402–2406.

165. H. Kobayashi, M. Nishikawa, C. Sakamoto, T. Nishio, H. Kanazawa and T. Okano, *Anal. Sci.*, 2009, **25**, 1043–1047.

166. S. Uchiyama, K. Kimura, C. Gota, K. Okabe, K. Kawamoto, N. Inada, T. Yoshihara and S. Tobita, *Chem.- Eur. J.*, 2012, **18**, 9552–9563.

167. C. Li and S. Liu, *J. Mater. Chem.*, 2010, **20**, 10716–10723.

168. P. Schattling, F. D. Jochum and P. Theato, *Chem. Commun.*, 2011, **47**, 8859–8861.

169. G. Pasparakis, M. Vamvakaki, N. Krasnogor and C. Alexander, *Soft Matter*, 2009, **5**, 3839–3841.

Organic–Inorganic Hybrids Thermometry

ANGEL MILLÁN,*[a] LUÍS D. CARLOS,*[b] CARLOS D. S. BRITES,[b] NUNO J. O. SILVA,[b] RAFAEL PIÑOL[a] AND FERNANDO PALACIO[a]

[a] Instituto de Ciencia de Materiales de Aragón, CSIC–Universidad de Zaragoza, 50009 Zaragoza, Spain; [b] Departamento de Física and CICECO, Universidade de Aveiro, 3810–193 Aveiro, Portugal
*Email: amillan@unizar.es; lcarlos@ua.pt

8.1 Introduction

In *strictu sensu* an organic–inorganic hybrid material is a combination of organic and inorganic components related by chemical bonds. This concept includes a large variety of materials that can be classified accordingly to the:

1. *type of bond*: ionic, covalent, coordination bond, hydrogen bond, van der Waals;
2. *size of the inorganic and organic domains*: molecular, nanoscopic, mesoscopic microscopic, or macroscopic (and therefore the synergic effects are less determinant in the resulting physical properties of the material). One of the components, *i.e.*, the inorganic one, can be in a macromolecular form (cluster) that is sustained by the moulding action of the other component (organic ligands);
3. *arrangement of organic and inorganic domains*: discrete, linear, laminar, tri-dimensional frameworks, or disordered. One of the components can be nanometric (inorganic nanoparticles) embedded in the other

RSC Nanoscience & Nanotechnology No. 38
Thermometry at the Nanoscale: Techniques and Selected Applications
Edited by Luís Dias Carlos and Fernando Palacio
© The Royal Society of Chemistry 2016
Published by the Royal Society of Chemistry, www.rsc.org

component (organic) that act as the matrix to form a nanocomposite, or they can be arranged in a core–shell structure; and

4. *nature of the compounds*: the inorganic and the organic compounds.

Among the molecular hybrids linked by covalent bonds, of particular interest are those derived from sol–gel chemistry routes, because they are the most commonly used hybrids and therefore will occupy a large part of this chapter. The mixing of organic and inorganic components may bring special features to the resulting material in such a way that:

1. one component (or both) achieves a new structure by modulation due to the other component;
2. the new material displays the physical properties of each of the components; and
3. the hybrid material has new physical properties that arise from synergic effects between the components.

The characteristics of organic–inorganic hybrids can be very useful in the design of thermometric systems. Their excellent mechanical properties and processability make them suitable as hosts for luminescent organic and inorganic thermometric probes (*e.g.*, organic dyes, semiconductor quantum dots (QDs) and tri-valent lanthanide ions, Ln^{3+}). The combination of organic and inorganic compounds can also give rise to cooperative effects resulting in an enhancement of the thermometric efficiency, relative to that of the organic or inorganic thermometric probes alone. For instance in inorganic optical thermometers, combination with organic antennas can improve their quantum efficiency. Moreover, new thermometric systems can be designed based on the synergies between organic and inorganic components.[1–6]

This chapter is interconnected with other chapters of the book, especially with that devoted to polymer-based thermometers (Chapter 7), as the organic component in many hybrid materials, especially in organic–inorganic nanoparticles (NPs), will be a polymer. It is also connected with Chapters 4, 5 and 6, which describe in detail the use of QDs, lanthanide ions and organic dyes as thermometric probes, respectively. These kinds of probes are also frequent in hybrid systems, and therefore they will also be treated in this chapter, as long as their thermometric performance is controlled by their hybrid character.

8.2 Classification

There are a large variety of measurable physical properties that vary with temperature and, in principle, can be used for temperature probing in hybrid nanothermometers. In practice, most of the hybrid nanothermometers developed so far are based on an optical temperature signalling system,[7] and therefore we will first classify them as *optical* or *non-optical*. There are several optical features that may vary with temperature and are, therefore, useful for

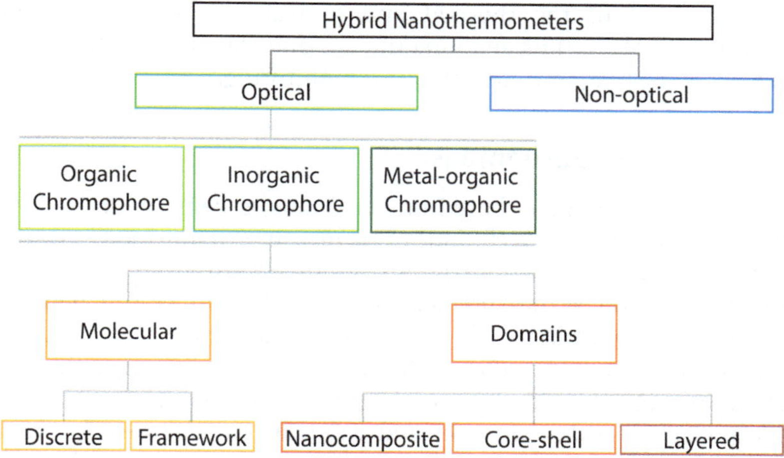

Scheme 8.1   Classification of hybrid nanothermometers.

temperature sensing, *i.e.*, emission intensity,[8] wavelength,[9] lifetime,[10] and anisotropy.[11] Among the types of chromophores, we can mainly distinguish: (1) organic or metallorganic dyes; (2) QDs; and (3) Ln^{3+} ions. Since Ln^{3+} ions present special features making them particularly useful in hybrid nano-thermometry, they will be explained in some detail in a separate section.

Finally, a rational criterion for the classification of hybrid thermometers can be the level of mixing between organic and inorganic components, in: (1) discrete, bi-dimensional or tri-dimensional molecular assemblies; or in (2) core–shell or nanocomposite domain nano-arrangements. Materials consisting of a matrix prepared by the sol–gel processing of metal alkoxides embedding Ln^{3+} luminescence coordination compounds are a special case of hybrid nanothermometers and they deserve a section in this chapter (Section 8.3.1.2). A representation of the most usual types of hybrid nano-thermometers is shown in Scheme 8.1.

8.3   Optical Thermometers

The main advantages of inorganic–organic hybrids in luminescence thermometry can be summarized as follows:

1. from the synthetic point of view, they offer a large versatility, simple procedures, and high purity;
2. they show high processability at low temperatures, which favours low-cost processing and miniaturization;
3. a large capacity for structuring that allows control of the density and separation of emitting centres, enabling the control of non-radiative decay pathways;
4. improved mechanical properties in the gap between organic and inorganic materials; and

5. amelioration of the thermal and optical stability of the isolated emitting centres, thus enabling the use of isolated dyes, QDs and Ln^{3+} chelates for long-term temperature monitoring.

8.3.1　Trivalent Lanthanides

Luminescence thermometry based on Ln^{3+} ions is being addressed as one of the most promising techniques to measure temperature with sub-micrometric resolution, because it can combine high spatial and temporal resolution with a robust measurement based on the temperature dependence of the steady-state emission,[12–14] lifetime[15–18] and rise time.[19] Complete tutorial reviews are well spread in the literature,[1,20] including examples devoted to organic–inorganic hybrids,[5,20] complexes with various chelating ligands,[21,22] ceramic phosphors that can withstand extreme temperatures,[23] and nano-thermometers[24–27] (which are additionally, reviewed in Chapter 5 of this book).

8.3.1.1　*Particular Characteristics of Ln^{3+} Ions*

The uniqueness of Ln^{3+} ions (the most prevalent oxidation state in which lanthanide elements are present in nature) lies in their electronic configuration corresponding to $4f^N$ ($N = 1$–14). The shielding of the 4f electrons from interaction with their surroundings (ligand–field interaction) by the filled 5s and 5p orbitals is responsible for the interesting chemical and photophysical properties of the Ln^{3+} ions,[1,22] which have been exploited to build molecular probes for pH,[27] oxygen[28] and temperature, among others.

As direct Ln^{3+} photo-excitation is not very efficient, limiting the light output, the design of lanthanide complexes, in which the ligands incorporate organic chromophores strongly bonded to the 4f metal centre increase substantially the Ln^{3+} luminescence intensity by:

1. shielding the metal ion from deleterious quenching interactions (*e.g.*, with solvent molecules);
2. raising the effective absorption cross-sections: 10^4–10^5 times higher and over a much broader spectral range (when compared with those of the corresponding Ln^{3+} ions); and
3. lanthanide luminescence sensitization, commonly named the *antenna effect*, which consists of transferring the energy absorbed by the ligands to nearby Ln^{3+} ions, which in turn undergo a radiative emitting process.

Ln^{3+} ions form complexes with various organic molecules, such as aromatic carboxylic acids, β-diketonates, calixarenes, cryptands, podands, and heterocyclic ligands, thus covering the near-UV (Ce^{3+} and Gd^{3+}), visible (blue, Tm^{3+}; green, Tb^{3+} and Er^{3+}; yellow, Dy^{3+}; orange, Sm^{3+}; and red, Eu^{3+}), and near-infrared (NIR, Nd^{3+}, Er^{3+}, Tm^{3+} and Yb^{3+}) spectral regions. The main limitations of Ln^{3+} complexes are their accentuated photo-bleaching (photostability under UV irradiation), and low thermal and mechanical stability.[29] From the materials science engineering point of view, these fundamental limitations can be overcome by encapsulating the lanthanide complexes in polymers, liquid crystals and sol–gel-derived organic–inorganic hybrids (mostly siloxane-based).[1,29]

8.3.1.2 Diureasils

Diureasils are sol–gel-derived organic–inorganic hybrids composed of a siliceous framework to which polyether chains of variable length are covalently bonded through urea linkages.[30] These hybrids present acceptable transparency, mechanical flexibility, and thermal stability to be processed as thick and thin films[31] and have been used by us as self-referenced and efficient luminescent probes to map temperature in micro-electronic circuits[12,13,32] and optoelectronic devices.[33]

Thermal managing, *e.g.*, control of the heat flux generated by the several million transistors that exist essentially on a single chip of few square millimetres, is essential for improving electronic performance and reliability, posing a limitation stronger than the downscaling itself. As explained in detail in Chapter 13, temperature resolving in integrated circuits with spatial resolution higher than 1 μm,[34,35] is nowadays recognized as one of the main challenges of the modern electronics industry, since the continuous miniaturization and integration, together with the increase of physical complexity of electronic devices and integrated circuits, results in extremely confined power dissipation and localized heating problems.[34,36]

Although infrared cameras are commonly used to map the surface temperature of microcircuits, they present the following disadvantages:

1. limited spatial resolution (>100 μm);
2. previous knowledge of the material surface emissivity (that is a function of wavelength and temperature) is required;
3. considerable temperature uncertainty ($\sim 2°$); and
4. strong dependence on the relative orientation (angle) between the camera and the measured surface.

An appropriate alternative is to deposit a thin film embedding a molecular luminescent thermometer on a microcircuit. The temperature probe is minimally invasive, as the temperature readout is fully remote, usually implemented by fibre optical technology. The thermal dynamics of the thermometric probe (*e.g.*, the thermalization time) are fully controllable by the careful deposition of the film (using well-implemented dip-coating and spin-coating techniques) and the acquisition rate is defined by the excitation and detection setups.

In 2010, the temperature of an integrated circuit was mapped by Brites *et al.*[12] using a di-ureasil film co-doped with [Eu(btfa)$_3$(MeOH)(bpeta)] and [Tb(btfa)$_3$(MeOH)(bpeta)] β-diketonate chelates (where btfa$^-$ is 4,4,4-trifluoro-l-phenyl-1,3-butanedionate, bpeta is 1,2-bis(4-pyridyl)ethane and

MeOH is methanol). The absolute emission quantum yield was 0.16 ± 0.02. The maximum relative sensitivity (S_m), defined as the maximum value of:

$$S_r = \frac{\partial \Delta / \partial T}{\Delta} \tag{8.1}$$

where S_r is the relative sensitivity, Δ is the thermometric parameter and T the absolute temperature,[37] is 1.9% K^{-1} at 201 K diminishing to 0.7–0.4% K^{-1} in the 300–350 K range. The sensitivity in this range is high enough to permit the use of commercial portable detectors to collect the emission spectra and to discriminate small temperature changes. The temperature mapping is achieved by linear scanning with a constant step, converting the emission spectra at each point into absolute temperature, using a calibration curve. The temperature uncertainty of the thermometer was estimated from the temperature fluctuations in the absence of temperature gradients as 0.5 K. Moreover, this proof-of-concept shows that temperature mapping under the same conditions by a state-of-the-art commercial infrared (IR) thermal camera has poorer spatial resolution, with no clear temperature discrimination of the narrower and closely spaced hotter tracks (Figure 8.1A).[12]

Defining the spatial resolution of the mapping as:

$$\delta x = \frac{\delta T}{|\nabla T|_{\max}} \tag{8.2}$$

where δT is the temperature uncertainty and $|\nabla T|_{\max}$ is the maximum measured temperature gradient,[38] it was possible to quantify the spatial resolution of the hybrid thermometer film as 35 μm.

The spatial resolution can be improved either by decreasing the temperature uncertainty or by increasing the measured temperature gradient. The first approach requires the use of lower noise detectors (*e.g.*, cooled detectors or photomultiplier tubes). It is known that the signal-to-noise ratio of non-cooled photodiode array detectors and photomultiplier tubes is on the order of 10^3 to 10^6 and using low noise detectors is expected to produce a decrease in the temperature uncertainty of $\sim 10^{-3}$ K. Increasing of the temperature gradient can be achieved either by increasing the value of the temperature step or by mapping with finer scanning steps.

In 2013, Brites *et al.*[13] reported the fabrication of diureasil films incorporating the same two Eu^{3+} and Tb^{3+} β-diketonate chelates and maghemite NPs. Relative to the first reported thermometer films,[12] the sensitivity was raised to 1.0% K^{-1} (for 300–350 K), and the gradient of the temperature profile increased 10-fold, with the concomitant improvement of the spatial resolution by an order of magnitude, $\delta x = 3.4$ μm.[13]

Using a Eu^{3+}/Tb^{3+} co-doped di-ureasil thermometric film[12] to cover a FR4 printed wiring board (with higher temperature gradients), new temperature profiles were measured (Figure 8.1B)[32] and the spatial resolution lower limit tested. Moreover, a more precise definition of the temperature uncertainty was adopted, based on the distribution of temperature

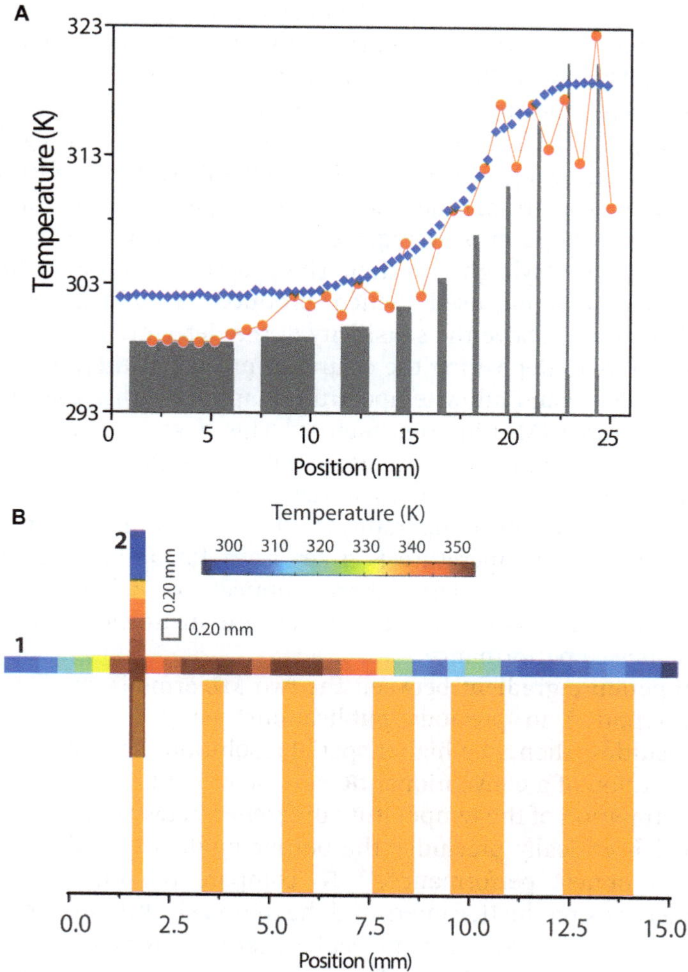

Figure 8.1 (A) Temperature profiles obtained with Eu^{3+}/Tb^{3+} co-doped di-ureasil (red circles, the size of which corresponds to a temperature uncertainly of 0.5 K) and with an IR camera (blue diamonds). The grey shadowed areas correspond to the position of the copper tracks. Reproduced with permission from ref. 12. © Wiley-VCH 2010. (B) Pseudocolour temperature maps reconstructed from the emission of Eu^{3+}/Tb^{3+} co-doped di-ureasil collected with a 200 μm core diameter fibre along two perpendicular directions (denoted by 1 and 2) of an FR4 printed wiring board, using a scanning step of 200 μm. The orange shadowed areas correspond to the positions of the copper tracks.

readouts in the absence of external heat sources. The stochastic nature of the temperature fluctuations results in a Gaussian distribution of temperature readouts, where the temperature uncertainty is given by its full-with-at-half maximum. A temperature uncertainty of 0.15 K was reported.[32]

The decrease in the scanning step from 800 to 50 µm (that is equivalent to an increase the temperature gradient) improved the spatial resolution up to 0.45 µm,[32] well below the Rayleigh limit of diffraction of the fibres employed, 2 µm. Below scanning steps that are of the order of the optical fibre radius (100 or 200 µm), the spatial resolution does not improved significantly. Furthermore, narrowing the optical fibre field-of-view by a factor of five produces only a marginal decrease in the spatial resolution. As the temperature at each point is averaged over the field-of-view of the fibre it is not limited by the Rayleigh criterion. The spatial resolution is limited by the experimental setup used, which produces a field-of-view-averaged temperature change above the sensitivity of the detector.[32]

The possibility of depositing the di-ureasil molecular thermometer using an inkjet printer would allow temperature mapping with a thermally actuated Mach–Zehnder (MZ) interferometer.[33] This is an uncommon example in which the same material that is used to fabricate the thermally actuated MZ interferometer is also used to measure one of the device's intrinsic properties (the operating temperature).[33] The determination of the interferometer operating temperature is a key issue for optimizing the device performance. Due to the large thermo-optical coefficient of diureasils ($ca.$ -10^{-4} K^{-1}) conventional IR cameras are insufficient to accurately predict the device performance.

The temperature gradient between the two MZ arms is smaller than the gradients reported in previous publications using the same di-ureasil film[13,32] resulting, then, in a higher spatial resolution, $\delta x = 28$ µm. The poor spatial resolution of a conventional IR camera, $\delta x = 128$ µm, corresponds to an underestimation of the temperature difference between the arms of up to 0.30 K, which critically precludes the output modelling and prediction of the device's optical performance.[33] To compare the uncertainty of the measurements using the IR camera and the di-ureasil emission, temperature maps were measured along the molecular thermometer layer region in the absence of a thermal gradient. The value estimated using the emission spectrum (0.1 K) is three times lower than that estimated using the camera (0.3 K).[33] Moreover, while for the luminescent molecular thermometer the temperature was recorded upon UV illumination, for the IR camera the temperature was determined with and without UV illumination. The same temperature mapping (within experimental error) was obtained, excluding, therefore, any contribution from the UV handheld lamp for the temperature measurements. This means that the Stokes shift (the difference between the excitation and emission energies that is converted in non-radiative deactivations) does not contribute to heating the film, within the experimental uncertainly of the camera.

8.3.2 Molecular Hybrids

In certain environments, the light emission of organic dyes is sensitive to temperature and therefore it can be used for temperature determination.

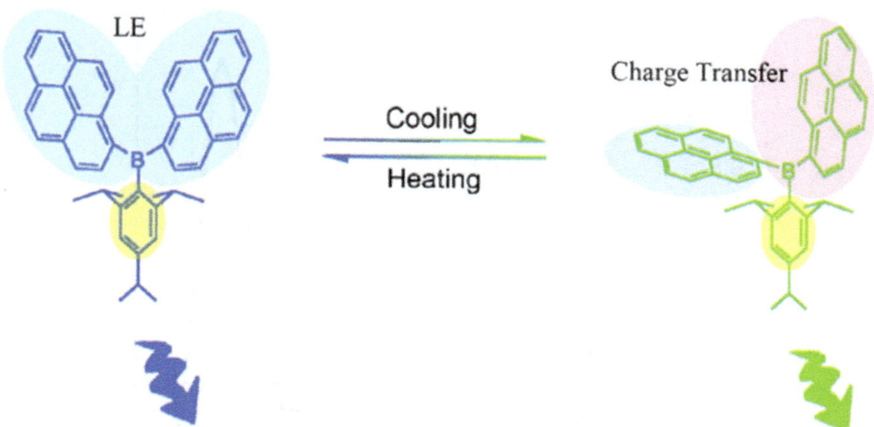

Scheme 8.2   Mechanisms of emission change with temperature of triarylboron-based thermometers.
Reproduced with permission from ref. 50. © Wiley-VCH 2011.

Typical organic dyes, such as Rhodamine B (RhB),[39–41] fluorescein iso-thiocyanates[42] and cyanines,[43] among others,[44,45] have been used for this purpose in biological media and in polymeric matrices.[46,47] A more advanced thermometric system is based on the ratio of the intensities of the monomer and excimer/exciplex[48] or on the decay of emission lifetimes. Moreover, the operating temperature range of monomer/excimer thermometers is improved in an ionic liquid (from 298 to 413 K).[49] A quite different dye-based thermometer[50] uses triarylboron molecules dissolved in an organic solvent, and shows temperature-dependent green-to-blue luminescence over the 223–373 K interval. In this case, the blue-shift is ascribed to the thermal equilibrium between twisted intramolecular charge transfer and the local excited state of the triarylboron molecule (Scheme 8.2). Some authors have suggested that organic dye solutions could be entrapped in sol–gel films to build temperature-sensitive paints.[49] In principle, they could also be implemented in other types of hybrid matrices, benefiting from the advantages that have been pointed out above.

8.3.2.1   *Discrete Metal–Organic Molecular Compounds*

Other kinds of dyes have a metal–organic composition, and therefore could be considered as molecular hybrids. An example is that of the platinum octaethyl porphyrin complex (Figure 8.2),[51] whose thermometric properties are based on the intensity ratios of two transitions: the first excited triplet level, at 650 nm, and one band, at 540 nm, whose origin is not entirely clear. Another kind of thermometric molecular hybrid is that of spin-cross-over compounds. The spin-cross-over process involves the switching between two molecular spin states (the so-called *high spin* and *low spin* configurations) in (pseudo)octahedral $3d^4$–$3d^7$ transition metal complexes. The combination of

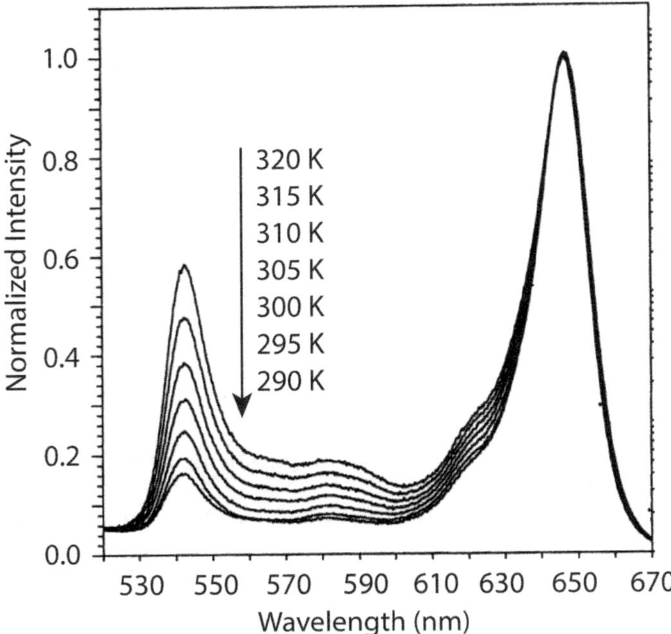

Figure 8.2   Photoluminescence spectra of PtOEP [2,3,7,8,12,13,17,18-octaethyl-21*H*,23*H*-porphyrin platinum(ii)] excited at 380 nm measured through a short-wavelength pass filter at the given temperatures. The spectra are normalized to the emission band at 650 nm.
Adapted with permission from ref. 51. © AIP Publishing 2002.

spin-cross-over and dye luminescence gives rise to the thermometric response. One of the first examples reported was that of Ni^{2+} cyclam (1,4,8,11-tetraazacyclotetradecane).[52] The intensity of cyclam emission is partially quenched by an energy transfer mechanism whose efficiency decreases with temperature between 300 and 370 K, with a maximum relative sensitivity of 3.6% K^{-1} at 300 K. Another example is that of Rh110-doped ultrasmall [Fe(NH$_2$triazole)$_3$](tosylate, NO_3^{-})$_2$ NPs,[53,54] in which the Rh110 luminescence intensity is modulated by the thermal cross-over process. The thermometer operates in the range 305–325 K with $S_m = 1.9\%$ K^{-1} at 309 K.

Photobleaching is the main drawback of dye-based thermometers, which limits their use for temperature changes at different time scales. On the other hand, their molecular nature permits a high spatial resolution, mostly limited by the optical detection system.

8.3.2.2   Layered Double Hydroxides

Layered double hydroxides (LDHs) are another class of organic–inorganic hybrids used in thermometric systems. LDHs show temperature-dependent changes in alignment, packing and conformational constraints. The thermometric operation can be enforced by a component with optical properties

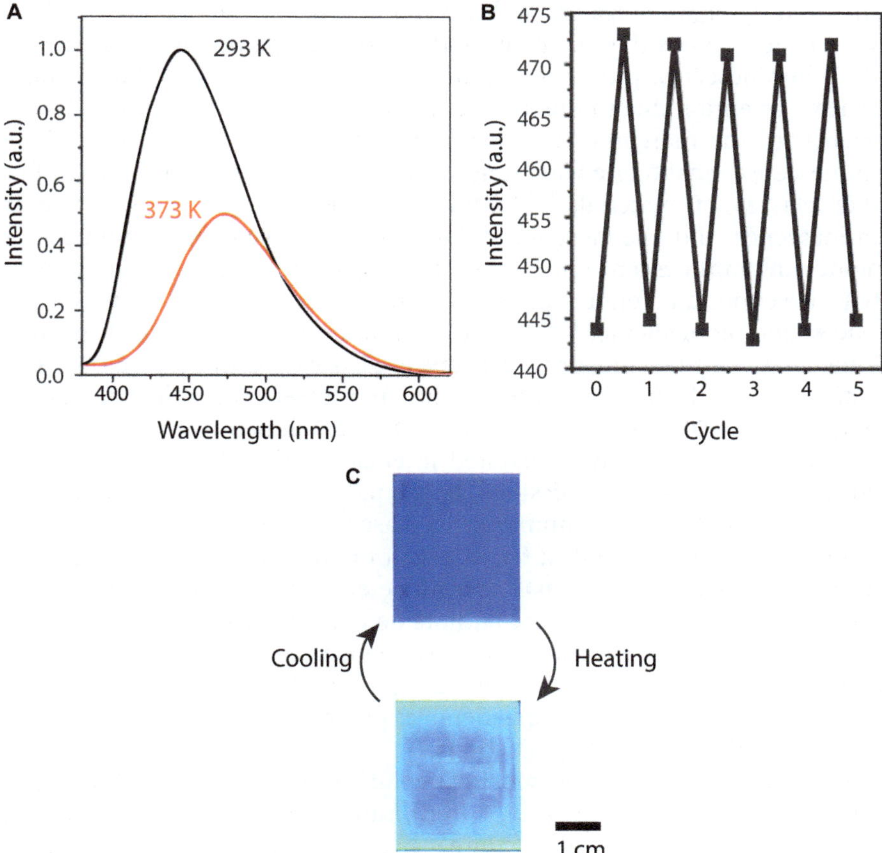

Figure 8.3 (A) Emission spectra of a BSB/LDH ultrathin film (eight layers) at 293 and 373 K. (B) Reversible fluorescence response over five consecutive cycles. (C) Photographs of the ultrathin film before and after heating). Adapted with permission from ref. 57. © Wiley-VCH 2011.

that responds to those changes.[55,56] For instance, LDH hybrids incorporating poly(diacetylenecarboxylates) develop colours ranging from yellow to blue as the temperature is increased from 293 to 353 K.[55] Inorganic–organic ultrathin films based on the layer-by-layer (LBL) assembly of the anionic stilbene derivative (bis(2-sulfonatostyryl)biphenyl, BSB) and LDH nanosheets change the peak of the emission band from 444 to 473 nm when the nanosheets are heated from 293 to 373 K (Figure 8.3).[57]

8.3.2.3   Metal–Organic Frameworks

Instead of single molecules, the combination of metals and ligands can give rise to coordination polymers with 1–3 dimensionality that may also be used as thermometric sensors. The thermometric process is based on the energy transfer between ions within the solid framework. The ion's emission is sensitized by organic ligands with a suitable triplet-state energy, which in this way determines the sensitivity and working range of the thermometer.

The luminescence properties of metal–organic frameworks (MOFs) have attracted attention due to the unique hybrid networks of these materials in which both the inorganic and organic moieties may be the source of the luminescence, enabling a wide range of emissive phenomena found in few other classes of materials.[58,59] Moreover, metal–ligand charge transfer luminescence and the host–guest interactions provide opportunities for engineering luminescence properties. In the past two decades, luminescent MOFs have found potential applications in chemical sensing, light-emitting devices, and biomedicine.[58–64] The use of luminescent MOF NPs in sensing, biomedical imaging and drug delivery is also well documented.[65–68] Cui et al.[69] reported the first ratiometric luminescent MOF thermometer, $Eu_{0.0069}Tb_{0.9931}$-DMBDC (where DMBDC = 2,5-dimethoxy-1,4-benzenedicarboxylate), based on the integrated intensity of the Tb^{3+} $(^5D_4 \rightarrow {}^7F_5, I_{Tb})$ and Eu^{3+} $(^5D_0 \rightarrow {}^7F_2, I_{Eu})$ emissions. In the past two years, several examples of mixed Ln-MOFs as luminescence-based thermometers have been reported.[70–73] However, only a few consist of nanosized particles. The most recent realization of an Ln-MOF thermometer at the sub-micron scale was reported by Cadiau et al.[74] These authors demonstrated that mixed Ln-MOF NPs made by reverse emulsion show excellent performance as ratiometric luminescent nanothermometers in the physiological temperature range (300–320 K), thus holding considerable potential as nanoplatforms for biological and biomedical applications.

In Figure 8.4C the relative sensitivities of the Ln^{3+}-based MOF thermometers reported by Cadiau et. al.[74] (red), Cui et al.[75,76] (orange and pink), Rao et al.[77] (blue), D'Vries et al.[71] (green), Zhou et al.[78] (yellow), Wei et al.[79] (black) and Wang et al.[80] (grey) are compared. The most obvious feature of the MOF hybrid thermometers based on Eu^{3+} and Tb^{3+} is that different host matrices enable a temperature range of 10–330 K to be covered with a relative sensitivity higher than 0.5% K^{-1}, reinforcing the importance of the interaction between the host and the emitting centres to tune the temperature range of maximum sensitivity.

8.3.3 Polymer Nanocomposites

Most optical thermometers are based on a luminescence emission that is directly dependent on temperature. However, the combination of inorganic and organic components in a hybrid composite opens up the possibility of a novel class of complex thermometric systems,[24] in which a variation of the luminescence response of one component is indirectly induced by a temperature-dependent process occurring in the second component. In many of these complex systems, the luminescent part is inorganic and the organic part, usually a polymer, undergoes structural and conformational changes with the temperature. Therefore, in this class of thermometers, the thermometric property is neither located in the organic or the inorganic

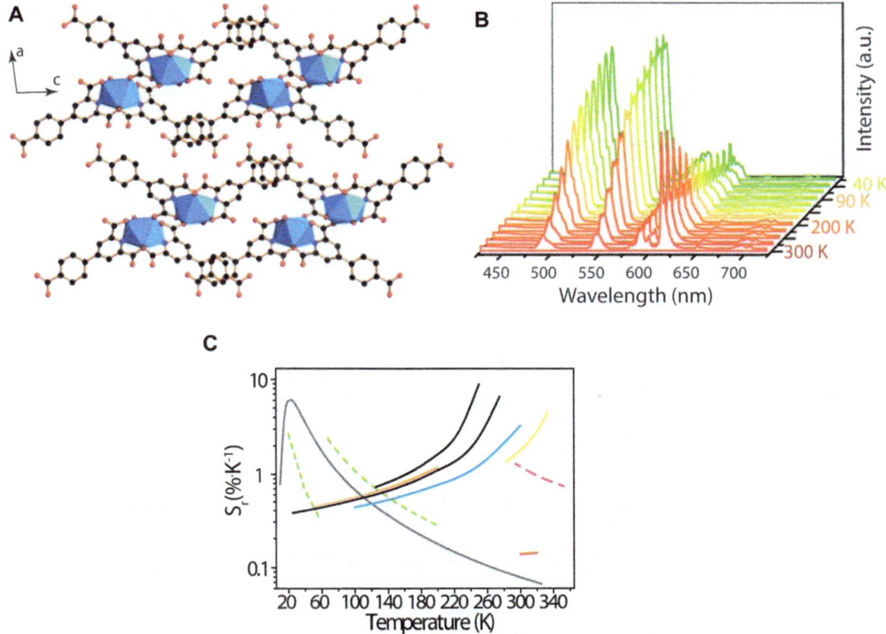

Figure 8.4 (A) Crystal packing viewed along the *b* crystallographic direction (Tb, blue polyhedra; C, black; O, red; N, blue; H atoms are omitted for clarity) of Tbcpda, isostructural with the MOF thermometer $Tb_{0.957}Eu_{0.043}$cpda (cpda = 5-(4-carboxyphenyl)-2,6-pyridinedicarboxylic acid). (B) Emission spectra of $Tb_{0.957}Eu_{0.043}$cpda recorded between 40 and 300 K excited at 335 nm. Adapted from ref. 72 with permission from the Royal Society of Chemistry. (C) Relative sensitivities of Ln^{3+}-based MOF thermometers. The full lines represent thermometers whose thermometric parameter is defined as I_{Tb}/I_{Eu}, whereas the dashed lines represent thermometers whose thermometric parameter is defined using other definitions.

parts but is a result of a cooperative effect between the two components. Polymers and inorganic emitters can be interconnected in several ways in order to generate a thermometer, as shown in Scheme 8.3.

Some polymers, such as poly(*N*-isopropylacrylamide) (PNIPAM)[19] and related polymers,[81] poly(ethylene glycol) (PEG)-based polymers,[82–86] poly(methyl methacrylate) (PMMA)[87] and helix-forming polysaccharides[88] show a conformational transition from a hydrophilic swollen-globule state to a hydrophobic collapsed-coil state over a range of temperatures close to physiological values.[89] When such polymers are interconnected with optical emitters, such as AuNPs, QDs and Ln^{3+} phosphors, temperature-dependent transitions can induce changes in the local environment of the emitters, producing variations in their absorption/emission features (intensity, energy or lifetime). An example of a complex hybrid thermometer is the assembly of AuNPs and PNIPAM.[90] The polymer shows a gradual conformational transition from 290 to 310 K inducing a corresponding shift of the

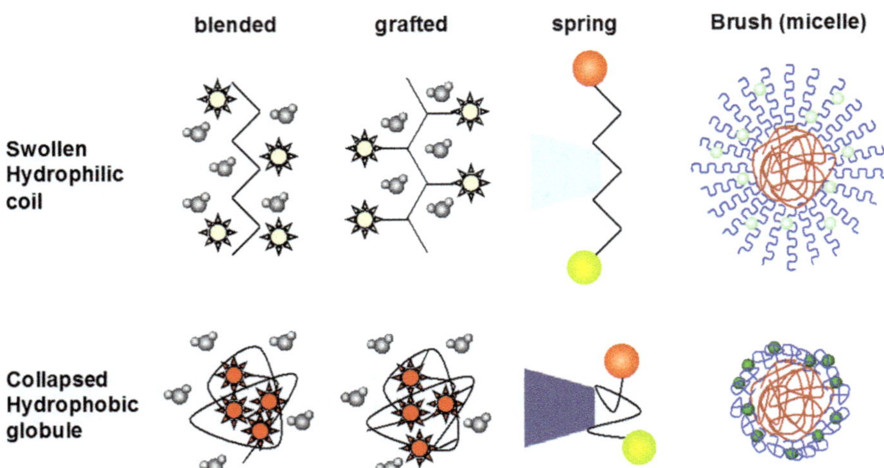

Scheme 8.3 Typical arrangements of chromophores in polymer nanocomposite nanothermometers.
Reproduced from ref. 24 with permission from The Royal Society of Chemistry.

absorption peak of the AuNPs from 540 to 546 nm that can be used as a thermometer with a maximum sensitivity of 0.1% K^{-1} at 295 K (Figure 8.5). In another example with the same components, the polymer PNIPAM was arranged as a NP coating.[91] The refractive index of the PNIPAM coating varies during the conformational transition inducing temperature-dependent changes in the Au plasmon emission wavelength.

The optical emitter can also be a QD,[92,93] as will be explained in more depth in Section 8.3.4. Then, the shrinking of the polymer during the conformational transition shortens the distance between CdTe QDs embedded in the polymer, which is translated into an increase in the intensity and wavelength of the QD emission.[93] By increasing both the level of complexity and the thermometric sensitivity, a truly hybrid nanothermometer is built with a thermosensitive polymer chain linking an AuNP on one end, and a QD on the other.[94,95] The optical emission of this system arises from plasmon-exciton interactions, which are highly dependent on the interparticle distance. The polymer works as a spring with a variable length over the temperature range of 293–333 K, which determines the distance-dependent emission. The high sensitivity of the optical output makes this system very appropriate for new sensing and optoelectronic devices.[95]

8.3.4 Semiconductor Quantum Dots

Semiconductor QDs have been successfully used for thermometry at sub-micrometre scales based on their temperature-dependent photoluminescence. In particular, QD thermometers explore temperature changes in the photoluminescence intensity, the wavelength of the maximum

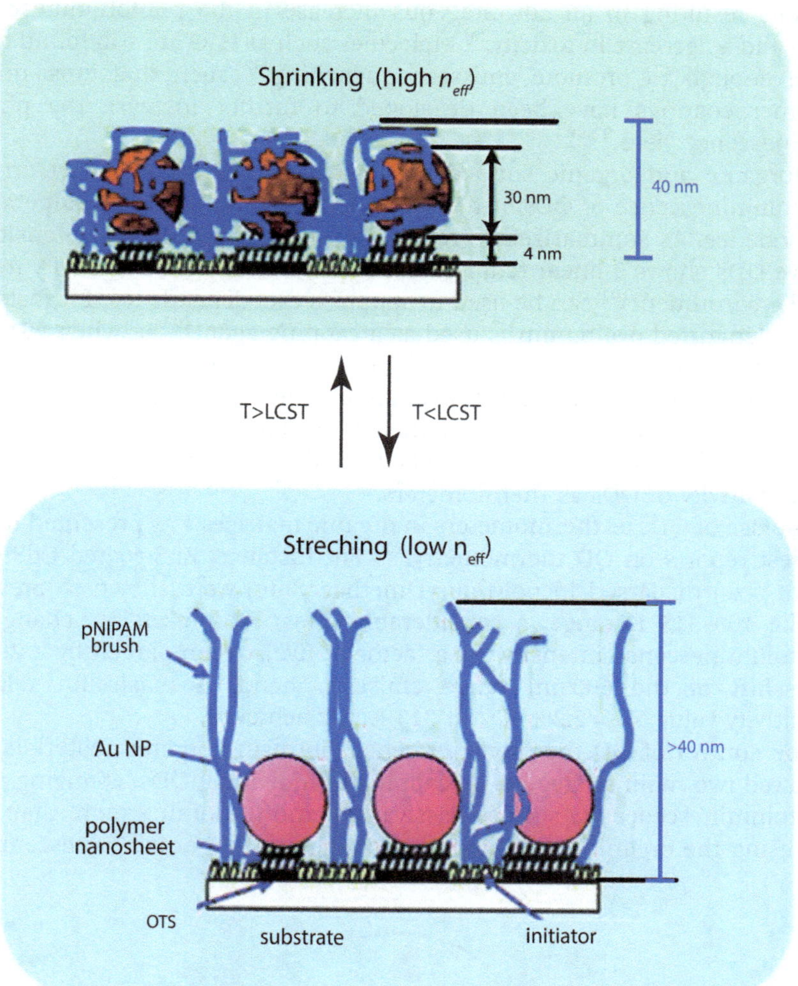

Figure 8.5 Schematic illustration of hybrid nano-assemblies with AuNPs and polymer brushes above and below a lower critical solution temperature (LCST).
Adapted with permission from ref. 90. © American Chemical Society 2007.

intensity peak (peak shift), and the lifetime. QD thermometry benefits from more than a decade of developments in routes for synthesis, capping and stabilization of QDs in polar and apolar fluids, leading to materials with a fine size tuning, a narrow distribution, high quantum yields and photostability that are already on the market for imaging, labelling and sensing. Size tuning and narrow distribution are key to achieving a tuneable emission from the UV to the IR, depending on composition. In QDs, surface control is of fundamental importance to achieving high quantum yields and photostability. An example is the protection of Cd-based cores with a ZnS shell capsule resulting in an advantageous increase in the photoluminescence yield and a decrease in toxicity.[96] Molecules such as H_2O are still found to be close enough to promote emission quenching,[97] such that cross-linked polymer coatings have been developed to further increase the photoluminescence yield.[97,98]

Inorganic and organic surface coatings have a dramatic effect on the photoluminescence of QDs at room temperature and on their temperature dependence. As summarized in ref. 99, while the fluorescence intensity of native QDs shows a linear temperature dependence,[100] its sensitivity to the local environment[101] can be used to suppress this dependence, for instance when denatured ovalbumin is used as a capping agent[102] or when QDs are embedded in polymer particles.[103,104] Capping and conjugation with organic materials (leading to the formation of an organic–inorganic hybrid material) (Figure 8.6) is also a way of enhancing the temperature dependence of photoluminescence intensity (see refs. 99 and 105 for instance), controlling the sensitivity of QDs as thermometers.

The use of QDs as thermometers in organic matrices was presented in the earliest reports on QD thermometry.[106] For instance, ZnS-coated CdS QDs (5 nm size) dispersed in poly(lauryl methacrylate) were shown to present, in the 100–315 K range, a considerable linear and reversible change in photoluminescence intensity (by a factor of five), accompanied by a 20 nm blue-shift of the 600 nm range emission band. A maximum relative sensitivity value $S_m = 2.2\%$ K^{-1} at 313 K was achieved.

The control of QD thermometer sensitivity using organic materials has followed two main strategies: (1) capping of individual QDs, changing their photoluminescence *via* surface passivation; and (2) joining more than one QD using the organic component and changing the photoluminescence of

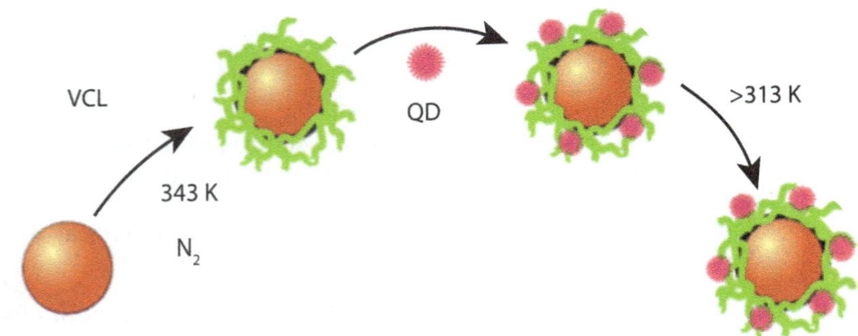

Figure 8.6 Engineering of thermosensitive polymer particles. Poly(acrolein-*co*-styrene) particles were used as solid cores (red spheres). A thermosensitive shell (green) around the solid core was obtained *via* radical polymerization of vinylcaprolactam (VCL). Embedding QDs (pink) in this thermosensitive shell resulted in fluorescent particles whose fluorescence changed due to variations in the distance between QDs as a result of changes in the PVCL conformation at the LCST.
Adapted with permission from ref. 99. © Elsevier 2013.

the composite by changing the inter-QD interactions. Systems using principle (1) include QDs stabilized with capping agents ranging from low-molecular-weight molecules, such as thiocarboxylic acid[107,108] to polymers[109,110] and proteins.[9,111,112] While the use of low-molecular-weight molecules and polymers is aimed primarily at stabilization in solution and control of photoluminescence (as mentioned), the use of proteins allows targeting and intracellular thermometry.[111,112]

In strategy (2) the alteration of inter-QD interactions is achieved *via* a change in the QD average distance mediated by polymers having a phase transition at a given temperature. For instance, CdTe QDs embedded in PNIPAM hydrogels show an increase in the quantum yield, the fluorescence intensity and the emission wavelength with the temperature, during the coil-to-globule transition.[93] PNIPAM-methylacrylic acid co-polymers were further used to promote self-assembly between surface-modified CdTe/ZnS QDs and the thermosensitive polymer, resulting in water-soluble $\sim$300 nm core–shell composites.[105] This concept of coupling a QD to a polymer undergoing a phase transition as a way of enhancing sensitivity was further applied in a different architecture. In this case, the central core (radius of about 90 nm) is a first polymer without a phase transition in the range of interest, with its surface decorated with a second thermosensitive polymer shell containing the QDs (radius of about 135 nm below the transition temperature that decrease to about 100 nm above the transition temperature) (Figure 8.6).[99] The maximum relative sensitivity (estimated here) is about 6.5% K^{-1} in the 300–310 K range. As in the previous examples, the thermopolymer controls the inter-QD distance and its fluorescence intensity. A possible advantage of this system is the improved homogeneity of the distance between the QDs and the external environment whose temperature is to be sensed, when compared to a gel[93] having surface and inner QDs.

The transition of PNIPAM occurs with a dramatic change of its surface hydrophobicity and, thus, stability in a given solvent both below and above the transition may be at risk. A more complex architecture aimed at overcoming this issue was obtained by the introduction of a surface linear thermosensitive polymer, such as PEG[92] linking CdTe NPs ($\sim$4 nm diameter) to a larger central AuNP ($\sim$20 nm diameter) (Figure 8.7). The underlying optical mechanism involves plasmon resonance and exciton–plasmon interactions. The variation of the length of the polymer in the temperature range of 293 to 333 K changes the inter-Au/CdTe distance-dependent resonance conditions of the CdTe emission.

8.3.5 Core–Shell Nanoparticles

An obvious candidate for nanothermometry is a luminescent NP with a temperature-dependent emission, such as a QD,[113] nanodiamond[114,115] or Ln^{3+}-doped NP.[116] For this purpose, organic–inorganic structures can be advantageous over bare inorganic NPs. A simple coating with an organic shell may confer stability in liquid suspensions, biocompatibility, chemical

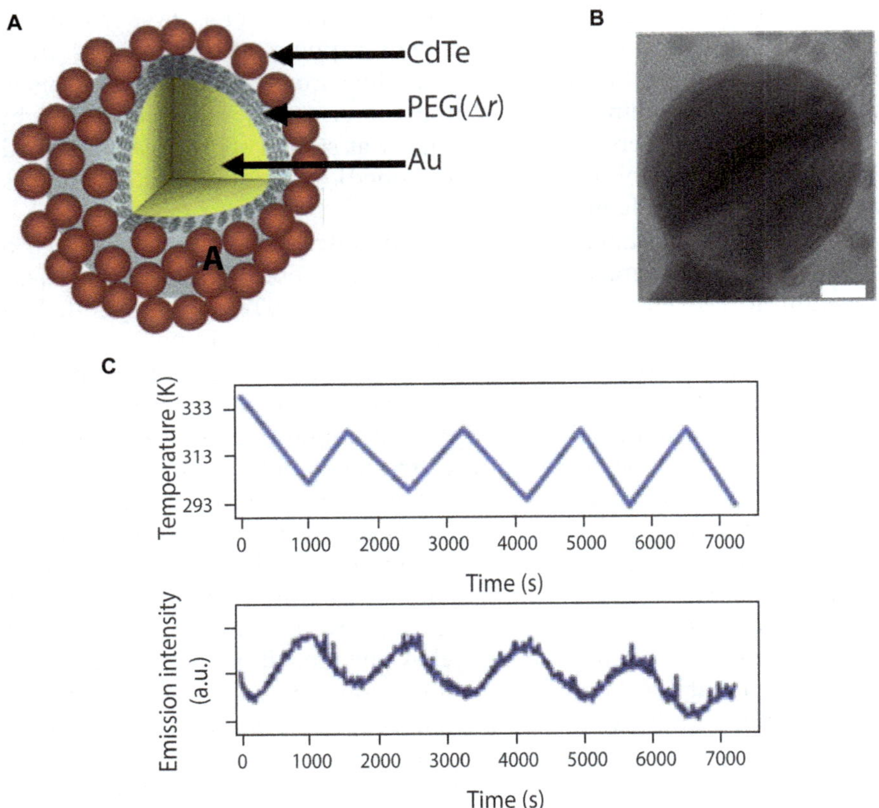

Figure 8.7   (A) Scheme of a hybrid nanothermometer based on two types of light-emitting NPs linked by a thermosensitive polymer, PEG, acting as a spring. (B) Electron microscope image of the nanothermometer (scalebar is 50 nm), and (C) temperature and experimental emission from the superstructure as a function of time.
Adapted with permission from ref. 94. © Wiley-VCH 2005.

reactivity and multifunctionality. Moreover, the organic shell may also provide enhanced sensitivity and tunability to the thermometric nanosensor. That is the case for QDs decorated with macrocyclic compounds, which improve the thermometric sensitivity of QDs by 2.4-fold, and increase the temperature working limit from 333 to 363 K.[117] It is also the case for ruthenium(II) tris(bipyridyl) ($Ru(bpy)_3^{2+}$), which show an orange-red emission that can be thermally deactivated through crossing to a non-radiative level.[118] As the emission is easily quenched, the molecule has to be protected, for instance by encapsulation in a transparent support such as a silica NP. The hybridization of silica-based NPs can be achieved by coating them with a poly(L-lysine) biocompatible polymer shell.[106] The system shows a linear variation of the emission intensity at 630 nm, under 450 nm excitation, in the 293–323 K temperature range. The combination of inorganic and organic components is particularly interesting for the construction of dual-emission ratiometric thermometers.[119] In ref. 120 for example, the thermosensitive emission of a metallorganic complex, β-diketonate chelate Eu^{3+} thenoyltrifluoroacetonate (Eu-TTA), is complemented with a thermo-insensitive organic fluorophore, such as Rh101, to build a ratiometric thermometer. Both fluorophores are encapsulated in a PMMA polymer NP, which is further coated with a protective layer of a poly(allylamine) hydrochloride (PAH, Figure 8.8). The working range is 299–313 K, the accuracy 1 K, and it has been proved useful for measuring local heating in cells in order to follow the Ca^{2+} dynamics.

There is another class of ratiometric luminescent thermometers with hybrid composition that are based on a cooperative effect between inorganic QDs and fluorescence resonance energy transfer (FRET) acceptor organic dyes attached to the NP surface.[121] The thermometric response is based on partial energy transfer from the photoexcited QD to the organic dye. An example is CdSe/CdS NPs modified with Alexa-647. The luminescence of the

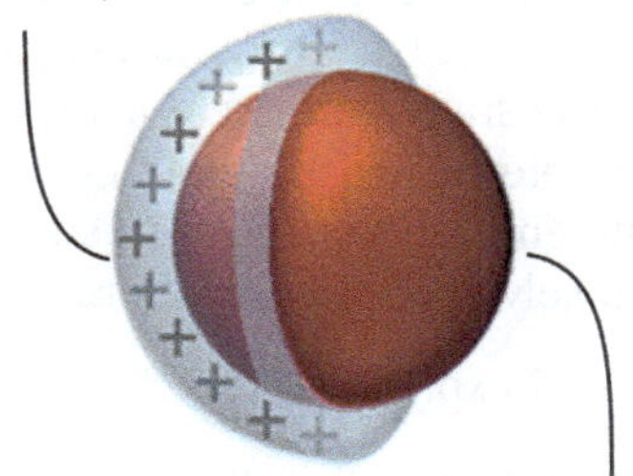

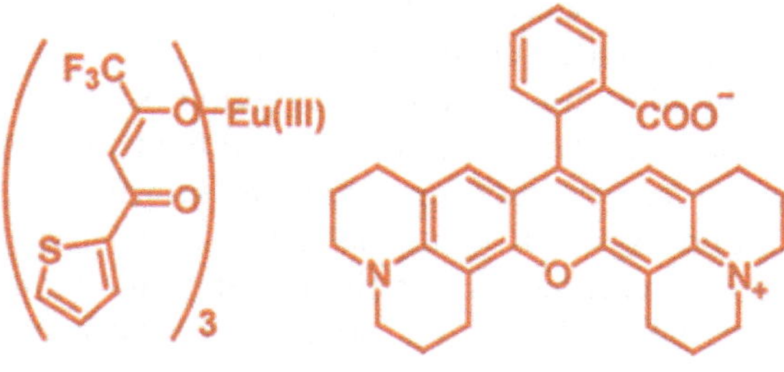

Figure 8.8 Schematic illustration of a ratiometric fluorescent core–shell nanothermometer. The thermosensitive Eu-TTA (excitation and emission wavelengths of 341 nm and 612 nm, respectively) and thermo-insensitive Rh101 (excitation and emission wavelengths of 560 nm and 587 nm, respectively) fluorophores are embedded in a hydrophobic PMMA core. A hydrophilic polymer shell composed of positively charged PAH, isolates the core from the exterior environment.
Adapted with permission from ref. 120. © American Chemical Society 2014.

QD around 615 nm decreases and red-shifts with temperature. The dye emission at 647 nm also decreases with increasing temperature in the range 293–313 K. The temperature can be obtained from the ratio of the QD and dye emission intensities. In some extraordinary cases, both the optical emission and the thermometric response originate from a synergic effect, as is the case for silicon NPs modified with diphenylamine.[122] In this case, the surface modification of SiNPs with the organic ligand improves their photoluminescence quantum yield from 0.08 to 0.75 and makes them water-dispersible and suitable for temperature mapping by fluorescence lifetime imaging. It is quite common for luminescent markers for biological applications to use organic–inorganic hybrid structures, such as inorganic NP emitters coated with a biopolymer with protein anti-adherent properties, like PEG.[123]

Due to the significant absorption by biological tissues in the visible range, caused by scattering and specific absorption of tissue components, such as melanin and haemoglobin, the ideal luminescent thermometers for biological applications should be excited and emit within the biological wavelength window for tissues (700–900 nm). QDs (which are considered in Section 8.3.4), Nd^{3+}-doped NPs and upconverting (UC) lanthanide-doped NPs[123,124] (addressed in Chapter 5) can combine such features. For UC NPs, the excitation and emission spectral range is determined by their Ln^{3+} dopant ions. The temperature is inferred from the emission spectrum using the ratio between the emission intensity of two closely spaced energy levels (1 and 2):

$$I_1/I_2 = C \, \exp(-\Delta E/k_B T), \tag{8.3}$$

where C is a pre-exponential constant depending on the degeneracies of the two levels, the spontaneous emission rates and frequencies of the I_1 and I_2 transitions, ΔE is the energy gap between the levels, k_B is the Boltzmann constant and T is the absolute temperature. The Er^{3+}/Yb^{3+} pair is, by far, the most widely reported due to the efficient $Yb^{3+} \rightarrow Er^{3+}$ energy transfer together with the high absorption cross-section of Yb^{3+} at 980 nm. Noticeably, this physical mechanism allows NIR-to-visible optical conversion through a two-photon excitation mechanism. However the Er^{3+} transitions $(^2H_{11/2} \rightarrow {}^4I_{15/2}$ and ${}^4S_{3/2} \rightarrow {}^4I_{15/2})$ in the visible range restrict its application, due to the above-mentioned short tissue penetration depth. To overcome this limitation, other emitting Ln^{3+} ions can be used with emission within the biological window. A seminal example of NIR-to-NIR conversion used CaF_2 co-doped with Yb^{3+} and Tm^{3+} (instead of Er^{3+}).[125] The maximum sensitivity (0.17% K^{-1} over a room temperature range), however, is not very competitive when compared to examples involving Yb^{3+}/Er^{3+}. An interesting example was reported by Dong *et al.*[125] using the judicious combinations of $Er^{3+}/Yb^{3+}/MoO_4^{2-}$ (visible emission) and $Tm^{3+}/Yb^{3+}/MoO_4^{2-}$ (NIR emission) for doping oxide materials (Al_2O_3, TiO_2, Gd_2O_3 or $Yb_3Al_5O_{12}$). As an example of emission in the visible, the authors reported co-doped $Yb_3Al_5O_{12}$ NPs working in the 295–973 K range with resolution up to 0.3 K and a $S_m = 1.0\%$ K^{-1} at 300 K (Figure 8.9).[125] Additionally the NPs have been used

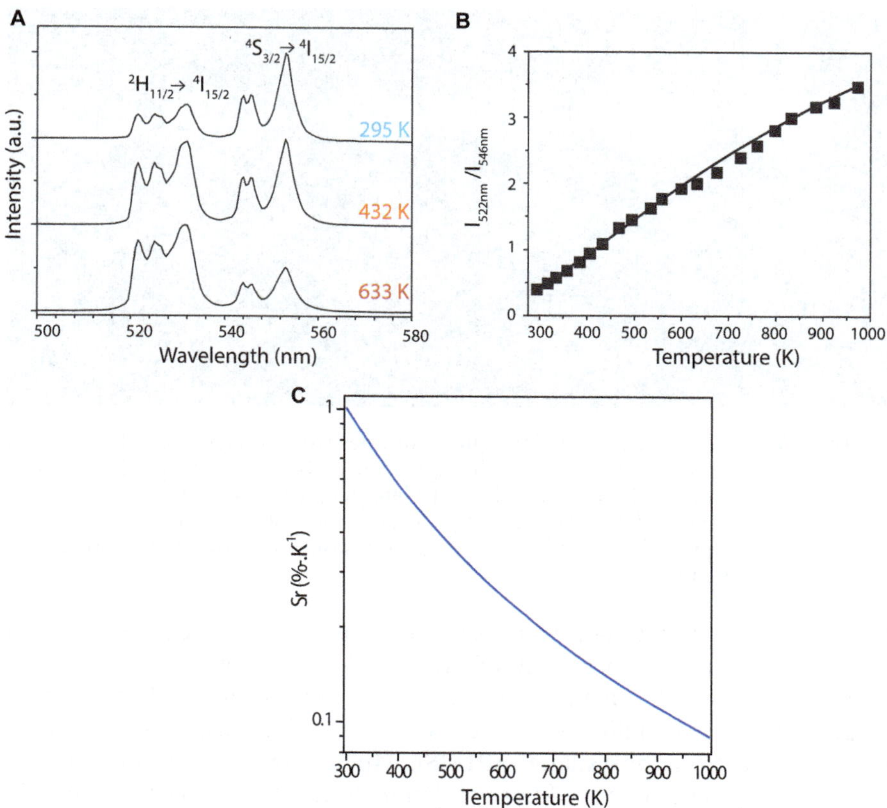

Figure 8.9 Temperature sensing based on Er-Mo:YbAG. (A) Green UC emission spectra at different temperatures. (B) Fluorescence intensity ratio relative to the temperature. Adapted with permission from ref. 125. © Wiley-VCH 2012. (C) Relative sensitivity computed from published data over the reported temperature range.

for *in vivo* UC colour optical imaging after coating with poly(acrylic acid) (PAA) with the successful localization of their emission inside the body.

An interesting example combining excitation and emission in the NIR was reported by Rocha *et al.* using the downshifting of Nd^{3+}-doped LaF_3 NPs.[126] The authors follow the change in the emission maximum or the intensity ratio of two Stark components (at 863 and 885 nm) for sub-tissue NIR thermal sensing. The main advantage of the system is the control in a single-beam plasmonic-mediated heating experiment. As gold nanorods can be efficiently excited by 808 nm radiation, the authors prepared aqueous solutions containing both Nd^{3+}-doped LaF_3 NPs (thermometers) and gold nanorods (heaters) at a concentration level of 0.3 and 0.01% by mass, respectively. The solution was placed underneath a 1 mm thick tissue phantom (a scattering medium that mimics the absorption of the human skin), and the plasmonic excitation heat release was followed by the luminescent thermometer. The main drawback was the low thermal

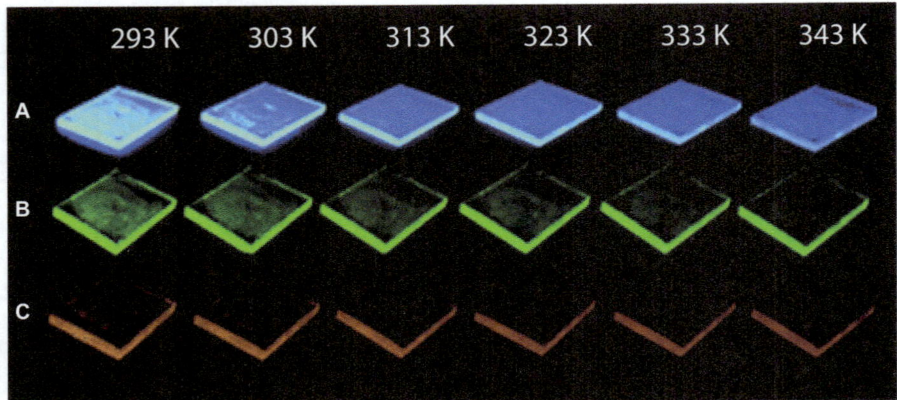

Figure 8.10   Temperature-dependent photoluminescence images of SiC-dot films at given temperatures. The light source is an Hg lamp with suitable excitation filters. Excitation wavelengths are in the (A) UV (360–370 nm), (B) blue (460–490 nm), and (C) green (530–550 nm) range.
Adapted from ref. 127 with permission from the Royal Society of Chemistry.

sensitivity of the thermometer (0.1% K^{-1} using the intensity ratio, or 0.001% K^{-1} using the maximum of the emission peak shift).

Another case of organic-modified NPs that exhibit thermometric properties[127] is that of SiC NPs (SiC-dots) coated with an organosilica layer of 3-aminopropyltrimethoxysilane (APTMS), which shows a quantum yield of 0.426. This system shows a linear variation of the luminescence emission intensity in the 293–343 K range under 360 nm excitation (Figure 8.10).

Although, in common with most thermometric hybrid NPs the emitter is the inorganic component, there are also examples of inorganic NPs encapsulating organic dyes to ensure the stability of the emission and chemical protection from the environment. Silica is an adequate material for this purpose because it is transparent, inert, and thermostable. Therefore it has been used to encapsulate organic dyes, such as RhB, to fabricate temperature nanosensors that show a 94% drop in fluorescence intensity in the 298–418 K interval when coupled to a MEMS (Microelectromechanical System) device.[128]

8.4   Non-optical Thermometers

Although any temperature-dependent property can be used to create a thermometer, most hybrid nanothermometers are based on luminescence emission. Nevertheless, there are a few examples of hybrid nanothermometers that are based on alternative properties.

8.4.1   Magnetic Resonance Imaging

Interventional magnetic resonance imaging (MRI) is becoming an important tool in thermal therapy due to its cost-effective and minimally invasive characteristics. Thus, optimization of local hyperthermia therapies can be better achieved by continuous thermometry. Moreover, local differences in heat absorption and heat conductivity render 3D real-time temperature mapping necessary. Based on the temperature dependence of several proton relaxation parameters, such as proton density, T_1 and T_2 relaxation times, the diffusion coefficient D, magnetization transfer, proton resonance frequency and proton spectroscopic imaging, the MRI technique allows both continuous thermometry and the 3D mapping of temperature changes.[129,130] Accurate use of the technique, however, based on intrinsic MRI thermometric parameters, is limited by its high sensitivity to patient motion and low sensitivity at low magnetic field strengths and in fatty tissues. Moreover, only proton spectroscopic MR imaging can provide absolute temperature values, while the other temperature-sensitive parameters can only provide temperature changes. Comparison of the thermal sensitivities of T_1, D and the chemical shift of proton resonance frequency in phantoms showed sensitivities better than 1 K for proton resonance frequency and diffusion methods, and about 2 K for T_1 methods.[131] Although such sensitivities decrease in homogeneous media containing fat,[132] values of ± 0.5 K *in vivo* have reported recently.[133]

The use of temperature-sensitive contrast agents can further increase the sensitivity and accuracy of the technique. Examples include paramagnetic thermosensitive liposomes,[134] lanthanide complexes,[135,136] multifunctional NPs,[137] and spin-transition molecular materials.[138] In most cases, these materials can provide absolute temperature values but, as thermometers, they are limited to the transition temperatures at which molecular changes occur. Application of some of these techniques extends beyond their use in MRI and will be described in separate sections.

8.4.2 Thermosensitive Liposomes

Paramagnetic thermosensitive liposomes encapsulate molecular contrast agents such as Gd- or Mn-complexes in lipid NPs that melt at a certain temperature liberating the contrast agent. When, in addition, these complexes have been previously incorporated into a polymer hydrogel, such as NIPAM, which shows swollen–collapsed transitions at a different temperature, the system is able to yield two temperature values (Figure 8.11). Based on this principle, a system has been proposed consisting of a gadolinium-*N,N,N',N'',N''*-diethylenetriaminepentaacetic acid (Gd-DTPA) complex attached to hydrogel NPs by chelation with *N,N'*-bisallylamido-diethylenetriamine-*N,N',N''*-triacetic acid (BADTTA).[139] Paramagnetic thermosensitive liposomes for MRI have been reviewed by Lindner *et al.*,[134] and *in vivo* application has been demonstrated in animals.[140,141]

In addition, hybrid thermosensitive liposomes are also applied for drug delivery in combination with local hyperthermia.[142,143] Liposomes are particularly suited to retaining a chemotherapeutic drug inside the phospholipid bi-layer membrane below a triggering temperature, above which the membrane leaks the drug. There are a rather large number of

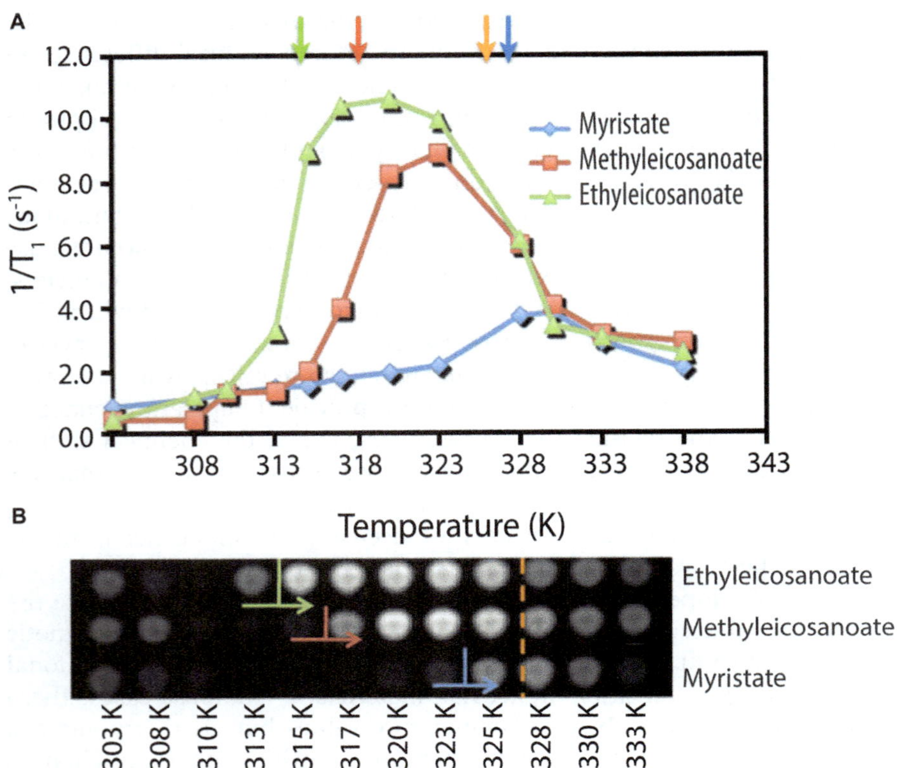

Figure 8.11 HLN provides an off–on signal for MRI thermometry *in vitro*. (A) The water proton relaxation rate $(1/T_1)$ as a function of temperature was determined, *in vitro*, for HLN (hydrogel–lipid hybrid nanoparticle) formulations with different matrix lipids, each with different melting temperatures indicated by the solid arrows above the curve. The arrows are colour matched to the curves they relate to, and the orange arrow indicates the T_{tr} (transition temperature) of the common hydrogel composition (65:35 mol NIAPM:AM) used. The T_1-weighted contrast signal enhancement provided by each HLN formulation as a function of temperature is shown in (B). The solid lines indicate the lipid melting temperatures of ethyleicosanoate (green), methyleicosanoate (red), and myristate (blue), the arrows indicate a signal enhancement transition, and the dashed orange line indicates the hydrogel *Ttr*.
Adapted with permission from ref. 139. © Elsevier 2012.

formulations already described, differences in their thermal properties being in the triggering temperature ranges and release times.[144] A temperature-sensitive liposome based on 1,2-dipalmitoyl-*sn*-glycero-3-phosphocholine (DPPC) and 1,2-distearoyl-*sn*-glycero-3-phosphocholine (DSPC) containing doxorubicin and originally proposed by Needham *et al.*[145] has been shown to release 100% of its contents through stabilized membrane pores within 10–20 s at 314 K. This formulation has exhibited dramatic improvements in pre-clinical drug delivery and tumour regression and is now in clinical trials for liver and prostate cancers.[146,147] Another formulation encapsulates cisplatine in the liposomes developed by Needham *et al.*;[145] at 310 K the release of cisplatine was less than 5% while more than 95% of the encapsulated cisplatine was released at 315 K.[148]

A temperature threshold can be achieved by combining thermally sensitive liposomes with magnetic NPs,[149,150] although a simple water bath has also been used for low-temperature thermosensitive liposomes.[144] Magnetic NPs, normally iron oxide particles of sizes ranging between 8 and 15 nm, can be induced to generate heat when exposed to an alternating magnetic field in the frequency range of radiowaves (500 kHz to 2 MHz).[151–153] The technique is called magnetic fluid hyperthermia (MFH) and the design of magnetic NPs for efficient loading and heating of host thermosensitive liposomes can be rather complex.[154] However, the thermometric system is always related to the composition of the liposome formulation.

8.4.3 Spin-transition Molecular Materials

Six-fold coordination compounds of $3d^n$ ($n = 4$ to 7) transition metals can have in their ground state as either a low spin (LS) or a high spin (HS) configuration depending, to a first approximation, on the relative magnitudes of their ligand field and spin-pairing energies. For a few compounds these energies are close enough to each other to be on the order of thermal energy. Consequently, the spin states of the compound can be reversibly switched between HS and LS by an external stimulus such as temperature, pressure or irradiation.[155–157] In many cases the transition between the electronic states occurs at two different values of the external stimulus depending on whether the stimulus increases its action or decreases it. These compounds show hysteresis. In the case of temperature-driven transitions, a spin-transition compound is characterized by two critical temperatures, $T_c(\text{up})$ and $T_c(\text{down})$ corresponding respectively to the warming up and cooling down pathways.[158]

Although molecular in nature, the phenomenon also depends in a co-operative way on intermolecular interactions within the crystal lattice. Thus, a variety of spin transitions have been observed, *e.g.*, abrupt, smooth, in one step, in two steps, *etc.*[156,159] In addition, compounds showing an abrupt spin transition in the room temperature region have been reported.[160] The possibility of using thermal spin-cross-over compounds possessing an abrupt transition and hysteresis for data recording,[161] memory devices,[162] temperature sensors,[163] and thermotherapy[164,165] has been explored and reviewed.

However, from the point of view of the scope of this book, space resolution achieved with this technique is at the sub-millimetre scale at best. Thus, polymeric structures of $[\text{Fe}(\text{Htrz})_{3-3x} (4\text{-NH}_2\text{trz})_{3x}](\text{ClO}_4)_2 \cdot n\text{H}_2\text{O}$, where Htrz = 1,2,4-triazole and 4-NH$_2$trz = 4-amino-1,2,4-triazole, can be tuned to the desired temperature transitions by adjusting the amount of 4-NH$_2$trz substituting Htrz.[166] In these compounds the Fe(II) ions have a transition from $S = 2$ in the paramagnetic state above the transition temperature T_c to $S = 0$ in the diamagnetic state below T_c. In the paramagnetic state the Fe(II) ions will introduce field inhomogeneities, which will alter the NMR signal in spectroscopy and in imaging. This has been used to investigate the possi-bilities of these materials for thermosensitive imaging by NMR in phan-toms.[164] Although optimising space resolution was not the aim of this study, it was not higher than millimetres. The possibilities of Fe(II) spin-transition systems have been also investigated as temperature-responsive paramagnetic chemical exchange saturation transfer (PARACEST) contrast agents in MRI thermometry. As PARACEST-based MRI thermometry is very sensitive to small changes of the electronic spin state of the contrast agent, two Fe(II) spin-cross-over systems, $[Fe(3\text{-}bpp)_2](BF_4)_2 \cdot 3Et_2O$ (3-bpp = 2,6-di(pyrazol-3-yl)pyridine) and $[(Me_2NPY_5Me_2)Fe(H_2O)](BF_4)_2 \cdot H_2O$ (Me$_2$NPY$_5$Me$_2$ = 4-dimethylamino-2,6-bis(1,1-bis(2-pyridyl)ethyl)pyridine), with an abrupt spin transition were considered ideal agents for these studies.[165] Moreover, the possibilities of $[(Me_2NPY_5Me_2)Fe(H_2O)](BF_4)_2 \cdot H_2O$ for temperature mapping were also explored in phantoms. The images were decomposed in pixels such that the pre-saturation frequency giving the minimum intensity was converted to a temperature using the linear relationship determined from the NMR Z-spectrum. The pixel surface was 0.234×0.234 mm.[165]

Temperature-driven spin-transition systems have also been prepared as NPs. Linter cellulose fiber nanocomposites were prepared by functionalizing the fibres with $[Fe(hptrz)_3](OTs)_2$ spin-cross-over NPs and acridine orange dye molecules. Cotton linters are fine, silky fibers which adhere to the seeds of the cotton plant after ginning. The cellulose sheets prepared from the nanocomposite show significant thermochromic and thermofluorescent effects associated with the spin-transition phenomenon, providing scope for paper thermometry and anti-counterfeit applications.[167]

8.4.4 Miscellaneous Systems

Examples of electrical resistance hybrid thermometers can also be found. In particular, metal–organic composites made by intercalation of Sn^+ ion beams in polyetheretherketone (PEEK) have been proposed as high-performance, low-cost thermometers, with a tuneable temperature range.[168]

Using the most typical thermal indicator, the expansion of a liquid in a tube, a hybrid nanothermometer was built by filling a carbon nanotube with liquid Ga.[169] This thermometer could work over a 323–773 K range, but it would be necessary to use a scanning electron microscope to do the reading.

Magnetic moment is also a thermosensitive property, thus a nanotherm-ometer based on magnetic NPs with an organic coating has also been pro-posed.[170] The accuracy can be as good as 0.0055%. The main inconvenience is that the reading is done with a SQUID (Superconducting Quantum Inter-ference Device) sensor with limited spatial resolution.

8.5 Perspectives: Hybrids Joining Nanothermometry and Nanoheating

Among the many possibilities that hybrid organic–inorganic materials offer in the field of nanothermometry one of them is now attracting much interest. That is joining thermometry and heating in a single nano-object. Inorganic NPs such as magnetic NPs or AuNPs can become heat sources by irradiation with alternating magnetic fields or laser light. The combination of these inorganic NPs with organic components in a hybrid structure may allow the incorporation of non-contact thermometers with these nanoheaters so the temperature of the nanoheater can be monitored as a function of the heater–thermometer distance. This nanothermometric heater could be beneficial in many technological applications from biomedicine to microelectronics. One of the most interesting is the application of hyperthermia to the therapy of cancer[171] and other diseases.[172] Heating tumours above 315–316 K with radiofrequency irradiation of magnetic NPs that have been injected into the tumour has already been used in clinical trials. The use of nanothermometric heaters can be very useful for developing more efficient hyperthermia therapies based on local intracellular heating.

Developments in this direction are very recent, and are based both on purely inorganic materials,[173,174] and on hybrid structures. A hybrid approach[175] consists of binding an organic fluorophore, such as fluorescein amine, to the shell of a magnetic heating core with an azo bond that breaks at a certain temperature. The measurement of the fluorescent emission of the fluorophore detached from the NP gives an estimation of the temperature reached in the NP. A further improvement is achieved by attaching the fluorophore to a DNA strand that is placed on the NP surface by hybridization with another DNA strand on the surface of the NP. The temperature is followed after the detachment of the fluorophore by the thermal denaturation of double-stranded DNA at a temperature that can be fine-tuned (see Figure 8.12).[176] A third approach uses a heating core of either Au or Fe_2O_3 coated with polymer and RhB (that is entrapped in the polymer surface by electrostatic interactions).[177] The temperature is monitored through the excited-state lifetime of the dye.

The development of local heating techniques and nanoheat diffusion mechanisms requires an adequate monitoring of the local temperature of the nanoheaters with a high time resolution and a low temperature uncertainty. Very recently,[178] the use of molecular luminescent probes, such as the lanthanide complexes described in Section 8.3.1, embedded in coreshell magnetic/co-polymer NPs has provided these features. The complexes are encapsulated in the hydrophobic inner part of the co-polymer shell, around the iron oxide magnetic core. The resulting nanoheater/nanothermometer shows outstanding performance in terms of: low thermometer heat capacitance (0.021 K^{-1}) and heater/thermometer resistance (1 K W^{-1}); high temperature sensitivity (5.8% K^{-1} at 296 K) and uncertainty (0.5 K); physiological working temperature range (295–315 K); readout reproducibility (>99.5%); and fast time response (0.250 s). Experiments of time-resolved thermometry under an ac magnetic field reveal the failure of using macroscopic thermal parameters to describe heat diffusion at the nanoscale. Very preliminary experiments on cells incubated with the NPs showed that temperature mapping of the cell's interior is possible by simple fluorescence microscopy at two different wavelengths with pixel-to-pixel resolution.

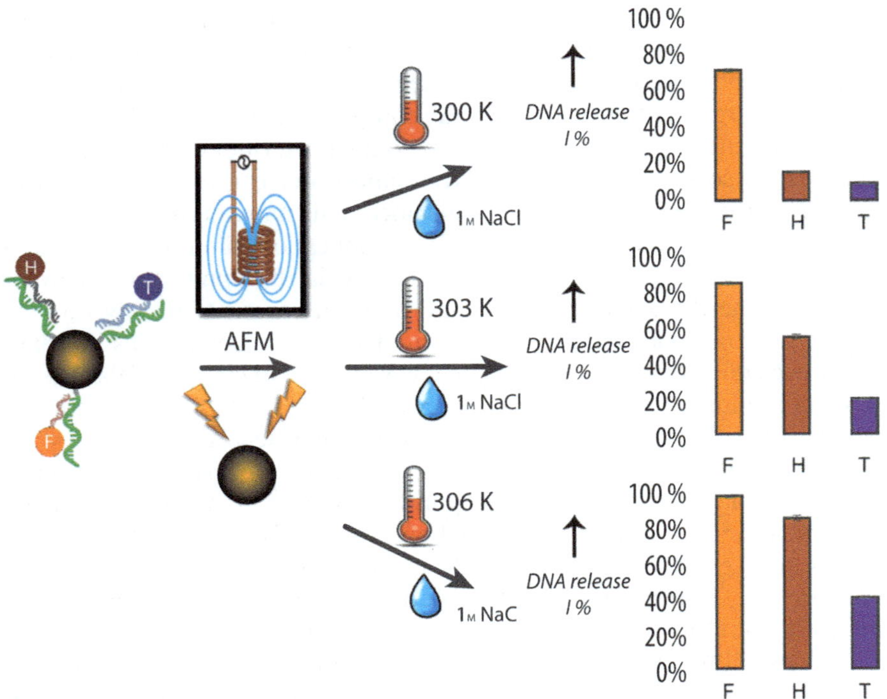

Figure 8.12 AMF (alternating magnetic field) heating of DNA-MNPs, followed by their precipitation by salt addition, and measurement of the corresponding fluorescence of the supernatants (F: DNA-F; H: DNA-H; T: DNA-T), for three of the final global temperatures reached. Release efficiencies were calculated after determination of the total DNA denatured after the samples were heated up to 368 K for 20 min. Adapted with permission from ref. 176. © Wiley-VCH 2013.

References

1. L. D. Carlos, R. A. S. Ferreira, V. de Zea Bermudez and S. J. L. Ribeiro, *Adv. Mater.*, 2009, **21**, 509–534.
2. C. Sanchez, B. Julián-López, P. Belleville and M. Popall, *J. Mater. Chem.*, 2005, **15**, 3559–3592.
3. P. Escribano, B. Julián-López, J. Planelles-Aragó, E. Cordoncillo, B. Viana and C. Sanchez, *J. Mater. Chem.*, 2008, **18**, 23.
4. C. Sanchez, P. Belleville, M. Popall and L. Nicole, *Chem. Soc. Rev.*, 2011, **40**, 696–753.
5. L. D. Carlos, R. A. Ferreira, V. de Zea Bermudez, B. Julián-López and P. Escribano, *Chem. Soc. Rev.*, 2011, **40**, 536–549.
6. B. Lebeau and P. Innocenzi, *Chem. Soc. Rev.*, 2011, **40**, 886–906.
7. X. D. Wang, D. E. Achatz, C. Hupf, M. Sperber, J. Wegener, S. Bange, J. M. Lupton and O. S. Wolfbeis, *Sens. Actuators, B*, 2013, **188**, 257–262.

8. P. Löw, B. Kim, N. Takama and C. Bergaud, *Small*, 2008, **4**, 908–914.

9. S. Li, K. Zhang, J. M. Yang, L. W. Lin and H. Yang, *Nano Lett.*, 2007, 7, 3102–3105.

10. E. M. Graham, K. Iwai, S. Uchiyama, A. P. de Silva, S. W. Magennis and A. C. Jones, *Lab-on-a-Chip*, 2010, **10**, 1267–1273.

11. G. Baffou, M. Kreuzer, F. Kulzer and R. Quidant, *Opt. Express*, 2009, **17**, 3291–3298.

12. C. D. S. Brites, P. P. Lima, N. J. O. Silva, A. Millán, V. S. Amaral, F. Palacio and L. D. Carlos, *Adv. Mater.*, 2010, **22**, 4499–4504.

13. C. D. S. Brites, P. P. Lima, N. J. O. Silva, A. Millán, V. S. Amaral, F. Palacio and L. D. Carlos, *J. Lumin.*, 2013, **133**, 230–232.

14. C. D. S. Brites, P. P. Lima, N. J. O. Silva, A. Millán, V. S. Amaral, F. Palacio and L. D. Carlos, *Nanoscale*, 2013, **5**, 7572–7580.

15. C. Gota, S. Uchiyama, T. Yoshihara, S. Tobita and T. Ohwada, *J. Phys. Chem. B*, 2008, **112**, 2829–2836.

16. J. B. Yu, L. N. Sun, H. S. Peng and M. I. J. Stich, *J. Mater. Chem.*, 2010, **20**, 6975–6981.

17. K. Okabe, N. Inada, C. Gota, Y. Harada, T. Funatsu and S. Uchiyama, *Nat. Commun.*, 2012, **3**, 705.

18. C. Paviolo, A. H. Clayton, S. L. McArthur and P. R. Stoddart, *J. Microsc.*, 2013, **250**, 179–188.

19. A. H. Khalid and K. Kontis, *Meas. Sci. Technol.*, 2009, **20**, 025305.

20. C. D. S. Brites, P. P. Lima, N. J. O. Silva, A. Millán, V. S. Amaral, F. Palacio and L. D. Carlos, *New J. Chem.*, 2011, **35**, 1177–1183.

21. S. Uchiyama, A. P. de Silva and K. Iwai, *J. Chem. Educ.*, 2006, **83**, 720–727.

22. J. C. Bunzli, *Acc. Chem. Res.*, 2006, **39**, 53–61.

23. A. Rabhiou, J. Feist, A. Kempf, S. Skinner and A. Heyes, *Sens. Actuators, A*, 2011, **169**, 18–26.

24. C. D. S. Brites, P. P. Lima, N. J. O. Silva, A. Millán, V. S. Amaral, F. Palacio and L. D. Carlos, *Nanoscale*, 2012, **4**, 4799–4829.

25. D. Jaque and F. Vetrone, *Nanoscale*, 2012, **4**, 4301–4326.

26. L. H. Fischer, G. S. Harms and O. S. Wolfbeis, *Angew. Chem., Int. Ed.*, 2011, **50**, 4546–4551.

27. X. D. Wang, O. S. Wolfbeis and R. J. Meier, *Chem. Soc. Rev.*, 2013, **42**, 7834–7869.

28. D. Parker, *Coord. Chem. Rev.*, 2000, **205**, 109–130.

29. P. P. Lima, M. M. Nolasco, F. A. A. Paz, R. A. S. Ferreira, R. L. Longo, O. L. Malta and L. D. Carlos, *Chem. Mater.*, 2013, **25**, 586–598.

30. V. de Zea Bermudez, L. D. Carlos and L. Alcácer, *Chem. Mater.*, 1999, **11**, 569–580.

31. E. Pecoraro, R. A. Sá Ferreira, C. Molina, S. J. L. Ribeiro, Y. Messsaddeq and L. D. Carlos, *J. Alloys Compd.*, 2008, **451**, 136–139.

32. C. D. S. Brites, P. P. Lima, N. J. O. Silva, A. Millán, V. S. Amaral, F. Palacio and L. D. Carlos, *Front. Chem.*, 2013, **1**, 9.

33. R. A. S. Ferreira, C. D. S. Brites, C. M. S. Vicente, P. P. Lima, A. R. N. Bastos, P. G. Marques, M. Hiltunen, L. D. Carlos and P. S. André, *Laser Photonics Rev.*, 2013, **7**, 1027–1035.
34. M. G. Burzo, P. L. Komarov and P. E. Raad, *IEEE Trans. Compon. Packag. Technol.*, 2005, **28**, 637–643.
35. H. X. Liu, W. Q. Sun, A. Xiang, T. W. Shi, Q. Chen and S. Y. Xu, *Nanoscale Res. Lett.*, 2012, **7**, 1–6.
36. J. Christofferson, K. Maize, Y. Ezzahri, J. Shabani, X. Wang and A. Shakouri, *J. Electron. Packag.*, 2008, **130**, 041101–041106.
37. V. K. Rai, *Appl. Phys. B: Lasers Opt.*, 2007, **88**, 297–303.
38. K. Kim, W. H. Jeong, W. C. Lee and P. Reddy, *ACS Nano*, 2012, **6**, 4248–4257.
39. D. Ross, M. Gaitan and L. E. Locascio, *Anal. Chem.*, 2001, **73**, 4117–4123.
40. W. Jung, Y. W. Kim, D. Yim and J. Y. Yoo, *Sens. Actuators, A*, 2011, **171**, 228–232.
41. R. Samy, T. Glawdel and C. L. Ren, *Anal. Chem.*, 2008, **80**, 369–375.
42. X. L. Guan, X. Y. Liu and Z. X. Su, *J. Appl. Polym. Sci.*, 2007, **104**, 3960–3966.
43. R. B. Mikkelsen and D. F. H. Wallach, *Cell Biol. Int. Rep.*, 1977, **1**, 51–55.
44. X. X. He, J. H. Duan, K. M. Wang, W. H. Tan, X. Lin and C. M. He, *J. Nanosci. Nanotechnol.*, 2004, **4**, 585–589.
45. F. Wang, W. B. Tan, Y. Zhang, X. P. Fan and M. Q. Wang, *Nanotechnology*, 2006, **17**, R1–R13.
46. M. E. Köse, B. F. Carroll and K. S. Schanze, *Langmuir*, 2005, **21**, 9121–9129.
47. M. E. Köse, A. Omar, C. A. Virgin, B. F. Carroll and K. S. Schanze, *Langmuir*, 2005, **21**, 9110–9120.
48. N. Chandrasekharan and L. A. Kelly, *J. Am. Chem. Soc.*, 2001, **123**, 9898–9899.
49. G. A. Baker, S. N. Baker and T. M. McCleskey, *Chem. Commun.*, 2003, 2932–2933.
50. J. Feng, K. J. Tian, D. H. Hu, S. Q. Wang, S. Y. Li, Y. Zeng, Y. Li and G. Q. Yang, *Angew. Chem., Int. Ed.*, 2011, **50**, 8072–8076.
51. J. M. Lupton, *Appl. Phys. Lett.*, 2002, **81**, 2478–2480.
52. M. Engeser, L. Fabbrizzi, M. Licchelli and D. Sacchi, *Chem. Commun.*, 1999, 1191–1192.
53. C. Quintero, G. Molnár, L. Salmon, A. Tokarev, C. Bergaud and A. Bousseksou, in *16th International Workshop on Thermal Investigations of ICs and Systems (THERMINIC)*, IEEE Components, Packaging, and Manufacturing Technology Society, Barcelona, 2010, pp. 1–5.
54. L. Salmon, G. Molnar, D. Zitouni, C. Quintero, C. Bergaud, J. C. Micheau and A. Bousseksou, *J. Mater. Chem.*, 2010, **20**, 5499–5503.
55. T. Itoh, T. Shichi, T. Yui, H. Takahashi, Y. Inui and K. Takagi, *J. Phys. Chem. B*, 2005, **109**, 3199–3206.

56. W. Li, D. P. Yan, R. Gao, J. Lu, M. Wei and X. Duan, *J. Nanomater.*, 2013, **2013**.

57. D. P. Yan, J. Lu, J. Ma, M. Wei, D. G. Evans and X. Duan, *Angew. Chem., Int. Ed.*, 2011, **50**, 720–723.

58. M. D. Allendorf, C. A. Bauer, R. K. Bhakta and R. J. T. Houk, *Chem. Soc. Rev.*, 2009, **38**, 1330–1352.

59. J. Rocha, L. D. Carlos, F. A. A. Paz and D. Ananias, *Chem. Soc. Rev.*, 2011, **40**, 926–940.

60. L. E. Kreno, K. Leong, O. K. Farha, M. Allendorf, R. P. Van Duyne and J. T. Hupp, *Chem. Rev.*, 2012, **112**, 1105–1125.

61. Y. Takashima, V. M. Martinez, S. Furukawa, M. Kondo, S. Shimomura, H. Uehara, M. Nakahama, K. Sugimoto and S. Kitagawa, *Nat. Commun.*, 2011, **2**.

62. B. V. Harbuzaru, A. Corma, F. Rey, P. Atienzar, J. L. Jorda, H. Garcia, D. Ananias, L. D. Carlos and J. Rocha, *Angew. Chem., Int. Ed.*, 2008, **47**, 1080–1083.

63. Y. J. Cui, Y. F. Yue, G. D. Qian and B. L. Chen, *Chem. Rev.*, 2012, **112**, 1126–1162.

64. B. V. Harbuzaru, A. Corma, F. Rey, J. L. Jorda, D. Ananias, L. D. Carlos and J. Rocha, *Angew. Chem., Int. Ed.*, 2009, **48**, 6476–6479.

65. W. J. Rieter, K. M. Pott, K. M. L. Taylor and W. B. Lin, *J. Am. Chem. Soc.*, 2008, **130**, 11584–11585.

66. H. Xu, F. Liu, Y. J. Cui, B. L. Chen and G. D. Qian, *Chem. Commun.*, 2011, **47**, 3153–3155.

67. J. Della Rocca, D. M. Liu and W. B. Lin, *Acc. Chem. Res.*, 2011, **44**, 957–968.

68. W. T. Yang, J. Feng and H. J. Zhang, *J. Mater. Chem.*, 2012, **22**, 6819–6823.

69. Y. J. Cui, H. Xu, Y. F. Yue, Z. Y. Guo, J. C. Yu, Z. X. Chen, J. K. Gao, Y. Yang, G. D. Qian and B. L. Chen, *J. Am. Chem. Soc.*, 2012, **134**, 3979–3982.

70. X. T. Rao, T. Song, J. K. Gao, Y. J. Cui, Y. Yang, C. D. Wu, B. L. Chen and G. D. Qian, *J. Am. Chem. Soc.*, 2013, **135**, 15559–15564.

71. R. F. D'Vries, S. Alvarez-Garcia, N. Snejko, L. E. Bausa, E. Gutierrez-Puebla, A. de Andres and M. A. Monge, *J. Mater. Chem. C*, 2013, **1**, 6316–6324.

72. Y. J. Cui, W. F. Zou, R. J. Song, J. C. Yu, W. Q. Zhang, Y. Yang and G. D. Qian, *Chem. Commun.*, 2014, **50**, 719–721.

73. Y. H. Han, C. B. Tian, Q. H. Li and S. W. Du, *J. Mater. Chem. C*, 2014, **2**, 8065–8070.

74. A. Cadiau, C. D. S. Brites, P. M. F. J. Costa, R. A. S. Ferreira, J. Rocha and L. D. Carlos, *ACS Nano*, 2013, **7**, 7213–7218.

75. Y. Cui, H. Xu, Y. Yue, Z. Guo, J. Yu, Z. Chen, J. Gao, Y. Yang, G. Qian and B. Chen, *J. Am. Chem. Soc.*, 2012, **134**, 3979–3982.

76. Y. Cui, R. Song, J. Yu, M. Liu, Z. Wang, C. Wu, Y. Yang, Z. Wang, B. Chen and G. Qian, *Adv. Mater.*, 2015, **27**, 1420–1425.

77. X. Rao, T. Song, J. Gao, Y. Cui, Y. Yang, C. Wu, B. Chen and G. Qian, *J. Am. Chem. Soc.*, 2013, **135**, 15559–15564.

78. Y. Zhou, B. Yan and F. Lei, *Chem. Commun.*, 2014, **50**, 15235–15238.

79. Y. Zhang, W. Wei, G. K. Das and T. T. Y. Tan, *J. Photochem. Photobiol. C*, 2014, **20**, 71–96.

80. Z. Wang, D. Ananias, A. Carné-Sánchez, C. D. S. Brites, D. Maspoch, J. Rocha and L. D. Carlos, *Adv. Funct. Mater.*, 2015, **25**, 2824–2830.

81. C. Gota, S. Uchiyama and T. Ohwada, *Analyst*, 2007, **132**, 121–126.

82. C. Pietsch, A. Vollrath, R. Hoogenboom and U. S. Schubert, *Sensors*, 2010, **10**, 7979–7990.

83. W. T. Wu, J. Shen, P. Banerjee and S. Q. Zhou, *Adv. Funct. Mater.*, 2011, **21**, 2830–2839.

84. S. Y. Lee, S. Lee, I. C. Youn, D. K. Yi, Y. T. Lim, B. H. Chung, J. F. Leary, I. C. Kwon, K. Kim and K. Choi, *Chem.–Eur. J.*, 2009, **15**, 6103–6106.

85. W. T. Wu, J. Shen, P. Banerjee and S. Q. Zhou, *Biomaterials*, 2010, **31**, 7555–7566.

86. Y. Chen and X. Li, *Biomacromolecules*, 2011, **12**, 4367–4372.

87. C. Pietsch, R. Hoogenboom and U. S. Schubert, *Polym. Chem.*, 2010, **1**, 1005–1008.

88. T. Shiraki, A. Dawn, Y. Tsuchiya and S. Shinkai, *J. Am. Chem. Soc.*, 2010, **132**, 13928–13935.

89. H. G. Schild, *Prog. Polym. Sci.*, 1992, **17**, 163–249.

90. M. Mitsuishi, Y. Koishikawa, H. Tanaka, E. Sato, T. Mikayama, J. Matsui and T. Miyashita, *Langmuir*, 2007, **23**, 7472–7474.

91. M. Honda, Y. Saito, N. I. Smith, K. Fujita and S. Kawata, *Opt. Express*, 2011, **19**, 12375–12383.

92. J. Lee and N. A. Kotov, *Nano Today*, 2007, **2**, 48–51.

93. J. Li, X. Hong, Y. Liu, D. Li, Y. W. Wang, J. H. Li, Y. B. Bai and T. J. Li, *Adv. Mater.*, 2005, **17**, 163–166.

94. J. Lee, A. O. Govorov and N. A. Kotov, *Angew. Chem., Int. Ed.*, 2005, **44**, 7439–7442.

95. S. M. Sadeghi, *IEEE Trans. Nanotechnol.*, 2011, **10**, 566–571.

96. I. L. Medintz, H. T. Uyeda, E. R. Goldman and H. Mattoussi, *Nat. Mater.*, 2005, **4**, 435–446.

97. H. Weller, J. Ostermann, C. Schmidtke and H. Kloust, *Zeitschrift Fur Physikalische Chemie-International Journal of Research in Physical Chemistry & Chemical Physics*, 2014, **228**, 183–192.

98. J. Dimitrijevic, L. Krapf, C. Wolter, C. Schmidtke, J. P. Merkl, T. Jochum, A. Kornowski, A. Schuth, A. Gebert, G. Huttmann, T. Vossmeyer and H. Weller, *Nanoscale*, 2014, **6**, 10413–10422.

99. A. N. Generalova, V. A. Oleinikov, A. Sukhanova, M. V. Artemyev, V. P. Zubov and I. Nabiev, *Biosens. Bioelectron.*, 2013, **39**, 187–193.

100. T. C. Liu, Z. L. Huang, H. Q. Wang, J. H. Wang, X. Q. Li, Y. D. Zhao and Q. M. Luo, *Anal. Chim. Acta*, 2006, **559**, 120–123.

101. G. Kalyuzhny and R. W. Murray, *J. Phys. Chem. B*, 2005, **109**, 7012–7021.

102. J. H. Wang, H. Q. Wang, Y. Q. Li, H. L. Zhang, X. Q. Li, X. F. Hua, Y. C. Cao, Z. L. Huang and Y. D. Zhao, *Talanta*, 2008, **74**, 724–729.

103. V. Stsiapura, A. Sukhanova, M. Artemyev, M. Pluot, J. H. Cohen, A. V. Baranov, V. Oleinikov and I. Nabiev, *Anal. Biochem.*, 2004, **334**, 257–265.

104. N. Joumaa, M. Lansalot, A. Theretz, A. Elaissari, A. Sukhanova, M. Artemyev, I. Nabiev and J. H. Cohen, *Langmuir*, 2006, **22**, 1810–1816.

105. R. J. Gui, X. Q. An, J. Gong and T. Chen, *Mater. Lett.*, 2012, **88**, 122–125.

106. G. Walker, V. Sundar, C. Rudzinski, A. Wun, M. Bawendi and D. Nocera, *Appl. Phys. Lett.*, 2003, **83**, 3555–3557.

107. D. Choudhury, D. Jaque, A. Rodenas, W. T. Ramsay, L. Paterson and A. K. Kar, *Lab-on-a-Chip*, 2012, **12**, 2414–2420.

108. P. Haro-Gonzalez, W. T. Ramsay, L. Martinez Maestro, B. del Rosal, K. Santacruz-Gomez, C. Iglesias-de la Cruz, F. Mdel, J. Y. Sanz-Rodriguez, P. Chooi, M. Rodriguez Sevilla, D. Bettinelli, A. K. Choudhury, J. G. Kar, D. Sole, Jaque and L. Paterson, *Small*, 2013, **9**, 2162–2170.

109. L. M. Maestro, C. Jacinto, U. R. Silva, F. Vetrone, J. A. Capobianco, D. Jaque and J. G. Sole, *Small*, 2013, **9**, 3198–3200.

110. L. M. Maestro, C. Jacinto, U. R. Silva, F. Vetrone, J. A. Capobianco, D. Jaque and J. G. Solé, *Small*, 2011, **7**, 1774–1778.

111. J. M. Yang, H. Yang and L. W. Lin, *ACS Nano*, 2011, **5**, 5067–5071.

112. L. M. Maestro, E. M. Rodríguez, F. S. Rodríguez, M. C. Iglesias-de la Cruz, A. Juarranz, R. Naccache, F. Vetrone, D. Jaque, J. A. Capobianco and J. G. Solé, *Nano Lett.*, 2010, **10**, 5109–5115.

113. L. M. Maestro, P. Haro-González, M. C. Iglesias-de la Cruz, F. S. Rodríguez, A. Juarranz, J. G. Solé and D. Jaque, *Nanomedicine*, 2013, **8**, 379–388.

114. D. M. Toyli, C. F. de las Casas, D. J. Christle, V. V. Dobrovitski and D. D. Awschalom, *Proc. Natl. Acad. Sci. U. S. A.*, 2013, **110**, 8417–8421.

115. G. Kucsko, P. C. Maurer, N. Y. Yao, M. Kubo, H. J. Noh, P. K. Lo, H. Park and M. D. Lukin, *Nature*, 2013, **500**, 54–58.

116. F. Vetrone, R. Naccache, A. Zamarron, A. J. de la Fuente, F. Sanz-Rodriguez, L. M. Maestro, E. M. Rodriguez, D. Jaque, J. G. Solé and J. A. Capobianco, *ACS Nano*, 2010, **4**, 3254–3258.

117. D. Zhou, M. Lin, X. Liu, J. Li, Z. Chen, D. Yao, H. Sun, H. Zhang and B. Yang, *ACS Nano*, 2013, **7**, 2273–2283.

118. Y. M. Yang, *Microchim. Acta*, 2014, **181**, 263–294.

119. E. J. McLaurin, L. R. Bradshaw and D. R. Gamelin, *Chem. Mater.*, 2013, **25**, 1283–1292.

120. Y. Takei, S. Arai, A. Murata, M. Takabayashi, K. Oyama, S. Ishiwata, S. Takeoka and M. Suzuki, *ACS Nano*, 2014, **8**, 198–206.

121. A. E. Albers, E. M. Chan, P. M. McBride, C. M. Ajo-Franklin, B. E. Cohen and B. A. Helms, *J. Am. Chem. Soc.*, 2012, **134**, 9565–9568.

122. Q. Li, Y. He, J. Chang, L. Wang, H. Chen, Y. W. Tan, H. Wang and Z. Shao, *J. Am. Chem. Soc.*, 2013, **135**, 14924–14927.

123. E. Hemmer, N. Venkatachalam, H. Hyodo, A. Hattori, Y. Ebina, H. Kishimoto and K. Soga, *Nanoscale*, 2013, **5**, 11339–11361.
124. F. Wang and X. G. Liu, *Chem. Soc. Rev.*, 2009, **38**, 976–989.
125. B. Dong, B. S. Cao, Y. Y. He, Z. Liu, Z. P. Li and Z. Q. Feng, *Adv. Mater.*, 2012, **24**, 1987–1993.
126. U. Rocha, C. Jacinto da Silva, W. Ferreira Silva, I. Guedes, A. Benayas, L. Martínez Maestro, M. Acosta Elias, E. Bovero, F. C. van Veggel, J. A. García Solé and D. Jaque, *ACS Nano*, 2013, **7**, 1188–1199.
127. P. C. Chen, Y. N. Chen, P. C. Hsu, C. C. Shih and H. T. Chang, *Chem. Commun.*, 2013, **49**, 1639–1641.
128. V. M. Chauhan, R. H. Hopper, S. Z. Ali, E. M. King, F. Udrea, C. H. Oxley and J. W. Aylott, *Sens. Actuators, B*, 2014, **192**, 126–133.
129. V. Rieke and K. Butts Pauly, *J. Magn. Reson. Imaging*, 2008, **27**, 376–390.
130. B. Quesson, J. A. de Zwart and C. T. W. Moonen, *J. Magn. Reson. Imaging*, 2000, **12**, 525–533.
131. W. Wlodarczyk, R. Boroschewski, M. Hentschel, P. Wust, G. Monich and R. Felix, *J. Magn. Reson. Imaging*, 1998, **8**, 165–174.
132. F. Bertsch, J. Mattner, M. K. Stehling, U. Muller-Lisse, M. Peller, R. Loeffler, J. Weber, K. Messmer, W. Wilmanns, R. Issels and M. Reiser, *Magn. Reson. Imaging*, 1998, **16**, 393–403.
133. L. Ludemann, W. Wlodarczyk, J. Nadobny, M. Weihrauch, J. Gellermann and P. Wust, *Int. J. Hyperthermia*, 2010, **26**, 273–282.
134. L. H. Lindner, H. M. Reinl, M. Schlemmer, R. Stahl and M. Peller, *Int. J. Hyperthermia*, 2005, **21**, 575–588.
135. S. K. Pakin, S. K. Hekmatyar, P. Hopewell, A. Babsky and N. Bansal, *NMR Biomed.*, 2006, **19**, 116–124.
136. S. K. Hekmatyar, R. M. Kerkhoff, S. K. Pakin, P. Hopewell and N. Bansal, *Int. J. Hyperthermia*, 2005, **21**, 561–574.
137. Y. Jin, C. Jia, S.-W. Huang, M. O'Donnell and X. Gao, *Nat. Commun.*, 2010, **1**, 41.
138. D. E. Woessner, S. R. Zhang, M. E. Merritt and A. D. Sherry, *Magn. Reson. Med.*, 2005, **53**, 790–799.
139. A. J. Shuhendler, R. Staruch, W. Oakden, C. R. Gordijo, A. M. Rauth, G. J. Stanisz, R. Chopra and X. Y. Wu, *J. Controlled Release*, 2012, **157**, 478–484.
140. L. Frich, A. Bjornerud, S. Fossheim, T. Tillung and I. Gladhaug, *Magn. Reson. Med.*, 2004, **52**, 1302–1309.
141. N. McDannold, S. L. Fossheim, H. Rasmussen, H. Martin, N. Vykhodtseva and K. Hynynen, *Radiology*, 2004, **230**, 743–752.
142. D. Needham and M. W. Dewhirst, *Adv. Drug Delivery Rev.*, 2001, **53**, 285–305.
143. B. Kneidl, M. Peller, G. Winter, L. H. Lindner and M. Hossann, *Int. J. Nanomed.*, 2014, **9**, 4387–4398.
144. G. Kong, G. Anyarambhatla, W. P. Petros, R. D. Braun, O. M. Colvin, D. Needham and M. W. Dewhirst, *Cancer Res.*, 2000, **60**, 6950–6957.

145. D. Needham, G. Anyarambhatla, G. Kong and M. W. Dewhirst, *Cancer Res.*, 2000, **60**, 1197–1201.

146. M. L. Hauck, S. M. LaRue, W. P. Petros, J. M. Poulson, D. Yu, I. Spasojevic, A. F. Pruitt, A. Klein, B. Case, D. E. Thrall, D. Needham and M. W. Dewhirst, *Clin. Cancer Res.*, 2006, **12**, 4004–4010.

147. A. M. Ponce, Z. Vujaskovic, F. Yuan, D. Needham and M. W. Dewhirst, *Int. J. Hyperthermia*, 2006, **22**, 205–213.

148. J. Woo, G. N. C. Chiu, G. Karlsson, E. Wasan, L. Ickenstein, K. Edwards and M. B. Bally, *Int. J. Pharm.*, 2008, **349**, 38–46.

149. M. Babincova, P. Cicmanec, V. Altanerova, C. Altaner and P. Babinec, *Bioelectrochemistry*, 2002, **55**, 17–19.

150. P. Pradhan, J. Giri, R. Banerjee, J. Bellare and D. Bahadur, *J. Magn. Magn. Mater.*, 2007, **311**, 208–215.

151. Q. A. Pankhurst, J. Connolly, S. K. Jones and J. Dobson, *J. Phys. D: Appl. Phys.*, 2003, **36**, R167–R181.

152. Q. A. Pankhurst, N. T. K. Thanh, S. K. Jones and J. Dobson, *J. Phys. D: Appl. Phys.*, 2009, **42**.

153. R. Bustamante, A. Millan, R. Pinol, F. Palacio, J. Carrey, M. Respaud, R. Fernandez-Pacheco and N. J. O. Silva, *Phys. Rev. B*, 2013, **88**, 184406.

154. L.-A. Tai, P.-J. Tsai, Y.-C. Wang, Y.-J. Wang, L.-W. Lo and C.-S. Yang, *Nanotechnology*, 2009, **20**.

155. P. Gutlich, A. Hauser and H. Spiering, *Angew. Chem., Int. Ed. Engl.*, 1994, **33**, 2024–2054.

156. P. Gutlich, *Struct. Bonding*, 1981, **44**, 83–195.

157. E. Konig, G. Ritter and S. K. Kulshreshtha, *Chem. Rev.*, 1985, **85**, 219–234.

158. Y. Garcia, V. Niel, M. C. Munoz and J. A. Real, in *Spin Crossover in Transition Metal Compounds i*, ed. P. Gutlich and H. A. Goodwin, 2004, vol. 233, pp. 229–257.

159. O. Kahn, *Molecular magnetism*, VCH, New York, 1993.

160. J. F. Letard, P. Guionneau, L. Rabardel, J. A. K. Howard, A. E. Goeta, D. Chasseau and O. Kahn, *Inorg. Chem.*, 1998, **37**, 4432–4441.

161. O. Kahn, J. Krober and C. Jay, *Adv. Mater.*, 1992, **4**, 718–728.

162. O. Kahn and C. J. Martinez, *Science*, 1998, **279**, 44–48.

163. A. Bousseksou, G. Molnar, L. Salmon and W. Nicolazzi, *Chem. Soc. Rev.*, 2011, **40**, 3313–3335.

164. R. N. Muller, L. Vander Elst and S. Laurent, *J. Am. Chem. Soc.*, 2003, **125**, 8405–8407.

165. I.-R. Jeon, J. G. Park, C. R. Haney and T. D. Harris, *Chem. Sci.*, 2014, **5**, 2461–2465.

166. J. Krober, E. Codjovi, O. Kahn, F. Groliere and C. Jay, *J. Am. Chem. Soc.*, 1993, **115**, 9810–9811.

167. V. Nagy, K. Halasz, M. T. Carayon, I. A. Gural'skiy, S. Tricard, G. Molnar, A. Bousseksou, L. Salmon and L. Csoka, *Colloids Surf., A*, 2014, **456**, 35–40.

168. A. P. Stephenson, A. P. Micolich, K. H. Lee, P. Meredith and B. J. Powell, *Chemphyschem*, 2011, **12**, 116–121.

169. Y. H. Gao and Y. Bando, *Appl. Phys. Lett.*, 2002, **81**, 3966–3968.

170. J. Zhong, W. Z. Liu, L. Kong and P. C. Morais, *Sci. Rep.*, 2014, **4**.

171. L. H. Reddy, J. L. Arias, J. Nicolas and P. Couvreur, *Chem. Rev.*, 2012, **112**, 5818–5878.

172. S. A. Stanley, J. E. Gagner, S. Damanpour, M. Yoshida, J. S. Dordick and J. M. Friedman, *Science*, 2012, **336**, 604–608.

173. M. L. Debasu, D. Ananias, I. Pastoriza-Santos, L. M. Liz-Marzan, J. Rocha and L. D. Carlos, *Adv. Mater.*, 2013, **25**, 4868–4874.

174. J. Dong and J. I. Zink, *ACS Nano*, 2014, **8**, 5199–5207.

175. A. Riedinger, P. Guardia, A. Curcio, M. A. Garcia, R. Cingolani, L. Manna and T. Pellegrino, *Nano Lett.*, 2013, **13**, 2399–2406.

176. J. T. Dias, M. Moros, P. del Pino, S. Rivera, V. Grazu and J. M. de la Fuente, *Angew. Chem., Int. Ed.*, 2013, **52**, 11526–11529.

177. S. Freddi, L. Sironi, R. D'Antuono, D. Morone, A. Dona, E. Cabrini, L. D'Alfonso, M. Collini, P. Pallavicini, G. Baldi, D. Maggioni and G. Chirico, *Nano Lett.*, 2013, **13**, 2004–2010.

178. R. Piñol, C. D. S. Brites, R. Bustamante, A. Martínez, N. J. O. Silva, J. L. Murillo, R. Cases, J. Carrey, C. Estepa, C. Sosa, F. Palacio, L. D. Carlos and A. Millán, *ACS Nano*, 2015, **9**, 3134–3142.

Section III
Non-luminescence-based Thermometry

CHAPTER 9

Scanning Thermal Microscopy

SÉVERINE GOMÈS,* ALI ASSY AND PIERRE-OLIVIER CHAPUIS

CETHIL UMR5008, Université de Lyon, CNRS, INSA-Lyon, Univ. Lyon 1,
F-69621, Villeurbanne, France
*Email: severine.gomes@insa-lyon.fr

9.1 Introduction

While the lateral spatial resolution of far-field optical techniques is limited
by diffraction at a few hundreds of nanometres, Scanning Thermal Microscopy (SThM) allows real nanoscale thermal imaging. SThM techniques
are based on Scanning Probe Microscopy (SPM) methods, with spatial
resolutions depending on the characteristic lengths associated with the heat
transfer between the small probes and the samples to be characterized. The
probes can be tailored with tips of curvature radii in the range of a few tens
of nanometres. The first instrument was invented in 1986, shortly after the
discovery of the Scanning Tunnelling Microscope[1] (STM) by Williams and
Wickramasinghe, and was termed the Scanning Thermal Profiler[2] (STP).
Its goal was to extend the possibilities of imaging topography because
the STM was limited to insulators. Although the STP was not intended
for mapping temperature distributions of surfaces, it stimulated intense
efforts to develop SPM-based techniques in the thermal area. *Scanning
Thermal Microscopy* is now an integral part of the experimental landscape in
sub-micron heat-transfer studies and has found a wide range of
applications.

This chapter focuses on the fundamentals and selected applications of
SThM methods. It includes a review of the main SPM-based techniques

RSC Nanoscience & Nanotechnology No. 38
Thermometry at the Nanoscale: Techniques and Selected Applications
Edited by Luís Dias Carlos and Fernando Palacio
© The Royal Society of Chemistry 2016
Published by the Royal Society of Chemistry, www.rsc.org

proposed for thermal imaging with nanoscale spatial resolution. It also underlines fundamental works performed to improve our understanding of the heat-transfer mechanisms at the micro- and nanoscales. The chapter is organized in three sections, namely, on instrumentation and associated setups, on SThM measurement approaches, and on SThM applications.

The first section reviews research on the design of experiments involving thermal probes for thermal imaging and the characterization of nanosystems and materials. Note that it does not discuss the related research on the mechanical and electrical designs of probes, which are also of significant practical importance.

The second section starts with a reminder of the fundamentals of thermal metrology by contact. It then presents the approaches currently used to calibrate SThM probes. In many cases, the link between the nominal measured signal and the investigated parameter has not yet been fully established due to the complexity of the micro-/nanoscale interaction between the probe and the sample. As a consequence, special attention is paid to this interaction, which is critical for many applications.

The third section gives some examples of applications of SThM. This includes the characterization of operating devices, the measurement of the thermal conductivity of nanomaterials and the determination of local phase-transition temperatures. Finally, this chapter briefly mentions the areas where progress could be made in the future, in order to develop novel opportunities for SThM.

9.2   Instrumentation and Associated Setups

9.2.1   General Layout

Various types of SThM have been developed. However, techniques based on Atomic Force Microscopy (AFM) have been mainly used, because AFM enables a wider variety of samples to be probed, in contrast to STM. Except for Tunnelling Thermometry[3,4] and the Point Contact Thermocouple method,[5] AFM-based SThM systems do not rely on the presence of an electronic tunnelling current from a conductive surface. Furthermore, AFM-based systems are very versatile systems. Beyond imaging, they allow measurements under various experimental configurations of probe–sample interaction (as a function of the tip–sample force and distance for example) and various types of sensors can be placed at the probe tip. Figure 9.1 illustrates the general layout of an AFM-based SThM system.

The operation consists of determining the temperature or temperature-dependent characteristics of the sample surface through the analysis of the near-field interaction between the probe and the sample. This may be performed when the probe and the sample are in contact or separated by a distance of a few nanometres to several micrometres. Therefore an effective anti-vibration mounting system is always essential, as is a fine positioning

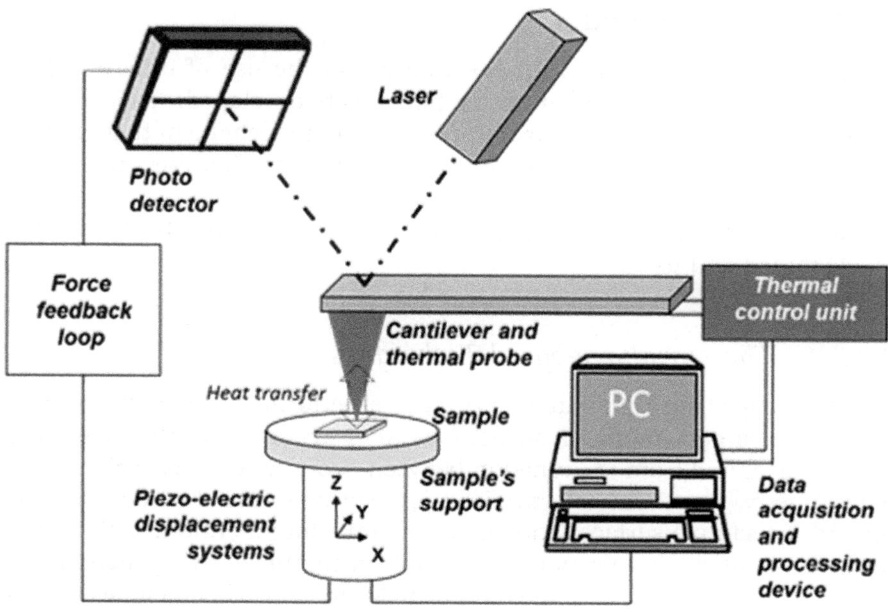

Figure 9.1 General layout of an SThM AFM-based system. This comprises the main following components: the probe operating as a thermal sensor, a source of heat which can be external or internal to the sample and/or can be incorporated into the probe itself, a cantilever on which the probe is mounted and whose deflection senses the force that acts between the probe and the sample, a piezo-ceramic system associated with an electronic circuit which servo controls the probe–sample near-field interaction intensity, and a second piezo-ceramic system for enabling the scanning of the sample surface by the probe. Not to scale.

system to place the probe tip close to the surface. Cantilever deflections are measured by reflecting a laser beam off the cantilever and sensing the beam using a photodiode. Other systems involving piezo-resistive or piezo-electric cantilevers, where deflections generate an electrical signal in the cantilever may also be used.[6,7]

In the imaging mode, the deflection signal is used in a feedback control loop to maintain a constant tip–sample contact force while the tip is scanned laterally. Piezo-electric scanners are used to move either the probe or the sample vertically (in the Z direction in Figure 9.1) and to scan the sample surface (in the X–Y direction). The combination of the X–Y scan position data, the feedback signal and the probe temperature-related signal gives the raw data for a 3D topographical image and a 'thermal' image of the surface. The thermal image contrast reflects the change in the amount of heat locally exchanged between the tip and the sample. The feedback control system, which maintains a constant force between the probe and surface, operates simultaneously, but quite independently, of the process of thermal measurement.

9.2.2　Main Techniques

Since 1993, various thermal methods based on the use of different thermosensitive sensors or phenomena have been developed. They can be classified according to the temperature-dependent mechanism that is used: thermovoltage,[8–16] electrical resistance change,[7,16–22] fluorescence,[23–25] or thermal expansion.[26]

Thermovoltage and electrical resistance based techniques have been the most studied and used.

9.2.2.1　*Thermovoltage-based Techniques*

Thermovoltage-based measurements can be performed either in the non-contact or contact mode of an STM or AFM system. The thermoelectric voltage generated from the junction between two electrodes is exploited for thermometry. The thermoelectric junction can be established between the tip and the sample surface. This includes tunnelling thermometry[3,4,27] and the point contact thermocouple method.[28] A thermoelectric junction can also be deposited on or integrated at the SPM tip apex. This includes probes with a built-in thermal sensor such as a thermocouple or a Schottky diode.

9.2.2.1.1　Tunnelling Thermometry.　Tunnelling thermometry is based on the principle of a *tunnelling thermocouple* and uses the temperature dependence of the tunnelling electronic current between two electrodes without contact. The technique was first proposed by Weaver *et al.*[27] in 1989 under the configuration of an STM-based technique, and first implemented in the configuration of an AFM by Nonnenmacher and Wickramasinghe in 1992.[3] In 2002, Zhou *et al.*[4] used an AFM-based method to map the temperature distribution of electronic devices. Compared with thermocouple probes with a built-in thermal sensor, the thermoelectric voltage measured by the tunnelling thermocouple has been shown to be significantly less affected by the tip–sample contact thermal resistance and air conduction. This allowed quantitative temperature imaging with nanometric spatial resolution, limited however by the metal film of 10–20 nm in thickness that is required by the measurement principle.[4]

In 2004, Pavlov[28] claimed the direct measurement of the absolute temperature using the tunnelling of free electrons. A tunnelling nanojunction was established between the copper oxide covering a copper substrate and the platinum tip of an STM. The method is based on the linear temperature dependence of the relative Fermi level shift. Measuring the tunnelling current at zero voltage between the tip and the substrate, Pavlov has shown that the Fermi level shift provides a measure of the substrate from room temperature up to 1250 K.

9.2.2.1.2　Point Contact Thermocouple Method.　The measurement of the thermoelectric voltage generated at a point junction between a metal tip and a sample covered with a metal seems to allow quantitative thermal imaging with nanoscale spatial resolution. The thermoelectric signal is not distorted by air heat conduction and is not affected by the tip–sample contact thermal resistance (see next section). Using this method, Sadat *et al.*[5] recently reported the direct mapping of topography and temperature fields of metal surfaces with a temperature resolution of 0.01 K and a spatial resolution of ∼100 nm.

For both of the above-mentioned methods, the probe and the sample must generally have an electrically-conducting surface or a surface covered with a metallic film. This limits their application.

9.2.2.1.3 Thermocouple Probes with a Built-in Thermal Sensor.

Since 1986, advances in microfabrication and characterization technologies have enabled a progressive and significant improvement in the design, operation and use of thermocouple probes. These advances have, in turn, enhanced our understanding of cantilever and tip heat-transfer fundamentals.[29] Miniaturization of the cantilever, the tip and the junction at the tip end has enabled decreased thermal time constants and improved spatial resolution.[13,29–31] Figure 9.2(a) shows an example of a nanojunction fabricated at the end of an AFM tip.[32] The diameter of the thermocouple junction is about 100 nm, and the radius of the tip is about 50 nm.

Until 2002, thermocouple SThM probes were used in passive mode and nanoscale thermal imaging was essentially qualitative with these probes. Different factors limit quantitative characterization in this configuration. Shi *et al.*[33] have shown that the temperature rise locally measured by a probe depends on the size of the heated zone at the sample surface due to heat transfer through the surrounding gas. The heat transfer through the surrounding gas may then be seen as a perturbative effect on the measured thermal signal, limiting its local character. This applies to all the methods involving a built-in thermal sensor. Furthermore, a large temperature drop can occur at the tip–sample junction due to a thermal contact resistance at the tip–sample contact. As discussed in the Section 9.3 of this chapter, the value of this thermal contact resistance depends on various physical properties and on the surface of the sample. It is often unknown and difficult to determine. Additionally the sample temperature can be modified due to the presence of the probe. Even if this perturbation is often neglected in SThM, it should be specified for each technique.

For quantitative temperature measurement using thermocouple probes, Nakabeppu *et al.*[34] proposed in 2002 measurements performed in vacuum (under 0.1 Pa) with an active thermal feedback scheme allowing it to maintain the tip temperature equal to the sample surface temperature. The tip was a single-wire thermocouple AFM probe and the probe mount was integrated with an additional thermocouple and a heater. These measurements suggested that maintaining zero heat flux between the tip and the sample may be an alternative for quantitative temperature measurement and profiling. The idea is that the temperature of the sample surface is not

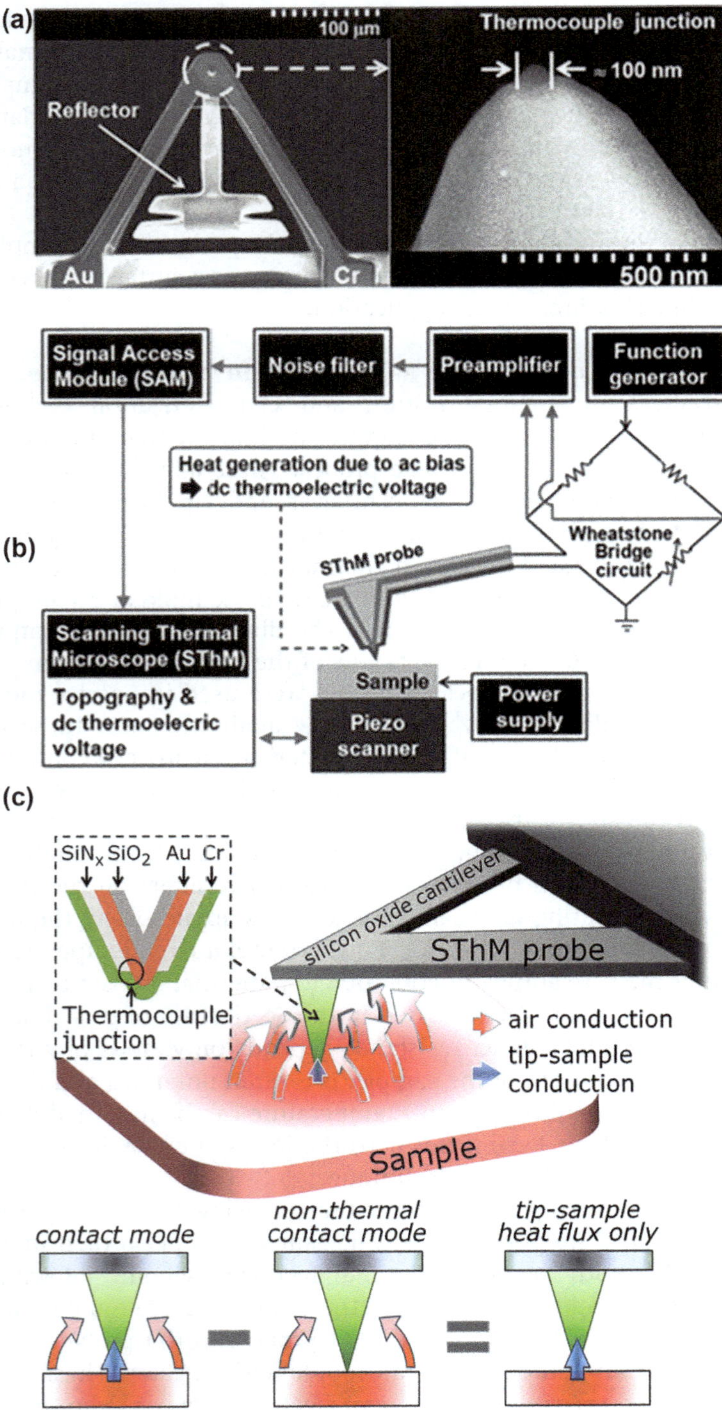
(a)
100 μm
Thermocouple junction
Reflector
≈ 100 nm
Au
Cr
500 nm
(b)
Signal Access Module (SAM)
Noise filter
Preamplifier
Function generator
Heat generation due to ac bias
➡ dc thermoelectric voltage
SThM probe
Wheatstone Bridge circuit
Scanning Thermal Microscope (SThM)
Topography & dc thermoelecric voltage
Sample
Power supply
Piezo scanner
(c)
SiNx SiO2 Au Cr
Thermocouple junction
silicon oxide cantilever
SThM probe
air conduction
tip-sample conduction
Sample
contact mode
non-thermal contact mode
tip-sample heat flux only

modified by the presence of the probe. The temperature of the sample surface can then be measured, so that the actual surface temperature profile is better reconstructed, even if the tip–sample contact resistance is unknown or perturbations occur during the scan due to changes in surface properties or topography. However, only point measurements were performed because of the large thermal time constant (~ 0.5 s) of the probe used, which had a large thermal mass.

Based on the above-mentioned thermodynamic principle and combined with SThM using a double scan technique,[35] the null-point SThM (NP SThM) method was recently proposed by Chung *et al.*[32,36,37] (see Figure 9.2). Figure 9.2(b) is a schematic diagram of the experimental setup for NP SThM.[32] The thermocouple junction of the probe is heated by an ac current of frequency (>100 kHz) high enough so that its corresponding period is much smaller than the thermal time constant of the thermocouple junction (>1 ms). The junction temperature is monitored by measuring the dc thermoelectric voltage from the small junction located at the tip end. A Wheatstone bridge circuit is used to remove the influence of the unwanted dc drift voltage due to ac bias.[32]

The method enables quantitative thermal profiling at the nanoscale for experiments under ambient conditions. NP SThM was demonstrated by measuring the temperature of an electrically heated multiwall carbon nanotube in a point-by-point fashion. A continuous temperature profile for a 5 mm wide aluminium line heater of monitored temperature, deposited on Pyrex glass was also obtained. NP SThM has furthermore shown promise for quantitative thermal conductivity profiling.[37] However, NP SThM requires the use of a nanothermocouple probe in active mode and two temperature profiles must be obtained by scanning the probe at two different heights above the sample at each position on the sample. This adds to the complexity of the technique.

Figure 9.2   Schematic diagram of the experimental setup and principle for an NP SThM. (a) SEM images of an SThM thermocouple nanotip. The thermocouple junction integrated at the apex of the tip has a diameter of about 100 nm, and the tip radius is about 50 nm. (b) Setup for the active-mode operation of the thermocouple SThM probe. The signal from the bridge circuit is amplified by a pre-amplifier. A noise filter is used to maximize the signal-to-noise ratio by reducing the 60 Hz harmonic noise. The filtered signal is fed into the signal access module (SAM) and becomes simultaneously available with the topography signal of the AFM. A dc power supply is used to heat the sample. Reproduced with permission from ref. 32. © Elsevier 2012. (c) Principle of quantitative thermal profiling. The SThM probe scans the same line in both contact and non-contact modes of AFM. The difference between the thermal signals obtained in the two modes is due only to the heat flux through the tip–sample thermal contact.
Reproduced with permission from ref. 36. © American Chemical Society 2010.

Recently, Kim *et al.*[38] described an ultrahigh vacuum (UHV) based SThM technique that is capable of quantitatively mapping temperature fields with ~ 15 mK temperature resolution and ~ 10 nm spatial resolution.

The application of thermocouple probes for thermal conductivity measurements in active mode was extended with the proposition of the 2ω method in 2006.[39,40] Thermal conductivity contrast imaging with a nanoscale spatial resolution was then reported. The 2ω method consists of monitoring the amplitude of the 2ω signal from the thermocouple junction of the probe, which is heated *via* the Joule effect by an ac current. The first measurements were performed in contact mode.[38,39]

9.2.2.1.4 Schottky Diode Probes with a Built-In Thermal Sensor. A Schottky diode made of a metal–semiconductor junction can be used as a sensitive temperature sensor as it offers high sensitivity (much larger than that of a typical thermocouple probe), and can be fabricated straightforwardly on a Si probe. In 1998, Leinhos *et al.*[41] developed and tested Si cantilever probes with a built-in metal–Si Schottky diode at the end of the Si tip. However the sensor was not thermally isolated from the environment due to the very high thermal conductivity of Si. As the sensor temperature could be completely different to the sample surface temperature in this case, this concept of an SThM probe was abandoned.

9.2.2.2   Resistive Probes

Resistive probes are based on resistance thermometry. The temperature dependence of the electrical resistivity $\rho(T)$ has been used to develop different kinds of SThM resistive probes, in particular, metallic probes and doped Si probes.

9.2.2.2.1 Metallic Probes. The first SThM metallic resistive probe was the Wollaston wire probe proposed in 1994 by Pylkki *et al.*[17,42] As its name suggests, this probe is a thermal sensor fabricated from a Wollaston wire. The cantilever arms are made of such wire, as shown in Figure 9.3(a). This wire encloses a core of diameter 5 μm made of platinum/10% rhodium. The wire is etched by electrochemically removing the outer silver layer of the Wollaston wire over a length of 200 μm. The liberated core is the thermally sensitive element. Its bending allows the formation of a V-shaped probe. A mirror is stacked on the cantilever arms to allow control of the cantilever deflection by optical methods.

Because of its high endurance, its time response, estimated to 200 μs in air,[43–45] and the high temperature coefficient of platinum/10% rhodium ($\alpha = 0.00166$ K^{-1}), the Wollaston probe is attractive. It has been commercialized for more than 15 years, and its demonstrated applications encompass microsystem diagnostics[46–49] and the local thermophysical characterization of various materials.[50–60] However its large active area limits thermal investigations at the nanoscale.

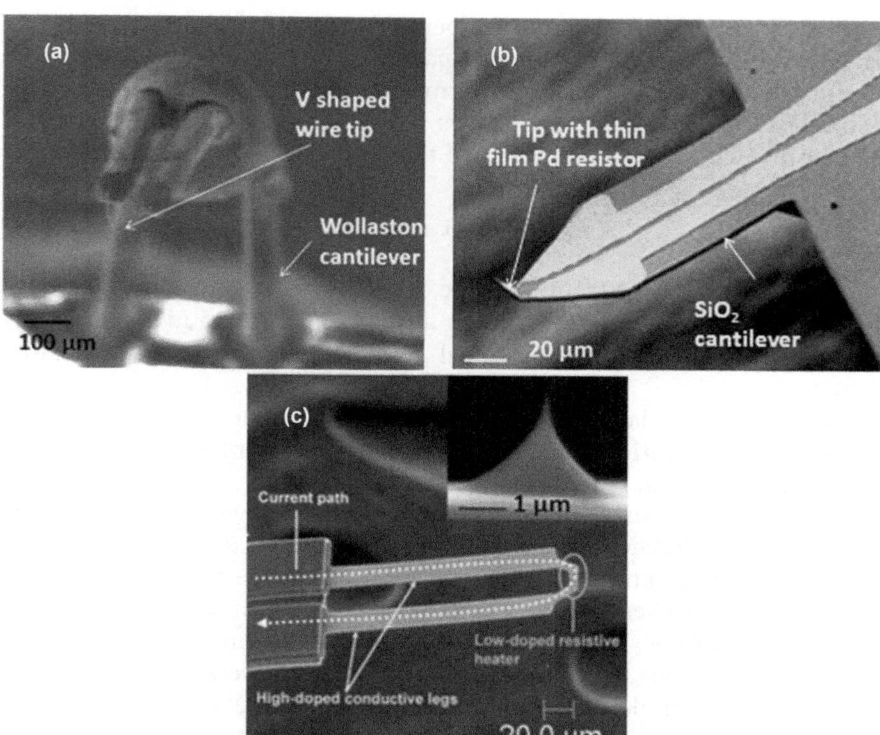

Figure 9.3 SEM images of SThM resistive probes: (a) a Wollaston wire microprobe, (b) a palladium probe, and (c) a doped silicon probe.
Adapted with permission from ref. 136 and 64. © AIP Publishing 2007, 2012.

To improve this method (in terms of thermal response time, spatial resolution and thermal sensitivity) using a miniaturized resistive metallic element, several probe designs based on the integration of a resistive element into an AFM cantilever have been proposed.[7,18–22,61–63]

As an example, Figure 9.3(b)[64] shows a probe that combines a thin Pd film resistor positioned at the very end of a flat tip at the end of a thin cantilever. The cantilever was initially made of SiO_2 but it has been recently extended to Si_3N_4 as reported, for example, by Weaver *et al.*[21] The cantilever design was such that it has a low spring constant (0.35 N m^{-1}). This is desirable for conventional SThM scans obtained with an AFM tip in constant contact with the sample and to ensure that the probe is minimally invasive. The tip is tall (around 10 µm) to maximize the cantilever–sample separation, minimizing the heating of the cantilever by the hot sample in the case of thermometry measurements.[21] This tip has a curvature radius of around 50 nm[64] and its electrical resistance has a temperature coefficient α of about 0.0012 K^{-1}.[65] Its time response has been estimated to be about a few tens of µs.[65,66]

All resistive metallic probes can operate in passive and active modes. In passive mode, a very small current is passed through the probe, resulting in minimal Joule heating whilst still permitting the measurement of electrical resistance. During a scan, heat flows from the sample to the probe and changes the electrical resistance R_p of the probe, according to the following equation:

$$R_p(T) = R_{p0}(1 + \alpha(T - T_0)) \tag{9.1}$$

where $R_p(T)$ is the electrical resistance of the probe at temperature T, R_{p0} is the electrical resistance of the probe at temperature T_0 and α is the temperature coefficient of the electrical resistivity of the probe's sensitive element.

In active mode, a large current is passed through the probe, resulting in significant Joule heating. The probe temperature is monitored by modifying the electrical current.

Passive mode is used in thermometry, while active mode is used to measure thermophysical properties of materials such as thermal conductivity. Active mode can also be used to locally heat the sample in order to induce and study thermally-dependent phenomena. In particular, thermal expansion of polymer materials can be investigated with *Scanning Thermal Expansion Microscopy* or with the dynamic localized thermomechanical analysis method, both developed by Hammiche *et al.* in 2000.[67]

Under both passive and active modes, dc, ac or both measurements can be performed. Different configurations of electrical bridges have been used to measure the electrical resistance of the probe and deduce its temperature.[47,61,62,68,69] Figure 9.4 shows a typical setup to investigate the surface

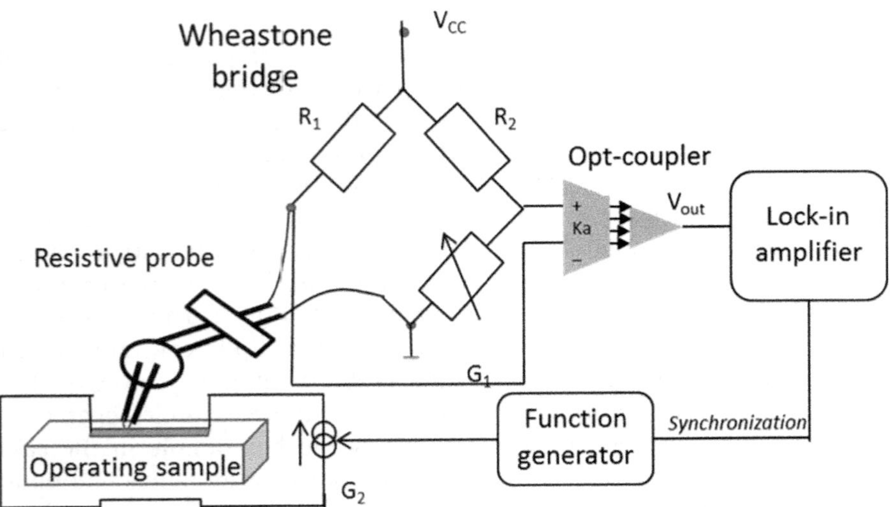

Figure 9.4   Schematic diagram of a typical setup for investigating the surface temperature of a sample operating in the ac regime. An opt-coupler ensures that the probe is indifferent to the sample potential.

temperature of an operating sample in the ac regime.[45] An opt-coupler ensures that the probe is indifferent to the sample potential. Another solution is to perform the measurements using a transformer-isolated bridge.[21]

9.2.2.2.2 Doped Si Probes. SThM doped Si resistor SPM probes are based on a modification of those initially developed in the Millipede project for data storage applications.[70,71] As represented in Figure 9.3(c), the cantilevers are U-shaped. They have a heater element at the cantilever end, placed above a sharp Si tip of micrometric height, the curvature radius of which can reach 10 nm. The electric current flowing through the resistive element causes resistive heating and a temperature rise of the tip.

Operating in active mode, this kind of probe has mainly been used for nanothermal analysis, thermomechanical actuation, nanolithography, and data storage. It has also found different applications in SThM. Some examples are given in Section 9.4. A review of its applications is given in ref. 72.

9.2.3 Other Probes

9.2.3.1 Fluorescent-particle-based Probes

The SThM method developed by Aigouy *et al.*[23–25] works as follows: the sensor is a fluorescent particle whose emission depends on temperature. An Er^{3+}/Yb^{3+} co-doped fluoride amorphous glass particle or a PbF_2:Er^{3+}/Yb^{3+} nanocrystal of diameter of 100–300 nm, which exhibits a stable fluorescence in the visible range, is glued at the end of a sharp AFM tip. The temperature of the particle is determined by comparing the intensity of two peaks emitted from a thermal equilibrium distribution. During a scan, variations in the sample temperature surface result in temperature variations of the Er^{3+} particle emission, giving the thermal contrast. Fluorescence images are then obtained and converted to temperature maps. Although luminescence is affected both by the near-field optical distribution and by temperature variations, the thermal contribution can be treated separately by comparing the obtained images with reference ones recorded when the device is not heated. The technique can operate in dc mode, or ac mode at low frequencies (<500 Hz). The method presents a spatial resolution in the range of the fluorescent particle size (<300 nm) and enables thermal imaging in liquids.[73]

9.2.3.2 Bimorph Sensors

Thermal bimorph sensors are composite cantilevers. They exploit the bending moment, due to a mismatch in thermal expansion coefficients of metal-coated SiN_x cantilever materials, induced by the temperature change of the cantilever.[74] Cantilever temperature change results in a cantilever deflection that can be measured by an AFM. The optical detection scheme in an AFM has a resolution of ~ 0.1 Å. A temperature resolution of 10^{-5} K for

SiN$_x$ cantilevers covered with Al was reported. Such a temperature resolution is far better than that of a thermocouple or an electrical resistance thermometer.

The method was applied for imaging the topography and thermal expansion of an indium-tin-oxide ac heater containing two defects.[26] An ac measurement was then required to separate the cantilever bendings due to topography or temperature, and measurements were performed under ambient air. Bimorphs that twist instead of bend have recently been proposed to circumvent this issue. The approach was demonstrated to have superior properties.[75]

Other interesting applications of such sensors include the study of the radiative near-field heat transfer between a micrometric sphere (used as a probe tip) and a sample surface,[76–78] and the near-field thermal imaging of nanostructured surfaces.[78] These measurements were performed under vacuum conditions.

9.2.4   Other Methods

9.2.4.1   Scanning Joule Expansion Microscopy

Conventional AFM measurements of the local thermal expansion of a sample heated by Joule effect by a sinusoidal or pulsed voltage have been used for thermal imaging. This constitutes the basic principle of the *Scanning Joule Expansion Microscopy* (SJEM) technique developed in 1998 by Varesi and Majumdar.[79,80] The SJEM technique was applied to image thermally interconnect structures in very large-scale integrated (VLSI) circuits.[81] A thin polymer film covered the sample surface to allow a large expansion signal. The spatial resolution was thus limited by the thickness of the polymer film to about 50 nm. Recently, Grosse *et al.*[82,83] reported the measurement by SJEM of the amplitude distribution of the steady periodic temperature field at the contacts of working graphene transistors (heating frequency of 65 kHz) with a spatial resolution of ∼10 nm. They also claimed the mapping of single-walled carbon nanotube devices with spatial resolution and temperature precision as good as ∼100 nm and ∼0.7 K, respectively. One of the major issues of SJEM remains that it is difficult to obtain the dc sample temperature from the expansion signal.

9.2.4.2   AFM-IR Technique

A local transient deformation of a sample can also be induced by an infrared (IR) pulsed laser tuned to the sample absorbing wavelength. Absorption of IR radiation by the sample leads to a rapid thermal expansion that excites the resonant oscillations of the cantilever.[84,85] Measuring the amplitude of the cantilever oscillation as a function of the source wavelength creates local absorption spectra. Based on this principle, Dazzi *et al.*[86,87] proposed a new method of IR microspectroscopy (the AFM-IR technique), with a sub-wavelength lateral resolution, which has opened the way to measure and identify spectroscopic contrasts not accessible by far-field or near-field optical methods. Applications of the technique have mainly concerned life sciences and the polymer materials field.[87]

9.2.4.3 *Thermal Radiation Scanning Tunnelling Microscopy*

SPM methods include also *Scanning Near-field Optical Microscopy* (SNOM). Its use for thermometric purposes was mentioned by Cahill *et al.*[88] in 2003 and Christofferson *et al.*[89] in 2008. Among the SNOM methods recently proposed, *Thermal Radiation Scanning Tunnelling Microscopy* (TRSTM) developed by De Wilde *et al.*[90] is promising for thermometry. The technique operates without any external illumination: it is a near-field analogue of a night-vision camera, making use of the thermal IR evanescent fields emitted by the surface. As it is based on the scattering by a probe (akin to a dipole) of the near-field thermal radiation of a material at a given temperature,[90,91] TRSTM does not require the contact of a tip with the sample surface. The method has a spatial resolution on the order of 100 nm.

The authors demonstrated that the TRSTM signal is a measurement related to the electromagnetic local density of states (EM-LDOS) at the tip position. Electromagnetic modes are associated with energy levels that are populated according to Bose–Einstein statistics. The temperature dependence of the TRSTM signal is known and obeys the relation giving the EM-LDOS $U(r,\omega)$ at the thermodynamic equilibrium:

$$U(r,\omega) = \rho(r,\omega)\hbar\omega/(\exp(\hbar\omega/k_{\mathrm{B}}T) - 1) \tag{9.2}$$

where $\rho(r,\omega)$ is the EM-LDOS for a photon of energy $\hbar\omega$, k_{B} is the Boltzmann constant, and T is the temperature. Through knowledge of the intensity $S(T_1)$ of the TRSTM signal measured at a given point, at a given wavelength λ and at a given temperature T_1 (as an example $T_1 = 300$ K and $\lambda = 10$ μm), it should be possible to determine the temperature T_2 at the same point by measuring the signal $S(T_2)$ at the same λ:

$$S(T_1)/S(T_2) = (\exp(\hbar\omega/k_{\mathrm{B}}T_1 - 1)/(\exp(\hbar\omega/k_{\mathrm{B}}T_2) - 1) \tag{9.3}$$

The TRSTM signal should thus give access to the determination of the local absolute temperature on the surface of a sample.[92]

The main methods that allow thermal imaging with a nanoscale spatial resolution have been presented. The following section focuses on the measurement of sample temperature and thermal conductivity. It will mainly deal with widely spread sensors.

9.3 SThM Measurement Approaches

9.3.1 Probe Calibration

The calibration method consists of performing measurements with a reference sample and comparing the determined value to the expected one. If not, one can correct the setting of the unit. In SThM, the calibration consists mainly of linking the thermovoltage (thermocouple junction) or the electrical resistance (resistive probe) measured to either the sample temperature or the sample thermal conductivity.

9.3.1.1 *Temperature Measurement*

As in any thermal measurement method involving a sensor in contact with, or in proximity to, the sample surface to be characterized, the quantity of heat exchanged between the sample and the tip Q_{s-t} depends on the energy balance of the tip–sample system. The system includes the sensor in contact with the sample and their surrounding environment (see Figure 9.5). Q_{s-t} is then a function of the effective thermal properties and temperatures of the sample, the probe and their surrounding environment, including the probe holder, *i.e.*, the probe cantilever in SThM.

In steady-state thermometry, a heat quantity is exchanged between the hot sample and the probe that is initially at room temperature. To simplify the description of the system, the thermal interaction between the probe and the sample is described by the thermal resistance network represented in Figure 9.6.

The corresponding expression of Q_{s-t} can then be written as:

$$Q_{s-t} = (T_s - T_{t,c})/(R_{th,s} + R_{th,c}) = (T_p - T_a)(R_{th,pe} + R_{th,cant})/R_{th,pe}R_{th,cant}$$
$$+ (T_{t,c} - T_a)/R_{th,env} \tag{9.4}$$

where T_s is the sample temperature to be determined and T_p is the probe temperature that is measured. The contact of the hot sample with the probe initially at ambient temperature T_a leads to a decrease of the temperature within the sample under the probe–sample contact of temperature $T_{s,c}$, and an increase of the temperature at the tip apex $T_{t,c}$. $R_{th,s}$, $R_{th,c}$ and $R_{th,t}$ are the thermal resistances associated respectively to the heat transfer within the sample at the level of the constriction near the contact (sample thermal

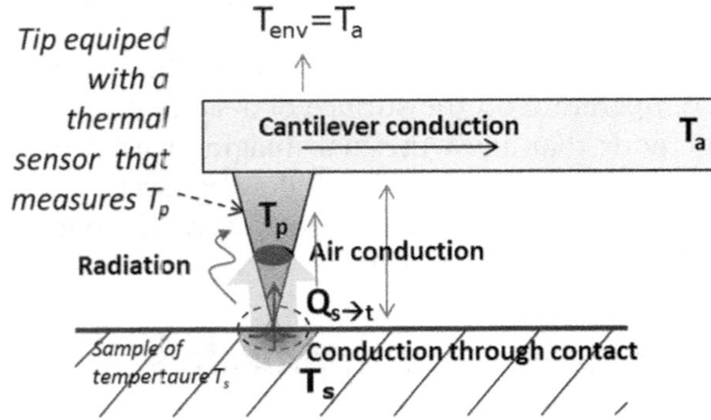

Figure 9.5 Schematic diagram of the heat flow within the system (probe–sample–surrounding environment) for a probe used in passive mode and in contact with a hot sample. Not to scale.

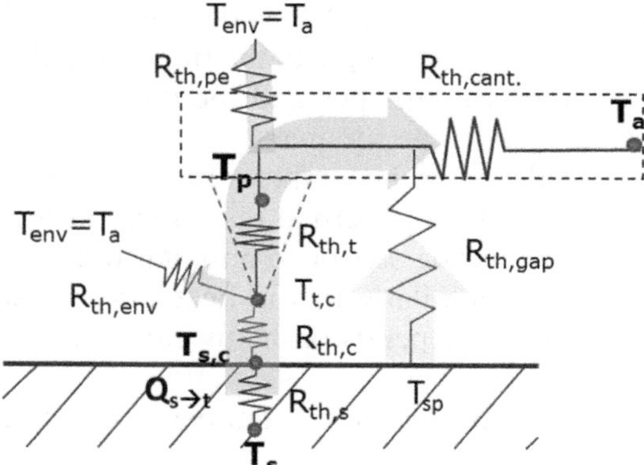

Figure 9.6   Thermal resistance network model for a probe used in passive mode.

spreading resistance), to the heat transfer from the sample to the tip, and to the heat transfer between the tip apex and the thermosensitive element of temperature T_p. The heat losses to the environment are included in three thermal resistances: $R_{th,env}$ describes the heat losses to the surrounding environment between the probe apex and the sensitive element; $R_{th,pe}$ represents the probe heat losses from the sensitive element by convection and radiation to the environment; and $R_{th,cant}$ corresponds to the heat losses by conduction in the probe support or cantilever. Let $R_{th,p}$ be the equivalent for the last two thermal resistances.

In the simple case of a sensitive element at the tip apex, which is the case for almost all the thermovoltage-based SThM probes ($T_{t,c} = T_p$ and $R_{th,env} = R_{th,t} = 0$), the heat transfer between the cantilever and the sample surface, Q_{s-t}, may be written as:

$$Q_{s-t} = (T_s - T_p)/(R_{th,s} + R_{th,c}) = (T_p - T_a)(R_{th,pe} + R_{th,cant})/R_{th,pe}R_{th,cant}$$
$$= (T_p - T_a)/R_{th,p} \tag{9.5}$$

Here, $R_{th,p}$ describes the resistance between the tip apex and the probe support assumed to be a heat sink (see Figure 9.6).

The value of the correction to be applied at the nominal measurement of the instrument T_p may then be written as:

$$\delta T_p = T_s - T_p = (T_p - T_a)(R_{th,s} + R_{th,c})/R_{th,p} \tag{9.6}$$

This last expression clearly shows pathways that allow optimization of the design of SThM probes considered for thermometry. The minimization of δT_p requires the maximization of $R_{th,p}$. It also requires the maximization of the thermal resistance of the cantilever $R_{th,cant}$ and the thermal isolation of the cantilever from its surrounding environment to increase $R_{th,pe}$. The optimization of the thermal design of the cantilever in terms of dimensions, shape and materials, and experiments under vacuum environment have been based on these requirements.

Furthermore, eqn (9.6) shows that δT_p is dependent on the heat transfer within the sample and from the sample to the tip. In addition, it depends on the heat transfer between the tip apex and the thermosensitive element, if this element is not located at the tip apex.

If δT_p is neglected, only the probe temperature T_p is measured. Practically, the error δT_p due to the different thermal resistances of the network shown in Figure 9.6 can vary from one sensor to another and is related to the experimental configuration. Similarly to the heat rate transferred from the sample to the probe, Q_{s-t}, it depends on many parameters:

(1) the surrounding gas (pressure, temperature, relative degree of humidity);
(2) the tip–sample mechanical contact *i.e.*, the mechanical properties of the tip and surface, the tip–sample force, the surface roughness and topography; and
(3) the thermophysical properties of the probe and sample.

As discussed in the following section, this estimation through modelling based on the conventional approach of thermometry mentioned before is not trivial, and is still one of the main limitations of SThM. The description of heat transfer at the micro- and nanoscales within the tip and sample, exchanged between the probe and sample through surrounding gas and radiation, and through nanoscale contacts must be considered to estimate δT_p.

The calibration methods implemented for the determination of δT_p have mainly dealt with laboratory self-heating samples. They are based on comparisons of SThM measurements with either measurements obtained by optical thermometry methods, or results of simulation of the sample surface temperature (or both). However, optical thermometry methods have a spatial resolution limited to few hundred nanometres and simulations at micro- and nanoscales are often dependent on simplifications or critical parameters. These comparisons are then not strictly applicable to SThM temperature measurements with a spatial resolution of a few tens of nanometres.

One regularly evoked solution would be to exploit the NP SThM method where:

$$Q_{s-t} = (T_s - T_p)/(R_{th,s} + R_{th,c}) = 0 \tag{9.7}$$

Self-heating samples that have been used or fabricated for SThM calibration include instruments that are specifically designed for absolute temperature measurements on the scale of one micron. They are based on the measurement of the Johnson noise in a small metallic resistor,[93] or instrumented membrane.[94] Other samples have been based on hot sources implemented in sub-surface volumes with a metallic line heated through the Joule effect.[25,33,35,37,38,47,95,96] The samples are generally heated in an ac regime to demonstrate thermal mapping with low signal-to-noise ratios.

They have also been used to characterize the dynamic response of sensors, which is also an important parameter to be considered. Accurate temperature measurements after calibration are only possible on samples that hold almost the same surface properties as the calibration samples.

9.3.1.2 Thermal Conductivity Measurement

Thermal conductivity measurement requires the heating of the sample by the probe operating in active mode. The temperature sensor is heated and plays the role of heat source for the sample. Under this condition, the rate of the heat transferred by the probe to the sample Q_{t-s} may be written as a function of the thermal power P used for the heating of the probe (see Figure 9.7) and the measured probe temperature T_p:

$$Q_{t\text{-}s} = (T_{t,c} - T_a)/(R_{th,s} + R_{th,c}) = P - ((T_{t,c} - T_a)/R_{th,env} + (T_p - T_a)/R_{th,p})$$

$$(9.8)$$

where $R_{th,p}$ includes the possible thermal losses from the cantilever to the sample.

In the case of a homogeneous and semi-infinite sample, the sample thermal conductivity k_s is included in the expression for the sample thermal spreading resistance $R_{th,s}$:

$$R_{th,s} = 1/4k_s b \qquad (9.9)$$

where eqn (9.9) assumes an isothermal circular area of effective radius b at the probe–sample interface.[97]

From eqn (9.8) and (9.9), the ideal situation for an easy estimation of the sample thermal conductivity should be that: (1) k_s is included in the expression of $R_{th,s}$ only; and (2) the tip apex temperature $T_{t,c}$ equals the averaged probe temperature T_p that is measured by the thermal sensor through the measurement of its electrical resistance, in the case of a resistive probe.

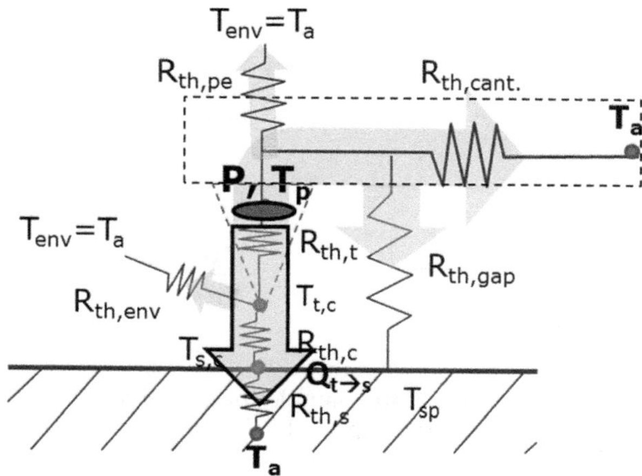

Figure 9.7 Thermal resistance network model for a probe used in active mode.

However, as discussed by Majumdar,[29] even assuming the thermal resistance associated with the heat transfer from the tip to the sample, $R_{\text{th,c}}$, b and $T_{\text{p}} = T_{\text{t,c}}$ ($R_{\text{th,t}} = 0$) are constant, whatever the sample is,

$$Q_{\text{t–s}} = 4k_{\text{s}}b(T_{\text{t,c}} - T_{\text{a}})/(1 + 4k_{\text{s}}bR_{\text{th,c}}) \tag{9.10}$$

For $Q_{\text{t–s}}$ to depend primarily on λ_{s}, $4bk_{\text{s}}R_{\text{th,c}} \ll 1$ must be satisfied, that is, the dominant thermal resistance should be that of the sample.

In the case of the Wollaston probe [Figure 9.3(a)], the order of magnitude of $R_{\text{th,c}}$ and b have been estimated to be about 2.10^{6}–1.10^{5} m K^{-1} and 1 μm,[44,57,98,99] respectively. These data were estimated from the fitting of theoretical curves to the experimental data obtained with known samples (see below). Consequently, SThM measurements with this probe would mainly be sensitive to k_{s} for materials of low thermal conductivity. It has been demonstrated that this is effectively the case.[98–101]

As shown in Figure 9.8, a purely experimental calibration of the Wollaston probe however can be used. The calibration can be performed with a set of experiments involving flat bulk samples of well-known thermal conductivities in a range that covers the expected value of the thermal conductivity k_{s} to be measured. Practically, the tip is usually heated with an increase of temperature larger than 80 K to ensure a good signal-to-noise ratio and avoid issues related to the presence of a water meniscus.[102]

A stable dc current heats up the tip through the Joule effect, and the electrical resistance of the tip is constantly monitored. The probe is measured with a balanced Wheatstone bridge that includes a feedback loop, allowing a constant value of its electrical resistance to be set

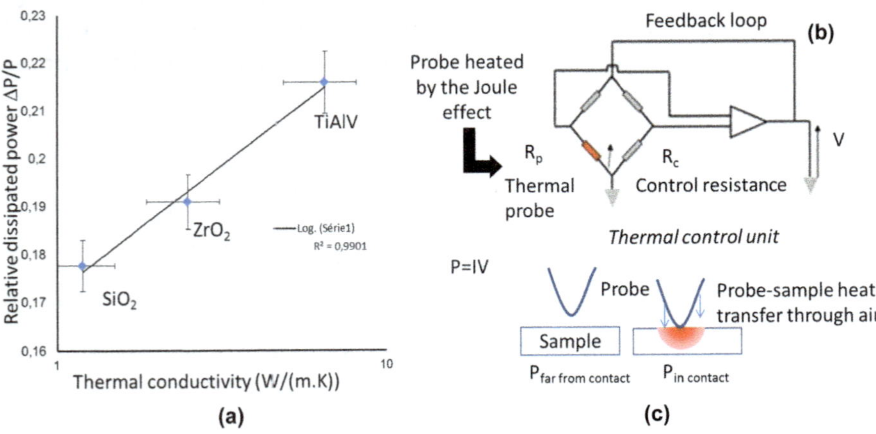

Figure 9.8   (a) Calibration curve. Three samples of known thermal conductivity over a range that covers the expected value of k_{s} to be measured are used. (b) Schematic diagram of a typical balanced Wheatstone bridge used for measurement with the Wollaston probe for investigating the thermal conductivity of samples. (c) Principles of measurements for a probe free in air, and in contact with the sample.

[see Figure 9.8(b)]. The average probe temperature is then kept constant during the measurement, and this is the electrical current that can vary.

When the tip is brought into contact, the value of the current is modified. If a possible heating of the connecting wires due to internal Joule effects and thermoelectric effects at the contacts are neglected, the electrical resistance is always constant. The exciting Joule power P_J is then $P_J(I) = (R_{c0} + R_{p0}(\bar{\theta}))I^2$, where R_{c0} is the electrical resistance due to the connection wires and $R_{p0}(\bar{\theta})$ is the probe electrical resistance at the set value $\bar{\theta} = T_p - T_a$. The relative dissipated probe power can then be computed as:

$$\Delta P/P = (R_{p0}(\bar{\theta}))I^2_{\text{in contact}} - R_{p0}(\bar{\theta}))I^2_{\text{far from contact}})/R_{p0}(\bar{\theta}))I^2_{\text{far from contact}} \quad (9.11)$$

In principle, this experimental calibration is possible only for materials with comparable surface states (roughness, mechanical properties *etc.*). Fortunately most of the heat is transferred through the air, not through the mechanical contact, when in ambient conditions. As a consequence, the experimental measurement is robust against probe and/or surface defects and irregularities in the case of low thermal conductivity samples. The method could furthermore be performed using measurements in an ac regime.

A quite similar calibration methodology has recently been demonstrated comparing measurements performed with a Wollaston probe far from contact, and without contact but at a small distance (<100 nm) from the sample surface, confirming that heat transfer is mainly exchanged through air with the Wollaston probe.[58] It has also been demonstrated applicable to the resistive palladium probe [see Figure 9.3(b)] operated in air.[103]

9.3.1.3 Calibration and Specification of SThM Measurements through Modelling

Beyond the purely experimental calibration of probes, the modelling of SThM measurements can allow a link to the nominal signal measured and the parameters to be determined. For this purpose, the modelling of SThM measurements has particularly been developed for resistive probes for which the thermal sensor cannot be assumed to be located at the probe apex. As shown by eqn (9.4) and (9.8), the temperature of the probe at the probe apex $T_{t,c}$ should be known to rigorously establish the expression of the heat rate exchanged between the probe and the sample. Some studies have proposed similar modelling for thermocouple probes that are also resistive and can then be used in active mode[35,104]

Analytical and numerical models have been proposed.[11,37,44,45,59,64,65,99,105–108] The majority of them have considered the geometrical and dimensional parameters and the physical properties of materials allowing the description of the probe. They also include effective parameters such as:

(1) an effective coefficient h of heat loss by the whole probe surface to its environment, allowing the expression of the thermal resistance $R_{\text{th,pe}}$ in eqn (9.5) and (9.8); and

(2) an effective thermal resistance $R_{\text{th,c}}$ to describe the probe–sample thermal interaction at the level of the probe–sample contact. Thermal interaction is then assumed to be across an area generally described as a disc of effective radius b at the sample surface.

The fitting of simulated measurements with experimental data in well-known conditions can allow the modelling parameters to be identified. The usual approach is given here for the Wollaston probe, and then discussed for smaller resistive probes.

9.3.1.3.1 Probe Parameters. Usual analytical thermal modelling of the probe is based on the resolution of the heat equation along the resistive wire of the Wollaston probe. The wire is then described as two thermal fins heated by the Joule effect.[44] This modelling allows the determination of the mean temperature of the wire tip whatever the heating regime (dc or ac).[44,108] It can allow the determination of probe parameters such as the time response constant and the sensor dimensional parameters from measurements performed in an ac regime.

As an example, Figure 9.9 shows data obtained from experiments performed in ac mode without contact with a sample under ambient air.[44]

In the corresponding experiments, an alternating current $I = I_0\cos(\omega t)$ is sent into the tip, which heats it with two contributions due to the Joule effect $P_{\text{J}} = R_{\text{p}0}^2 I_0^2 \left[\dfrac{1}{2} + \cos(2\omega t)\right]$. This leads to an increase of temperature

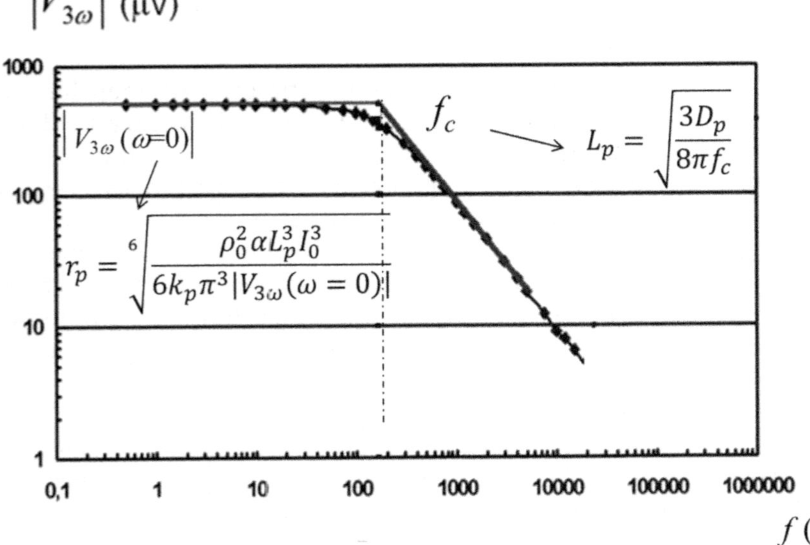

Figure 9.9 Modulus of the 3ω voltage $V_{3\omega}$ for a Wollaston probe as a function of the excitation frequency, $f = \omega/2\pi$, of the probe heating current. Fitting allows the determination of the probe's effective parameters.

$\theta = \theta_{dc} + \theta_{2\omega}\cos(2\omega t + \phi_{2\omega})$ where the subscript dc indicates the continuous component of the temperature increase. Alternating current measurements are performed with a Wheatstone bridge without a feedback loop. The bridge is balanced when the probe is free in air and the imbalance of the bridge is monitored when the tip is brought into contact with the sample. Remembering that the electrical resistivity ρ depends linearly on temperature through the coefficient of temperature α, the voltage signal due to the imbalance of the bridge is:

$$V = R_0(1 + \alpha\theta_{dc})I_0 \cos(\omega t) + \frac{\alpha R_0 I_0}{2} \theta_{2\omega} \cos(\omega t - \phi_{2\omega})$$

$$+ \frac{\alpha R_0 I_0}{2} \theta_{2\omega} \cos(3\omega t + \phi_{2\omega}) \tag{9.12}$$

Interestingly, $V_{3\omega}$ is directly proportional to $\theta_{2\omega}$, akin to a thermometer. Figure 9.9 shows that the tip can be assumed to operate as a low-pass filter of cut-off frequency about $f_c = 200$ Hz. Once f_c is experimentally determined, the effective radius r_p and the effective half-length L_p of the tip can be determined from the model. Let us note that the length and diameter of the tip can vary from one tip to another, since Wollaston probes are fabricated manually.

9.3.1.3.2 Probe Losses to the Environment: the h Coefficient.

A fraction of the heat is lost from the probe–sample system to the environment while working under ambient conditions. When the dimensions go down to the microscale, the heat-transfer coefficient due to convection becomes smaller as the buoyancy forces decrease.[109] The heat losses to the air are suggested to be mainly due to heat conduction.[110] The heat-transfer coefficient h equivalent to these losses depends on the probe size and is larger than those at the macroscale due to the larger surface-to-volume ratio in SThM.

The analytical model reported previously was also used to specify the heat loss coefficient h for the Wollaston probe.[111] The measurements of the amplitude of the component $\theta_{2\omega}$ of the probe ac temperature were performed with a probe free in air at different pressures P. The obtained value of h was approximately 3500 W m^{-2} K^{-1} for experiments under ambient conditions. The experimental graph $h = f(P)$ could be analysed in light of the theory of Lees[112] based on the resolution of the Boltzmann equation for the heat transfer between two cylinders. Other values of h are available in the literature.[108,113] The obtained differences can be attributed to the fact that the model sensitivity to this parameter is rather low.[114] Furthermore, h should vary when the tip is close to the surface of the sample.

9.3.1.3.3 Effective Parameters Describing the Probe–Sample Thermal Interaction.

Measuring the amplitude of the component $\theta_{2\omega}$ of the probe ac temperature, as a function of the angular frequency ω while the probe contacts a sample can also allow the determination of the parameters $R_{th,c}$

and b that describe the probe–sample thermal interaction. Thus, with an appropriate description of the sample resistance $R_{\text{th,s}}$, one can determine the contact conductance $R_{\text{th,c}}$ for a sample of well-known thermal conductivity.

Results obtained in this way under ambient air on a set of samples with various thermal conductivities have suggested that $R_{\text{th,c}}$ and b are two parameters which depend on the thermal conductivity of sample.[114] Values of $R_{\text{th,c}}$ and b were respectively found to vary from 2.10^6 K W^{-1} to 1.10^5 K W^{-1}, and from 550 nm to 150 nm when the sample thermal conductivity of the sample increases.[115] Assy *et al.* have recently shown that this dependence is not negligible, from similar measurements performed in a dc regime.[116] Furthermore other works[99,100] suggested a non-negligible contribution of the thermal resistance of the solid–solid contact between the probe and the sample. This contribution depends on the sample thermal conductivity but also on the sample mechanical properties and roughness.[100]

To date, no simple analytical expression of $R_{\text{th,c}}$ according to the sample's physical properties is available. It seems that numerical simulation is more appropriate to describe the probe–sample interaction through air. As a consequence, the purely experimental method previously described is currently preferable.

9.3.1.3.4 Approaches for Smaller Probes.

Puyoo *et al.*[117] have proposed a similar model and approach for the palladium probe [see Figure 9.3(b)]. A value of h of 6100 W m^{-2} K^{-1} was identified for a probe free in air (far from the sample). Like for the Wollaston probe, this value has been used to simulate the measurement performed when the probe contacts the sample.

Recently, Tovee *et al.*[54] also modelled the heat transfer within the palladium probe based on the diffusive regime through finite element simulations. The heat transfer through the air was modelled as diffusive conduction in their work. Experimental measurements performed under ambient conditions, and results of simulations for the same environmental conditions were in good agreement. The thermal resistance of the probe that the authors determined in a vacuum was however larger than expected.[64] This could be related to boundary scattering within the probe apex, as such phenomena induce another thermal resistance to be added to the one described by the diffusive regime.[118] Since the dimensions at the tip apex can be lower than the average mean free path of energy carriers within the tip, the thermal conductivity of the tip apex can differ from the one estimated in the diffusive regime.[118] This size effect must be taken into account for the smallest probes such as the resistive Si tips [see Figure 9.3(b)].

At room temperature the phonon averaged mean free path Λ_{ph} in pure crystalline Si is of the order of 300 nm,[119] which exceeds the actual size of the tip apex. As a consequence, a continuous description of heat transfer does not accurately model the heat conduction at the tip apex. Instead, heat conduction through the tip is modelled as quasi-ballistic transport. For example, following the expression of thermal conductivity from kinetic theory:

$$k = Cv\Lambda_{\mathrm{ph}}/3 \tag{9.13}$$

where k is the thermal conductivity, C is the heat capacity per unit volume and v *is* the phonon velocity. Mathiessen's rule can be applied to calculate a mean free path modified from the bulk value through boundary scattering. Analytical expressions have already been proposed for simple geometries.[120]

Through numerical simulations, Nelson and King[106] have modelled the thermal interaction between Si nanotips and samples. In their modelling, the thermal resistance in the tip is divided into two parts: conical resistance R_{cone} and hemi-spherical resistance R_{sphere}. The radius of the sphere was assumed to be the same as the tip apex, and the heat transfer in the hemisphere was assumed to be one directional. These conditions would include fewer errors while working under a UHV environment where the conduction through the air is eliminated.[106] The thermal resistance at the tip–sample contact has been given as a function of the boundary resistance. However, this approach encounters some uncertainties and, as discussed in the following section, also depends on the sample material.[106]

It appears through this discussion that a better understanding of the thermal interaction between an SThM tip and the sample is crucial for a good understanding of measurement, and a proper interpretation of the contrast of thermal images. The following section focuses on the current description of the different heat transfers contributing to the probe–sample thermal interaction.

9.3.2 Heat Transfer between the Probe and the Sample

Various mechanisms of heat transfer may coexist between the probe and the sample. Figure 9.10 shows the heat flow paths from a hot probe and within the sample for experiments performed under ambient air.

The heat transfer channels between the probe and the sample include radiative heat transfer, thermal conduction through the surrounding gas, heat conduction through the liquid meniscus formed at the tip–sample junction and heat conduction through the mechanical contacts between both objects. Very schematically, and as proposed by Majumdar,[29,33] the effective thermal resistance $R_{\mathrm{th,c}}$ describing the probe–sample thermal interaction at the level of the probe–sample interface may be simply written as:

$$R_{\mathrm{th,c}} = 1/G_{\mathrm{th,c}} = 1/(G_{\mathrm{rad}} + G_{\mathrm{gas}} + G_{\mathrm{w}} + G_{\mathrm{mc}}) \tag{9.14}$$

where G_{rad}, G_{gas}, G_{w} and G_{mc} are the thermal conductances used to describe the probe–sample heat transfer through radiation, gas conduction, liquid meniscus conduction and conduction through the mechanical contacts, respectively.

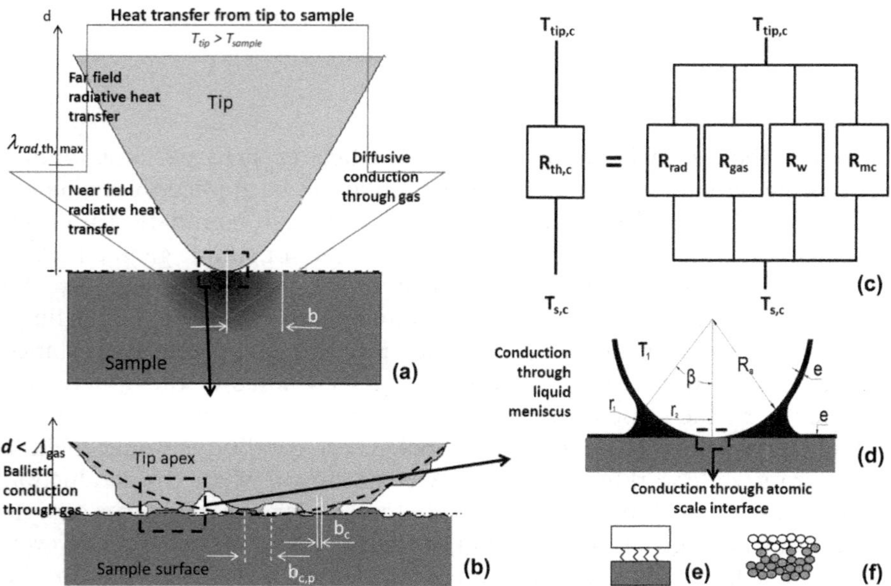

Figure 9.10   Schematic representations of the heat flow paths from a heated tip to a sample at different length scales (not to scale). (a) Microscopic probe–sample contact. (b) Microscopic multi-asperity contact. (c) Thermal resistance network model of the thermal resistance of the probe–sample thermal interaction $R_{\mathrm{th,c}}$. (d) Water meniscus model. (e) Atomic-scale interface. (f) Atomic-scale roughness.

9.3.2.1   *Thermal Radiation*

To date the probe–sample heat transfer through radiation has been totally neglected in SThM, excepted in the SPM works specifically developed for its investigation[77,78,90,91,121] at the micro- and nanoscales and/or SThM works based on the exploitation of near-field thermal radiation for thermometry measurements.[31,76,122]

The thermal radiation contribution to the probe–sample interaction has been estimated as being much smaller than the other heat-transfer mechanisms based on various far-field macroscopic considerations.[29,72,121] The participation of radiation in this interaction is still to be properly evaluated by accounting for the near-field regime. Its clean detection requires a vacuum environment where conduction through the water meniscus and air could both be suppressed. The probe should not be in contact with the sample to avoid conduction through the mechanical contact. Near-field thermal radiation occurs when the separation distance d between the probe and the sample becomes smaller than $\lambda_{\mathrm{rad,th,max}}$, the maximum wavelength of thermal radiation ($\simeq 10\ \mu\mathrm{m}$ at 300 K). We note that materials, and the shape of the sample and tip should therefore be well chosen.[106,111]

Measured values[31,76,78] of the probe–sample thermal conductance G_{rad} are smaller by two orders of magnitude or more than the values measured for G_{gas}.

For most practical operating conditions, radiation may be ignored in the interpretation of SThM measurements in contact under ambient conditions.[72] This may however be seen as a strong assumption for experiments performed under vacuum.

9.3.2.2 Air Heat Transfer

As already shown throughout this chapter, the probe–sample heat transfer through air cannot be neglected when experiments are performed under ambient air. It is then a dominant heat-transfer mechanism. This is the case for large tips such as Wollaston wire microprobes.[111] For smaller probes, this transfer greatly contributes to the heat transfer between the probe cantilever and the sample.[106] Due to its influence, temperature or thermal property distributions measured by SThM for nanoscale devices or structures are distorted and can greatly deviate from the true distributions.[33,37] This makes it necessary to correct the thermal distribution obtained from SThM by thermal modelling.

As represented in Figure 9.10, the regime of the heat transfer between a probe and its gaseous surroundings depends on the separation distance d between the probe and the sample. Let us consider Λ_{gas} the mean free path of the gas and the Knudsen number, $Kn_{gas} = \Lambda_{gas}/d$. When $d \gg \Lambda_{gas}$ ($Kn_{gas} < 1$), thus the conduction regime is diffusive and dependent on d. When $d < \Lambda_{gas}$ ($Kn_{gas} > 1$), the continuum assumption is not appropriate. This concerns the very end of the tip where d is smaller than micrometres,[123,124] or when the probe is operated under very low vacuums ($P < 1$ mBar).[125]

The different regimes of conduction through air have often been accounted for through local unidirectional thermal conductance expressions[33,121] where a geometry factor is used to be close to the shape of the tip. The conductance through air G_{gas} can then be estimated as:

$$G_{gas} = \iint h_{gas(\text{diffusive, ballistic or slip regime})}\, dxdy \qquad (9.15)$$

The order of magnitude of the identified values of G_{gas} for experiments performed under ambient air can vary from 10^{-8} W K^{-1} to 10^{-6} W K^{-1} depending on the size of the tip.

Direct simulation Monte Carlo (DSMC)[123,126] and a quasi-ballistic heat-transfer model[124] have more recently been developed to better assess the sub-continuum conduction. We note that for the largest distances, the air heat transfer between the probe and the sample can also involve heat convection.[127]

9.3.2.3 *Water Meniscus*

In humid air, a meniscus forms on surfaces from water molecules. This especially takes place at the tip–sample contact, where contaminants adsorbed on the sample and tip surfaces can also be present. This phenomenon depends on the surface roughness, the interaction geometry and the hydrophilicity of the materials.[128] To date little information on heat transfer through the water meniscus is available.

Luo *et al.*[11] described the probe–sample heat transfer by a network of thermal conductances as shown in Figure 9.10(c). The effective thermal conductance G_{eff} between a thermocouple nanoprobe and a heated sample was determined experimentally. The thermal conductances for each of the heat-transfer mechanisms of the probe–sample interaction were calculated. By approximating the probe apex to a spherical cap of radius R_{a} in contact with the meniscus, the water meniscus thermal conductance G_{w} was given as:

$$G_{\mathrm{w}} = 2\pi k_{\mathrm{w}} R_{\mathrm{a}} \int_0^{\beta} R_{\mathrm{a}} \sin\varphi \cos\varphi \, \mathrm{d}\varphi / (R_{\mathrm{a}}(1 - \cos\varphi) + a) \qquad (9.16)$$

where k_{w} and β are, respectively, the thermal conductivity and opening angle of the meniscus, and a is the distance between the apex of the tip and the sample [see Figure 9.10(d)]. The authors found that the effective thermal conductance G_{eff} corresponds better to the conductance of the meniscus. It was then suggested that the heat transfer through water meniscus is the dominant heat mechanism that operates between the thermocouple probe and the sample.

However, few years later, the same group found a remarkable difference in the experimental determination of G_{w} with similar equipment. The thermal conduction due to air appeared to dominate the thermal interaction between the probe and the heated sample, and was much larger than the heat transfer through water ($G_{\mathrm{w}} = 6.7$ nW K^{-1}). They suggested that such a difference between the two studies is related to the conductances at the probe–water and water–sample interfaces, that were not taken into account in the modelling of their previous work.[33]

9.3.2.4 *Heat Conduction at the Tip–Sample Mechanical Contact*

The radius of the mechanical contact $b_{\mathrm{c,p}}$ is generally estimated from mechanical contact theories such the ones developed by Hertz, Johnson–Kendall–Roberts (JKR)[129] or Derjaguin–Müller–Toporov (DMT)[130–132] for elastic deformation. These theories apply to contacts between a sphere and a flat surface and their use leads to considering an ideal contact between surfaces of perfect quality.

Whatever the considered SThM tip is, $b_{\mathrm{c,p}}$ can be estimated to be lower than a few tens of nanometres. For such a nanoscale contact, both phonon mismatch and mechanical contact geometry have to be taken into account for estimating the contribution of the solid–solid interface to the measurement.

Even for interfaces of perfect quality, an interfacial contact thermal resistance occurs at the tip–sample interface due to the acoustic property mismatch, *i.e.*, the difference in phonon dispersion, between the two materials in contact.[88,118,133–135] The interfacial resistance can then be estimated as:

$$R_{\text{th, contact, B}} = R_{\text{th,B}} / \pi b_{\text{c,p}}^2 \tag{9.17}$$

where $R_{\text{th,B}}$ is a thermal boundary resistance that has the same units as the bulk thermal contact resistance.

The values of $R_{\text{th,B}}$ experimentally determined for solid–solid contacts near room temperature do not vary significantly with the contacting materials. $R_{\text{th,B}}$ values typically range from 5×10^{-9} to 5×10^{-7} m^2 K W^{-1}.[134,136,137] In reality, however, the tip–sample interface is not perfect. As represented in Figures 9.10(b) and (d), water, contaminants or oxide layers can cover the probe and sample surfaces and surface roughness or weak coupling bonds between the atoms of solids lead to a non-continuous contact area.

Advanced contact models have been developed to account for the weak coupling. The transmission probability is related to a theoretical mechanical coupling spring between the two solids.[138,139] Furthermore, as represented in Figure 9.10(b), the apparent contact area may be divided into smaller contact spots [of length scale b_{c} in Figure 9.10(b)] due to the roughness of the tip and sample surfaces. Appropriate modelling of this effect will depend on the ratio of the averaged mean free path of heat carriers Λ_{c} to the length scale b_{c}.

If $b_{\text{c}} \gg \Lambda_{\text{c}}$, diffusive transport applies. Solutions of type of eqn (9.9) are applicable:

$$G_{\text{th,contact}} = K k_{\text{s}} b_{\text{c}} \tag{9.18}$$

where K is a geometrical factor describing the heat spreading within the sample. On the other hand, if $b_{\text{c}} < \Lambda_{\text{c}}$, ballistic solutions must be considered.[140]

When $b_{\text{c}} \ll \Lambda_{\text{c}}$, the notion of finite contact spots may be extrapolated to the atomic scale.[141,142] For this extreme case, Gotsmann[142] has recently suggested that a quantization of conductance may occur when b_{c} is lower than λ_{coh}, the phonon coherence length, while the distance between individual contact spots may exceed λ_{coh}. Related experimental works were performed under UHV with doped Si probes with end tips specifically prepared for this purpose.

The above discussion of a roughness size effect may apply to the expression of the sample thermal resistance $R_{\text{th,s}}$ and its related effective radius b [(see eqn (9.9)]. $R_{\text{th,s}}$ dependence is strongly changed if $b_{\text{c}} \ll \Lambda_{\text{c}}$, and it no longer depends on the sample thermal conductivity. The contact becomes the essential parameter. This seems to be an intrinsic limitation to SThM: the sensitivity to the thermal conductivity decreases as the characteristic size of the probed volume approaches the value of the averaged mean free path. If it becomes much smaller, the SThM measurement becomes completely independent of thermal conductivity. The measurement then provides direct information on the energy carrier properties.[121] Furthermore, if SThM should enable the investigation of thermal phenomena with a resolution of <10 nm (as has already been the case in tunnelling thermo-voltage-based SThM[27]), it is unlikely that temperature can vary within a distance of atomic spacing. Statistical thermodynamics states that temperature can be defined only when thermodynamic equilibrium is established, *i.e.*, on a given minimal volume.

9.4 Applications

Although the tip–sample interaction is not yet fully understood, SThM has been applied in many diverse fields of science and engineering since the 1990s due to its high spatial resolution. These fields include:

(1) the thermal characterization of operating devices;
(2) the measurement of the thermal conductivity of nanostructured materials and nanomaterials; and
(3) polymer sciences.[50,56,67,136,143]

9.4.1 Characterization of Devices

For sample surface temperature investigation, the tip acts as a thermometer. The Wollaston probe in passive mode has been used for microsystem diagnostics.[46–49] It was successfully used to characterize the temperature profile measurements of a semiconductor thermoelectric couple[46] and shown to be useful for failure localization and analysis of integrated circuits.[47,144,145] These studies were made in an ac regime. However, the active area, the size and shape of the Wollaston probe, limited the thermal investigations of integrated circuits (ICs) due to topography related artefacts in the thermal images,[47] and the analysis of nanoscale structures. As such, applications of SThM to active microdevices and nanodevices have mainly been performed with smaller probes.

The first thermocouple probes with a built-in thermal sensor[146] were used to measure the temperature distribution inside a vertical-cavity surface-emitting laser (VCSEL).[146] SThM images of the VCSEL showed that the peak temperature occurred at the intersection of the optical axis and the active quantum wells, and increased with input power at a rate of $0.74\,\text{K}\,\text{mW}^{-1}$. Comparisons of the results with model predictions showed that the *n* mirrors and the substrate produce higher heat generation rates, possibly due to Joule heating and/or the absorption of spontaneous emissions, which are often neglected in models.[146]

Since then, SThM using small probes (thermocouple and resistive probes) has been used to characterize the heat dissipation and transport pathways in various nanocomponents:

(1) multiwall-carbon nanotubes (MWNTs generally about 10 nm in diameter) and single-wall carbon nanotubes (SWNTs about 1–2 nm in diameter).[147–149] The temperature profiles obtained with SThM showed, for example, that when electron current flows in MWNTs a uniform Joule heating and then diffusive electron transport occurs. In SWNTs, electron flow was shown to be ballistic at low voltages (near room temperature);

(2) Joule self-heated graphene nanoribbons (GNRs).[150] Hot spot formation at well-defined and localized sites was observed;

(3) self-heating Si nanowires and self-heating nanowire diodes (with doped Si probes).[96]

While a homogeneous nanowire shows a bell-shaped temperature profile, a nanowire diode exhibits a hot spot centred near the junction between the two doped segments;nanoscale constrictions in metallic microwires deposited on an oxidized Si substrate.

(4) Nanoscale constrictions were tuned in terms of temperature and confinement size[95] (see Figure 9.11); and

(5) conjugated-polymer light-emitting diodes (LEDs).[148] Polymer LEDs with thermally conductive substrates and Al/Cu double cathodes were shown to enhance the thermal stability of the organic device.[151]

The heat dissipation characteristics in substrate-supported and suspended (with asymmetric-type contacts) current-carrying GaN nanowires with diameters of *ca.* 40–60 nm were also investigated.[152] Using an approach combining SThM and spatially resolved Raman spectroscopy, direct measurements of the nanowire–substrate/electrode interface thermal resistance, and the nanowire thermal conductivity were performed. On the basis of these results, the relative significance of nanowire–substrate/electrode interfaces in dissipating heat was demonstrated in suspended nanowire devices.[152]

Kar-Narayan *et al.*[153] also used SThM for measuring the electrocaloric (EC) temperature change in a multilayer capacitor (MLC) based on $BaTiO_3$. EC effects are thermal changes that arise in electrically insulating materials due to changes in the applied electric field. They showed that SThM can be used to measure giant EC effects in thin films with substrates present.

9.4.2 Measurement of the Thermal Conductivity

For thermal conductivity measurements, the tip acts as a thermometer and delivers heat to the sample simultaneously (conventional SThM in active mode). The thermal investigation of matter by the use of a very localized heat source allows materials to be probed at the level of very small sub-surface

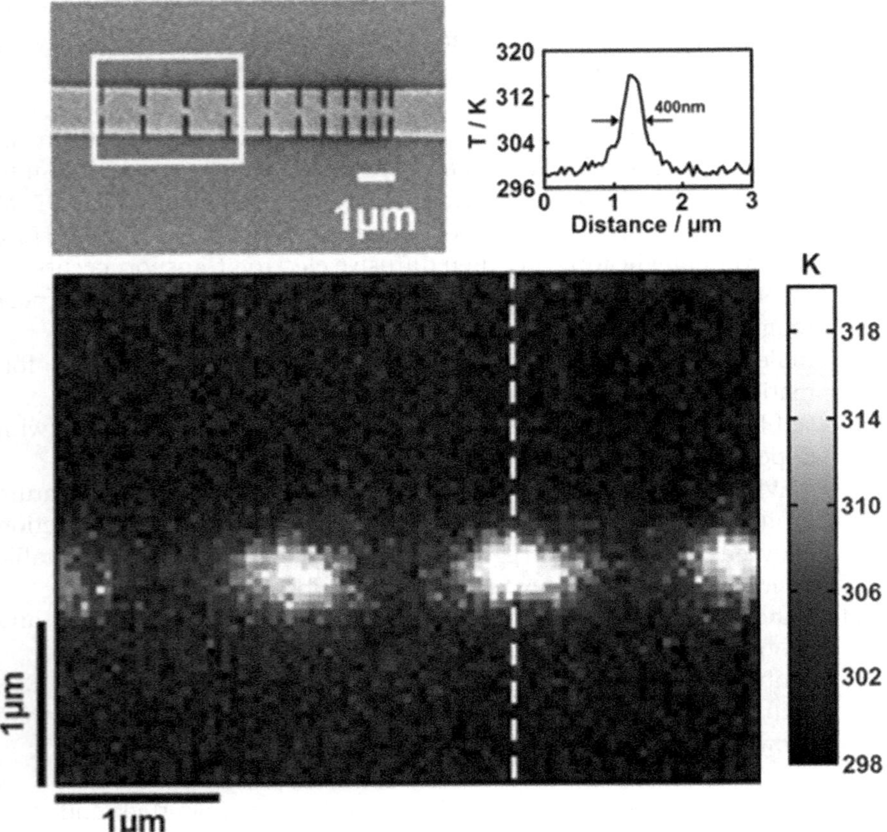

Figure 9.11   Thermal image of sub-micrometre hot spots showing nanoscale constrictions in a metallic microwire (heated with a 1 mA current) deposited on an oxidized silicon substrate. This image was obtained by the SThM method using a fluorescent nanocrystal glued to the apex of an AFM probe. The SEM image indicates the observed constrictions. Reproduced with permission from ref. 95. © Wiley-VCH 2011.

volumes. Therefore, the technique has quickly appeared as a promising method to study the thermal conductivity of thin films.

The Wollaston wire probe was the leading probe for many years. It has been applied to the characterization of various bulk materials and thin films, mainly insulating materials, such as:

(1) bulk ZnO (0001),[51] and Ba_8Si_{46} which is a simple binary representative of intermetallic clathrates, guest–host systems with a high potential for thermoelectric applications due to their ultralow thermal conductivity and even their perfect crystal structure;[154]

(2) porous[54,58,105,155] and mesoporous[57] bulks or thin films of sub-micrometric thickness.[99,152] Some results obtained with SThM were

compared with values measured on the same material samples by Raman thermometry and were in good accordance.[57,99,155,156] The measurement of the rate of reduction of the thermal conductivity in porous Si by irradiation with swift heavy ions was demonstrated;[155] and

(3) thin films on substrates, including SiO_2 thin films on Si substrates,[59] and fully and partially coalesced lateral epitaxial overgrown GaN/sapphire (0001).[52]

The smallest details thermally imaged with the Wollaston probe were found to be about a few tens of nanometres in size. In these cases, however, topography related artefacts may be suspected,[101] carbon contamination-assisted sharpening of the probe was shown,[157] or probes were modified.[158,159] In the two last cases, a better sensitivity to sample thermal properties[157,158] was obtained due to the decrease in the probe–sample thermal resistance at the probe–sample contact. As a result, micrometric spatial resolution has generally been agreed by SThM users for the Wollaston probe. For a more complete overview of applications of the Wollaston probe, readers are referred to the reviews in ref. 50 and 160.

Regarding the other resistive metallic probes with smaller tips, the thermal conductivity of single Si nanowires was studied[65] with the palladium probe operating *via* the 3ω method. A spatial resolution of around 100 nm was achieved in thermal imaging performed under vacuum. No significant reduction in comparison with bulk Si was found for nanowires with diameters ranging from 200 to 380 nm.[65] The method was recently applied to SiGe nanowires[161] and to characterize a Sb_2Te_3 phase-change nanowire.[162]

A modified SThM technique has been proposed to image thermally multilayered periodic photonic structures.[163] With an analysis of the topographic and thermal signals the thermal boundaries between the layers were revealed and the periodicity of the structure was analyzed. A spatial resolution of at least about 70 nm was found.

Doped Si probes allowed the imaging of the thermal conductivity contrast of biological materials with a spatial resolution of 10 nm and thermal resolution of 50 nW.[164] They furthermore enabled the estimation of the thermal conductivity of a 3 nm thick HfO_2 film on a Si substrate with a spatial resolution of around 25 nm. In the last case, experiments were also performed under vacuum.[165]

9.4.3 Phase-transition Temperature Measurement

SThM probes operating in active mode have also been used to characterize polymer materials. Ref. 50, 56, 67, 136 and 143 give the origin and review this application of SThM. As an example, Figure 9.12(a) provides one of the principles of the *Local Thermal Analysis* (LTA) method for the local measurement of the melting temperature of polymers.

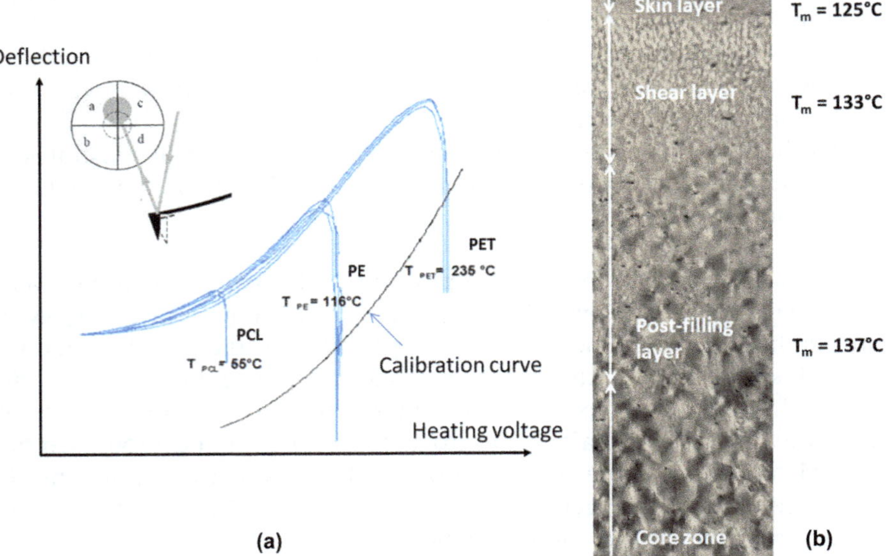

Figure 9.12 Measurement principles in LTA and an example. (a) The calibration of a probe is obtained from measurements of the cantilever deflection performed for three samples of known melting temperature T_m. These measurements allow the determination of the calibration curve giving the correlation between the voltage applied to the probe to heat the tip, and the corresponding temperature of the tip. (b) Measured T_m of an isotactic polypropylene (iPP), according to its local microstructure.

In practice, the probe is used as a local heating element and detector simultaneously. In imaging mode, topography images are obtained by scanning the probe over the surface while maintaining it at a constant temperature in the tapping mode of AFM. LTA measurements are performed on selected spots from topographical images by following the system's response [the vertical displacement of the probe, see Figure 9.12(a)] during a controlled fast heating while the tip remains in contact with the surface. The calibration of a probe is obtained from measurements of the cantilever deflection performed for three samples of known melting temperature T_m.

Figure 9.12(b) shows results obtained with the method for a semicrystalline polymer sample drawn from an injection moulding plate, according to its local microstructure. The probe used was a doped Si probe [see Figure 9.3(b)]. The presented results allow the verification of the increase of T_m with the increase in the crystallite size from the skin layer to the core zone within the thickness of a moulding plate.

9.5   Conclusions and Perspectives

This review of the research on the design of experiments and probes for thermal characterization at the nanoscale by SPM methods emphasizes that there have been several important developments and breakthroughs in SThM since the 1990s. Thermal imaging and measurements at spatial resolutions in the sub-100 nm regime have become achievable not only through the technological development of new miniaturized sensors but also by exploiting experiments under vacuum.

SThM measures local temperature or thermal properties of a sample with a sharp tip. Purely experimental calibration methods are available. The interpretation of the measurements will however always depend on the description of the heat flows within the probe–sample system and its surrounding environment. This includes the description of the macroscopic heat flows through the cantilever as well as the microscopic heat flows through the tip and substrate. This description is indeed required to determine the heat flow exchanged between the probe and the sensor to assess the temperature at the tip–sample interface and the error on the thermal parameters to be identified.

Although progress has been made in the understanding of heat transfer at point contacts, the key challenge is still to understand the heat flows at the nanoscale interface between the tip and a substrate. Many fundamental issues are still unresolved. Quantitative comparisons between experimental observations and theory are needed. For that, improved models for probe–sample heat transfer have to be developed.

Many newly-developed nanomaterials and devices are composed of heterogeneous materials, and it is critical to understand the nanoscale transport mechanism at the interfaces.[166] SThM should enable the characterization of these new devices and materials and contribute to elucidating the nanoscale transport mechanisms at interfaces.

Acknowledgements

The research leading to this chapter has received funding from the European Union Seventh Framework Programme FP7-NMP-2013-LARGE-7 under grant agreement no. 604668.

References

1. G. Binnig, H. Rohrer, C. Gerber and E. Weibel, *Appl. Phys. Lett.*, 1982, **40**, 178–180.
2. C. Williams and H. Wickramasinghe, *Appl. Phys. Lett.*, 1986, **49**, 1587–1589.
3. M. Nonnenmacher and H. Wickramasinghe, *Appl. Phys. Lett.*, 1992, **61**, 168–170.
4. J. Zhou, C. Yu, Q. Hao, D. Kim and L. Shi, ASME 2002 International Mechanical Engineering Congress and Exposition, 2002.
5. S. Sadat, A. Tan, Y. J. Chua and P. Reddy, *Nano Lett.*, 2010, **10**, 2613–2617.

6. H. Lang, M. Hegner and C. Gerber, *Springer Handb. Nanotechnol.*, 2007, 443–460.

7. I. Rangelow, T. Gotszalk, P. Grabiec, K. Edinger and N. Abedinov, *Microelectron. Eng.*, 2001, **57**, 737–748.

8. A. Majumdar, J. Lai, M. Chandrachood, O. Nakabeppu, Y. Wu and Z. Shi, *Rev. Sci. Instrum.*, 1995, **66**, 3584–3592.

9. Y. Suzuki, *Jpn. J. Appl. Phys.*, 1996, **35**, L352.

10. K. Luo, Z. Shi, J. Lai and A. Majumdar, *Appl. Phys. Lett.*, 1996, **68**, 325–327.

11. K. Luo, Z. Shi, J. Varesi and A. Majumdar, *J. Vac. Sci. Technol., B: Microelectron. Nanometer Struct.-Process., Meas., Phenom.*, 1997, **15**, 349–360.

12. M. Lederman, D. Richardson and H. Tong, *IEEE Trans. Magn.*, 1997, **33**, 2923–2925.

13. G. Mills, H. Zhou, A. Midha, L. Donaldson and J. Weaver, *Appl. Phys. Lett.*, 1998, **72**, 2900–2902.

14. L. Shi, O. Kwon, C. Miner and A. Majumdar, *J. Microelectromech. Syst.*, 2001, **10**, 370–378.

15. K. Hwang, K. Kim, J. Chung, O. Kwon, B. Lee, J. S. Lee, S. Park and Y. K. Choi, ASME 2009 Second International Conference on Micro/ Nanoscale Heat and Mass Transfer, 2009.

16. S. Heisig, H.-U. Danzebrink, A. Leyk, W. Mertin, S. Münster and E. Oesterschulze, *Ultramicroscopy*, 1998, **71**, 99–105.

17. R. J. Pylkki, P. J. Moyer and P. E. West, *Jpn. J. Appl. Phys.*, 1994, **33**, 3785.

18. G. Mills, J. Weaver, G. Harris, W. Chen, J. Carrejo, L. Johnson and B. Rogers, *Ultramicroscopy*, 1999, **80**, 7–11.

19. K. Edinger, T. Gotszalk and I. Rangelow, *J. Vac. Sci. Technol., B: Microelectron. Nanometer Struct.-Process., Meas., Phenom.*, 2001, **19**, 2856–2860.

20. A. Gaitas and Y. B. Gianchandani, *Ultramicroscopy*, 2006, **106**, 874–880.

21. P. S. Dobson, J. M. Weaver and G. Mills, Sensors, 2007 IEEE, 2007.

22. Y. Zhang, P. Dobson and J. Weaver, *Microelectron. Eng.*, 2011, **88**, 2435–2438.

23. L. Aigouy, G. Tessier, M. Mortier and B. Charlot, *Appl. Phys. Lett.*, 2005, **87**, 184105.

24. L. Aigouy, E. Saïdi, L. Lalouat, J. Labéguerie-Egéa, M. Mortier, P. Löw and C. Bergaud, *J. Appl. Phys.*, 2009, **106**, 074301.

25. E. Saïdi, B. Samson, L. Aigouy, S. Volz, P. Löw, C. Bergaud and M. Mortier, *Nanotechnology*, 2009, **20**, 115703.

26. O. Nakabeppu, M. Chandrachood, Y. Wu, J. Lai and A. Majumdar, *Appl. Phys. Lett.*, 1995, **66**, 694–696.

27. J. Weaver, L. Walpita and H. Wickramasinghe, *Nature*, 1989, **342**, 783–785.

28. A. Pavlov, *Appl. Phys. Lett.*, 2004, **85**, 2095–2097.

29. A. Majumdar, *Annu. Rev. Mater. Sci.*, 1999, **29**, 505–585.

30. H. Zhou, A. Midha, G. Mills, S. Thoms, S. Murad and J. Weaver, *J. Vac. Sci. Technol., B: Microelectron. Nanometer Struct.-Process., Meas., Phenom.*, 1998, **16**, 54–58.

31. A. Kittel, W. Müller-Hirsch, J. Parisi, S.-A. Biehs, D. Reddig and M. Holthaus, *Phys. Rev. Lett.*, 2005, **95**, 224301.
32. J. Chung, K. Kim, G. Hwang, O. Kwon, Y. K. Choi and J. S. Lee, *Int. J. Therm. Sci.*, 2012, **62**, 109–113.
33. L. Shi and A. Majumdar, *J. Heat Transfer*, 2002, **124**, 329–337.
34. O. Nakabeppu and T. Suzuki, *J. Therm. Anal. Calorim.*, 2002, **69**, 727–737.
35. K. Kim, J. Chung, J. Won, O. Kwon, J. S. Lee, S. H. Park and Y. K. Choi, *Appl. Phys. Lett.*, 2008, **93**, 203115.
36. J. Chung, K. Kim, G. Hwang, O. Kwon, S. Jung, J. Lee, J. W. Lee and G. T. Kim, *Rev. Sci. Instrum.*, 2010, **81**, 114901.
37. K. Kim, J. Chung, G. Hwang, O. Kwon and J. S. Lee, *Acs Nano*, 2011, **5**, 8700–8709.
38. K. Kim, W. Jeong, W. Lee and P. Reddy, *Acs Nano*, 2012, **6**, 4248–4257.
39. H. H. Roh, J. S. Lee, D. L. Kim, J. Park, K. Kim, O. Kwon, S. H. Park, Y. K. Choi and A. Majumdar, *J. Vac. Sci. Technol., B: Microelectron. Nanometer Struct.-Process., Meas., Phenom.*, 2006, **24**, 2398–2404.
40. H. H. Roh, J. S. Lee, D. L. Kim, J. Park, K. Kim, O. Kwon, S. H. Park, Y. K. Choi and A. Majumdar, *J. Vac. Sci. Technol., B: Microelectron. Nanometer Struct.-Process., Meas., Phenom.*, 2006, **24**, 2405–2411.
41. T. Leinhos, M. Stopka and E. Oesterschulze, *Appl. Phys. A: Mater. Sci. Process.*, 1998, **66**, S65–S69.
42. T. W. Tong, *Thermal Conductivity 22*, CRC Press, 1994.
43. A. Buzin, P. Kamasa, M. Pyda and B. Wunderlich, *Thermochim. Acta*, 2002, **381**, 9–18.
44. S. Lefèvre, J.-B. Saulnier, C. Fuentes and S. Volz, *Superlattices Microstruct.*, 2004, **35**, 283–288.
45. Y. Ezzahri, L. Patiño Lopez, O. Chapuis, S. Dilhaire, S. Grauby, W. Claeys and S. Volz, *Superlattices Microstruct.*, 2005, **38**, 69–75.
46. L. D. P. Lopez, S. Grauby, S. Dilhaire, M. Amine Salhi, W. Claeys, S. Lefèvre and S. Volz, *Microelectron. J.*, 2004, **35**, 797–803.
47. S. Gomès, P.-O. Chapuis, F. Nepveu, N. Trannoy, S. Volz, B. Charlot, G. Tessier, S. Dilhaire, B. Cretin and P. Vairac, EEE Trans. *Compon. Packag. Technol.*, 2007, **30**, 424–431.
48. J. Altet, S. Dilhaire, S. Volz, J.-M. Rampnoux, A. Rubio, S. Grauby, L. D. Patino Lopez, W. Claeys and J.-B. Saulnier, *Microelectron. J.*, 2002, **33**, 689–696.
49. J. Altet, W. Claeys, S. Dilhaire and A. Rubio, *Proc. IEEE*, 2006, **94**, 1519–1533.
50. H. Pollock and A. Hammiche, *J. Phys. D: Appl. Phys.*, 2001, **34**, R23.
51. D. I. Florescu, L. Mourokh, F. H. Pollak, D. C. Look, G. Cantwell and X. Li, *J. Appl. Phys.*, 2002, **91**, 890–892.
52. D. Florescu, V. Asnin, F. H. Pollak, A. Jones, J. Ramer, M. Schurman and I. Ferguson, *Appl. Phys. Lett.*, 2000, **77**, 1464–1466.
53. E. R. Meinders, *J. Mater. Res.*, 2001, **16**, 2530–2543.
54. S. Volz, X. Feng, C. Fuentes, P. Guérin and M. Jaouen, *Int. J. Thermophys.*, 2002, **23**, 1645–1657.

55. V. Gorbunov, N. Fuchigami, J. Hazel and V. Tsukruk, *Langmuir*, 1999, **15**, 8340–8343.
56. A. Hammiche, M. Reading, H. Pollock, M. Song and D. Hourston, *Rev. Sci. Instrum.*, 1996, **67**, 4268–4274.
57. S. Gomes, L. David, V. Lysenko, A. Descamps, T. Nychyporuk and M. Raynaud, *J. Phys. D: Appl. Phys.*, 2007, **40**, 6677.
58. Y. Zhang, E. E. Castillo, R. J. Mehta, G. Ramanath and T. Borca-Tasciuc, *Rev. Sci. Instrum.*, 2011, **82**, 024902.
59. S. Callard, G. Tallarida, A. Borghesi and L. Zanotti, *J. Non-Cryst. Solids*, 1999, **245**, 203–209.
60. J. Pelzl, M. Chirtoc and R. Meckenstock, *Int. J. Thermophys.*, 2013, **34**, 1353–1366.
61. G. Wielgoszewski, P. Sulecki, T. Gotszalk, P. Janus, D. Szmigiel, P. Grabiec and E. Zschech, *J. Vac. Sci. Technol., B: Microelectron. Nanometer Struct.–Process., Meas., Phenom.*, 2010, **28**, C6N7–C6N11.
62. G. Wielgoszewski, P. Sulecki, P. Janus, P. Grabiec, E. Zschech and T. Gotszalk, *Meas. Sci. Technol.*, 2011, **22**, 094023.
63. Y. Zhang, P. Dobson and J. Weaver, *J. Vac. Sci. Technol., B: Nanotechnol. Microelectron.: Mater., Process., Meas., Phenom.*, 2012, **30**, 010601.
64. P. Tovee, M. Pumarol, D. Zeze, K. Kjoller and O. Kolosov, *J. Appl. Phys.*, 2012, **112**, 114317.
65. E. Puyoo, S. Grauby, J.-M. Rampnoux, E. Rouvière and S. Dilhaire, *J. Appl. Phys.*, 2011, **109**, 024302.
66. E. Puyoo, S. Grauby, J.-M. Rampnoux, E. Rouvière and S. Dilhaire, *Rev. Sci. Instrum.*, 2010, **81**, 073701.
67. A. Hammiche, D. Price, E. Dupas, G. Mills, A. Kulik, M. Reading, J. Weaver and H. Pollock, *J. Microsc.*, 2000, **199**, 180–190.
68. J. Lee and Y. B. Gianchandani, *J. Microelectromech. Syst.*, 2005, **14**, 44–53.
69. H. H. WAH, Ph. D Thesis, University of Singapore, 2012.
70. G. Binnig, M. Despont, U. Drechsler, W. Haeberle, M. Lutwyche, P. Vettiger, H. Mamin, B. Chui and T. Kenny, *Appl. Phys. Lett.*, 1999, **74**, 1329–1331.
71. P. Vettiger, M. Despont, U. Drechsler, U. Durig, W. Haberle, M. I. Lutwyche, H. E. Rothuizen, R. Stutz, R. Widmer and G. K. Binnig, *IBM J. Res. Dev.*, 2000, **44**, 323–340.
72. W. P. King, B. Bhatia, J. R. Felts, H. J. Kim, B. Kwon, B. Lee, S. Somnath and M. Rosenberger, *Annu. Rev. Heat Transfer*, 2013, **16**.
73. L. Aigouy, L. Lalouat, M. Mortier, P. Löw and C. Bergaud, *Rev. Sci. Instrum.*, 2011, **82**, 036106.
74. J. Gimzewski, C. Gerber, E. Meyer and R. Schlittler, *Chem. Phys. Lett.*, 1994, **217**, 589–594.
75. M. E. McConney, D. D. Kulkarni, H. Jiang, T. J. Bunning and V. V. Tsukruk, *Nano Lett.*, 2012, **12**, 1218–1223.
76. A. Narayanaswamy, S. Shen and G. Chen, *hys. Rev. B: Condens. Matter Mater. Phys.*, 2008, **78**, 115303.

77. S. Shen, A. Narayanaswamy and G. Chen, *Nano Lett.*, 2009, **9**, 2909–2913.
78. E. Rousseau, A. Siria, G. Jourdan, S. Volz, F. Comin, J. Chevrier and J.-J. Greffet, *Nat. Photonics*, 2009, **3**, 514–517.
79. J. Varesi and A. Majumdar, *Appl. Phys. Lett.*, 1998, **72**, 37–39.
80. A. Majumdar and J. Varesi, *J. Heat Transfer*, 1998, **120**, 297–305.
81. M. Igeta, K. Banerjee, G. Wu, C. Hu and A. Majumdar, *IEEE Electron Device Lett.*,, 2000, **21**, 224–226.
82. X. Xie, K. L. Grosse, J. Song, C. Lu, S. Dunham, F. Du, A. E. Islam, Y. Li, Y. Zhang and E. Pop, *ACS Nano*, 2012, **6**, 10267–10275.
83. K. L. Grosse, M.-H. Bae, F. Lian, E. Pop and W. P. King, *Nat. Nanotechnol.*, 2011, **6**, 287–290.
84. K. Yamanaka and S. Nakano, *Appl. Phys. A: Mater. Sci. Process.*, 1998, **66**, S313–S317.
85. U. Rabe, J. Turner and W. Arnold, *Appl. Phys. A: Mater. Sci. Process.*, 1998, **66**, S277–S282.
86. A. Dazzi, R. Prazeres, F. Glotin and J. Ortega, *Opt. Lett.*, 2005, **30**, 2388–2390.
87. C. Prater, K. Kjoller, D. Cook, R. Shetty, G. Meyers, C. Reinhardt, J. Felts, W. King, K. Vodopyanov and A. Dazzi, *Microsc. Anal. UK*, 2010, 5.
88. D. G. Cahill, W. K. Ford, K. E. Goodson, G. D. Mahan, A. Majumdar, H. J. Maris, R. Merlin and S. R. Phillpot, *J. Appl. Phys.*, 2003, **93**, 793–818.
89. J. Christofferson, K. Maize, Y. Ezzahri, J. Shabani, X. Wang and A. Shakouri, *J. Microelectron. Electron. Packag.*, 2008, **130**, 041101.
90. Y. De Wilde, F. Formanek, R. Carminati, B. Gralak, P.-A. Lemoine, K. Joulain, J.-P. Mulet, Y. Chen and J.-J. Greffet, *Nature*, 2006, **444**, 740–743.
91. K. Joulain, R. Carminati, J.-P. Mulet and J.-J. Greffet, *Phys. Rev. B: Condens. Matter Mater. Phys.*, 2003, **68**, 245405.
92. P.-A. Lemoine, *Ph.D Thesis, ESPCI Paris*, 2008.
93. P. Dobson, G. Mills and J. Weaver, *Rev. Sci. Instrum.*, 2005, **76**, 054901.
94. L. Thiery, S. Toullier, D. Teyssieux and D. Briand, *J. Heat Transfer*, 2008, **130**, 091601.
95. E. Saïdi, N. Babinet, L. Lalouat, J. Lesueur, L. Aigouy, S. Volz, J. Labéguerie--Egéa and M. Mortier, *Small*, 2011, 7, 259–264.
96. F. Menges, H. Riel, A. Stemmer and B. Gotsmann, *Nano Lett.*, 2012, **12**, 596–601.
97. M. Yovanovich and E. Marotta, *Heat Transfer Handb.*, 2003, **1**, 261–394.
98. S. Lefevre, S. Volz, J.-B. Saulnier, C. Fuentes and N. Trannoy, *Rev. Sci. Instrum.*, 2003, **74**, 2418–2423.
99. L. David, S. Gomes and M. Raynaud, *J. Phys. D: Appl. Phys.*, 2007, **40**, 4337.
100. H. Fischer, *Thermochim. Acta*, 2005, **425**, 69–74.
101. S. Gomès, *Ph.D Thesis, University of Reims*, 1999.

102. S. Gomès, N. Trannoy and P. Grossel, *Meas. Sci. Technol.*, 1999, **10**, 805.
103. J. Juszczyk, M. Wojtol and J. Bodzenta, *Int. J. Thermophys.*, 2013, **34**, 620–628.
104. L. Thiery, E. Gavignet and B. Cretin, *Rev. Sci. Instrum.*, 2009, **80**, 034901.
105. T. Borca-Tasciuc, *Annu. Rev. Heat Transfer*, 2013, **16**.
106. B. A. Nelson and W. P. King, *Nanoscale and Microscale Thermophys. Eng.*, 2008, **12**, 98–115.
107. P. Grossel, O. Raphaël, F. Depasse, T. Duvaut and N. Trannoy, *Int. J. Therm. Sci.*, 2007, **46**, 980–988.
108. F. Depasse, P. Grossel and N. Trannoy, *Superlattices Microstruct.*, 2004, **35**, 269–282.
109. Z.-Y. Guo and Z.-X. Li, *Int. J. Heat Mass Transfer*, 2003, **46**, 149–159.
110. X. Hu, A. Jain and K. E. Goodson, ASME 2005 Summer Heat Transfer Conference collocated with the ASME 2005 Pacific Rim Technical Conference and Exhibition on Integration and Packaging of MEMS, NEMS, and Electronic Systems, 2005.
111. P.-O. Chapuis, E. Rousseau, A. Assy, S. Gomès, S. Lefèvre and S. Volz, *MRS Proc.*, 2013, **1543**, 159–164.
112. L. Lees and C.-Y. Liu, *Phys. Fluids*, 1962, **5**, 1137–1148.
113. M. Chirtoc and J. Henry, *Eur. Phys. J.:Spec. Top.*, 2008, **153**, 343–348.
114. S. Lefèvre, Ph.D Thesis, University of Poitiers, 2004.
115. S. Lefèvre and S. Volz, *Rev. Sci. Instrum.*, 2005, **76**, 033701–033701-033706.
116. A. Assy, S. Gomès, S. Lefèvre and P.-O. Chapuis, **1557**, mrss13-1557-y1505-1510, *MRS Online Proc. Libr.*, 2013.
117. E. Puyoo, Ph.D Thesis, University of Bordeaux 1, 2010.
118. R. Prasher, *Nano Lett.*, 2005, 5, 2155–2159.
119. F. Yang and C. Dames, *Phys. Rev. B: Condens. Matter Mater. Phys.*, 2013, **87**, 035437.
120. B. Gotsmann, M. A. Lantz, A. Knoll and U. Dürig, *Nanotechnology*, 2010.
121. P.-O. Chapuis, Ph.D Thesis, Ecole Centrale Paris, 2007.
122. A. Kittel, U. F. Wischnath, J. Welker, O. Huth, F. Rüting and S.-A. Biehs, *Appl. Phys. Lett.*, 2008, **93**, 193109.
123. N. D. Masters, W. Ye and W. P. King, *Phys. Fluids*, 2005, **17**, 100615.
124. P.-O. Chapuis, J.-J. Greffet, K. Joulain and S. Volz, *Nanotechnology*, 2006, **17**, 2978.
125. J. Lee, T. L. Wright, M. R. Abel, E. O. Sunden, A. Marchenkov, S. Graham and W. P. King, *J. Appl. Phys.*, 2007, **101**, 014906.
126. X. Liu, Y. Yang and J. Yang, *J. Appl. Phys.*, 2009, **105**, 013508.
127. S. Lefèvre, S. Volz and P.-O. Chapuis, *Int. J. Heat Mass Transfer*, 2006, **49**, 251–258.
128. H.-J. Butt, B. Cappella and M. Kappl, *Surf. Sci. Rep.*, 2005, **59**, 1–152.
129. K. Johnson, K. Kendall and A. Roberts, *Proc. R. Soc. London, Ser. A*, 1971, **324**, 301–313.
130. B. Derjaguin, V. Muller and Y. P. Toporov, *J. Colloid Interface Sci.*, 1975, **53**, 314–326.

131. V. Muller, V. Yushchenko and B. Derjaguin, *J. Colloid Interface Sci.*, 1980, **77**, 91–101.
132. V. Muller, B. Derjaguin and Y. P. Toporov, *Colloids Surf.*, 1983, **7**, 251–259.
133. M. Kelkar, P. E. Phelan and B. Gu, *Int. J. Heat Mass Transfer*, 1997, **40**, 2637–2645.
134. R. Stoner and H. Maris, *Phys. Rev. B: Condens. Matter Mater. Phys.*, 1993, **48**, 16373.
135. E. T. Swartz and R. O. Pohl, *Rev. Mod. Phys.*, 1989, **61**, 605.
136. B. Nelson and W. King, *Rev. Sci. Instrum.*, 2007, **78**, 023702.
137. T. Beechem, S. Graham, P. Hopkins and P. Norris, *Appl. Phys. Lett.*, 2007, **90**, 054104.
138. R. Prasher, *Appl. Phys. Lett.*, 2009, **94**, 041905–041905–041903.
139. B. Persson, A. Volokitin and H. Ueba, *J. Phys.: Condens. Matter*, 2011, **23**, 045009.
140. R. S. Prasher and P. E. Phelan, *J. Appl. Phys.*, 2006, **100**, 063538.
141. Y. Mo, K. T. Turner and I. Szlufarska, *Nature*, 2009, **457**, 1116–1119.
142. B. Gotsmann and M. Lantz, *Nat. Mater.*, 2013, **12**, 59–65.
143. A. Hammiche, D. Hourston, H. Pollock, M. Reading and M. Song, *J. Vac. Sci. Technol., B: Microelectron. Nanometer Struct.–Process., Meas., Phenom.*, 1996, **14**, 1486–1491.
144. G. Fiege, V. Feige, J. Phang, M. Maywald, S. Gorlich and L. Balk, *Microelectron. Reliab.*, 1998, **38**, 957–961.
145. G. Fiege, F.-J. Niedernostheide, H.-J. Schulze, R. Barthelmess and L. Balk, *Microelectron. Reliab.*, 1999, **39**, 1149–1152.
146. K. Luo, R. Herrick, A. Majumdar and P. Petroff, *Appl. Phys. Lett.*, 1997, **71**, 1604–1606.
147. L. Shi, S. Plyasunov, A. Bachtold, P. L. McEuen and A. Majumdar, *Appl. Phys. Lett.*, 2000, **77**, 4295–4297.
148. F. Boroumand, M. Voigt, D. Lidzey, A. Hammiche and G. Hill, *Appl. Phys. Lett.*, 2004, **84**, 4890–4892.
149. L. Shi, J. Zhou, P. Kim, A. Bachtold, A. Majumdar and P. L. McEuen, *J. Appl. Phys.*, 2009, **105**, 104306.
150. Y.-J. Yu, M. Y. Han, S. Berciaud, A. B. Georgescu, T. F. Heinz, L. E. Brus, K. S. Kim and P. Kim, *Appl. Phys. Lett.*, 2011, **99**, 183105.
151. S. H. Choi, T. I. Lee, H. K. Baik, H. H. Roh, O. Kwon and D. hak Suh, *Appl. Phys. Lett.*, 2008, **93**, 183301.
152. A. Soudi, R. D. Dawson and Y. Gu, *ACS Nano*, 2010, **5**, 255–262.
153. S. Kar-Narayan, S. Crossley, X. Moya, V. Kovacova, J. Abergel, A. Bontempi, N. Baier, E. Defay and N. Mathur, *Appl. Phys. Lett.*, 2013, **102**, 032903.
154. S. Pailhès, H. Euchner, V. Giordano, R. Debord, A. Assy, S. Gomès, A. Bosak, D. Machon, S. Paschen and M. de Boissieu, *Phys. Rev. Lett.*, 2014, **113**, 025506.
155. P. J. Newby, B. Canut, J.-M. Bluet, S. Gomès, M. Isaiev, R. Burbelo, K. Termentzidis, P. Chantrenne, L. G. Fréchette and V. Lysenko, *J. Appl. Phys.*, 2013, **114**, 014903.

156. S. Gomès, P. Newby, B. Canut, K. Termentzidis, O. Marty, L. Fréchette, P. Chantrenne, V. Aimez, J.-M. Bluet and V. Lysenko, *Microelectron. J.*, 2013, **44**, 1029–1034.

157. M. Sano, M. Yudasaka, R. Kikuchi and S. Yoshimura, *Langmuir*, 1997, **13**, 4493–4497.

158. A. Altes, Ph.D Thesis, University of Wuppertal, 2004.

159. E. Brown, L. Hao, D. Cox and J. Gallop, J. Phys.: Conf. Ser., 2008.

160. B. Cretin, S. Gomes, N. Trannoy and P. Vairac, in *Microscale and Nanoscale Heat Transfer*, Springer, 2007, pp. 181-238.

161. S. Grauby, E. Puyoo, J.-M. Rampnoux, E. Rouvière and S. Dilhaire, *J. Phys. Chem. C*, 2013, **117**, 9025–9034.

162. A. Saci, J.-L. Battaglia, A. Kusiak, R. Fallica and M. Longo, *Appl. Phys. Lett.*, 2014, **104**, 263103.

163. J. Juszczyk, M. Krzywiecki, R. Kruszka and J. Bodzenta, *Ultramicroscopy*, 2013, **135**, 95–98.

164. W. Haeberle, M. Pantea and J. Hoerber, *Ultramicroscopy*, 2006, **106**, 678–686.

165. M. Hinz, O. Marti, B. Gotsmann, M. Lantz and U. Durig, *Appl. Phys. Lett.*, 2008, **92**, 043122–043122-043123.

166. D. G. Cahill, P. V. Braun, G. Chen, D. R. Clarke, S. Fan, K. E. Goodson, P. Keblinski, W. P. King, G. D. Mahan and A. Majumdar, *Appl. Phys. Rev.*, 2014, **1**, 011305.

Near-field Thermometry

KENNETH D. KIHM[*a] AND SEONGHWAN KIM[b]

[a] Mechanical, Aerospace, and Biomedical Engineering, The University of Tennessee, Knoxville TN 37996, USA; [b] Department of Mechanical and Manufacturing Engineering, University of Calgary, Calgary Alberta T2N 1N4, Canada
*Email: kkihm@utk.edu

10.1   Introduction

Various approaches to sub-micron scale thermometry have been attempted for over a decade as researchers have attempted to push the limits of spatial and temporal measurement resolution. Among them, scanning thermal microscopy (SThM) using a microfabricated thermocouple or platinum thermistor tip seems to be the best for achieving sub-micron spatial measurement resolution in the case of solid surfaces exposed in ambient air environments.[1–6] In the case of aqueous media, however, the SThM probe has not yet been reported to be successful. Recently, one research group demonstrated the potential of SThM as a thermometry tool in an aqueous medium, but was not able to show any real temperature measurement data, which is necessary to validate the probe's accuracy and resolution capability.[7] Another group tried to measure cellular thermal responses in a culture medium with a microthermocouple probe built on a glass micropipette, but no meaningful temperature signals were obtained.[8]

On the other hand, several optical techniques have been developed to measure sub-micron scale temperature distribution in an aqueous medium. The ratiometric laser-induced fluorescence (LIF) technique using fluorescent dye molecules achieved microscale spatial measurement resolution, but the

RSC Nanoscience & Nanotechnology No. 38
Thermometry at the Nanoscale: Techniques and Selected Applications
Edited by Luís Dias Carlos and Fernando Palacio

Published by the Royal Society of Chemistry, www.rsc.org

measurement uncertainties were excessive, particularly in the near-wall or near-meniscus region, because of the interfacial interference of the fluorescent emission light.[9] Similarly, the optical serial sectioning microscopy (OSSM) technique showed good potential as a sub-micron scale thermometry system for nanoparticle suspension fluids, where the Brownian motion of nanoparticles is related to the local temperature, but this technique also suffered from large measurement uncertainties in the near-field due to the near-wall-hindered and biased Brownian motion.[10] In addition to the excessive measurement uncertainties of these optical thermometry techniques, the presence of dye molecules or dye-coated nanoparticles may alter the thermal characteristics of an aqueous medium. To lessen the shortcomings that are associated with the foreign trace particles, a label-free, real-time, full-field surface plasmon resonance (SPR) reflectance sensing technique has been used to map sub-micron scale temperature distribution in the aqueous medium.[11] However, the detection range of SPR reflectance sensing is confined to the extremely small penetration depth of the SPR wave field, approximately on the order of 100 nm from the solid surface.

In this chapter, we describe nano-optical SPR imaging thermometry as a near-field temperature measurement technique in a liquid medium in Section 10.2, and present the nanomechanical thermometry concept as a potential tool for near-field thermometry in Section 10.3. This chapter is completed with pertinent conclusions and perspectives in Section 10.4.

10.2 Nano-optical Surface Plasmon Resonance Imaging Thermometry

SPR imaging thermometry is based on the well-known Kretschmann's analysis,[12] which describes the dependence of resonance of surface plasmon polaritons upon the near-field refractive index (RI) or temperature. SPR thermometry serves as a potential tool for full-field and real-time temperature mapping for thermally transient liquid media. Section 10.2.1 presents the basic principles of SPR and SPR thermometry, Section 10.2.2 describes the implementation of the SPR thermometry system, Section 10.2.3 discusses measurement uncertainties associated with the temperature dependence of RI values of the system components (the prism and the gold layers), and Section 10.2.4 presents an example application for transient temperature field developments in the near-wall region.

10.2.1 Working Principles of SPR Imaging Thermometry

At an incident angle larger than the critical angle [Figure 10.1(a)], the light incident onto the glass prism is completely reflected at the interface, *i.e.*, total internal reflection (TIR), and no significant amount of light is refracted into the bulk phase of the external contacting medium. This is true, at least from the macroscopic point of view. From the microscopic point of view,

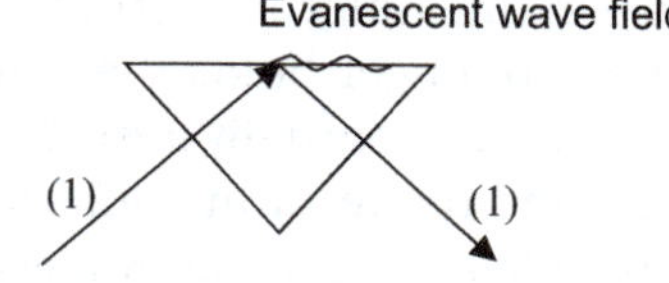

(a) Total Internal Reflection

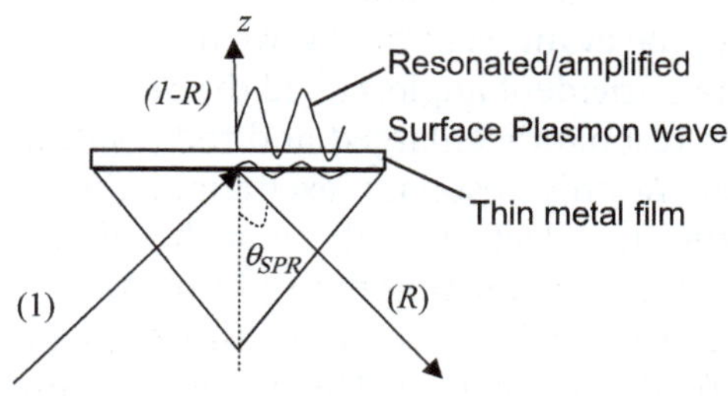

(b) Surface Plasmon Resonance

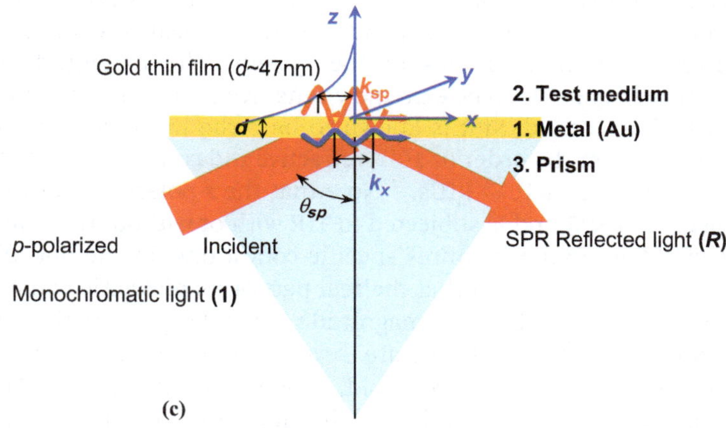

(c)

Figure 10.1   Schematic principles of TIR and SPR: (a) Total internal reflection (TIR); (b) SPR reflection; and (c) Kretschmann configuration of a three-layered SPR system consisting of a metallic gold film (1), the test fluid medium (2), and the glass prism (3). θ_{sp} is the SPR angle, d_2 is the thickness of the gold layer, k_x denotes the wave propagation number of the evanescent wave, and k_{sp} denotes the wave propagation number of the SPR wave.
Reproduced with permission from ref. 29. © Springer 2006.

however, the incident light penetrates the interface into the external medium and propagates parallel to the surface in the plane of incidence, creating a standing electromagnetic wave field while the reflectance remains unchanged from the incidence. This standing wave field is called an *evanescent wave field*, and decays substantially within an order of 100 nm from the interface.[13]

When a thin metal film intervenes between the internal (prism) and external media [Figure 10.1(b)], the reflectance no longer holds the conservation of intensity. The created evanescent wave field is partially absorbed by free electrons contained in the thin metal film, and the resulting reflectance is frustrated or reduced from that of TIR. In other words, the evanescent wave is partially attenuated to trigger coherent fluctuations of free electrons, and this coherent energy conversion of the photons into free electrons is called the *surface plasmon polariton* (SPP) phenomenon.[14]

When the *p*-polarized wave vector of the evanescent field wave matches the surface plasmon wave vector at a certain incident angle, called the *SPR angle*, the free electron absorption becomes resonantly enlarged and the resulting reflection intensity approaches zero. At this resonant excitation, the resulting reradiative emissions from the free electrons will be at the highest level at the upper surface of the metal film contacting the fluid medium.[15] The intensity of this reradiative SPP wave field is about 10 times greater than the ordinary evanescent wave field generated by TIR under the same illumination source strength.[16–19]

The resulting *p*-polarized SPR reflectance R is always less than one because of the absorbance, which depends on the optical, the material, and the contact medium properties of the system. The magnitude of the SPR reflectance is known to be extremely sensitive to the RI variation of the near-field region. Indeed, SPR is known to have the most sensitive RI-detection capabilities, on the order of 10^{-5} refractive index units (RIUs)[20,21] and more recently as fine as 10^{-8} RIUs.[22] Note that the *s*-polarized wave vector is not subjected to SPR and is subjected to TIR with or without the thin metal film.

The SPP excitation requires specific conditions for the optical properties of the thin metal film in that the real part of its dielectric constant must be negative, and its absolute magnitude must be greater than that of the imaginary part.[16,23,24] There are several noble metals available for SPR applications, including silver, gold, copper, and aluminium. Among them, gold is preferred because of its stability and superior performance in various environmental conditions.[25,26]

The penetration depth of the SPR wave field into the contact medium is defined as the depth at which the electric field of surface plasmon wave falls to $1/e$, which is equivalent to the reciprocal of the absolute value of the wave vector amplitude in the z-direction:[17,27]

$$z_{\text{penetration}} = \frac{\lambda}{2\pi} \text{Re} \left\{ \sqrt{\frac{\varepsilon'_{\text{met}} + \varepsilon_{\text{med}} + i\varepsilon''_{\text{met}}}{\varepsilon^2_{\text{med}}}} \right\} \quad (10.1)$$

where λ is the incident light wavelength, ε denotes the dielectric constant, and the sub-scripts met and med represent the metal layer and the external medium, respectively. For example, when $\lambda = 632.8\,\text{nm}$ is incident onto a gold metal film ($\varepsilon = -13.2 + 1.25i$), the SPR penetration depth is 192 nm for a water medium. Therefore, the near-field characterization herein refers to a thin layer within an order of a few hundred nanometres measured from the metal–medium interface.

The SPR reflectance R for the three-layer configuration [Figure 10.1(c)] of a metal film (1), dielectric medium (2), and prism substrate (3), is given by:[18]

$$R = R(n_1, n_2, n_3, d, \omega, \theta) \tag{10.2}$$

where n_i ($i = 1, 2, 3$) represents the RIs for the three media, d is the metal layer thickness, ω is the incident photon frequency, and θ is the incident ray angle. For p-polarized incident light, the reflectance R can be calculated by applying the Fresnel theory to the above-mentioned three-layer configuration. The resulting Fresnel equations are given as:

$$R = \left| r_p \right|^2 \tag{10.2a}$$

$$r_p = \frac{[r_{31} + r_{12}\exp(2ik_{z1}d)]}{[1 + r_{31}r_{12}\exp(2ik_{z1}d)]} \tag{10.2b}$$

$$r_{31} = \frac{Z_3 - Z_1}{Z_3 + Z_1}, r_{12} = \frac{Z_1 - Z_2}{Z_1 + Z_2} \tag{10.2c}$$

$$Z_i = \frac{\varepsilon_i}{k_{zi}}, \quad k_{zi} = \left[\varepsilon_i(\omega/c)^2 - k_{x3}^2\right]^{0.5}, \quad i = 1, 2, 3 \tag{10.2d}$$

$$k_{x3} = \varepsilon_3^{1/2} \cdot \frac{\omega}{c}\sin\theta \tag{10.2e}$$

$$n_i = \sqrt{\varepsilon_i} \tag{10.2f}$$

where r_p is the reflection coefficient[†] for the p-polarized incident light, ω is the angular frequency ($2\pi\nu$), c is the speed of light in a vacuum, k is the wave propagation number ($2\pi/\lambda$), and ε is the dielectric constant.

With given metal layer thickness (d), the incident wave angle (θ), and wavelength (λ), the reflectance R of eqn (10.2) varies exclusively with the values of the three RIs, n_1, n_2 and n_3. Furthermore, if the RI values of the experimental system are known, *i.e.*, the RI of the thin metal layer n_1 and the RI of the prism n_3 are specified, then the resulting R depends solely on the contacting medium RI n_2. Therefore, the measured R determines n_2, and the corresponding temperature of the near-field medium can be determined from the temperature dependence of $n_2 = n_2(T)$, which is available for most media from a reference such as the *CRC Handbook*.[28]

[†]The reflection coefficient is defined as the magnitude ratio of the reflection electric field to the incident electric field.

In summary, SPR thermometry is first aligned for the SPR resonance angle at a base temperature and this provides a dark image of the reflecting wave resulting from the resonant absorption of almost the entire incident wave. If the temperature field deviates from the base temperature, the resonance condition will be biased and the corresponding non-zero reflectance will appear upon the dark background. Analysis of this grey image field $R(n(T))$ will determine the near-field temperature distributions.

10.2.2 Experimental Setup for Near-field SPR Imaging Thermometry

While a variety of SPR detection systems are commercially available world-wide, virtually all of them are designed for very specific purposes, mostly for biomedical applications; none of them allows enough flexibility to be modified for the thermometry applications at hand. Therefore, a new SPR configuration is necessary for near-field fluidic characterization.

Figure 10.2(a) shows the optical arrangement of the first SPR thermometry system developed at the University of Tennessee.[29] This setup uses a white light source that is mono-chromatized by a narrow bandpass filter in order to minimize the diffraction interference noise. The randomly polarized white light from a 100 W tungsten–halogen lamp collimates onto the *p*-polarizer, the *E*-field of which is parallel to the incident plane. The rotating mirror adjusts the incident angle to be optimized for the SPR angle, and the incident ray illuminates the 47.5 nm thick gold film deposited on the 2.5 nm thick titanium adhesion layer that is coated on the 160 μm thick cover glass substrate. The cover glass RI matches the prism RI, and both are interfaced by a thin index-matching oil layer.

Figure 10.2(b) shows the more recently developed system at Seoul National University a part of the Korean Government sponsored World Class University (WCU) Program.[30] This system is more integrated in that an LED white light source replaces the white light of the original system, and this new system rotates the entire incident optical train to align the SPR angle, while the white light system uses a rotating mirror for beam steering. Although both systems work equally well, this new system requires fewer optical elements; thus, its SPR angle alignment is relatively easy and is more accurate than the original system. The bandpass filter narrows the LED light spectra to be centred at 625 nm, with a full width half maximum (FWHM) of 10 nm. A state-of-the-art Hamamatsu 14-bit cooled EM-CCD camera (Model C9100-02) uses a long focal distance lens of Mitutoyo M Plan APO 59 ($NA = 0.14, f = 40$ mm) to record SPR images at an extremely high spatial resolution.

10.2.3 Temperature Dependence of Refractive Index Values

The RI values n_1, n_2, and n_3 in eqn (10.2) vary with temperature. In most cases, the temperature dependence of n_2 is known for a specified near-field

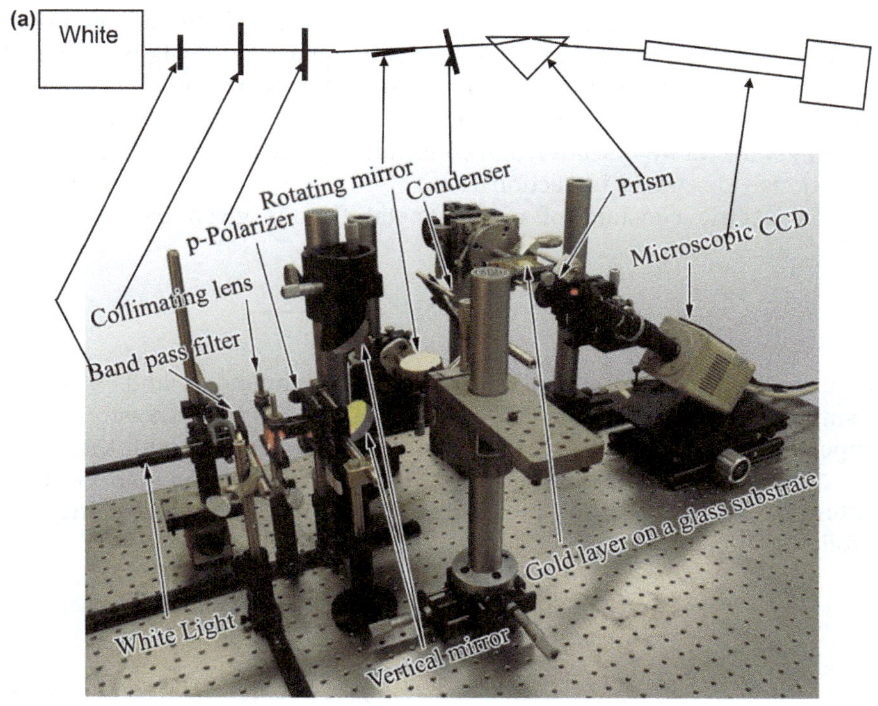

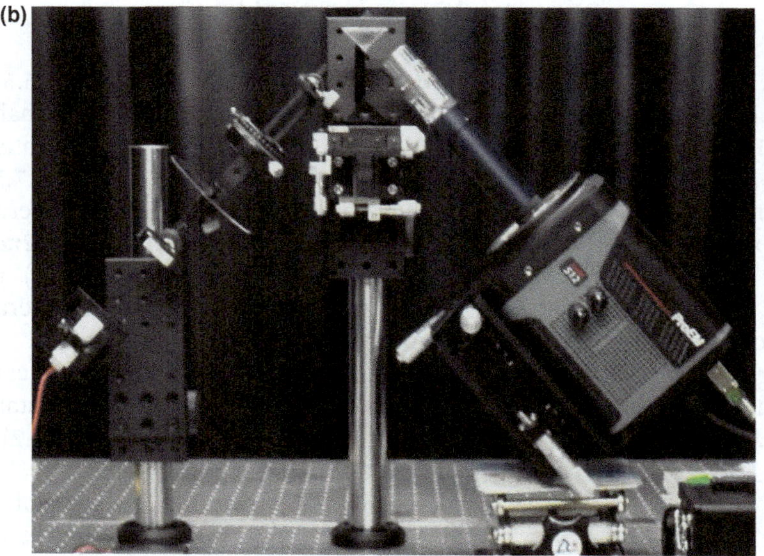

Figure 10.2 The layout of SPR imaging thermometry systems: (a) The first system, using a filtered white light source at 632.8 nm at the University of Tennessee.[29] (b) The improved and integrated system, using an LED light source at 625 nm at Seoul National University.
Reproduced with permission from ref. 30. © Springer 2011.

dielectric medium. The RI of the prism, n_3, is specified by the prism material, and its temperature dependence is generally negligibly small.[31] The temperature dependence of the thin metal (Au) film $n_1(T)$, on the other hand, is small but not completely negligible. More importantly, the measured data for $n_1(T)$ scatter, and consequently, the R–T correlation varies, as previously discussed in Section 10.2.1.

The dielectric constant of a thin metal film is given by the Drude model[32] as:

$$\varepsilon \equiv (n_R + in_I)^2 = 1 - \frac{\omega_p^2}{\omega(\omega + i\omega_c)} \tag{10.3}$$

where ω is the angular frequency of the incident light and ω_p and ω_c are the plasmon and collision frequencies, respectively. It has been shown that the temperature dependence of ω_p is negligibly small compared with that of ω_c (ref. 33). The collision frequency ω_c consists of the phonon–electron scattering frequency ω_{cp} and the electron–electron scattering frequency ω_{ce} as follows:[34]

$$\omega_c = \omega_{cp} + \omega_{ce}$$

$$\omega_{cp}(T) = \omega_0\left[\frac{2}{5} + 4\left(\frac{T}{T_D}\right)^5\right]\int_0^{T_D/T} \frac{z^4}{e^z - 1}\,dz \tag{10.4}$$

$$\omega_{ce}(T) = \frac{1}{6}\pi^4 \frac{\Gamma\Delta}{hE_F}\left[(k_BT)^2 + \left(\frac{h\omega}{4\pi^2}\right)^2\right]$$

where, the Debye temperature $T_D = 170$ K, the Fermi Energy $E_F = 5.53$ eV, the Boltzmann constant $k_B = 1.387\times10^{-23}$ J K^{-1}, the scattering probability $\Gamma = 0.55$, the fractional scattering $\Delta = 0.77$, and the Plank constant $h = 1.0546\times10^{-34}$ J s. Using $n = n_R + in_I = 0.1718 + i3.637$ for the 47.5 nm Au thin film at $\lambda = 632.8$ nm, the above equations are solved to determine that $\omega_p = 3.7826\omega$, and $\omega_c = 0.08802\omega$ where the specified $\omega = 2\pi c/\lambda = 2.9788\times10^{12}$ rad s^{-1}. Substituting these results into eqn (10.4) gives $\omega_0 = 1.171\times10^{14}$ rad s^{-1} and provides completion of the temperature dependence of the dielectric constant ε of the Au film.

Therefore, the resulting correlation of SPR reflectance R with T can be calculated by incorporating the T-dependence of the dielectric constant of Au as well as the dielectric constant of the test medium into eqn (10.2). For the case of a water medium to be tested using a BK7 prism ($n_3 = 1.515$) coated with a 47.5 nm Au layer, for example, the resulting R–T correlation is shown in Figure 10.3(a), with and without accounting for the temperature effect on the dielectric constant of gold. Note that the SPR resonance condition ($R = 0$) is set at 90 °C for the water temperature. In order to ensure measurement accuracy, inclusion of the T-effect on the dielectric constant of the metal layer is strongly recommended. Figure 10.3(b) shows the evolution of near-wall temperature profiles when a hot water droplet at 80 °C falls into

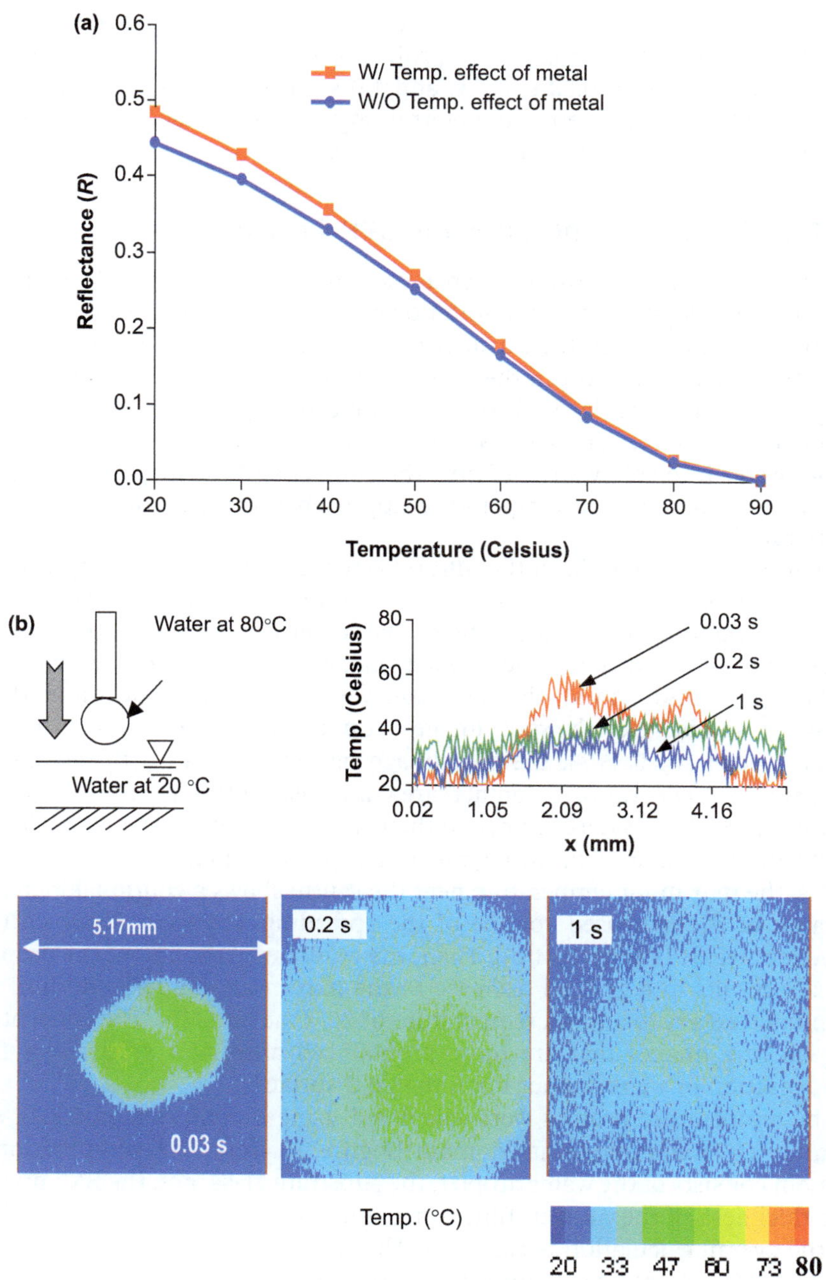

Figure 10.3 (a) Correlation of SPR reflectance intensity *R* with water temperature with and without accounting for the temperature dependence of the RI of the 47.5 nm thick gold film. (b) Full-field and real-time mapping of transient temperature fields when a hot water droplet (80 °C) falls on the cold water environment (20 °C) above the Au layer.
Reproduced with permission from ref. 11. © Optical Society of America 2007.

a shallow water pool at 20 °C and spreads in the near-field above the Au film.[11] The converted temperature profiles across the centreline cross-section are shown for $t = 0.03$ s, 0.2 s, and 1.0 s. This demonstrates the feasibility of SPR thermometry in determining time-dependent temperature profiles quantitatively as well as qualitatively.

10.2.4   Example Applications of SPR Thermometry

Figure 10.4 shows SPR thermometric measurements of near-wall temperature field development as a hot water drop of 2 mm diameter at 90 °C initial temperature spreads on the Au surface in the air environment. The gradual heat and energy transport in the air environment allows the contact surface shape to remain circular and to spread concentrically. In contrast, for the water environment as shown in Section 10.2.3, the more aggressive diffusion of hot water into cold water deforms the contact surface shape and spread, and the contact surface temperature approaches the environmental level more rapidly.

Figure 10.4(a) shows both the SPR reflectance images with the optimized SPR angle of 69.1°, which is set for the water medium at 90 °C and the corresponding temperature fields based on the above-mentioned R–T correlation. Prior to the liquid droplet making the initial contact with the surface ($t = 0$ and 0.14 s), the hot water vapour condenses on the surface showing as the greyish spot images. Upon the initial contact of the water droplet ($t = 0.15$ s), the spot image becomes darker with increasing temperature. The droplet contact area increases until $t = 1.55$ s as the spreading reaches its maximum, and thereafter, the droplet of water begins to shrink due to evaporation. During the initial spreading period from $t = 0.15$ to 0.45 s, the maximum temperature near the centre shows a continual increase because of the increased volume of the spreading hot water fed from the injection needle [Figure 10.4(b)]. Then, the evaporative cooling of the droplet surface results in a gradual decrease in the maximum temperature until its eventual recovery to a room temperature of 20 °C at $t = 142$ s. The concentric greyish ring images that are more pronounced after $t = 4$ s represent the condensation of water vapour upon the gold surface.

The temperature profiles were calculated using Fluent Version 6.2.16 assuming a 2D computational domain [Figure 10.4(c)]. The computational domain consists of the water droplet, the substrate glass, and the BK7 prism. The thickness of the Au thin film is negligibly small, and its thermal effect on the overall calculation is excluded. The total number of calculation cells is 12 615. The droplet is assumed to have an initial temperature of 90 °C and the contact surface is assumed to maintain a constant area with an initial temperature of 20 °C. For $t > 0$, the natural convection boundary condition with a typical heat convection coefficient $h = 10$ W m^2 K^{-1} is given at all surfaces interfacing the air.[35] Figure 10.4(d) shows fairly good agreement between the calculated profiles and the measured data using SPR thermometry at $t = 1.55$ and 22 s.

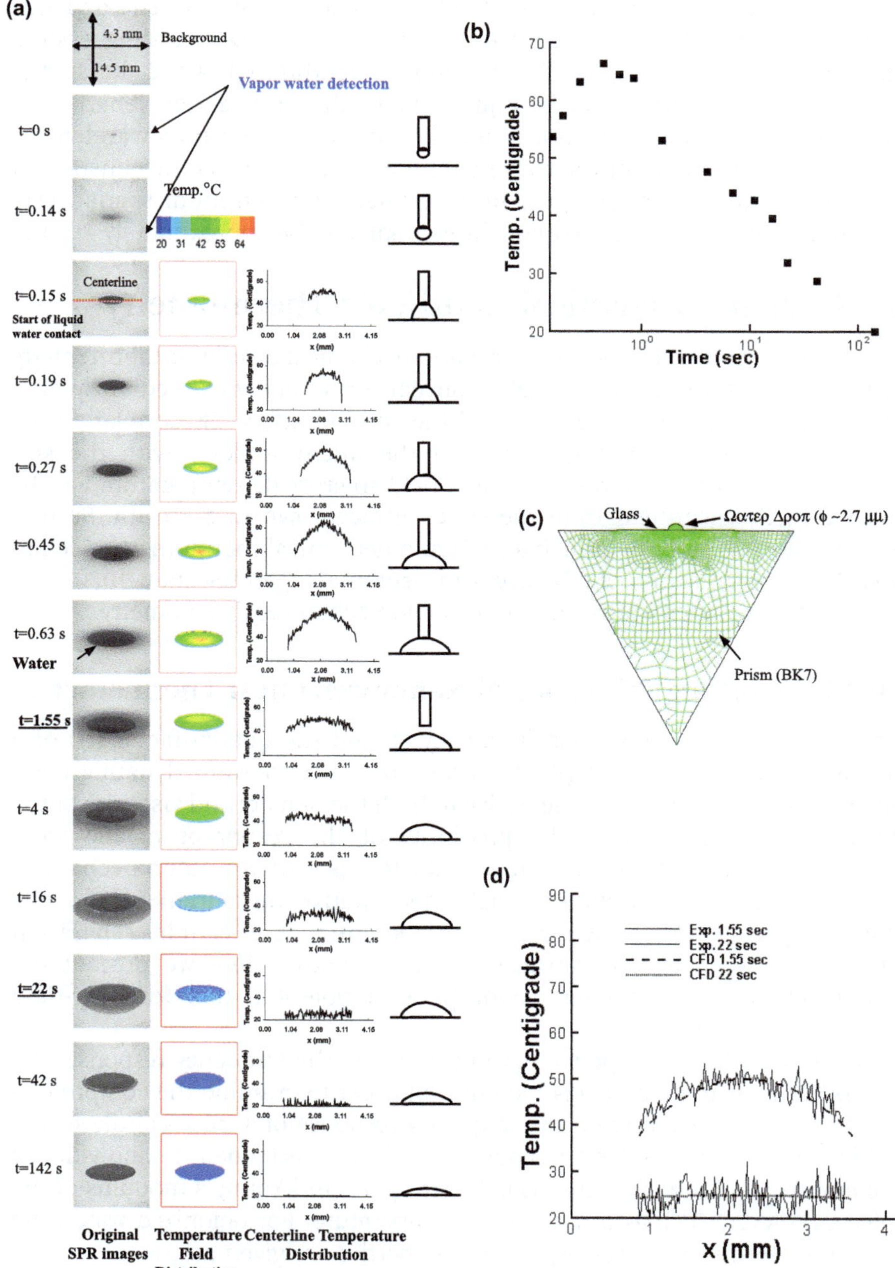

Figure 10.4 (a) Transient temperature change with time measured by SPR thermometry. (b) Temporal variation of the maximum temperature. (c) Grid generation for computational predictions. (d) Comparison between the experimental data and the CFD results for $t = 1.55$ s and 22 s.

The temporal resolution of near-field SPR imaging thermometry depends exclusively upon the data acquisition rate of the CCD recording system (6.4 ms or 156 frames per s for the present conditions), while the spatial resolution is determined to be equivalent to the propagation length of a surface plasmon wave, which is a function of the incident ray's wavelength, the dielectric constant of the thin metal film, and the dielectric constant of the test medium.[36] For the present experiment the minimum spatial resolution (in the xy plane) is theoretically estimated to be approximately 4.5 µm.

10.3   Nanomechanical Cantilever Thermometry

Employing a tipless microcantilever, nanomechanical cantilever thermometry is presented as a new thermometry concept for an aqueous medium within approximately one cantilever width of the solid interface. By correlating the thermal Brownian vibrating motion of the microcantilever with the surrounding liquid temperature, the near-field microscale temperature can be probed in the near-field from the solid surface down to $z = 5$ µm. Section 10.3.1 presents the basic principles of nanomechanical thermometry, Section 10.3.2 demonstrates proof-of-concept measurements, and Section 10.3.3 discusses the measurement resolution of nanomechanical thermometry.

10.3.1   Working Principles of Nanomechanical Thermometry

The peak resonance response frequency (f) and the quality factor (Q) of a nanomechanical resonator (Figure 10.5) are two important dynamic characteristics that are extremely sensitive to both the density and viscosity of the surrounding fluid and to the proximity of the resonator to the solid surface.[37–44] Thus, these dynamic characteristics of the nanomechanical resonator can be calibrated to yield quantitative measurements for the fluid properties which depend on temperature at a certain separation distance between a solid surface and a resonator. Here we present the nanomechanical thermometry concept as a potential tool for near-field nanothermometry.

For the proof of concept, we have investigated the frequency response of a microcantilever in an aqueous medium. In order to examine the comprehensive effect of temperature on the frequency response of a microcantilever immersed in a fluid, Sader's viscous model[44] was extended to individually account for the fluid viscosity, $\eta(T)$, density, $\rho(T)$, and Young's modulus of the microcantilever, $E(T)$, as a function of temperature. For deionized water, the functional forms of the corresponding properties are given as:

$$\eta(T) = \eta_0 e^{-1.704 - 5.306z + 7.003z^2},$$

$$z = \frac{273}{T} \quad \eta_0 = 1.788 \times 10^{-3} \, \mathrm{kg \, m^{-1} \, s^{-1}}$$

$$\rho(T) = 1000 - 0.0178 |T - 273|^{1.7} \, \mathrm{kg \, m^{-3}} \tag{10.6}$$

(10.5)

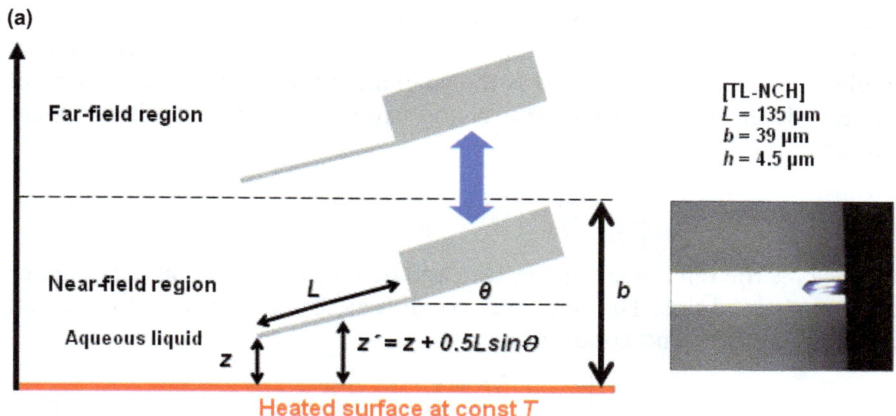

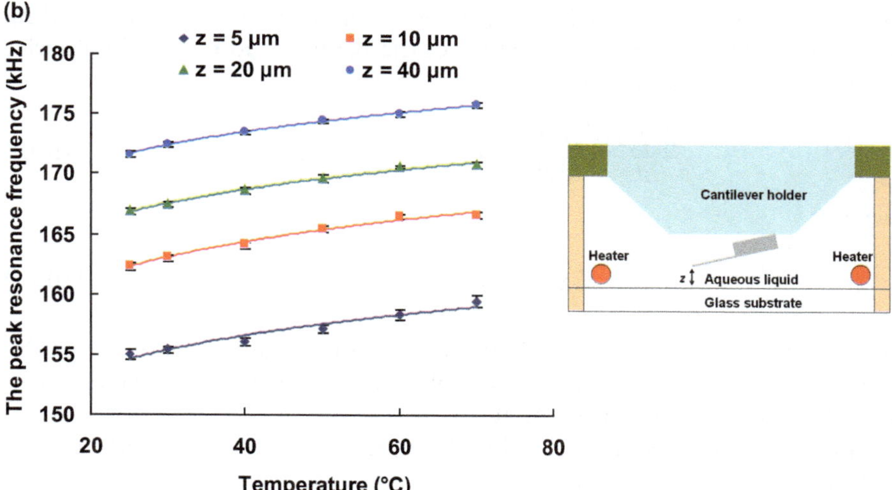

Figure 10.5 (a) A schematic illustration of near-field microscale thermometry using a tipless microcantilever for the case of a liquid near-field environment and a microphotograph of the microline heater and cantilever. (b) Experimental correlation of the microcantilever peak resonance frequency *f* with the medium temperature at four different separation distances ($z = 5$, 10, 20 and 40 μm) between the lower end of the cantilever and the glass substrate surface. Each symbol represents the averaged resonance frequencies from 10 measurements with the error bar of a 95% confidence interval. The corresponding power-fitting curves at each separation distance correlate fairly well with the measured data.
Reproduced with permission from ref. 51. © MDPI 2007.

The governing equation for the dynamic deflection function $w(x, t)$ of a microcantilever[45] is given by:

$$E(T)I\frac{\partial^4 w(x,t)}{\partial x^4} + \mu\frac{\partial^2 w(x,t)}{\partial t^2} = F(x,t) \tag{10.7}$$

where E is Young's modulus, T is temperature, I is the moment of inertia of the cantilever, μ is the mass per unit length of the cantilever, F is the external applied force per unit length, x is the spatial coordinate along the length of the cantilever, and t is time. The Young's modulus of the cantilever beam can be approximated[46] by:

$$E(T) = E(T_0) + \frac{\mathrm{d}E}{\mathrm{d}T}(T - T_0) \tag{10.8}$$

where T_0 is the reference temperature and $\mathrm{d}E/\mathrm{d}T$ is the thermal dependence of Young's modulus. The boundary conditions for eqn (10.7) are the usual clamped and free end conditions

$$\left[w(x,t) = \frac{\partial w(x,t)}{\partial x} \right]_{x=0} = \left[\frac{\partial^2 w(x,t)}{\partial x^2} = \frac{\partial^3 w(x,t)}{\partial x^3} \right]_{x=L} = 0 \tag{10.9}$$

where L is the length of the beam. Following Sader's previous work,[44] scaling the spatial variable x with the length of the beam L and taking the Fourier transform of eqn (10.7) obtains:

$$\frac{E(T)I}{L^4} \frac{\mathrm{d}^4 \hat{W}(x|\omega)}{\mathrm{d}x^4} - \mu\omega^2 \hat{W}(x|\omega) = \hat{F}(x|\omega), \tag{10.10}$$

where

$$\hat{X}(x|\omega) = \int_{-\infty}^{\infty} X(x|\omega) e^{i\omega t} \mathrm{d}t \tag{10.11}$$

for any function of time X, ω is the characteristic radial frequency of the vibration, or $2\pi f$, and the spatial variable x in eqn (10.10) refers to its scaled quantity for simplicity of notation.

For a cantilever beam moving in a fluid, the external applied load can be separated into two contributions:

$$\hat{F}(x|\omega) = \hat{F}_{\text{hydro}}(x|\omega) + \hat{F}_{\text{drive}}(x|\omega) \tag{10.12}$$

where the first component in eqn (10.12) is a hydrodynamic loading component due to the motion of the fluid around the beam, and the second term is a driving force that excites the beam. To proceed with the analysis, the general form of the first term is required. Therefore, the Fourier-transformed equations of motion for the fluid are examined.

$$\nabla \hat{u} = 0, \qquad -\nabla \hat{P} + \eta \nabla^2 \hat{u} = -i\rho\omega\hat{u}, \tag{10.13}$$

where $\hat{u}$ is the velocity field and $\hat{P}$ is the pressure. From eqn (10.13), the general form of the hydrodynamic loading component is given by

$$\hat{F}_{\text{hydro}}(x|\omega) = \frac{\pi}{4}\rho\omega^2 b^2 \Gamma(\omega)\hat{W}(x|\omega), \tag{10.14}$$

where b is the nominal width and $\Gamma(\omega)$ is the dimensionless hydrodynamic function, which is obtained from the solution of eqn (10.13) for a rigid beam.

For a beam that is circular in cross-section, the exact analytical result for $\Gamma_{\text{circular}}(\omega)$ is well known,[47] and is given by

$$\Gamma_{\text{circular}}(\omega) = 1 + \frac{4iK_1\left(-i\sqrt{i\text{Re}}\right)}{\sqrt{i\text{Re}}K_0\left(-i\sqrt{i\text{Re}}\right)}, \qquad (10.15)$$

where $\text{Re} = \rho(T)\omega b^2/(4\eta(T))$ and the functions K_0 and K_1 are modified Bessel functions of the third kind.[48] For a beam that is rectangular in cross-section, the hydrodynamic function is formulated with an approximate empirical correction function for eqn (10.15) as:

$$\Gamma_{\text{rec}}(\omega) = \Omega(\omega)\Gamma_{\text{circular}}(\omega), \qquad (10.16)$$

where $\Omega(\omega)$ is the correction function which can be found in the Sader's viscous model.[44] This approximate hydrodynamic function $\Gamma_{\text{rec}}(\omega)$ is accurate to within 0.1% over the entire range $\text{Re} \in [10^{-6}, 10^4]$.

For the case that the cantilever beam is excited by thermal Brownian motion of the molecules in the fluid, the expectation value of the potential energy for each mode of the cantilever must be identically equal to the thermal energy $\frac{1}{2}k_{\text{B}}T$, where k_{B} is Boltzmann's constant and T is absolute temperature. To evaluate the magnitudes of each of these driving forces, the modes of the damped cantilever beam must be decomposed into the modes of the undamped beam, since any damping will couple all modes. The undamped modes of a cantilever beam are given by:

$$\varphi_n(x) = (\cos C_n x - \cosh C_n x) + \frac{\cos C_n + \cosh C_n}{\sin C_n + \sinh C_n}(\sinh C_n x - \sin C_n x), \quad (10.17)$$

where C_n is a solution of $1 + \cos C_n \cosh C_n = 0$. The deflection function of the cantilever beam can be expressed as:

$$\hat{W}(x|\omega) = \sum_{n=1}^{\infty} \hat{F}_n(\omega)\alpha_n(\omega)\varphi_n(x), \qquad (10.18)$$

where

$$\alpha_n(\omega) = \frac{2\cos C_n \tan C_n}{C_n\left(C_n^4 - B^4(\omega)\right)(\sin C_n + \sinh C_n)}, \qquad (10.19)$$

$$\left|\hat{F}_n(\omega)\right|_s^2 = \frac{3\pi k_B T}{kC_n^4\int_0^\infty |\alpha_n(\omega')|^2 d\omega'}, \qquad (10.20)$$

where the subscript s refers to the spectral density. $B(\omega)$ is given as:

$$B(\omega) = C_1\sqrt{\frac{\omega}{\omega_{\text{vac},1}}}\left(1 + \frac{\pi\rho b^2}{4\mu}\Gamma(\omega)\right)^{1/4}, \qquad (10.21)$$

and the spring constant of the beam k is given by:

$$k(T) = \frac{3E(T)I}{L^3}. \qquad (10.22)$$

By integrating both sides of the thermal sensitivity equation for a single-layer cantilever,[49]

$$\frac{\mathrm{d}f}{f\mathrm{d}T} = \frac{\alpha}{2} + \frac{\mathrm{d}E}{2E(T)\mathrm{d}T}, \qquad (10.23)$$

the fundamental resonance frequency in vacuum $\omega_{\mathrm{vac},1}$ is formulated as:

$$\omega_{\mathrm{vac},1}(T) = \omega_{\mathrm{vac}}(T_0)\sqrt{\frac{E(T_0) + \dfrac{\mathrm{d}E}{\mathrm{d}T}(T - T_0)}{E(T_0)}}\, e^{\frac{\alpha}{2}(T - T_0)}, \qquad (10.24)$$

where T is the surrounding medium temperature in which the cantilever is submerged, T_0 is the reference temperature, and α is the thermal expansion coefficient of the cantilever. The fundamental resonant frequency at the reference temperature $\omega_{\mathrm{vac}}(T_0)$ is provided by the Simple Harmonic Oscillator (SHO) model using the readily measured peak resonance response frequency and the Q-factor in air.[37]

Taking all of these equations into consideration along with the original Sader's viscous model equations, the extended Sader's viscous model is obtained as:

$$\left|\hat{W}(x|\omega)\right|_s^2 = \frac{3\pi k_{\mathrm{B}}T}{k}\sum_{n=1}^{\infty}\frac{|\alpha_n(\omega)|^2}{C_n^4\int_0^{\infty}|\alpha_n(\omega')|^2 d\omega'}\varphi_n^2(x), \qquad (10.25a)$$

$$\left|\frac{\partial\hat{W}(x|\omega)}{\partial x}\right|_s^2 = \frac{3\pi k_{\mathrm{B}}T}{k}\sum_{n=1}^{\infty}\frac{|\alpha_n(\omega)|^2}{C_n^4\int_0^{\infty}|\alpha_n(\omega')|^2 d\omega'}\left(\frac{d\varphi_n(x)}{dx}\right)^2. \qquad (10.25b)$$

Equation (10.25a) gives the frequency response of the square of the magnitude of the displacement function at all positions along the beam, whereas eqn (10.25b) gives the corresponding result for the slope. It is noted that in the optical lever readout, the magnitude of the slope is typically measured [*i.e.*, the square root of eqn (10.25b)]. Eqn (10.25b) then gives the relationship between the measured resonance frequency of a micro-cantilever and the surrounding aqueous medium temperature.[50]

10.3.2 Proof-of-concept Measurements using Nanomechanical Thermometry

In order to prove the nanomechanical thermometry concept, well-controlled thermal experiments using a tipless microcantilever were performed.[51] A TL-NCH cantilever (length $L = 135$ μm, width $b = 39$ μm, and thickness

$h = 4.5$ µm; Nanosensors Inc.) was mounted on a solution-compatible cantilever holder attached to the head unit of an MFP-3D-BIO™ atomic force microscope (AFM) (Asylum Research Inc.) and the peak resonance frequencies of the TL-NCH immersed in deionized water were measured as a function of temperature at four specified separation distances $z = 5$, 10, 20 and 40 µm, which were measured between the lower end of the cantilever and the glass substrate surface. Figure 10.5(a) illustrates a schematic drawing of such a nanomechanical thermometry setup using a micro-cantilever working for both the near-field and far-field regions in the aqueous medium. Here, the near-field corresponds to the region that is one cantilever width from the solid interface, where the no-slip hydrodynamic boundary conditions effectively quench the thermal vibration of the microcantilever. The inset photograph shows the TL-NCH cantilever aligned parallel to the 100 µm wide gold line heater, which is shown as a thick horizontal white line. The separation distance z is controlled with 100 nm resolution by the piezoelectric actuator attached to the AFM head. The peak resonance response frequencies are measured, and the amount of data is reduced by the use of a thermal frequency fitting function, available from the MFP-3D-BIO AFM data acquisition software.

It is essential that a microcantilever of an AFM be aligned at a small angle of inclination to the horizon for efficient and effective surface imaging purposes. The present cantilever has an inclination angle θ of 11° and the probing site is widened to $L\sin\theta$. Assuming a linear temperature variation along the probing site, the modified separation distance $z' = z + 0.5L\sin\theta$ is taken to the midpoint of the cantilever. Note that the modified distance z' is used to calculate the temperature distribution while the nominal separation distance z is measured from the lower end of the cantilever.

The calibrations were carried out with the BioHeater™ system (Asylum Research Inc.), providing a uniform fluid temperature environment within ± 0.1 °C in a steady state as shown in Figure 10.5(b). The inset schematic illustrates the experimental setup using the BioHeater™ system. The peak resonance frequencies (f) quenched due to the nearby no-slip solid wall interference, should be taken into consideration in order to correlate the measured resonance frequency of the microcantilever with the near-wall aqueous medium temperature.[42,52]

To validate the accuracy of the nanomechanical thermometry, two carefully designed experiments were performed to measure either point-wise transient temperature or steady-state temperature fields in a controlled microscale thermal environment. Figure 10.6(a) shows the schematic of a microscale line heater also working as a temperature sensor based on the resistance–temperature relation of a thin gold film.[53] The microline heater consists of a vapour-deposited gold line which is 38 mm long, 100 µm wide, and 0.5 µm thick, and two 3 mm square and 0.5 µm thick soldering pads [Figure 10.6(a)]. A 500 nm thick gold layer is deposited with a 10 nm thick chromium adhesion layer by e-beam evaporation onto the Borofloat glass substrate. The microline heater patterns are deposited onto the glass

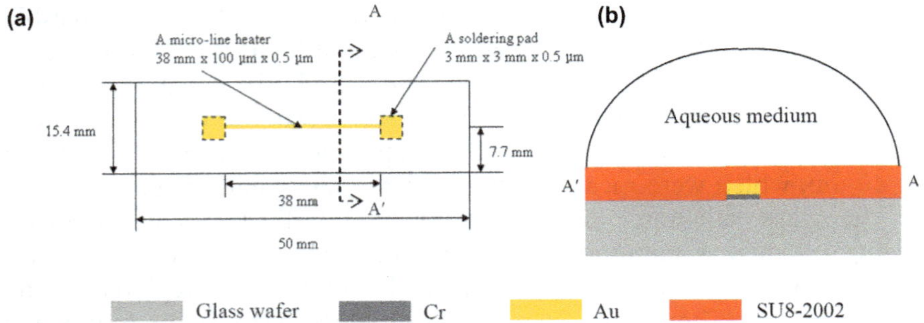

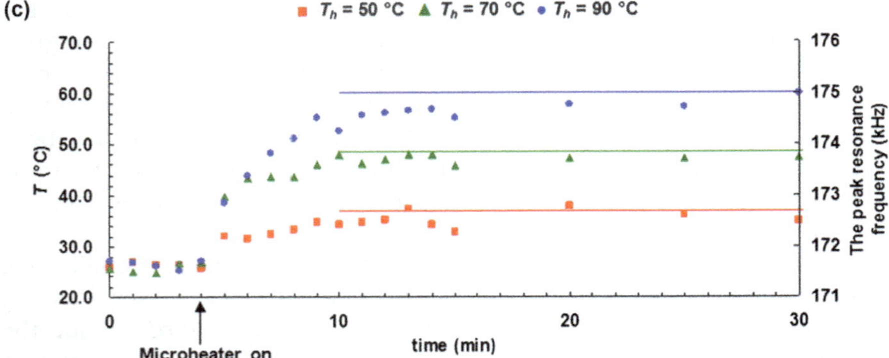

Figure 10.6 (a) A microline heater design and dimensions. (b) Schematic illustration of the microscale thermal environment with a microline heater. (c) Transient temperature measurements at $z = 40$ µm while the microline heater temperature T_h was maintained at 50, 70 or 90 °C. Reproduced with permission from ref. 51. © MDPI 2007.

substrate by a lift-off technique. To physically protect and electrically insulate the microline heater, the top layer is coated with SU8-2002 (MicroChem Inc.), except for the soldering pads. The coating thickness is uniform at 2 µm: a 1.5 µm coating on top of the 0.5 µm heater line [Figure 10.6(b)]. A feedback circuit based on the constant-temperature hot wire anemometer[54–56] is connected to provide a control for the microline heater power for a constant-temperature environment.

Figure 10.6(c) shows point-wise transient temperature measurements at $z = 40$ µm when the microline heater temperature T_h was maintained at 50, 70 or 90 °C, respectively. In the approximately 10 min after the heater was turned on, the test field temperature reached a steady state for each of the three cases and these agree well with the corresponding calculated steady-state temperature shown by the three horizontal lines. The measured steady-state temperature remains constant for a sufficiently long period of time under the feedback circuit control to ensure steady-state thermal conditions.

Using the f–T correlation functions at four specific separation distances, microscale steady-state temperature profiles were measured and compared

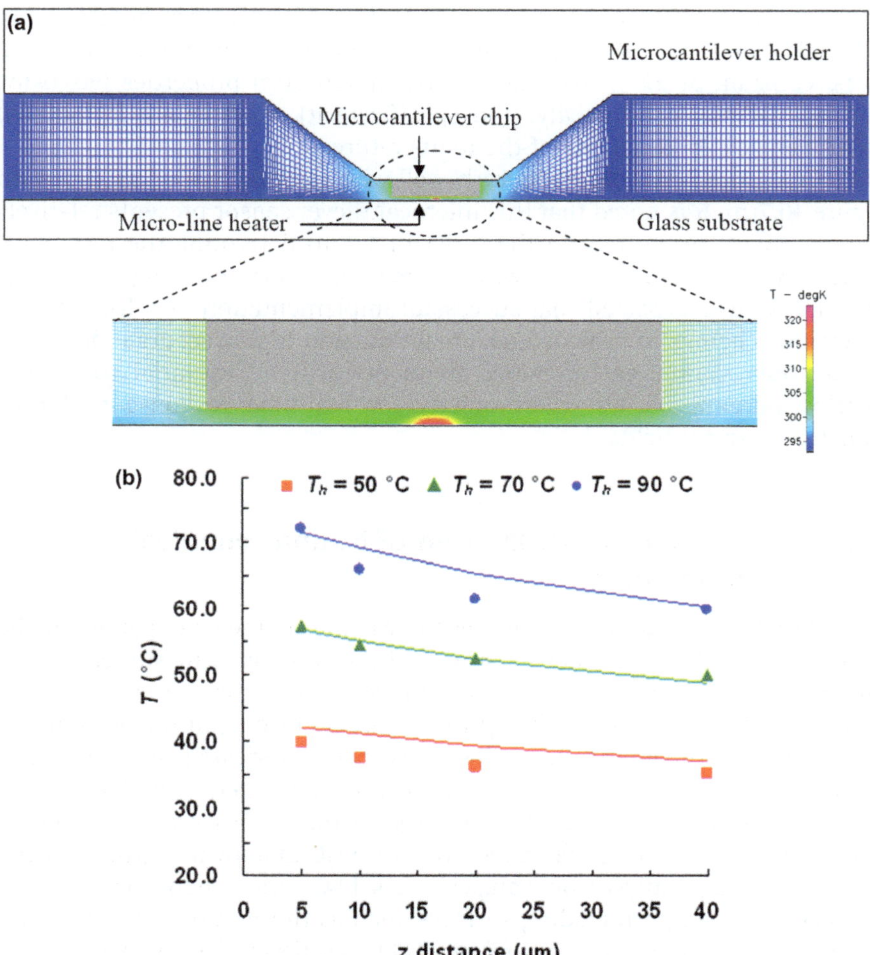

Figure 10.7 (a) Two-dimensional computational simulation domain of the microscale thermal environment and the detailed prediction of the temperature distribution. (b) Measured and calculated steady-state temperature profiles of an aqueous medium in the vicinity of the microline heater surface at four different z locations (5, 10, 20 and 40 µm) for each of three different heater temperatures (50, 70 or 90 °C).
Reproduced with permission from ref. 51. © MDPI 2007.

with the 2D numerical calculation results as shown in Figure 10.7. Figure 10.7(a) presents a calculated temperature field for the 2D computational domain that is consistent with the cross-section of the test medium. A room temperature (20 °C) is imposed for all outer boundary conditions, and a constant-temperature condition (50 °C) is specified for the gold heater surface. The computational domain to solve for the heat conduction equation consists of the microcantilever holder ($k = 1.38$ W m^{-1} K^{-1}), the silicon cantilever chip ($k = 124$ W m^{-1} K^{-1}), the aqueous medium

($k = 0.613$ W m^{-1} K^{-1}), SU8-2002 ($k = 0.17$ W m^{-1} K^{-1}), and the glass substrate ($k = 1.2$ W m^{-1} K^{-1}). All other thermophysical properties are assumed to be constant at 27 °C, whereas the thermophysical properties (viscosity, density, thermal conductivity, and specific heat) of water are formulated using polynomial functions of the temperature.

The measurements correlate fairly well with the predictions as shown in Figure 10.7(b). It is noted that this microcantilever sensor presents relatively slow temporal resolution on the order of 1 s to determine the resonance frequency of a microcantilever with thermal frequency scanning, averaging, and fitting. It is expected that successful implementation of self-excitation detection schemes to provide quasi-real-time and highly accurate detection of the microcantilever resonance frequency will be essential to further develop nanomechanical thermometry for real-time response to highly transient thermal fields.

10.3.3 Measurement Resolution of Nanomechanical Thermometry

The spatial resolution of a nanomechanical thermometer depends on the dimensions and the thermal sensitivity of the resonator. By sufficiently reducing the dimensions of a nanomechanical resonator, we expect to enhance the spatial resolution by up to two orders of magnitude. Recently, a variety of nanotubes and nanowires have been fabricated and characterized.[57–61] For example, Ag$_2$Ga nanowires can be fabricated with diameters of 25–500 nm and lengths of 1–110 µm.[58] Figure 10.8 shows scanning electron microscopy (SEM) images of state-of-the-art Ag$_2$Ga nanowires grown on the tip of an AFM cantilever by NaugaNeedles LLC.; these nanowires can be fabricated on sharp protruding surfaces such as those at the end of etched tungsten wires or on the tip of an AFM cantilever.[59] Considering that the critical distance between the probe and the surface, which induces the near-surface hydrodynamic interference effect, is the nominal width of a nanomechanical resonator,[42] near-field nanomechanical thermometry on a sub-micron scale seems to be within reach by correlating the peak resonance frequencies of nanowires to the local surrounding fluid temperature. In addition, our theoretical predictions based on the extended Sader's model,[44,50] suggest higher thermal sensitivity when the nanowire material has a positive thermal dependence of Young's modulus dE/dT and its dimensions are reduced.

Although successful measurements of microscale viscosity, density, and temperature of a simple fluid have been achieved by analysing the frequency response of a cantilever, the interpretation of the measured frequency response becomes more complicated when we are dealing with complex chemical and biological fluids. Both flexural rigidity changes due to the adsorbate and adsorption-induced surface stress and added mass due to the adsorbate and surface condition change should be additionally considered

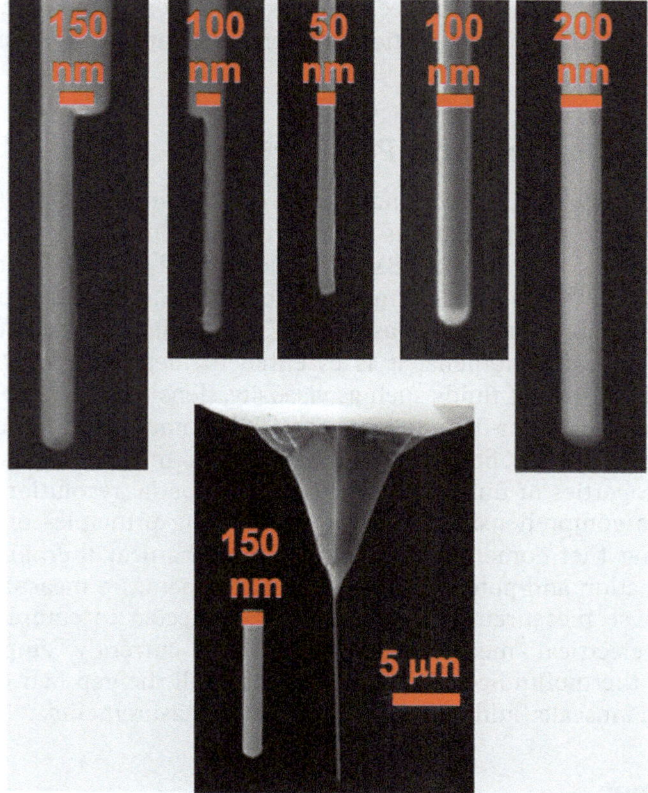

Figure 10.8 SEM images of typical Ag$_2$Ga nanowires grown on the tip of an AFM cantilever by NaugaNeedles LLC (Image courtesy of M. M. Yazdanpanah).

in order to correctly analyse the measured frequency response of a micro-cantilever in viscous fluids.[62–71]

Assuming evenly distributed adsorbate on the microcantilever surface, the governing Euler–Bernoulli beam equation for the dynamic deflection function $w(x,t)$ of the cantilever is modified to the following:[69]

$$(EI + \delta EI)\frac{\partial^4 w(x,t)}{\partial x^4} + (\mu + \delta\mu)\frac{\partial^4 w(x,t)}{\partial t^4} = F(x,t) \tag{10.26}$$

where δEI is the flexural rigidity change of the cantilever due to adsorbate and adsorption-induced surface stress and $\delta\mu$ is the adsorbate mass per unit length of the cantilever. However, the variation of the hydrodynamic loading due to the adsorption of biological molecules on the cantilever surface in a biological fluid needs to be properly modelled and combined in the externally applied force function $F(x, t)$ for accurate determination of adsorbed molecular mass[67,71,72] and correct interpretation of the frequency response of a nanomechanical resonator as a function of the surrounding fluid temperature. Anti-fouling surface coatings on nanomechanical resonators may alleviate surface contamination problems in near-field nanomechanical thermometry.

10.4 Conclusion and Perspectives

The marked progress in micro/nano-electromechanical systems (M/NEMS) technology and its applications to micro total analysis system (µTAS), lab-on-a-chip, and BioMEMS/NEMS has demanded the development of a fundamental understanding of micro/nanoscale fluidics and energy transport phenomena. To reveal the basic physics of micro/nanoscale fluidics and energy transport phenomena, it is essential to measure micro/nanoscale transport properties of fluids such as viscosity, density, and temperature. It is anticipated that near-field SPR imaging thermometry and nanomechanical thermometry will provide real-time, *in situ* measurements of these physical properties of fluids in sufficiently fine spatial resolution. We have presented a comprehensive review of the working principles of near-field SPR imaging thermometry as well as nanomechanical thermometry and their application and potential for near-field temperature measurements in liquids. These measurement techniques are expected to complement the advanced electrical measurement techniques currently employed for microscale thermofluidic sensors, as well as to fill the gap between micro-scale and nanoscale fluidic transport property measurements.

References

1. L. Shi and A. Majumdar, *Microscale Thermophys. Eng.*, 2001, **5**, 251–265.
2. D. G. Cahill, K. Goodson and A. Majumdar, *J. Heat Transfer*, 2002, **124**, 223–241.
3. L. Shi and A. Majumdar, *J. Heat Transfer*, 2002, **124**, 329–337.
4. H. H. Roh, J. S. Lee, D. L. Kim, J. Park, K. Kim, O. Kwon, S. H. Park, Y. K. Choi and A. Majumdar, *J. Vac. Sci. Technol. B.*, 2006, **24**, 2398–2404.
5. H. H. Roh, J. S. Lee, D. L. Kim, J. Park, K. Kim, O. Kwon, S. H. Park, Y. K. Choi and A. Majumdar, *J. Vac. Sci. Technol. B.*, 2006, **24**, 2405–2411.
6. S. Sadat, A. Tan, Y. J. Chua and P. Reddy, *Nano Lett.*, 2010, **10**, 2613–2617.
7. M. H. Li and Y. B. Gianchandani, *Sens. Actuators A-Phys*, 2003, **104**, 236–245.
8. M. S. Watanabe, N. Kakuta, K. Mabuchi and Y. Yamada, *Proc. 27th Ann. Conf. 2005 IEEE EMB.*, 2005, 4858–4861.
9. H. J. Kim, K. D. Kihm and J. Allen, *Int. J. Heat Mass Transfer*, 2003, **46**, 3967–3974.
10. J. S. Park, C. K. Choi and K. D. Kihm, *Meas. Sci. Technol.*, 2005, **16**, 1418–1429.
11. I. T. Kim and K. D. Kihm, *Opt. Lett.*, 2007, **32**, 3456–3458.
12. E. Z. Kretschmann, *Physik*, 1971, **241**, 313–324.
13. E. Hecht, *Optics*, Addison and Wesley, New York, 4th edn, 2002.

14. R. H. Richie, *Phys. Rev.*, 1957, **106**, 874–881.
15. T. L. Ferrell, T. A. Callcott and R. J. Warmack, *Am. Sci.*, 1985, **73**, 344–353.
16. J. R. Lakowicz, *Anal. Biochem.*, 2004, **324**, 153–169.
17. B. Rothenhausler and W. Knoll, *Nature*, 1988, **332**, 615–617.
18. T. Libermann and W. Knoll, *Colloids Surf.*, 2000, **171**, 115–130.
19. T. Neumann, M.-L. Johansson, D. Kambhampati and W. Knoll, *Adv. Funct. Mater.*, 2002, **12**, 575–586.
20. H. P. Ho and W. W. Lam, *Sens. Actuators, B*, 2003, **96**, 554–559.
21. Y. Xinglong, W. Dingxin, W. Xing, D. Xiang, L. Wei and Z. Xinsheng, *Sens. Actuators, B*, 2005, **108**, 765–771.
22. R. Slavik and J. Homola, *Sens. Actuators, B*, 2007, **123**, 10–12.
23. P. B. Johnson and R. W. Christy, *Phys. Rev. B*, 1972, **6**, 4370–4379.
24. E. Kryukov, Y.-K. Kim and J. B. Kettersonb, *J. Appl. Phys.*, 1997, **82**, 5411–5415.
25. D.-L. Hornauser, *Opt. Commun.*, 1976, **16**, 76–79.
26. T. Inagaki, K. Kagami and E. T. Arakawa, *Phys. Rev. B.*, 1981, **24**, 3644–3646.
27. K. A. Peterlinz and R. Georgiandis, *Langmuir*, 1996, **12**, 4731–4740.
28. D. R. Lide, *CRC Handbook of Chemistry and Physics*, CRC Press, 86th edn, 2005.
29. I. T. Kim and K. D. Kihm, *Exp. Fluids*, 2006, **41**, 905–916.
30. K. D. Kihm, *Near-Field Characterization of Micro/Nano-Scaled Fluid Flows*, Springer, 2011.
31. A. K. Sharma and B. D. Gupta, *Opt. Fiber Technol.*, 2006, **12**, 87–100.
32. H. P. Chiang, P. T. Leung and W. Tse, *J. Chem. Phys.*, 1998, **108**, 2659–2660.
33. H. P. Chiang, H. T. Yeh, C. M. Chen, J. C. Wu, S. Y. Su, R. Chang, Y. J. Wu, D. P. Tsai, S. U. Jen and P. T. Leung, *Opt. Commun.*, 2004, **241**, 409–418.
34. A. A. Kolomenskii, P. D. Gershon and H. A. Schuessler, *Appl. Opt.*, 1997, **36**, 6539–6547.
35. F. P. Incropera and D. P. DeWitt, *Fundamentals of Heat Transfer and Mass transfer*, John Willey & Sons, New York, 5th edn, 2002.
36. C. E. H. Berger, R. P. H. Kooyman and J. Greve, *Rev. Sci. Instrum.*, 1994, **65**, 2829–2835.
37. J. W. M. Chon, P. Mulvaney and J. E. Sader, *J. Appl. Phys.*, 2000, **87**, 3978–3988.
38. W. Y. Shih, X. Li, H. Gu, W.-H. Shih and I. A. Aksay, *J. Appl. Phys.*, 2001, **89**, 1497–1505.
39. M. A. Hopcroft, B. Kim, S. Chandorkar, R. Melamud, M. Agarwal, C. M. Jha, G. Bahl, J. Salvia, H. Mehta, H. K. Lee, R. N. Candler and T. W. Kenny, *Appl. Phys. Lett.*, 2007, **91**, 1–3.
40. A. Roters and D. Johannsmann, *J. Phys.: Condens. Matter.*, 1996, **8**, 7561–7577.
41. T. Naik, E. K. Longmire and S. C. Mantell, *Sens. Actuators, A*, 2003, **102**, 240–254.
42. C. P. Green and J. E. Sader, *J. Appl. Phys.*, 2005, **98**, 1–12.

43. R. J. Clarke, O. E. Jensen, J. Billingham, A. P. Pearson and P. M. Williams, *Phys. Rev. Lett.*, 2006, **96**, 1–4.
44. J. E. Sader, *J. Appl. Phys.*, 1998, **84**, 64–76.
45. L. D. Landau and E. M. Lifshitz, *Theory of Elasticity*, Pergamon, Oxford, 1970.
46. R. Sandberg, W. Svendsen, K. Mølhave and A. Boisen, *J. Micromech. Microeng.*, 2005, **15**, 1454–1458.
47. L. Rosenhead, *Laminar Boundary Layers*, Clarendon, Oxford, 1963.
48. M. Abramowitz and I. A. Stegun, *Handbook of Mathematical Functions*, Dover, New York, 1972.
49. J. Mertens, E. Finot, T. Thundat, A. Fabre, M.-H. Nadal, V. Eyraud and E. Bourillot, *Ultramicroscopy*, 2003, **97**, 119–126.
50. S. Kim and K. D. Kihm, *Appl. Phys. Lett.*, 2006, **89**, 1–3.
51. S. Kim, K. C. Kim and K. D. Kihm, *Sensors*, 2007, 7, 3156–3165.
52. S. Kim and K. D. Kihm, *Appl. Phys. Lett.*, 2007, **90**, 1–3.
53. S. W. Paik, K. D. Kihm, S. P. Lee and D. M. Pratt, *J. Heat Transfer*, 2007, **129**, 966–976.
54. H. Bruun, *Hot-wire Anemometry*, Oxford University Press, London, 1995.
55. R. J. Goldstein, *Fluid Mechanics Measurements*, Hemisphere Publishing Co., Washington D.C., 1983.
56. S. P. Lee and S. K. Kauh, *Exp. Fluids*, 1997, **22**, 212–219.
57. M. M. Yazdanpanah, S. A. Harfenist, A. Safir and R. W. Cohn, *J. Appl. Phys.*, 2005, **98**, 1–7.
58. L. B. Biedermann, R. C. Tung, A. Raman and R. G. Reifenberger, *Nanotechnology*, 2009, **20**, 035702.
59. L. B. Biedermann, R. C. Tung, A. Raman, R. G. Reifenberger, M. M. Yazdanpanah and R. W. Cohn, *Nanotechnology*, 2010, **21**, 305701.
60. E. Gil-Santos, D. Ramos, J. Martínez, M. Fernández-Regúlez, R. García, Á. S. Paulo, M. Calleja and J. Tamayo, *Nat. Nanotechol.*, 2010, 5, 641–645.
61. D. R. Kiracofe, M. M. Yazdanpanah and A. Raman, *Nanotechnology*, 2011, **22**, 295504.
62. G. Y. Chen, R. J. Warmack, T. Thundat, D. P. Allison and A. Huang, *Rev. Sci. Instrum.*, 1994, **65**, 2532–2537.
63. G. Y. Chen, T. Thundat, E. A. Wachter and R. J. Warmack, *J. Appl. Phys.*, 1995, 77, 3618–3622.
64. S. Cherian and T. Thundat, *Appl. Phys. Lett.*, 2002, **80**, 2219–2221.
65. P. Lu, H. P. Lee, C. Lu and S. J. O'Shea, *Phys. Rev. B*, 2005, 72, 1–5.
66. J. Tamayo, D. Ramos, J. Mertens and M. Calleja, *Appl. Phys. Lett.*, 2006, **89**, 1–3.
67. T. Y. Kwon, K. Eom, J. H. Park, D. S. Yoon, T. S. Kim and H. L. Lee, *Appl. Phys. Lett.*, 2007, **90**, 1–3.
68. M. J. Lachut and J. E. Sader, *Phys. Rev. Lett.*, 2007, **99**, 1–4.
69. S. Kim and K. D. Kihm, *Appl. Phys. Lett.*, 2008, **93**, 1–3.
70. S. Kim, K. D. Kihm and T. Thundat, *Exp. Fluids*, 2010, **48**, 721–736.
71. S. Kim, D. Yi, A. Passian and T. Thundat, *Appl. Phys. Lett.*, 2010, **96**, 1–3.
72. K. Rijal and R. Mutharasan, *Anal. Chem.*, 2007, 79, 7392–7400.

Nanotube Thermometry

KOJI TAKAHASHI[a,b]

[a] Department of Aeronautics and Astronautics, Kyushu University, Fukuoka 819-0395, Japan; [b] International Institute of Carbon-Neutral Energy Research (WPI-I2CNER), Kyushu University, Fukuoka 819-0395, Japan
Email: takahashi@aero.kyushu-u.ac.jp

11.1 Introduction

Carbon nanotubes (CNTs) are composed of seamless cylinders of honeycomb graphite lattice and are known to have unique properties.[1] Since they were first reported in 1991,[2] a wide variety of researchers have devoted themselves to understanding and applying CNTs. One of the most promising applications for CNTs is scanning probe microscopy (SPM) because their thin diameter and high aspect ratio are ideal for exploring steep features with high spatial resolution. In addition, their stiffness and flexibility due to their defect-free structure of carbon atoms linked by strong sp^2 bonding are advantageous for tip lifetime, by preventing tip wear and crashing in comparison with normal Si probes. Additionally, their well-characterized structure allows us to accurately reconstruct the surface morphology. Probe microscopy using a CNT as a tip was first reported in 1996:[3] a single CNT of diameter 5 nm extended from CNT bundles that had been attached to the Si pyramid tip of an AFM cantilever with acrylic adhesive. Stiffness in the axial direction and flexibility in the lateral were confirmed, and a trench of width 0.4 µm and depth 0.8 µm was successfully traced, which is a difficult target for conventional Si tips. Currently, CNT AFM probes are commercially available.

Following that, efforts were made to use a CNT as the tip for scanning thermal microscopy (SThM). A recent numerical study[4] claimed that CNTs

RSC Nanoscience & Nanotechnology No. 38
Thermometry at the Nanoscale: Techniques and Selected Applications
Edited by Luís Dias Carlos and Fernando Palacio
© The Royal Society of Chemistry 2016
Published by the Royal Society of Chemistry, www.rsc.org

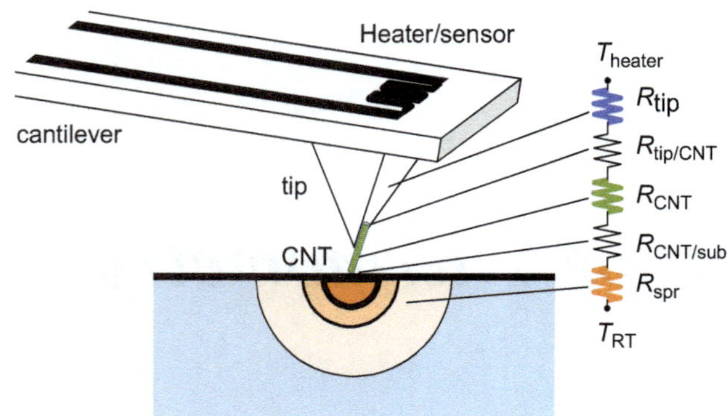

Figure 11.1   Schematic of CNT thermometry based on normal SThM. The heat goes through five thermal resistances including the thermal interface resistance and the thermal spreading resistance.

are beneficial not only for spatial resolution, but also for the thermal sensitivity of SThM required to obtain thermal conductivity mapping. A fundamental structure and heat conduction model for CNT thermal probes is shown in Figure 11.1, and is similar to normal SThM,[5] where a CNT is bonded to an SThM tip and heat flows from a heater on the cantilever to the sample substrate. To analyze the heat conduction, we have to consider the total thermal resistance from heater to sample, which consists of probe tip resistance, R_{tip}, interfacial resistance between the tip and CNT, $R_{tip/CNT}$, the CNT itself, R_{CNT}, the CNT–substrate interface, $R_{CNT/sub}$ and spreading resistance in the sample, R_{spr}. The low thermal resistance of CNTs is attractive when we increase the tip aspect ratio keeping the nanoscale spatial resolution.

11.2   Structure and Mechanical Properties of CNTs

CNTs are categorized into single-walled nanotubes (SWNTs) and multiwalled nanotubes (MWNTs), both discovered by Iijima *et al.* in 1993[6] and 1991,[2] respectively. The former tubes are single cylinders of graphene, whose properties are defined by a pair of integers (n, m) from the roll-up vector $C = na_1 + ma_2$, where a_1 and a_2 are the graphene lattice vectors. The diameter of a SWNT is around 1 nm and can be calculated precisely from the chiral index as $d = 3^{1/2}a(n^2 + nm + m^2)^{1/2}/\pi$, where a is the C–C bond length (0.144 nm). The lengths of SWNTs were initially on the order of micrometres but recently a new chemical vapour deposition (CVD) technique[7] called the *super growth method* has been invented to synthesize SWNT forests longer than 1 mm. MWNTs are made of co-axial graphene cylinders with 5–100 nm diameter and up to 20 μm length. CVD methods for MWNT synthesis have been developed for low-cost mass production. As well as other carbon materials, the high temperature of the laser ablation method or the carbon arc method yields a high-quality lattice of MWNTs. Raman spectroscopy can estimate the number of defects that induce large degradations in mechanical properties. Excellent mechanical properties of CNTs have been reported, although there is much variation in the data.[1] For example, Young's modulus in the axis direction is about 1.8 TPa, which is similar to or higher than diamond, but decreases as the diameter increases. The flexibility and fracture toughness are unique characteristics of CNTs due to their defect-free structure and definitely contribute to their long lifetimes as scanning probes.

11.3   Thermal Conductivity of CNTs

CNTs are expected to have thermal conductivities as high as diamond due to their strong sp^2 bonding and defect-free structures. This conjecture is based on the phonon gas model,[8] which predicts the lattice thermal conductivity of a bulk material as $k = CvL/3$, where C is the specific heat, v is the phonon velocity, and L is the phonon mean free path (MFP). The seamless and atomically perfect structure of CNTs can produce much longer L values than other bulk materials. For SWNTs, the first reported thermal conductivity[9] of 6600 $\mathrm{W\,m^{-1}\,K^{-1}}$ at room temperature might have been overestimated, but a recent study[10] also reported thermal conductivities higher than 3000 $\mathrm{W\,m^{-1}\,K^{-1}}$ for a long SWNT, which is much higher than crystalline silicon (140 $\mathrm{W\,m^{-1}\,K^{-1}}$) and even diamond (*ca.* 2000 $\mathrm{W\,m^{-1}\,K^{-1}}$). For heat conduction in CNTs, the contribution of the electrons is relatively low, and around 10% of the phonon thermal conductivity even for metallic SWNTs at room temperature. So far, several experiments[11–19] have tried to confirm the high thermal conductivity of CNTs by treating individual samples. Some of them obtained much lower data, probably due to the effects of contact resistance and lattice quality. By using careful methods, however, the reported data for SWNTs is around 3000 $\mathrm{W\,m^{-1}\,K^{-1}}$ at room temperature[11–12] and that of MWNTs ranges from over 2000 to under 100 $\mathrm{W\,m^{-1}\,K^{-1}}$ depending on their diameter[13,19] as shown in Figure 11.2(a).

One of the major reasons for the wide range of measured thermal conductivities for MWNTs is their anisotropic heat conduction. As described above, MWNTs consist of multiple cylinders or shells, and the intershell heat conduction *via* van der Waals interactions is much weaker than the intrashell heat conduction *via* the stiff sp^2 bonds as well as the bulk graphite.[20] Consequently, the heat flow in the inner shells is restrained. Due to this low contribution of the inner shells, MWNTs with large diameters show low thermal conductivities, by the way they are measured using a sensor/heater on the outermost shell, as shown in Figure 11.2(b). The details of this anisotropic heat conduction are still unknown but a difference of more than four orders of magnitude was reported when large gaps exist between the shells.[21]

When estimating the thermal conductivity of CNTs, ballistic heat conduction is another inevitable issue. Several theoretical studies[22–24] unveiled the length dependence of the thermal conductivity of SWNTs. If the length of a SWNT is shorter than, or of the same order as the phonon MFP, the

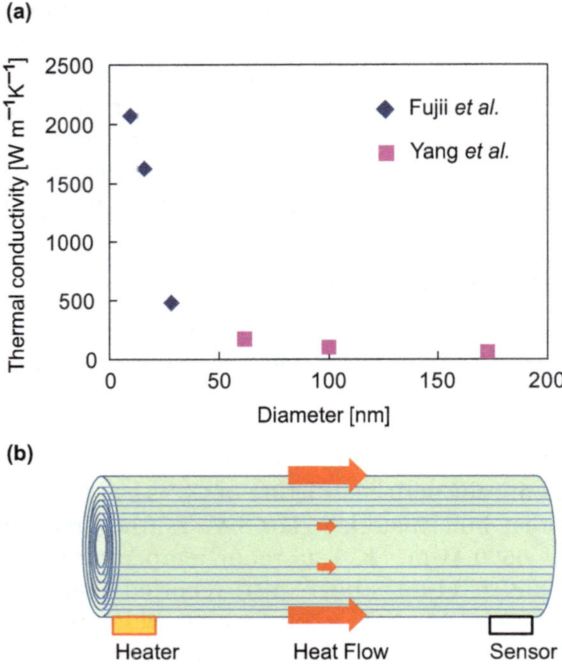

Figure 11.2   (a) Experimentally obtained thermal conductivity as a function of MWNT diameter at room temperature.[14,19] (b) Schema of heat conduction in a MWNT, where heat flow in the inner shells is restrained by the weak intershell heat conduction.

thermal conductivity decreases as the length decreases because the phonons are scattered at both ends of the sample. For example, an analysis[10] suggests that a SWNT of (10,0) chirality of length longer than 100 μm has a thermal conductivity of *ca.* 5000 $W\,m^{-1}\,K^{-1}$ but, it is 1000 $W\,m^{-1}\,K^{-1}$ for a length of 1 μm, and 300 $W\,m^{-1}\,K^{-1}$ for a length of 0.1 μm. On the other hand, it is difficult to analyse the heat conduction in MWNTs. To understand the phonon transport in detail, we are required to obtain the precise dispersion relationship with the density of states of the phonon, but the number of atoms in a MWNT is too large to treat numerically. However, well-prepared experimental studies[25–26] have reported that the thermal conductivity increases with length for MWNTs longer than 1 μm. The phonon MFP in MWNTs is estimated to be much shorter than 1 μm, thus these experiments suggest that the phonons of free path longer than the MFP carry enough thermal energy. Recently, this trend has been confirmed for other materials, for example, graphene and Si thin films.[27–28]

11.4   Temperature Measurement using CNT Probes

Due to the unknown thermal resistances of the probe itself and at the interfaces, quantitative thermal characterization of sample surfaces on the nanoscale is a difficult task. Even though the thermal resistance of a probe can be determined, the interfacial resistance should change due to the nanoscale topography of the sample surface and the damage of the scanning tip. To overcome this problem, a couple of methods have been proposed and demonstrated. Nakabeppu and Suzuki[29] developed an SThM cantilever in which thermocouples monitor the temperature and the amount of heat flow, and an electric heater ensures that the cantilever and the sample surface are at the same temperature. In contrast, Chung *et al.*[30] built another SThM cantilever with a thermocouple at the very end of the probe tip with a heating system to adjust the probe temperature to be the same as the sample surface. Their method is based on the null-point method, which compares the contact-mode temperature to the non-contact one and gives the sample temperature, removing the effect of air. In both methods, because no heat flows between the sample and cantilever, we do not have to consider the thermal contact resistance, which is the biggest issue causing uncertainty in the contact-type thermal measurements. The thermocouple built on the end of the probe tip was developed by Shi *et al.*[31] using Pt and Cr thin films and is useful for measuring the local surface temperature. However, they also found that wear of the probe tip occurs at contact forces over 30 nN.[32]

In this chapter, nanoscale thermometry using a CNT probe based on the T-type method proposed by Fujii *et al.*[14] is explained, which uses a single MWNT as a probe to enhance the spatial resolution and lifetime, as shown in Figure 11.3 together with a schema of its structure. The thermal model of this CNT thermometry is described in Figure 11.4. A CNT is fixed at one end to the Pt hot film sensor, and the other end can be in contact with the sample surface. The hot film controls the amount of Joule heating and

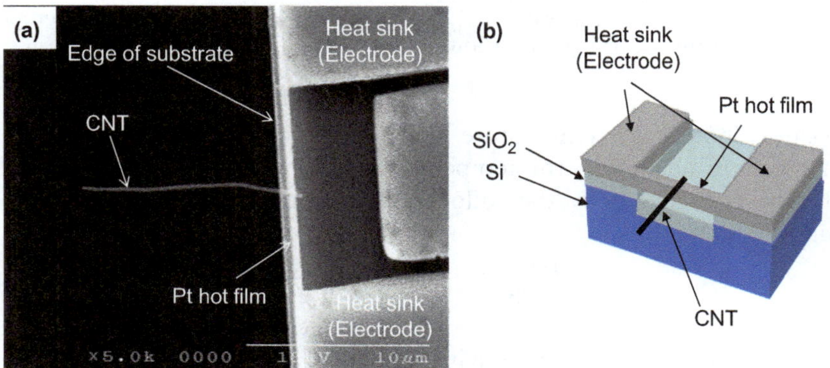

Figure 11.3 (a) SEM image of a CNT thermal probe where a suspended Pt hot film has been deposited within a few micrometres of the substrate edge to make the CNT probe protrude from the substrate. (b) Schematic of a MEMS-based structure of a suspended Pt film with a bonded CNT.

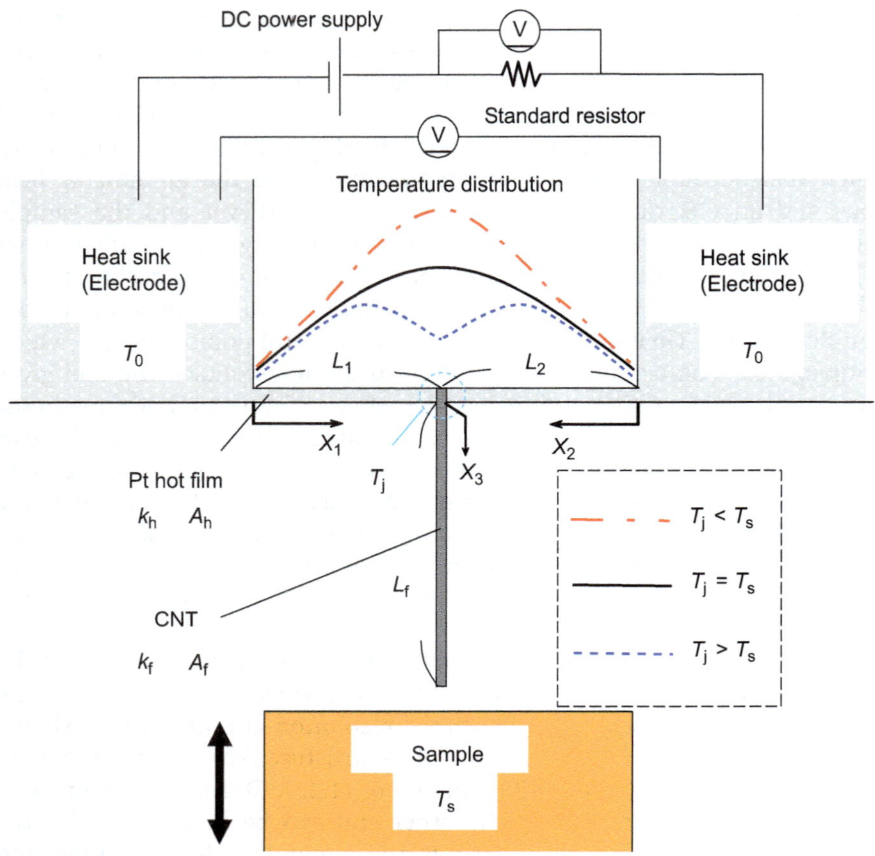

Figure 11.4   Heat transfer model of CNT thermometry based on T-type method.[14] A CNT probe is bonded to a suspended Pt hot film. The temperature distribution along the Pt film for three cases is shown: hot film–CNT junction temperature, T_j, is the same as the sample surface temperature, T_s, or higher, or lower.

measures its average temperature at the same time. When convection and radiation are negligible, the temperature distribution T of this suspended hot film is analysed by the following one-dimensional heat conduction equation:

$$k_h \frac{\partial^2 T_i(x_i)}{\partial x_i^2} + q_h = 0 \quad \text{for } i = 1, 2 \tag{11.1}$$

where k_h is the thermal conductivity of hot film and $q_h = Q_h/(A_h(L_1 + L_2))$. Q_h is the total Joule heat of the hot film given by $Q_h = IV$, where I is the electrical current and V is the voltage. The solution of eqn (11.1) is:

$$T_1(x_1) = -\frac{Q_h}{2k_h A_h(L_1 + L_2)} x_1^2 + C_1 x_1 + C_2 \tag{11.2}$$

When the CNT probe is not in contact with the surface, $T_1(0) = T_2(0) = T_0$, $T_1(L_1) = T_2(L_2)$, and $dT_1(L_1)/dx = dT_2(L_2)/dx$. By applying these boundary conditions, T_1 is obtained by,

$$T_1(x_1) = -\frac{Q_h}{2k_hA_h(L_1 + L_2)}x_1^2 + \frac{Q_h}{2k_hA_h}x_1 + T_0 \qquad (11.3)$$

and the temperature at the hot film–CNT junction, T_j, is given by,

$$T_j = T_1(L_1) = T_2(L_2) = \frac{Q_h}{2k_hA_h}\frac{L_1L_2}{L_1 + L_2} + T_0. \qquad (11.4)$$

By using eqn (11.3), the thermal conductivity, k_h, is estimated from the average temperature of the hot film, T_{ave}, which is associated with its electrical resistance, R_h as $T_{ave} = (R_h - R_{ref})/\alpha + T_{ref}$. Here α is the pre-measured temperature coefficient of the resistance of the hot film, and R_{ref} is the electrical resistance at the reference temperature T_{ref}. As long as Pt is used as the hot film, the linearity holds over a wide temperature range. When an end of the CNT is in contact with the sample of temperature T_s, the temperature distribution T_3 in the CNT probe of thermal conductivity k_f has to be considered as,

$$k_f\frac{\partial^2 T_3(x_3)}{\partial x_3^2} = 0 \qquad (11.5)$$

$$-k_hA_h\left[\frac{\partial T_1(x_1)}{\partial x_1}\bigg|_{x_1=L_1} + \frac{\partial T_2(x_2)}{\partial x_2}\bigg|_{x_2=L_2}\right] = -k_fA_f\frac{\partial T_3(x_3)}{\partial x_3}\bigg|_{x_3=0} \qquad (11.6)$$

$$-k_fA_f\frac{\partial T_3(x_3)}{\partial x_3}\bigg|_{x_3=0} = \frac{T_1(L_1) - T_3(0)}{R_j} \qquad (11.7)$$

$$-k_fA_f\frac{\partial T_3(x_3)}{\partial x_3}\bigg|_{x_3=L_f} = \frac{T_3(L_f) - T_s}{R_e} \qquad (11.8)$$

where T_s is the sample surface temperature, R is the thermal contact resistance, and the subscripts j and e denote the hot film/CNT and CNT–end/ sample interfaces, respectively. The subscript f means the CNT probe. By solving these equations, the temperature profiles along the hot film are obtained. If $T_j > T_s$, a portion of heat in the hot film goes into the sample and the temperature profile changes (see the dotted line in Figure 11.4). If $T_j < T_s$, heat comes from the sample surface through the CNT probe and the temperature distribution follows the dashed line shown in Figure 11.4. Only when $T_j = T_s$, does heat flow through the CNT not occur, and the temperature distribution is identical with the non-contact case of the same T_{ave}. The sample surface temperature can be determined from eqn (11.3) by using the value of Q_h that satisfies the last condition ($T_j = T_s$).

The procedure to measure the quantitative local temperature at the nanoscale contact point using CNT thermometry based on the above principle[33] is as follows. It is necessary to conduct the experiment in a vacuum to remove the convection effect. An electrical current must be applied to the Pt hot film as the probe is manipulated towards the sample surface. If the voltage of the Joule-heated hot film after contact, V_{after}, is not the same with that before contact, V_{before}, the CNT is detached from the sample surface and the electrical current is controlled to make V_{before} and V_{after} match. By repeating this process, the condition $V_{before} = V_{after}$ is obtained and the electrical resistance of the hot film, R_{ave} is calculated from $V_{before} = V_{after}$ and the applied electrical current I, to estimate the sample temperature, T_s. The demonstration study claimed that the temperature uncertainty is within 1 K, the spatial resolution is smaller than the diameter of the MWNT employed, and that no damage occurs at the end of MWNT probe, even after many contact tests.

11.5   Fabrication Techniques

The key to CNT thermometry is in the handling and bonding of the CNTs. In the earliest experiments,[34,35] an acrylic adhesive was used to bond CNTs to a Si cantilever tip *via* manual operation under the direct view of an optical microscope. Successful demonstration as an AFM tip was reported but its applicability as a thermal probe is questionable because the large thermal resistance of the adhesive existing between the CNT and the cantilever tip erases the advantage of the high thermal conductivity of the CNT probe. An alternative assembly method is the dielectrophoresis process,[36–38] which can set an individual CNT protruding from a fibril on the tip of a probe. This assembly process is explained in Figure 11.5. First, CNTs are dispersed in liquid media (water) with the aid of sonication. The tip and a flat counter-electrode are immersed in the CNT suspension and an alternating current (ac) is applied between them. Then, migration of CNTs starts towards the high-field region at the sharp tip because the dielectrophoresis force depends on the gradient of the square of the electrical field. Tang *et al.*[36] successfully demonstrated this alignment process in deionized water with MWNTs, at ac 10 V and 2 MHz. CNTs accumulate on the tip and form a fibril there. After withdrawal of the tip from the suspension, an individual CNT is seen to protrude from the fibril, which is more common for MWNTs than for SWNTs due to the difference in their flexibility. Because the van der Waals forces between the CNT and probe tip, and between CNTs are pretty strong, the obtained CNT tip can be used not only as an electron field emitter but also as an AFM probe.

Recent progress in CNT synthesis has enabled us to make CNTs grow directly on Si pyramidal tips by CVD.[39–42] Firstly, metal catalysts of Fe, Mo, Ni, their alloys *etc.* are deposited on the pyramidal AFM tips by dip/spin coating, electrophoretic methods, *etc.*, then CNTs are synthesized by CVD employing methane, acetylene, *etc.*, as the precursor on these catalytic

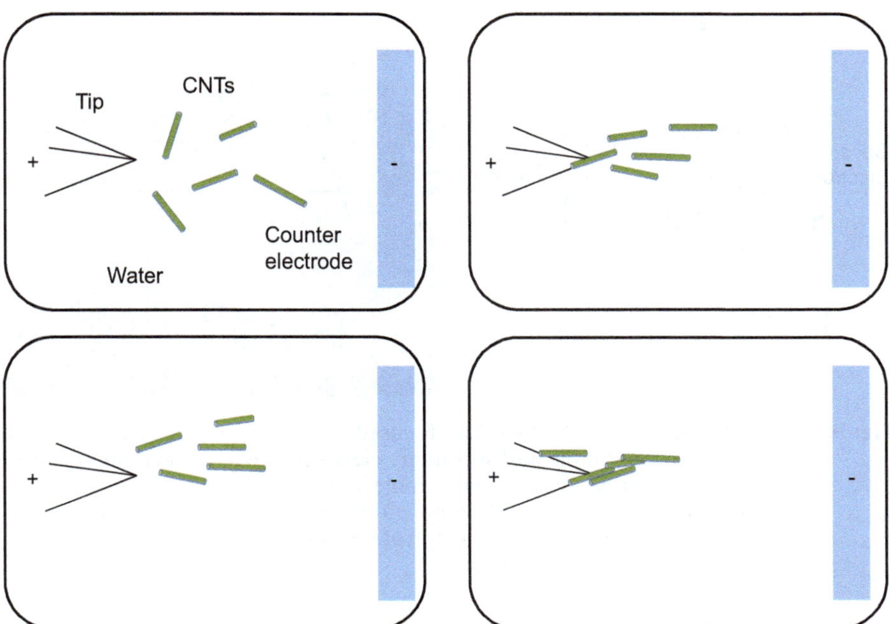

Figure 11.5 Dielectrophoresis method for CNTs to assemble into fibrils on a tip with a protruding CNT,[36] where CNTs migrate towards the higher electric field region around the tip.

particles. CNTs grow along the surface but some of them protrude from the tip apex. Too-lengthy CNTs can be shortened using an electrical etching method.[39] This straightforward growth method is applicable to commercially available AFM probes and the thermal resistance between the CNT and cantilever tip is expected to be lower than the adhesive method.

The most fragile point in the CNT probe is the junction between the CNT and the cantilever tip. Although the CNT itself has atomic scale robustness, van der Waals forces at the junction are not enough to guarantee the long lifetime required for contact scanning measurements. In recent years, electron beam-induced deposition (EBID) has been commonly used to enhance the bonding between a CNT and a cantilever tip, since the first successful report in 1999.[43] EBID was first discovered as contamination in an electron microscope in 1947,[44] and its physical mechanism is now understood, as shown in Figure 11.6. The irradiated electron beam induces the second electron in the sample, which is ejected and is usually detected during scanning electron microscopy (SEM). The interaction with the second electron decomposes the residual organic matter and amorphous carbon is deposited on the irradiated surface. This deposition works as an additional adhesive and the junction with this EBID becomes much stronger. For the above-mentioned CNT–hot film device, this EBID method is indispensable, and is used not only for the final bonding of the CNT to the hot film, but also for the temporal tacking of CNTs to a manipulator.

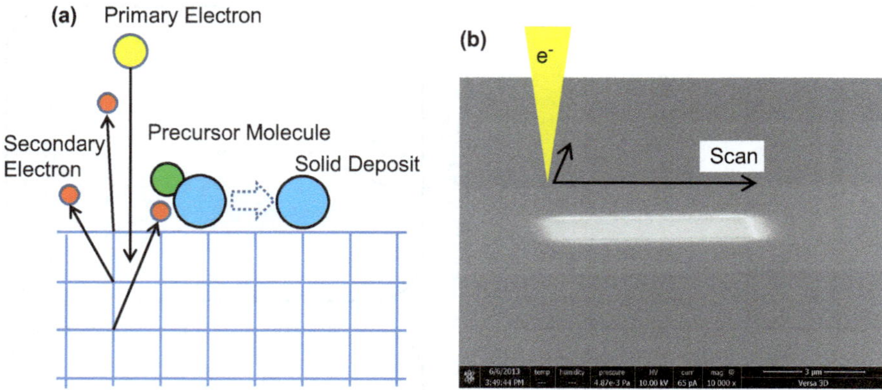

Figure 11.6   (a) Mechanism of EBID. The focused primary electron beam is irradiated on the sample and secondary electrons are generated and ejected. Their collisions with organic molecules absorbed on or close to the sample surface induce decomposition of the molecule and deposition of carbon atoms. (b) SEM image of deposited carbon (bright colour) by electron beam irradiation.

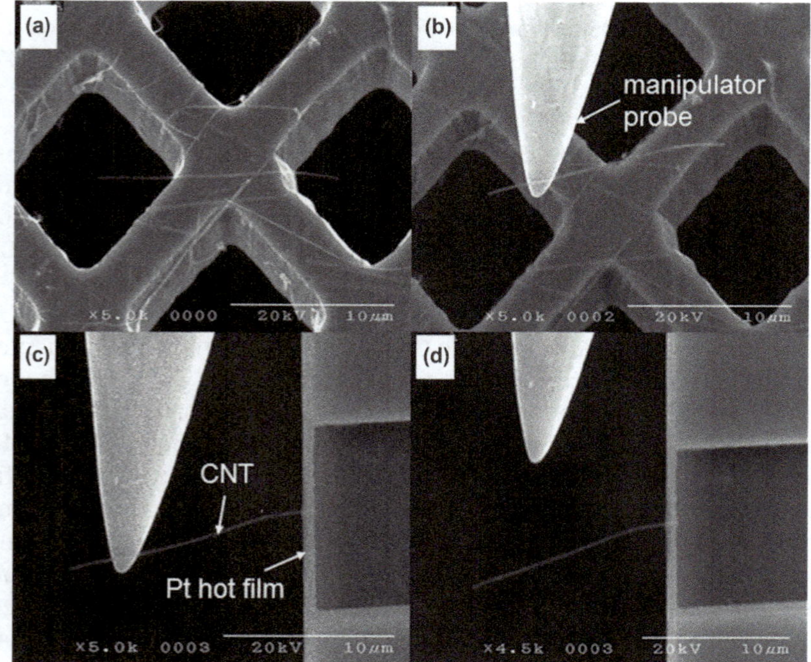

Figure 11.7   Fabrication procedure of an individual MWNT probe fixed on a suspended hot film sensor. (a) MWNTs are dropped onto a TEM grid. (b) Manipulator probe picks up a MWNT which is temporarily bonded to the probe by electron beam irradiation. (c) MWNT is set on a Pt hot film and EBID is conducted at their junction for permanent bonding. (d) Manipulator probe is finally detached from the MWNT.

Focused ion beam (FIB) deposition is also applicable but may degrade the CNT's properties.

The fabrication of a Pt hot film is based on a microelectromechanical systems (MEMS) technique consisting of electron beam (photo)lithography, physical vapour deposition of Pt, a lift-off technique for patterning the Pt film, and isotropic etching of the SiO_2 layer underneath the film. A suspended Pt hot film of 10 µm length and 0.5 µm width can be built closer than 2 µm from the edge of the Si substrate, as seen in Figure 11.3. It should be noted that the thermal and electrical properties of the thin film are different from the bulk material due to the grain and surface boundaries, which scatter the electrons and phonons.[45] In an SEM, a single CNT of appropriate length is picked up from bulk or isolated CNTs and bonded onto the hot film using the EBID method, the process of which is shown in Figure 11.7. In addition, it has to be considered that such a tiny suspended hot film may fail immediately due to the overcurrent from outside.

References

1. M. S. Dresselhaus and Ph. Avouris, *Carbon Nanotubes: Synthesis, Structure, Properties, and Applications*, ed. M. S. Dresselhaus, G. Dresselhaus and Ph. Avouris, Springer, 2001.
2. S. Iijima, *Nature*, 1991, **354**, 56.
3. H. Dai, J. H. Hafner, A. G. Rinzler, D. T. Colbert and R. E. Smalley, *Nature*, 1996, **384**, 147.
4. P. Tovee, M. Pumarol, D. Zeze, K. Kjoller and O. Kolosov, *J. Appl. Phys.*, 2012, **112**, 114317.
5. F. Menges, H. Riel, A. Stemmer, C. Dimitrakopoulos and B. Gotsmann, *Phys. Rev. Lett.*, 2013, **111**, 205901.
6. S. Iijima and T. Ichihashi, *Nature*, 1993, **363**, 603.
7. K. Hata, D. N. Futaba, K. Mizuno, T. Namai, M. Yumura and S. Iijima, *Science*, 2004, **306**, 1362.
8. C. Kittel, *Introduction to Solid State Physics*, Wiley, 8th edn, 2004.
9. S. Berber, Y.-K. Kwon and D. Tománek, *Phys. Rev. Lett.*, 2000, **84**, 4613.
10. N. Mingo and D. A. Broido, *Nano Lett.*, 2005, **5**, 1221.
11. C. Yu, L. Shi, Z. Yao, D. Li and A. Majumdar, *Nano Lett.*, 2005, **5**, 1842.
12. E. Pop, D. Mann, Q. Wang, K. Goodson and H. Dai, *Nano Lett.*, 2006, **6**, 96.
13. P. Kim, L. Shi, A. Majumdar and P. L. McEuen, *Phys. Rev. Lett.*, 2001, **87**, 215502.
14. M. Fujii, X. Zhang, H. Xie, H. Ago, K. Takahashi, T. Ikuta, H. Abe and T. Shimizu, *Phys. Rev. Lett.*, 2005, **95**, 065502.
15. T. Y. Choi, D. Poulikakos, J. Tharian and U. Sennhauser, *Appl. Phys. Lett.*, 2005, **87**, 013108.
16. T. Y. Choi, D. Poulikakos, J. Tharian and U. Sennhauser, *Nano Lett.*, 2006, **6**, 1589.

17. C. Dames, S. Chen, C. T. Harris, J. Y. Huang, Z. F. Ren, M. S. Dresselhaus and G. Chen, *Rev. Sci. Instrum.*, 2007, **78**, 104903.

18. M. T. Pettes and L. Shi, *Adv. Funct. Mater.*, 2009, **19**, 3918.

19. J. Yang, S. Waltermire, Y. Chen, A. A. Zinn, T. T. Xu and D. Li, *Appl. Phys. Lett.*, 2010, **96**, 023109.

20. G. A. Slack, *Phys. Rev.*, 1962, **127**, 694.

21. H. Hayashi, T. Ikuta, T. Nishiyama and K. Takahashi, *J. Appl. Phys.*, 2013, **113**, 014301.

22. S. Maruyama, *Physica B*, 2002, **323**, 193.

23. G. Zhang and B. Li, *J. Chem. Phys.*, 2005, **123**, 114714.

24. T. Yamamoto, S. Konabe, J. Shiomi and S. Maruyama, *Appl. Phys. Express*, 2009, **2**, 095003.

25. H. Hayashi, K. Takahashi, T. Ikuta, T. Nishiyama, Y. Takata and X. Zhang, *Appl. Phys. Lett.*, 2014, **104**, 113112.

26. C. W. Chang, D. Okawa, H. Garcia, A. Majumdar and A. Zettl, *Phys. Rev. Lett.*, 2008, **101**, 075903.

27. K. T. Regner, D. P. Sellan, Z. Su, C. H. Amon, A. J. H. McGaughey and J. A. Malen, *Nat. Commun.*, 2013, **4**, 1640.

28. X. Xu, L. F. C. Pereira, Y. Wang, J. Wu, K. Zhang, X. Zhao, S. Bae, C. T. Bui, R. Xie, J. T. L. Thong, B. H. Hong, K. P. Loh, D. Donadio, B. Li and B. Özyilmaz, *Nat. Commun.*, 2014, **5**, 3689.

29. O. Nakabeppu and T. Suzuki, *J. Therm. Anal. Calorim.*, 2002, **69**, 727.

30. J. Chung, K. Kim, G. Hwang, O. Kwon, S. Jung, J. Lee, J. W. Lee and G. T. Kim, *Rev. Sci. Instrum.*, 2010, **81**, 114901.

31. L. Shi, S. Plyasunov, A. Bachtold, P. L. McEuen and A. Majumdar, *Appl. Phys. Lett.*, 2000, **77**, 4295.

32. L. Shi, O. Kwon, C. C. Miner and A. Majumdar, *J. Microelectromech. Syst.*, 2001, **10**, 370.

33. J. Hirotani, J. Amano, T. Ikuta, T. Nishiyama and K. Takahashi, *Sens. Actuators, A*, 2013, **199**, 1.

34. H. Dai, J. H. Hafner, A. G. Rinzler, D. T. Colbert and R. E. Smalley, *Nature*, 1996, **384**, 147.

35. S. S. Wong, A. T. Woolley, T. W. Odom, J.-L. Huang, P. Kim, D. V. Vezenov and C. M. Lieber, *Appl. Phys. Lett.*, 1998, **73**, 3465.

36. J. Tang, B. Gao, H. Geng, O. D. Velev, L.-C. Qin and O. Zhou, *Adv. Mater.*, 2003, **15**, 1352.

37. J. Zhang, J. Tang, G. Yang, Q. Qiu, L.-C. Qin and O. Zhou, *Adv. Mater.*, 2004, **16**, 1219.

38. J. Tang, G. Yang, Q. Zhang, A. Parhat, B. Maynor, J. Liu, L.-C. Qin and O. Zhou, *Nano Lett.*, 2005, **5**, 11.

39. J. H. Hafner, C. L. Cheung and C. M. Lieber, *J. Am. Chem. Soc.*, 1999, **121**, 9750.

40. E. B. Cooper, S. R. Manalis, H. Fang, H. Dai, K. Matsumoto, S. C. Minne, T. Hunt and C. F. Quate, *Appl. Phys. Lett.*, 1999, **75**, 3566.

41. C. L. Cheung, J. H. Hafner, T. W. Odom, K. Kim and C. M. Lieber, *Appl. Phys. Lett.*, 2000, **76**, 3136.

42. E. Yenilmez, Q. Wang, R. J. Chen, D. Wang and H. Dai, *Appl. Phys. Lett.*, 2002, **80**, 2225.
43. H. Nishijima, S. Kamo, S. Akita, Y. Nakayama, K. I. Hohmura, S. H. Yoshimura and K. Takeyasu, *Appl. Phys. Lett.*, 1999, **74**, 4061.
44. J. H. L. Watson, *J. Appl. Phys.*, 1947, **18**, 153.
45. X. Zhang, H. Xie, M. Fujii, H. Ago, K. Takahashi, T. Ikuta, H. Abe and T. Shimizu, *Appl. Phys. Lett.*, 2005, **86**, 171912.

Section IV
Applications

Cellular Thermometry

SEIICHI UCHIYAMA*[a] AND NORIKO INADA*[b]

[a] Graduate School of Pharmaceutical Sciences, The University of Tokyo, 7-3-1 Hongo, Bunkyo-ku, Tokyo 113-0033, Japan; [b] Plant Global Education Project, Graduate School of Biological Sciences, Nara Institute of Science and Technology, 8916-5 Takayama-cho, Ikoma-shi, Nara 630-0192, Japan
*Email: seiichi@mol.f.u-tokyo.ac.jp; norikoi@bs.naist.jp

12.1   Background

The cell is the minimum unit of all living organisms. Each single cell contains all of the information that defines the species, and has the ability to reproduce itself. It requires its selective barrier, the plasma membrane, to not only protect cells from the environment but also concentrate nutrients, which are used to synthesise molecules for the cell function. Eukaryotic cells contain membrane-enclosed compartments, called *organelles*, each of which has its own specialised function. The nucleus contains the genome and is the primary site of both DNA and RNA synthesis; the endoplasmic reticulum (ER) is a factory that synthesises proteins and lipids, and the mitochondrion produces ATP, the energy currency utilised in many chemical reactions inside the cell.[1] In addition to mitochondria, the chloroplast in the plant cell also produces ATP; other important functions of the chloroplast are the production of O_2 and the assimilation of carbon atoms to sugars, which serve as food, biomass, and fossil fuels.[2] The activity of the cell could affect the cellular temperature, as cellular function is supported by numerous chemical reactions, which are either exothermic or endothermic.

As our body temperature is used as an indicator of our health, measuring the temperature of cells could provide valuable information on the cellular

RSC Nanoscience & Nanotechnology No. 38
Thermometry at the Nanoscale: Techniques and Selected Applications
Edited by Luís Dias Carlos and Fernando Palacio
© The Royal Society of Chemistry 2016
Published by the Royal Society of Chemistry, www.rsc.org

status. Earlier work using microcalorimetry revealed that tumourous tissues exhibited higher metabolism, thus producing more heat than non-tumourous tissues.[3] High metabolism is now regarded as one of the hallmarks of cancer cells.[4] Other human diseases, such as mitochondrial syndromes, could also exhibit abnormal metabolism and thus altered cellular temperature.[5] Temperature elevation is also observed for diseased plant tissues.[6] Cellular thermometry could be applied as a tool to detect diseased cells with high spatial resolution. As another study using microcalorimetry has indicated a good correlation between the degree of malignancy and heat production,[7] cellular thermometry could also assist in quantitative diagnosis or the analysis of the mechanism of the progression of malignancy.

Cellular thermometry can serve as a tool for quantitative analysis of cellular metabolism. When the ATP production of mitochondria in mammalian cells is inhibited by treatment with uncoupling reagents, the energy that was utilised for ATP synthesis is released as heat.[8] Thus, the mitochondrial activity can be quantitatively estimated based on the temperature changes. In addition to the mitochondria, the chloroplast could be another heat source inside plant cells. In the photosynthetic process, the light energy absorbed by chlorophyll is utilised to produce ATP and NADPH, which are used for carbon fixation. Plants have evolved an ability to balance the input and utilisation of light energy to avoid light-induced damage, because the intensity of sunlight fluctuates substantially depending on the weather and other environmental changes, and the absorption of excessive light by the chlorophyll pigments could result in the production of harmful reactive oxygen species. Thermal dissipation is one of the strategies used by plants for this purpose,[9] and thus, measuring the temperature of plant cells could provide valuable information to elucidate the mechanism of photosynthesis and environmental adaptation.

The cellular temperature is affected by not only the cellular metabolism but also changes in the ambient temperature. The sensation of ambient temperature is necessary for living organisms for survival, and cellular thermometry could serve as a tool to unravel the mechanism of thermal sensation at the cellular level. For example, animal thermo-TRP (transient receptor potential) cation channels are known as sensors of ambient temperature; however, the exact mechanism of how TRP channels respond to the temperature shift is unknown.[10] Simultaneous monitoring of the TRP activity and cellular temperature during an ambient temperature shift could help to elucidate the mechanism of thermal sensing with TRPs. The development and growth of plants, which do not have the ability to maintain a constant body temperature, are critically affected by the ambient temperature. Although the signalling downstream of temperature perception has been thoroughly studied in plants,[11,12] the temperature-sensing mechanism of plant cells remains unknown. With cellular thermometry, one could obtain a precise evaluation of changes in the cellular temperature affected by ambient temperature shifts, which could be used to specify the factors and pathways that are most sensitive to the temperature change in combination with other information, such as gene expression.

12.2 Tools for Measuring Cellular Temperature

12.2.1 Fluorescent and Luminescent Molecular Thermometers

The precedents regarding cellular thermometry in the recent literature have clearly indicated that fluorescent and luminescent molecular thermometers are promising tools for measuring the temperature of living cells. Compared to other thermometers, such as thermocouples and infrared radiation detectors (thermography), fluorescent and luminescent molecular thermometers are sufficiently small to enter into living cells. Based on the novel functional mechanisms of molecular thermometers proposed by many research groups in the past decade,[13–15] the temperature sensitivity of fluorescent and luminescent molecular thermometers has been significantly improved, now reaching the sub-degree level even within cells. Table 12.1 summarises works on cellular thermometry using fluorescent and luminescent molecular thermometers. Each case will be briefly outlined in the following sections in the order of size of the molecular thermometer, *i.e.*, small organic fluorophores, metal complexes with organic ligands, macromolecules (synthetic polymers, DNA and proteins) and inorganic substances (quantum dots and other materials).

12.2.1.1 *Organic Fluorophores*

A commercially available fluorescent compound, 2-(12-*N*-NBD-amino)dodecanoyl-1-hexadecanoyl-*sn*-glycero-3-phosphocholine (NBD-PC), exhibits a temperature-dependent change in its fluorescence properties due to the rotation of its amino substituent group.[16] Using NBD-PC as a fluorescent molecular thermometer, Chapman *et al.* attempted to measure cellular temperature for the first time in 1995.[17] NBD-PC was localised on the cellular membrane of CHO (Chinese hamster ovary) cells upon incubation of the cells in a solution containing NBD-PC and exhibited a decrease in its fluorescence lifetime with increasing medium temperature. The temperature resolution (*i.e.*, the minimum temperature that can be significantly distinguished) was evaluated to be *ca.* 2 °C.[17] The same report demonstrated that 6-dodecanoyl-2-dimethylaminonaphthalene (Laurdan) could be used for cellular thermometry, as the fluorescence properties of Laurdan located on the cellular membrane changed depending on environmental changes in hydrophobicity/hydrophilicity, which originated from a heat-induced gel-to-liquid crystalline phase transition of the membrane (Chart 12.1).[17]

Dylight549 (Dylight® is a trademark of Thermo Fisher Scientific Inc., USA) is another organic fluorophore that functions as a molecular thermometer; unfortunately, the chemical structure of this fluorophore remains unclear.

Table 12.1 Cellular thermometry using fluorescent and luminescent thermometers.

Fluorescent (luminescent) thermometer	Commercially available	Observed temperature-dependent parameter	Tested cell line	Introduction into cells	Distribution within cells	Temperature resolution/°C	Ref.
Organic fluorophores							
NBD-PC	Yes	FI (53 at 17 °C, 18 at 50 °C) or τ_f	CHO	Incubation 4 °C, 4 h	On a cell membrane	2	17
Laurdan	Yes	FI ratio (0.39 at 16 °C, 0.22 at 44 °C)	CHO	Incubation 37 °C, 35 min	On a cell membrane	0.1–1.0	17
Streptavidin-DyLight 549 coated nanoparticle	No	FI ($-1.5\%/°C$)	HEK293	Incubation room temperature, 1 min	On a cell membrane	—	18
Rhodamine B	Yes	FI (1.0 at 20 °C, 0.07 at 75 °C) or τ_f	Onion skin	Soak 4 °C, overnight	Whole cell	<1.8	22
Metal complexes with organic ligands							
Eu-TTA	Yes	Emission intensity (10 at 15 °C, 2.4 at 40 °C)	CHO	Incubation room temperature, 30 min	On a cell membrane	—	24
Eu-TTA containing nanoparticle	No	Emission intensity (10 at 15 °C, 2.4 at 40 °C)	HeLa	Incubation 37 °C, 1 h	Endosome/lysosome	0.3	26
Eu-TTA/Rhodamine 101 containing nanoparticle	no	Emission intensity ratio (4.4 at 26 °C, 2.8 at 40 °C)	HeLa	Incubation 37 °C, 2 h	Endosome	1.0	27
Ru(bpy)$_3{}^{2+}$-doped silica nanoparticle	no	Emission intensity (113 at 20 °C, 78 at 49 °C)	HepG2	Incubation 37 °C, 24 h	Within a cell	—	29

Thermoresponsive synthetic polymers							
NIPAM-DBD nanogel	No	FI (0.33 at 23 °C, 1.0 at 37 °C)	COS7	Microinjection	Dotted in cytoplasm	0.29–0.50	32
NIPAM-DBThD nanogel	Yes	FI (1.0 at 23 °C, 5.9 at 35 °C)	COS7	Microinjection	Dotted in cytoplasm	—	33
NNPAM-DBD polymer	Yes	τ_f (4.6 ns at 28 °C, 7.6 ns at 40 °C)	COS7, HeLa	Microinjection	Whole cell	0.18–0.58	34
NNPAM-DBD cationic polymer DNA	No	τ_f (6.2 ns at 15 °C, 8.6 ns at 35 °C)	Yeast, MOLT-4, HEK293T	Incubation 25 °C, 20 min	Cytoplasm or whole cell	0.09–0.78	35
L-MB	Yes	FI (1.0 at 20 °C, 2.0 at 37 °C)	HeLa	Liposome transfection	Mostly nucleus	0.2–0.7	37
GFP							
GFP	Yes	Fluorescence polarisation anisotropy (0.222 at 24 °C, 0.205 at 40 °C)	HeLa, U-87 MG	Genetically expressed	Whole cell	1.2	39
tsGFP1	No	FI ratio (1.7 at 20 °C, 1.1 at 50 °C)	HeLa, brown adipocyte, myotube	Genetically expressed	Cytoplasm, and targeted to mitochondria and endoplasmic reticulum	—	38
Quantum dots							
QD655	Yes	λ_{em} (+0.105 nm/°C)	NIH/3T3	Assisted by Qtracker cell labelling kit 37 °C, 1 h	Dotted within a cell	—	43
Quantum dot/ quantum rod complex	No	FI ratio (+2.4%/°C)	HeLa, NIH/3T3	Assisted by cationic polymer	Within a cell	0.2	44

Table 12.1 (*Continued*)

Fluorescent (luminescent) thermometer	Commercially available	Observed temperature-dependent parameter	Tested cell line	Introduction into cells	Distribution within cells	Temperature resolution/°C	Ref.
Inorganic materials							
$NaYF_4{:}Er^{3+},Yb^{3+}$ nanoparticle	No	FI ratio ($+1.1\%$/°C)	HeLa	Incubation 1.5 h	Dotted within a cell	—	45
Lipoic acid-protected gold nanocluster	No	τ_f (970 nm at 14 °C, 670 nm at 43 °C)	HeLa	Incubation 37 °C, 2 h	Dotted within a cell	0.3–0.5	46
Nitrogen-vacancy centres in diamond nanocrystal	No	FIs with microwave excitations	WS1	Assisted by nanowire	Dotted within a cell	0.044 ± 0.010	47

FI: fluorescence intensity; τ_f: fluorescence lifetime; λ_{em}: maximum emission wavelength; —: not available.

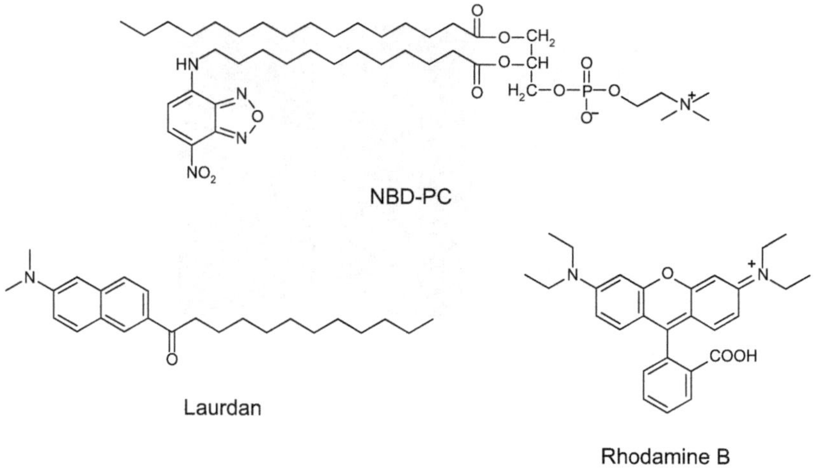

NBD-PC

Laurdan

Rhodamine B

Chart 12.1 The structures of NBD-PC, Laurdan and Rhodamine B.

Huang *et al.* prepared streptavidin-DyLight549-coated superparamagnetic MnF_2O_4 nanoparticles and located the nanoparticles on the cellular membrane, where an enzymatically biotinylated biotin acceptor peptide was present.[18] The fluorescence intensity of DyLight549 in the vicinity of MnF_2O_4 decreased upon heating the nanoparticles by application of a radiofrequency magnetic field, providing an estimate of the temperature rise over 15 °C (Figure 12.1).[18]

Rhodamine B (Chart 12.1) is perhaps the most common organic fluorophore that exhibits a temperature-dependent fluorescence change, which is ascribed to the rotation of the diethylamino group.[19] Although Rhodamine B has been widely utilised in the thermometry of extremely small spaces, such as microfluidic systems,[20,21] its use for cellular thermometry has been questioned. Paviolo *et al.* monitored the temperature-dependent fluorescence intensity and fluorescence lifetime of Rhodamine B in an onion skin and concluded that Rhodamine B was undesirably affected by environmental changes, such as pH change and protein denaturation, in addition to the temperature change, and that the encapsulation or immobilisation of Rhodamine B would be required in biological samples.[22]

12.2.1.2 Metal Complexes with Organic Ligands

Europium(III) thenoyltrifluoroacetonate (Eu-TTA) exhibits a temperature-dependent emission intensity because of the competition between the luminescence and non-radiative energy transfer from the Eu(III) ion to the ligand [Figure 12.2(a) and (b)].[23] Zohar *et al.* stained the cell membrane of CHO cells with Eu-TTA and monitored the temperature change after treatment with acetylcholine to activate the metabotropic m1-muscarinic receptors. They observed a bi-phasic heat wave [Figure 12.2(c)], which was

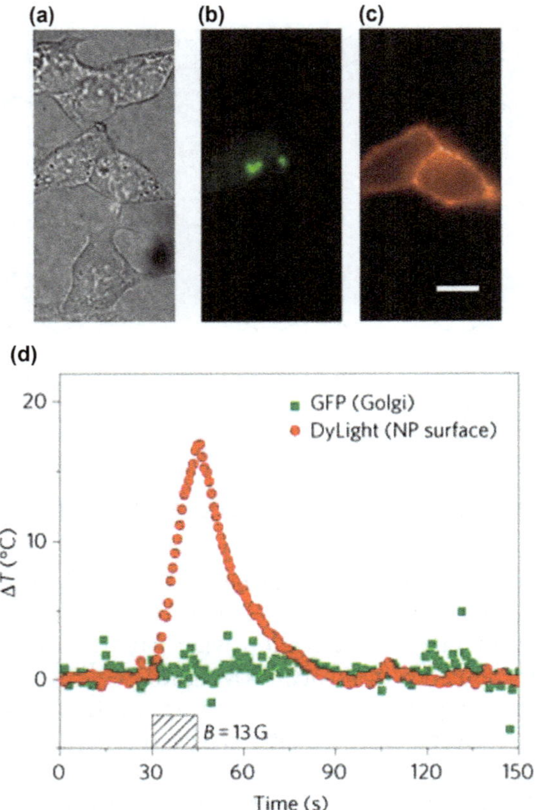

Figure 12.1 Temperature monitoring of the HEK293 cell membrane by using DyLight549-coated manganese ferrite ($MnFe_2O_4$) nanoparticles. (a) Differential interference contrast (DIC) image. (b) Fluorescence image of the Golgi body labelled with a green fluorescent protein (GFP). (c) Fluorescence image of DyLight549-coated nanoparticles. Scale bar = 20 μm. (d) Change in the local temperature of HEK293 cells after the application of the radiofrequency magnetic field (as indicated by the hatched box). The temperature was evaluated from the fluorescence intensity of DyLight549.
Adapted with permission from ref. 18. © Macmillan Publishers 2010.

blocked by pre-treatment with a muscarinic receptor antagonist.[24] Although not described in their study, Eu-TTA is easily decomposed in an aqueous environment. Suzuki *et al.* used a glass micropipette filled with Eu-TTA dissolved in non-aqueous dimethylsulfoxide (DMSO) and measured the change in cellular temperature of a single HeLa (human epithelial carcinoma) cell. Using this method, the heat production from a HeLa cell upon the ionomycin-induced calcium ion influx from the extracellular space was observed.[25] In updated work by the same group,[26,27] temperature-sensitive Eu-TTA and temperature-insensitive Rhodamine 101 were both encapsulated in a cationic nanoparticle [Figure 12.3(a) and (b)].[27] A cationic shell

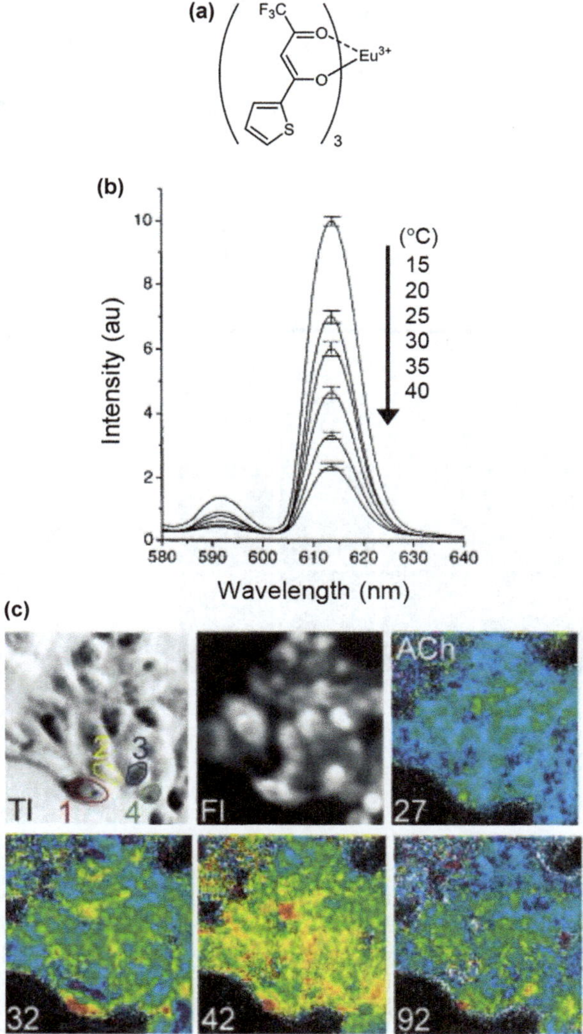

Figure 12.2 Thermal imaging of living CHO cells using Eu-TTA. (a) Chemical structure of Eu-TTA. (b) Temperature-dependent emission spectra of Eu-TTA integrated in liposomal membranes (pH 7.4, $\lambda_{ex} = 372$ nm). (c) Transmitted light (Tl), luminescence (Fl) and time lapse images of Eu-TTA-labelled CHO cells treated with acetylcholine (ACh). In the time lapse images, the fluorescence intensity is shown in pseudocolour (blue represents lower temperature, and red represents higher temperature). The number indicates the frame number (frame interval: *ca.* 1 s). Acetylcholine was added to the medium at frame 27. Four cells are indicated in the Tl image.
Adapted from ref. 24 with permission. © Elsevier 1998.

formed by protonated poly(allylamine) supported the incorporation of the nanoparticles into the cytoplasm of HeLa cells from the culture medium *via* an endocytic pathway. The change in the fluorescence ratio of two

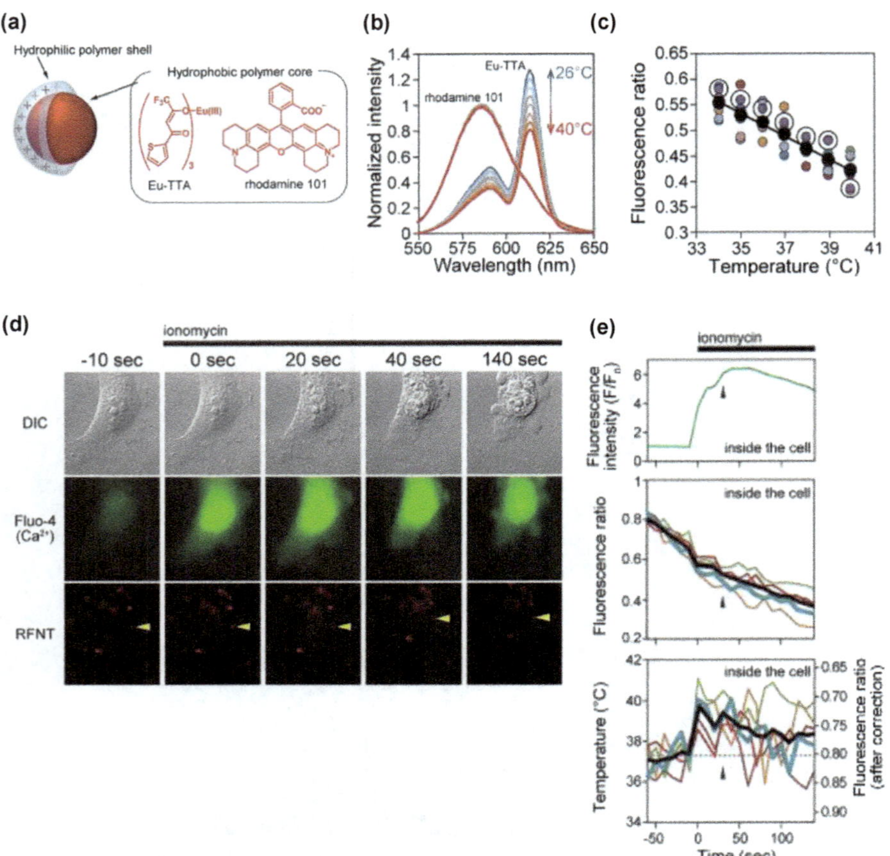

Figure 12.3 Intracellular thermometry of living HeLa cells using a ratiometric fluorescent nanothermometer (RFNT) containing Eu-TTA and Rhodamine 101. (a) Structure of the hydrophilic RFNT. (b) Fluorescence spectra of Eu-TTA ($\lambda_{ex} = 340$ nm) and Rhodamine 101 ($\lambda_{ex} = 530$ nm) in the RFNT. (c) Relationship between the fluorescence ratio (Eu-TTA/ Rhodamine 101) in a HeLa cell and the medium temperature. Each colour corresponds to a single nanoparticle, and the black plots indicate an average value. (d) Time lapse imaging of DIC (differential interference contrast), Fluo-4 (indicating Ca^{2+} ion) and Rhodamine 101 in the RFNT after the application of the ionomycin–Ca^{2+} complex. (e) Time courses of the fluorescence intensity of Fluo-4 (top), the fluorescence ratio (Eu-TTA/Rhodamine 101) of the RFNT (middle), and the evaluated temperature using the fluorescence ratio corrected by cancelling photobleaching effects.
Adapted with permission from ref. 27. © American Chemical Society 2014.

wavelengths, which corresponded to the emission maxima of Eu-TTA and Rhodamine 101, was independent of the concentration of fluorophores. In addition, the fluorescence ratio of the nanoparticles was not affected by other environmental changes, such as pH, ionic strength, coexisting proteins and viscosity, and thus, its change is highly temperature specific

[Figure 12.3(c)]. Using this method, heterogeneous heat production in HeLa cells was observed after a burst of calcium ion influx after an ionomycin treatment[27] [Figure 12.3(d) and (e)].

The ruthenium(II) tris(2,2′-bipyridyl) ion $Ru(bpy)_3^{2+}$ is another metal complex that exhibits temperature-dependent changes in its luminescence properties. The temperature sensitivity of $Ru(bpy)_3^{2+}$ is derived from non-emissive triplet metal-centred (MC) states existing at energies slightly above the emissive triplet metal-to-ligand charge transfer (MCLT) state.[28] Recently, Yang *et al.* prepared $Ru(bpy)_3^{2+}$-doped silica nanoparticles for intracellular thermometry.[29] Nanoparticles that were further coated with poly(L-lysine) spontaneously entered into HepG2 (human hepatocellular carcinoma) cells and were used to monitor the increase in cellular temperature during laser irradiation at 808 nm of gold nanorods internalised within the same cells.[29]

12.2.1.3   Thermoresponsive Synthetic Polymers

The combination of a thermoresponsive polymer with a water-sensitive fluorophore produced a highly sensitive fluorescent polymeric thermometer functioning in an aqueous environment.[30] Several groups, including ours, have exploited the remarkable sensitivity of the fluorescent polymeric thermometers to temperature variation and have developed thermometers for cellular thermometry.[31] The fluorescent nanogel based on the *N*-isopropylacrylamide structure that we developed was the first report of the accurate intracellular thermometry of a single live cell [Figure 12.4(a)].[32] The fluorescence of the fluorescent nanogels microinjected in COS7 (derived from African green monkey kidney) cells was quenched by interior water molecules at a lower temperature, but recovered at a higher temperature at which the water molecules were expelled from the interior of the nanogels [Figures 12.4(b) and (c)]. The evaluated temperature resolution of this fluorescent nanogel was 0.28–0.50 °C in the range of 27–33 °C.[32] The effects of an uncoupler, carbonyl cyanide 4-(trifluoromethoxy)phenylhydrazone [FCCP, Figure 12.4(d)], and an apoptosis inducer, camptothecin, on the average intracellular temperature of COS7 cells were successfully monitored.[32] We recently reported an improvement in the photostability of the fluorescent nanogel thermometer based on the development of a more stable water-sensitive fluorophore.[33]

Although the fluorescent nanogel was successful in monitoring temperature changes in living cells, it tended to aggregate at high temperatures and thus could not provide a heat distribution map inside the cell. To increase the spatial resolution in intracellular thermometry, we developed a fluorescent polymeric thermometer [Figure 12.5(a)][34] that consisted of the same units as the fluorescent nanogel but was considerably smaller (8.9 *vs.* 51.2 nm in diameter). Fluorescence lifetime imaging microscopy was used in this study to analyse the intracellular temperature distribution. As the fluorescence lifetime is less affected by the change in the concentration of the fluorescent polymeric thermometer, it is considered to be a particularly suitable parameter for the case in which the exact concentration

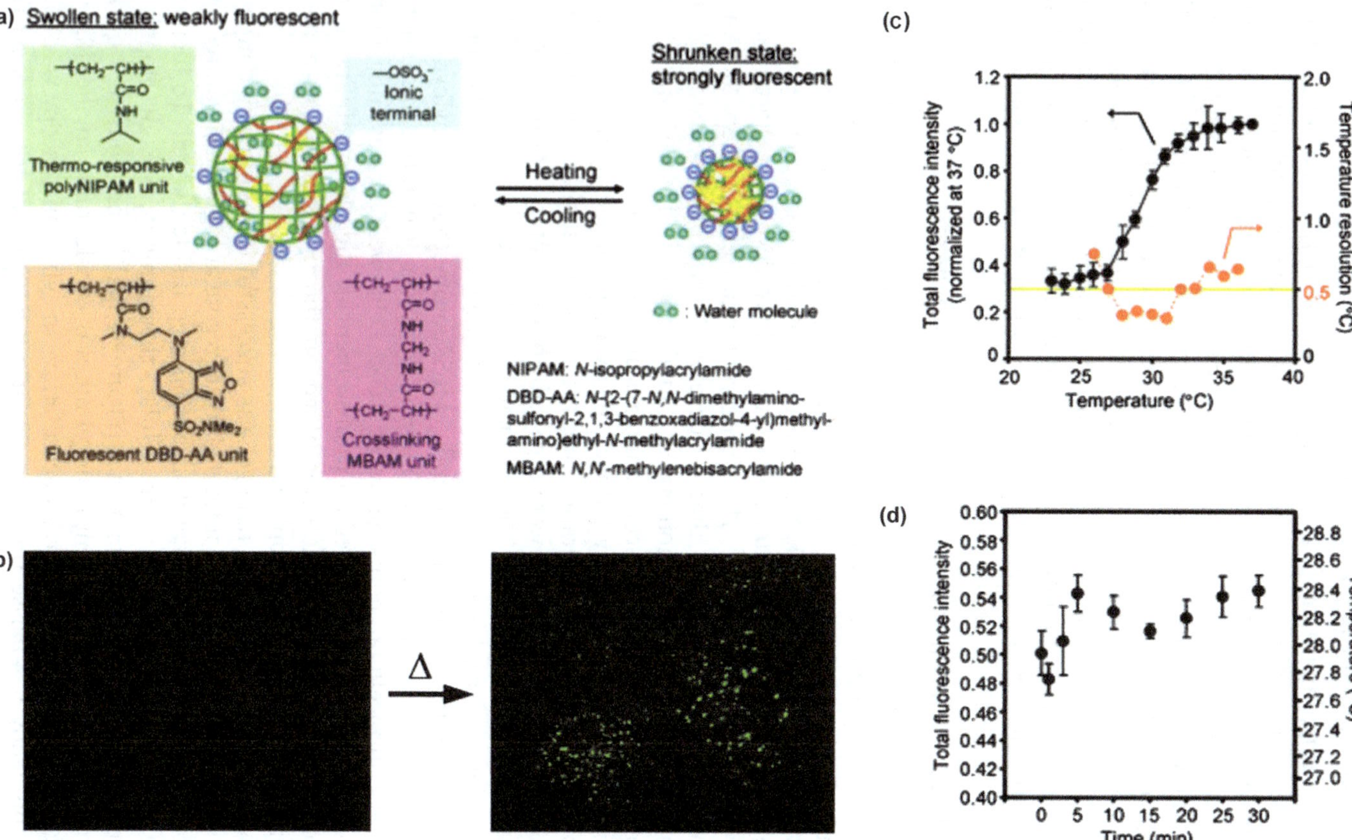

(a) Swollen state: weakly fluorescent
Thermo-responsive polyNIPAM unit
Fluorescent DBD-AA unit
Crosslinking MBAM unit
—OSO₃⁻ Ionic terminal
Shrunken state: strongly fluorescent
Heating
Cooling
: Water molecule
NIPAM: N-isopropylacrylamide
DBD-AA: N-{2-(7-N,N-dimethylamino-sulfonyl-2,1,3-benzoxadiazol-4-yl)methyl-amino}ethyl-N-methylacrylamide
MBAM: N,N-methylenebisacrylamide
(b)
Δ
(c)
Total fluorescence intensity (normalized at 37 °C)
Temperature resolution (°C)
Temperature (°C)
(d)
Total fluorescence intensity
Temperature (°C)
Time (min)

of fluorophores in local spaces cannot be determined. Based on the relationship between the fluorescence lifetime and the temperature, the temperature resolution of this fluorescent polymeric thermometer was evaluated to be 0.18–0.58 °C in the range of 29–39 °C [Figure 12.5(b)]. With the fluorescent polymeric thermometer, the nucleus and centrosome were clearly warmer than the cytoplasm in COS7 cells [Figure 12.5(c)]. In addition, local thermogenesis near mitochondria was captured [Figure 12.5(d)].

In the above works, the microinjection technique was adopted to introduce fluorescent thermometers into mammalian cells. However, microinjection cannot be applied to yeast cells because of the cell wall, which prevents the ingression of a glass needle. A cationic fluorescent polymeric thermometer NN-AP2.5 was developed [Figure 12.6(a)] to solve this problem and establish intracellular temperature measurement of yeast cells.[35] NN-AP2.5 rapidly (within 20 min) entered into cells of the yeast *Saccharomyces cerevisiae* strain SYT001 and remained in the cytoplasm [Figure 12.6(b)]. The fluorescence properties of NN-AP2.5 enabled sensitive intracellular thermometry of yeast cells [Figure 12.6(c)]. Spontaneous entry of the cationic thermometer was also observed in the mammalian MOLT-4 (human acute lymphoblastic leukaemia) and HEK293T (human embryonic kidney) cells.[35]

12.2.1.4 DNA

DNA has become an interesting and important macromolecule for both biology and nanotechnology.[36] Ke *et al.* reported an L-DNA (the enantiomeric form of natural D-DNA)-based molecular beacon (L-MB) as a fluorescent thermometer.[37] L-MB is a hairpin-structured dual-labelled oligonucleotide, and the distance between the fluorophore and quencher varies with temperature [Figure 12.7(a)]. L-MB transfected into HeLa cells accumulated in the nucleus and became highly fluorescent at higher temperatures [Figure 12.7(b)]. The utilisation of non-natural L-DNA is crucial, as the D-DNA-based molecular beacon (D-MB) did not exhibit any temperature-dependent changes, likely due to its rapid digestion by endogenous nucleases [Figure 12.7(c)].

Figure 12.4 Intracellular thermometry of living COS7 cells using a fluorescent nanogel thermometer. (a) Chemical structure and functional mechanism of a fluorescent nanogel thermometer. (b) Fluorescence images of a fluorescent nanogel thermometer in COS7 cells at a lower temperature (left) as well as at a higher temperature (right). (c) Response of the fluorescence intensity ($\lambda_{ex} = 488$ nm) of the fluorescent nanogel thermometer to the changes in temperature and its temperature resolution in COS7 cells. (d) Effect of treatment with an uncoupler reagent FCCP on the intracellular temperature of COS7 cells. Changes in the fluorescence intensity (left axis) and the corresponding temperature (right axis) after the addition of FCCP (100 µM, $n = 3$, mean ± s.d.). Adapted with permission from ref. 32. © American Chemical Society 2009.

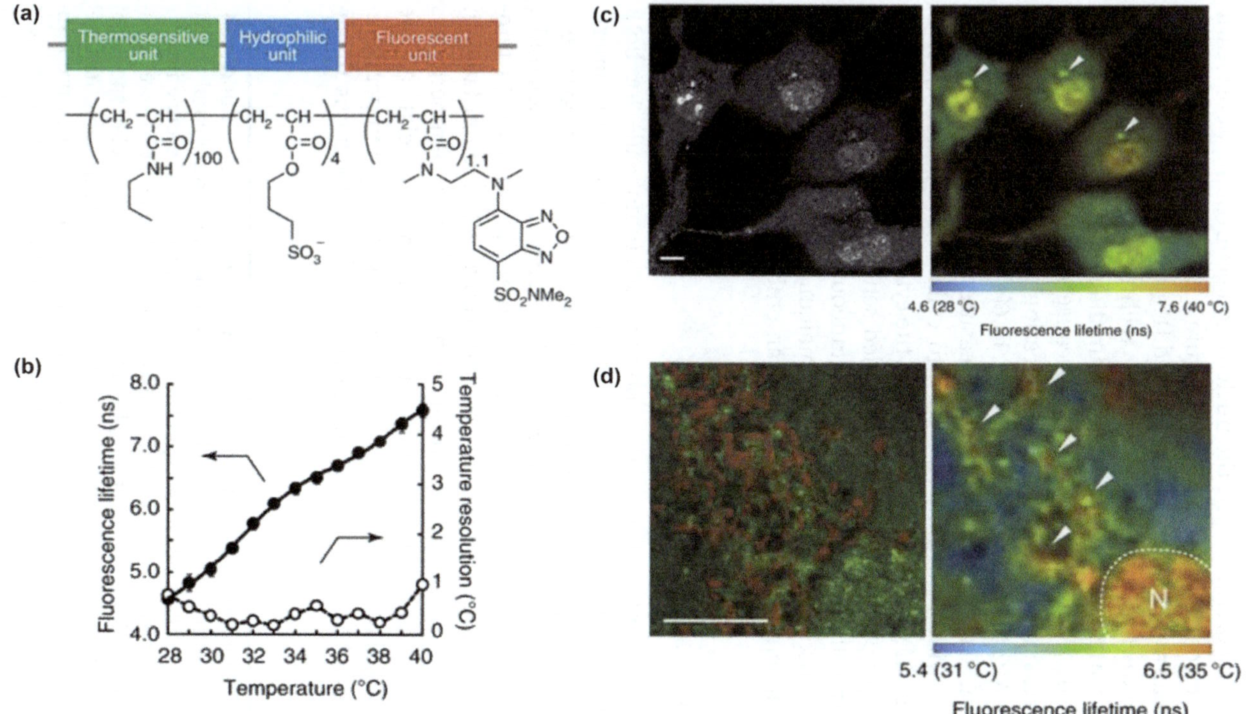

Figure 12.5 Intracellular temperature mapping of living COS7 cells using a fluorescent polymeric thermometer. (a) Chemical structure of the thermometer. (b) Response of the fluorescence lifetime ($\lambda_{ex} = 405$ nm) and the temperature resolution in a COS7 cell extract. (c) Confocal fluorescence image (left) and fluorescence lifetime image (right) of the fluorescent polymeric thermometer in COS7 cells. The centrosomes are indicated by arrowheads. (d) Enlarged confocal fluorescence image (left) of the thermometer (green) and mitochondria (red) and fluorescence lifetime image (right) of the thermometer in COS7 cells. Arrowheads indicate the local heat production near the mitochondria. N = nucleus. Adapted from ref. 34.

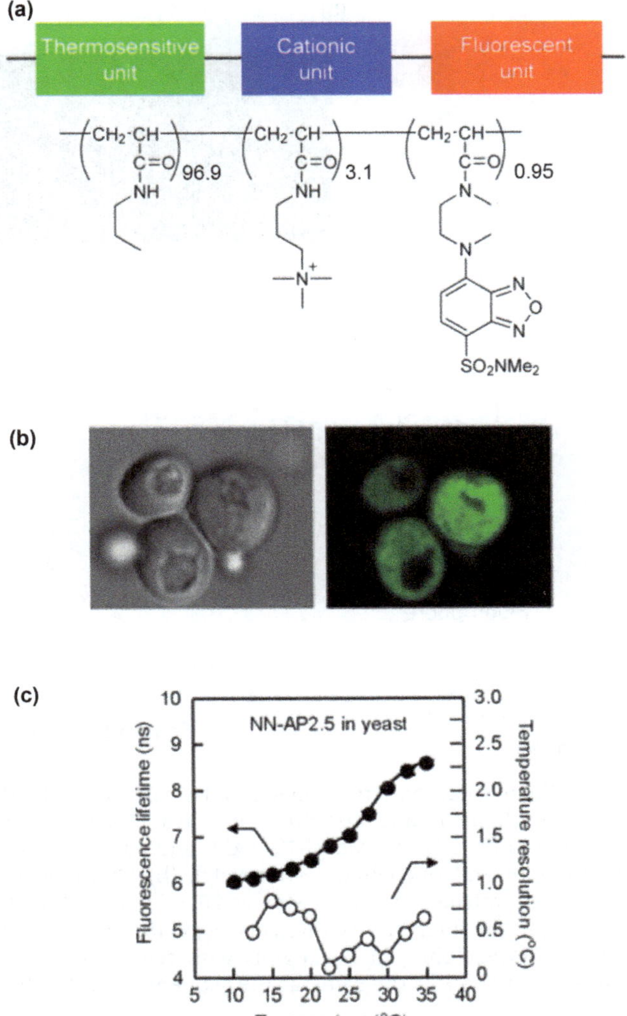

Figure 12.6 Intracellular thermometry of living yeast cells using a cationic fluorescent polymeric thermometer NN-AP2.5. (a) Chemical structure of NN-AP2.5. (b) DIC (left) and confocal fluorescence (right) images of yeast cells treated with NN-AP2.5. (c) Response of the fluorescence lifetime ($\lambda_{ex} = 456$ nm) and temperature resolution of NN-AP2.5 in yeast cells. Adapted with permission from ref. 35. © American Chemical Society 2013.

12.2.1.5 Green Fluorescent Protein

Although fluorescent thermometers based on green fluorescent protein (GFP) are highly desirable, as they are routinely used for cell biological analyses, GFPs generally exhibit only trivial sensitivity to temperature variation.[38] Donner *et al.* reported that the fluorescence polarisation anisotropy

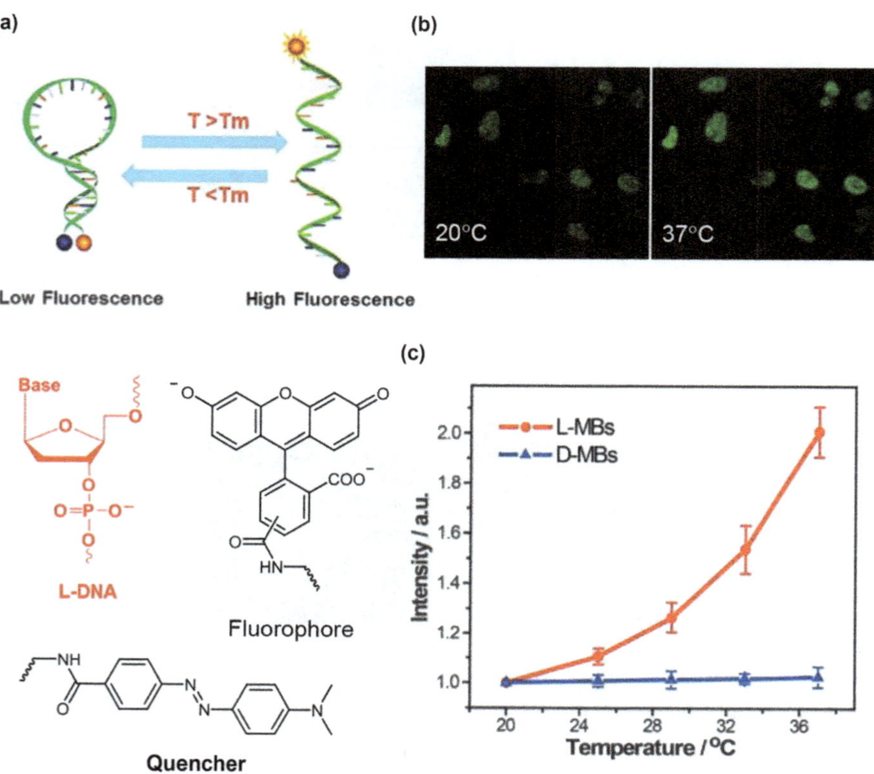

Figure 12.7 Intracellular thermometry of living HeLa cells using L-MB. (a) Functional mechanism and structure of L-MB (5′-FAM-CGAG TTT TTT TTT TTT TTT CTC G-Dabcyl-3′, FAM and Dabcyl units contain a fluorophore and quencher, respectively). (b) Fluorescence images of HeLa cells transfected with L-MBs at 20 °C (left) and 37 °C (right). (c) Changes in the fluorescence intensity (λ_{ex} = 488 nm) of L-MBs (red) and control D-MBs (blue) in HeLa cells with temperature variation. The sequence of D-MB is the same as the sequence of L-MB but composed of D-DNA.
Adapted with permission from ref. 37. © American Chemical Society 2012.

(FPA) of GFP was a measurable temperature-dependent parameter inside living HeLa cells, U-87 MG (human glioblastoma-astrocytoma) cells[39] and *Caenorhabditis elegans*[40] (Figure 12.8).

Kiyonaka *et al.* recently developed tsGFP1, a more sensitive GFP thermometer that consists of GFP and the temperature-sensitive protein TlpA derived from *Salmonella* bacterium. The temperature-dependent structural changes of TlpA enhance the temperature sensitivity of GFP (Figure 12.9).[38] They monitored changes in the temperature of the ER as well as of the mitochondria in HeLa cells by fusing organelle-targeting sequences to this GFP thermometer. As shown in Figures 12.9(d) and 12.9(e), thermogenesis by an uncoupler 3-chlorophenylhydrazone (CCCP) treatment was clearly

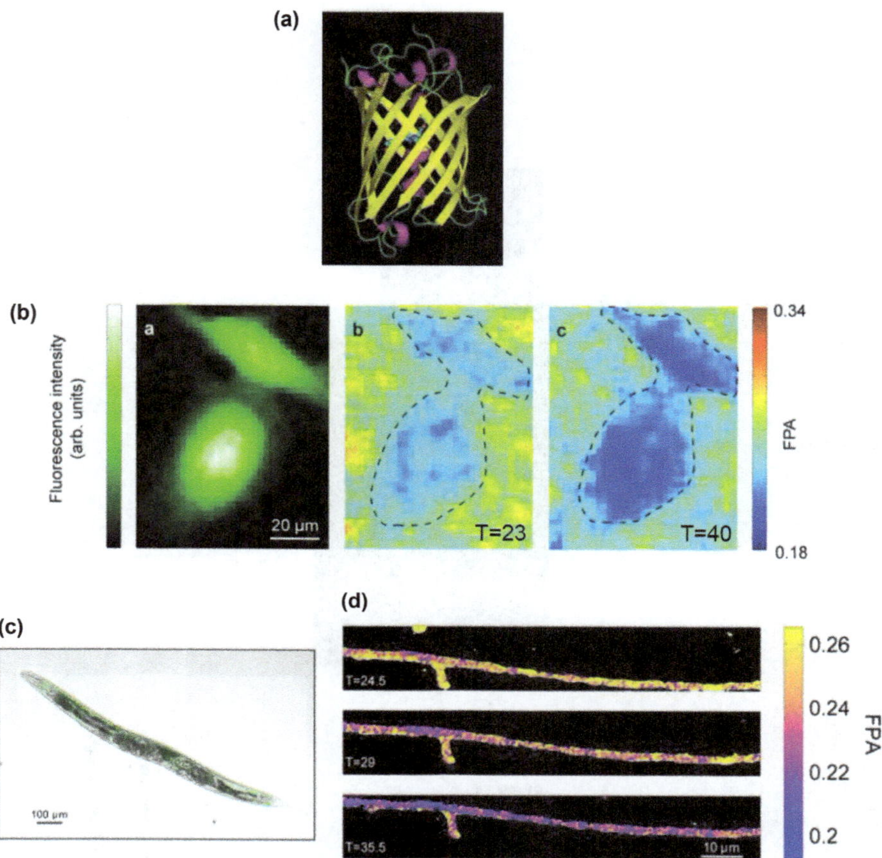

Figure 12.8 Thermometry of a living cell and living organism using GFP visualised by fluorescence anisotropic microscopy. (a) Structure of GFP. (b) Change in FPA of GFP in HeLa cells. a – Fluorescence image of GFP expressed in HeLa cells. b and c – FPA images at 23 °C and 40 °C, respectively. (c) A merged image of bright-field and GFP fluorescence expressed in GABAergic neurons of *C. elegans*. (d) FPA images of GFP expressed in *C. elegans* neurons at the different temperatures (24.5, 29 and 35.5 °C).
Adapted with permission from ref. 39 and 40. © American Chemical Society 2012, 2013.

observed near the mitochondria.[38] The intracellular temperatures of brown adipocytes and myotubes were also imaged.[38]

12.2.1.6 Quantum Dots

In general, the remarkable photostability of quantum dots is a considerable advantage for their use in bioimaging. Yang *et al.* used streptavidin-coated CdSe–ZnS core–shell quantum dots (QD655) as luminescent thermometers

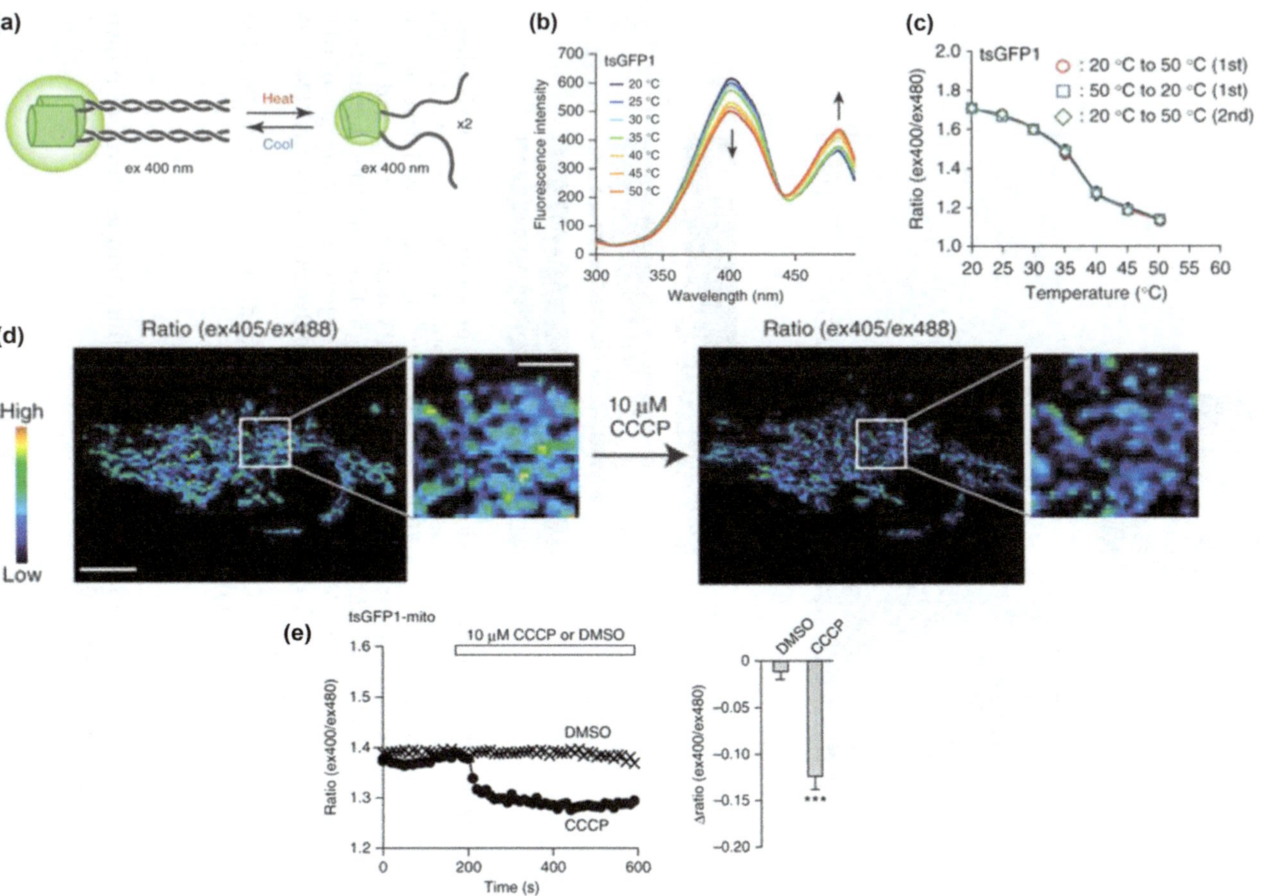
(a)
Heat
Cool
ex 400 nm
ex 400 nm
x2
(b)
tsGFP1
20 °C
25 °C
30 °C
35 °C
40 °C
45 °C
50 °C
Fluorescence intensity
700
600
500
400
300
200
100
0
300
350
400
450
Wavelength (nm)
(c)
tsGFP1
: 20 °C to 50 °C (1st)
: 50 °C to 20 °C (1st)
: 20 °C to 50 °C (2nd)
Ratio (ex400/ex480)
2.0
1.8
1.6
1.4
1.2
1.0
20 25 30 35 40 45 50 55 60
Temperature (°C)
(d)
Ratio (ex405/ex488)
Ratio (ex405/ex488)
High
Low
10 µM
CCCP
(e)
tsGFP1-mito
10 µM CCCP or DMSO
Ratio (ex400/ex480)
1.6
1.5
1.4
1.3
1.2
DMSO
CCCP
0
200
400
600
Time (s)
DMSO
CCCP
Δratio (ex400/ex480)
0
-0.05
-0.10
-0.15
-0.20
***

for intracellular thermometry.[41] QD655 exhibited a temperature-dependent shift in the maximum emission wavelength (*i.e.*, 652 nm at 20 °C and 653 nm at 40 °C) as the temperature affected the lattice of the quantum dots and the electron–lattice interactions.[42] The same research groups later reported the imaging of local heterogeneous thermogenesis in NIH/3T3 (mouse embryonic fibroblast) cells induced by a calcium ion stress (Figure 12.10) and cold-shock by using QD655.[43]

Albers *et al.* prepared a more complex luminescent thermometer based on a CdSe–CdS quantum dot–quantum rod, in which a Förster resonant energy transfer was involved with a cyanine dye as the acceptor [Figure 12.11(a)].[44] This quantum dot-based thermometer was introduced into the cytoplasm of HeLa and NIH/3T3 cells, supported by a cationic polymer colloid. Figure 12.11(b) and (c) display the response of the quantum dot-based thermometer in HeLa cells.

12.2.1.7 Inorganic Materials

The use of inorganic species as luminescent thermometers has also been reported. For example, $NaYF_4{:}Er^{3+},Yb^{3+}$ nanoparticles, for which temperature sensitivity is derived from the competition of two radiative excited states ($^2H_{11/2}$ and $^4S_{3/2}$), exhibit a temperature-dependent change in luminescent intensity in HeLa cells.[45] More recently, Shang *et al.* reported intracellular thermometry using fluorescent gold nanoclusters (diameter: 1.6 ± 0.3 nm).[46] The fluorescent gold nanoclusters, which were protected by lipoic acid to bear colloidal stability in biological media, exhibited a reduction in both the temperature intensity and fluorescence lifetime with increasing temperature [Figure 12.12(a)]. Intracellular temperature imaging of HeLa cells was performed with fluorescence lifetime imaging microscopy after the fluorescent gold nanoclusters were internalised by endocytosis [Figures 12.12(b) and 12.12(c)].

In 2013, three independent research groups consecutively reported thermometry using nitrogen-vacancy colour centres in diamond nanocrystals (nanodiamonds).[47–49] These nanodiamonds are potentially the most

---

Figure 12.9 Monitoring of intracellular thermogenesis using a GFP-based thermosensor tsGFP1. (a) Schematic diagram of tsGFP1, which consists of GFP (green) and coiled–coil regions of TlpA (grey). The tandem formation/deformation of the coiled–coil structure upon changes in the temperature resulting in the fluorescence change of tsGFP1. (b) Fluorescence excitation spectra of tsGFP1 ($\lambda_{em} = 510$ nm). (c) Reversibility in the fluorescence intensity ratio of tsGFP1. (d) Pseudocolour confocal images of the fluorescence intensity ratio in HeLa cells expressing mitochondria-targeted tsGFP1 after the treatment with the uncoupler CCCP. Scale bars = 10 μm (entire image) and 3 μm (inset). (e) Change in the fluorescence intensity ratio of mitochondria-targeted tsGFP1 after a treatment with CCCP or DMSO (control). Adapted with permission from ref. 38. © Macmillan Publishers 2013.

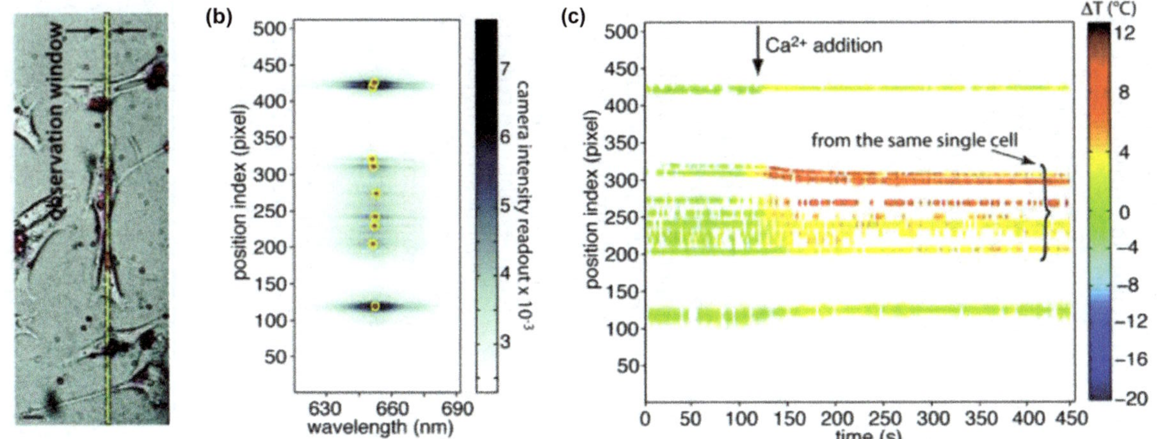

Figure 12.10 Intracellular thermometry of living NIH/3T3 cells using streptavidin-coated quantum dots QD655. (a) Merged image of bright-field and emission of QD655 in NIH/3T3 cells. The vertical lines indicate an experimental observation window. Scale bar = 20 μm. (b) Position-spectrum map of the sample shown in (a). (c) Location-dependent intracellular temperature progression after the addition of an ionomycin–Ca^{2+} complex.
Adapted with permission from ref. 43. © American Chemical Society 2011.

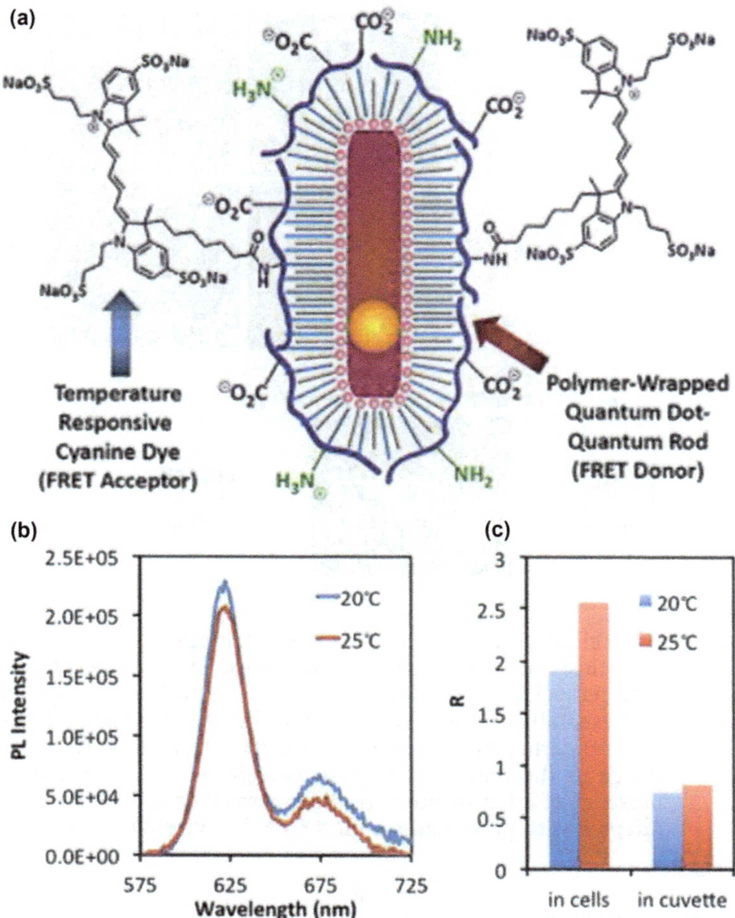

Figure 12.11 Quantum dot/quantum rod-based nanothermometer for intracellular thermometry. (a) Chemical structure of the nanothermometer. A red-emitting CdSe–CdS quantum dot–quantum rod semiconductor nanocrystal (centre) was passivated with an amphiphilic polymer shell and a far-red emitting cyanine dye. (b) Luminescence spectra in live HeLa cells. (c) Fluorescence ratios ($R = \mathrm{FI}_{630-640}/\mathrm{FI}_{664-674}$) in live HeLa cells and 100 mM bicarbonate buffer (pH 8.3).
Adapted with permission ref. 44. © American Chemical Society 2012.

sensitive luminescent thermometers ever reported, with a temperature resolution of 0.0018 °C.[47] In the nanodiamonds, the ground state is split into two energy levels, and the energy difference between the two levels is temperature dependent. One of the three research groups, Kucsko *et al.*, performed intracellular thermometry of a human embryonic fibroblast, in which the temperature at local points in living cells was increased by laser heating (Figure 12.13).[47]

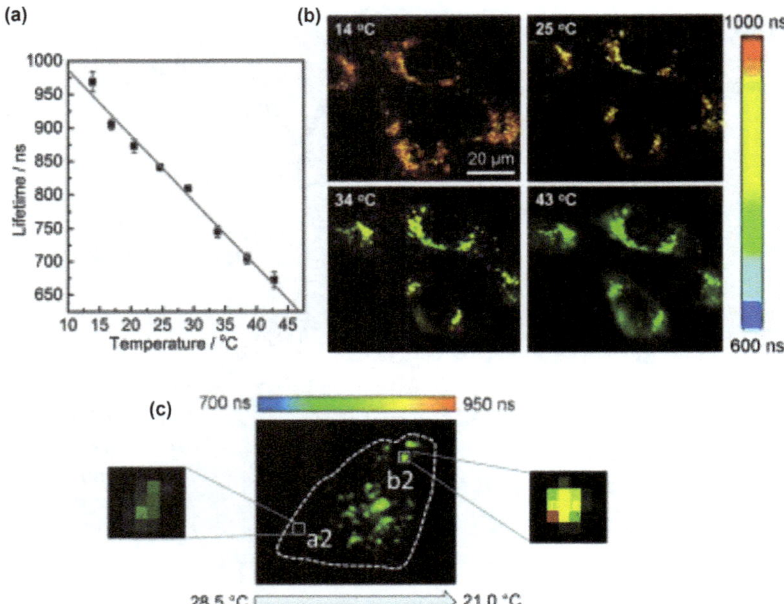

Figure 12.12 Intracellular thermometry of living HeLa cells using fluorescent gold nanoclusters. (a) Relationship between the fluorescence lifetime of the gold nanoclusters in HeLa cells ($\lambda_{\mathrm{ex}} = 470\,\mathrm{nm}$) and temperature. (b) Representative fluorescence lifetime images of HeLa cells internalising the fluorescent gold nanoclusters. (c) Fluorescence lifetime image of the fluorescence gold nanoclusters in a HeLa cell in the presence of a temperature gradient across the cell.
Adapted with permission from ref. 46. © Wiley-VCH 2013.

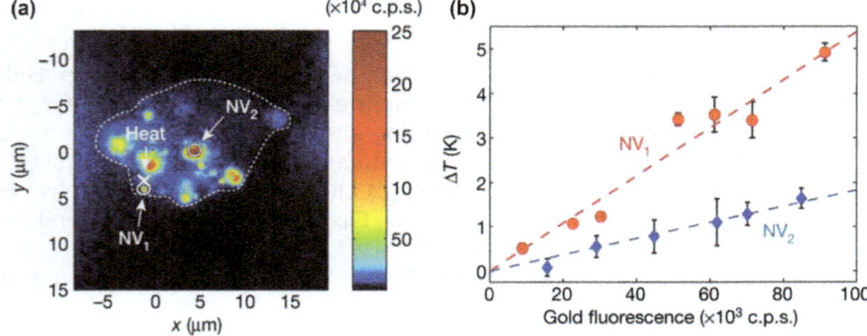

Figure 12.13 Nitrogen-vacancy centres in diamond nanocrystals (nanodiamonds) for intracellular temperature measurements. (a) A confocal image of the nanodiamonds in a WS1 cell (the cell shape is outlined with a dotted line). The cross indicates the position of a gold nanoparticle subjected to heating. The circles indicate the locations of the nanodiamonds (NV_1 and NV_2) used to measure the temperature. (b) Changes in the intracellular temperature at NV_1 and NV_2 with variations in the laser power applied to the gold nanoparticle.
Adapted with permission from ref. 47. © Macmillan Publishers 2013.

12.2.2 Others

In addition to fluorescent (or luminescent) thermometers, we can find few, but important additional technologies used to achieve cellular thermometry in the recent literature. Inomata *et al.* developed an extremely small calorimeter with a detection limit of 5.2 pJ, which can detect the heat released from a single cell [Figure 12.14(a) and (b)].[50] A heat release from a brown fat cell was observed after treatment of cells with norepinephrine [Figure 12.14(c)]. Wang *et al.* fabricated a sub-micron thermocouple probe consisting of tungsten, polyurethane, and a platinum film [Figure 12.15(a)].[51] This thermocouple probe can distinguish temperature differences of less than 0.1 °C. Using this thermocouple method, an increase in the U251 (human astrocytoma) cell temperature was observed upon

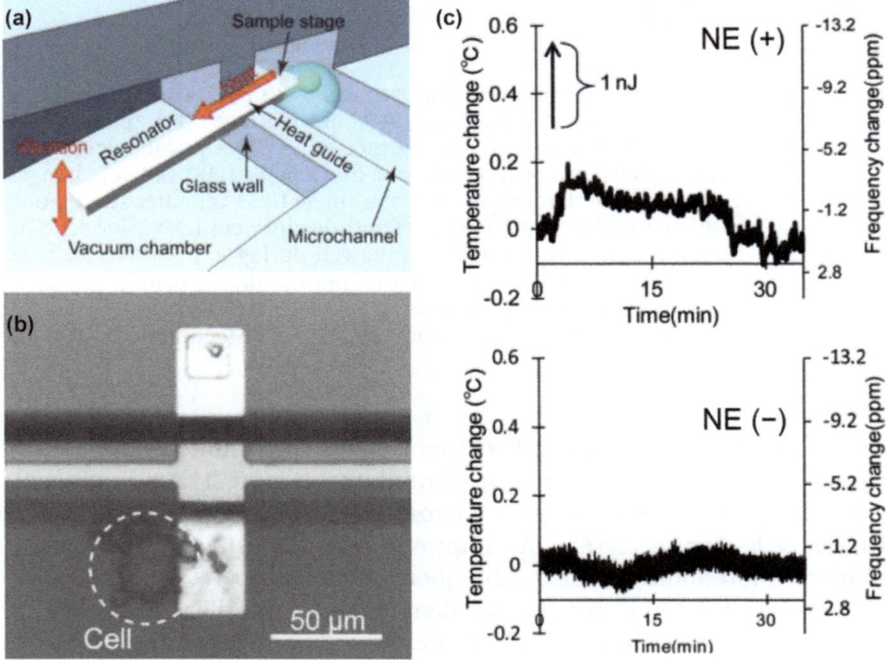

Figure 12.14 Cellular thermometry of a living brown fat cell by a pico calorimeter based on a resonant thermal sensor. (a) Schematic diagram of the resonant thermal sensor. The resonator (cantilevered Si sensor) is enclosed in the vacuum chamber, and the heat from the sample in a microchannel is conducted to the resonator through a sample stage. The resultant temperature change causes a shift in the resonant frequency of the resonator. (b) A brown fat cell attached to the sample stage. (c) Changes in the temperature of a brown fat cell by the stimulus with (upper, +) or without (lower, −) norepinephrine (NE).

Adapted with permission from ref. 50. © AIP Publishing 2012.

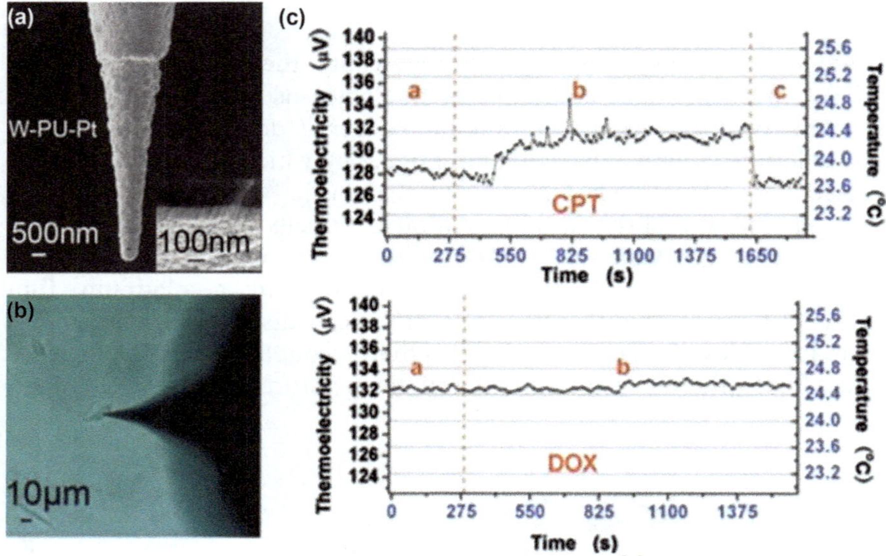

Figure 12.15   Intracellular thermometry of a U251 cell by a thermocouple consisting of tungsten (W), polyurethane (PU), and platinum (Pt). (a) Scanning electron micrograph of the thermocouple. (b) Optical microscopic image of the thermocouple inserted into a living U251 cell. (c) Changes in the intracellular temperature of a single U251 cell after stimulation with camptothecin (CPT, upper) and doxorubicin (DOX, lower). The thermocouple was inserted into the cell during a, b – CPT and DOX were added to the culture medium, and the thermocouple was withdrawn from the cell in duration c.
Adapted with permission from ref. 51. © Macmillan Publishers 2011.

treatment with camptothecin (a DNA topoisomerase I inhibitor) but not with doxorubicin (a DNA topoisomerase II inhibitor) [Figure 12.15(b) and (c)]. Gao *et al.* applied a photoacoustic microscopy for the temperature imaging of HeLa cells (Figure 12.16).[52] An improved photoacoustic method using a reference fluorophore was also subsequently developed.[53]

Environmental temperature-dependent changes in the production of biomolecules have been used to evaluate the cellular temperature. McCabe *et al.* introduced β-galactosidase, which was regulated by the temperature-dependent LacI(Ts) promoter in *Escherichia coli* bacterium and detected by a fluorescent substrate for β-galactosidase. Stronger fluorescence was observed as *E. coli* cells were incubated at a higher temperature.[54] Chiu *et al.* observed a temperature-dependent production of ergosterol in living fission yeast cells, which was monitored as a Raman peak at 1602 cm^{-1}.[55] Ergosterol was produced abundantly in the yeast cells incubated at a temperature higher than 35 °C, resulting in a strong Raman signal at 1602 cm^{-1}. Temperatures lower than 35 °C hindered the ergosterol biosynthesis, decreasing the peak.[55]

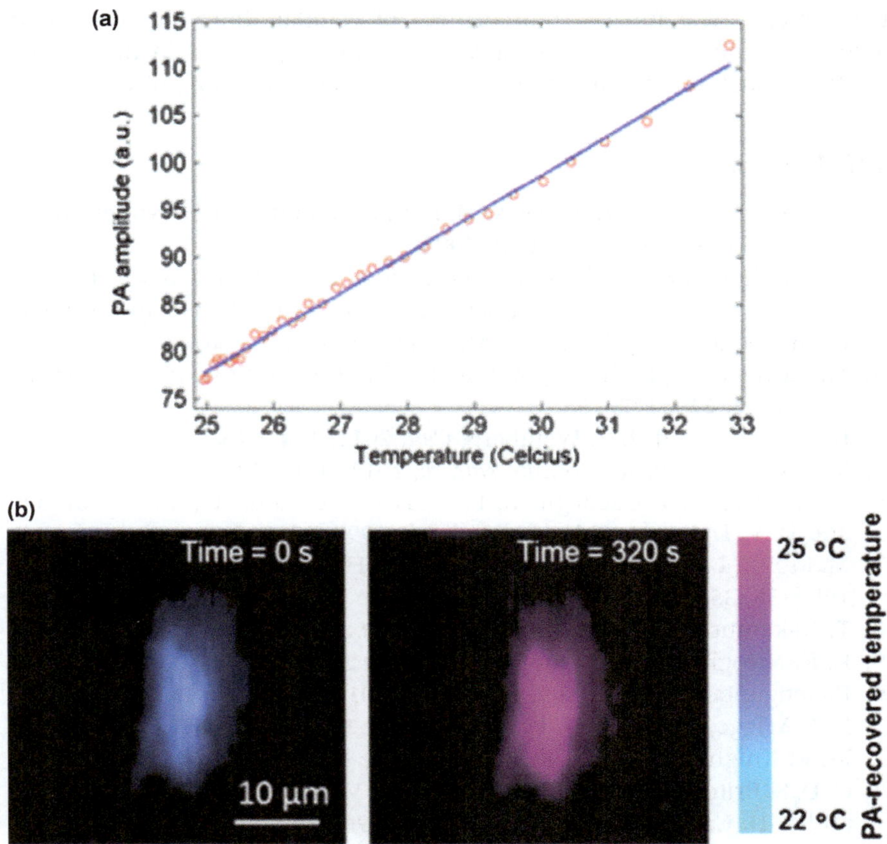

Figure 12.16 Intracellular thermometry with high-resolution photoacoustic microscopy. (a) Relationship between photoacoustic amplitude and temperature in a single HeLa cell. (b) Temperature sensing of a single HeLa cell during photothermal heating by laser irradiation (532 nm). Adapted from ref. 52.

12.3 Closing Remarks

As summarised in this chapter, the number of studies on the development of cellular thermometry has increased in recent years. This increase indicates that considerable attention has been paid to cellular thermometry. In fact, novel information, *i.e.*, the changes in the cellular temperature induced by chemical and physical stimuli, has been provided in numerous studies, although its biological importance has not been fully discussed. The use of cellular thermometry to solve biological questions in various samples will be expected in near future.

Current cellular thermometry already has sufficient spatial and temperature resolution, and some techniques display the ability to target sub-cellular areas in biological samples. Thus, the comparison of intracellular temperatures among different cell types is one of the next steps for cellular thermometry. The further improvement of cellular thermometry and its biological applications must greatly contribute to progress in life science and thus contribute to our health beyond the understanding of cells.

References

1. B. Alberts, A. Johnson, J. Lewis, M. Raff, K. Roberts and P. Walter, in *The Cell*, Garland Science, 5th edn, 2008.
2. R. Malkin and K. K. Niyogi, in *Biochemistry & Molecular Biology of Plants*, ed. B. B. Buchanan, W. Gruissem and R. L. Jones, American Society of Plant Physiologists, Rockville, Maryland, 2000, pp. 568–628.
3. M. Karnebogen, D. Singer, M. Kallerhoff and R.-H. Ringert, *Thermochim. Acta*, 1993, **229**, 147–155.
4. D. Hanahan and R. A. Weinberg, *Cell*, 2011, **144**, 646.
5. S. B. Vafai and V. K. Mootha, *Nature*, 2012, **491**, 374.
6. L. Chaerle, W. V. Caeneghem, E. Messens, H. Lambers, M. V. Montagu and D. V. D. Straeten, *Nat. Biotechnol.*, 1999, **17**, 813.
7. M. Monti, L. Brandt, J. Ikomi-kumm and H. Olsson, *Scand. J. Haematol.*, 1986, **36**, 353.
8. T. Nakamura and I. Matsuoka, *J. Biochem.*, 1978, **84**, 39.
9. K. K. Niyogi and T. B. Truong, *Curr. Opin. Plant Biol.*, 2013, **16**, 307.
10. P. Sengupta and P. Garrity, *Curr. Biol.*, 2013, **23**, R304.
11. P. A. Wigge, *Curr. Opin. Plant Biol.*, 2013, **16**, 661.
12. M. R. Knight and H. Knight, *New Phytol.*, 2012, **195**, 737.
13. C. D. S. Brites, P. P. Lima, N. J. O. Silva, A. Millán, V. S. Amaral, F. Palacio and L. D. Carlos, *Nanoscale*, 2012, **4**, 4799.
14. D. Jaque and F. Vetrone, *Nanoscale*, 2012, **4**, 4301.
15. X.-D. Wang, O. S. Wolfbeis and R. J. Meier, *Chem. Soc. Rev.*, 2013, **42**, 7834.
16. S. Fery-Forgues, J.-P. Fayet and A. Lopez, *J. Photochem. Photobiol. A*, 1993, **70**, 229.
17. C. F. Chapman, Y. Liu, G. J. Sonek and B. J. Tromberg, *Photochem. Photobiol.*, 1995, **62**, 416.
18. H. Huang, S. Delikanli, H. Zeng, D. M. Ferkey and A. Pralle, *Nat. Nanotechnol.*, 2010, **5**, 602.
19. M. J. Snare, F. E. Treloar, K. P. Ghiggino and P. J. Thistlethwaite, *J. Photochem.*, 1982, **18**, 335.
20. D. Ross, M. Gaitan and L. E. Locascio, *Anal. Chem.*, 2001, **73**, 4117.
21. R. K. Benninger, Y. Koç, O. Hofmann, J. Requejo-Isidro, M. A. A. Neil, P. M. W. French and A. J. deMello, *Anal. Chem.*, 2006, **78**, 2272.
22. C. Paviolo, A. H. A. Clayton, S. L. Mcarthur and P. R. Stoddart, *J. Microscopy*, 2013, **250**, 179.
23. G. A. Crosby, R. E. Whan and R. M. Alire, *J. Chem. Phys.*, 1961, **34**, 743.
24. O. Zohar, M. Ikeda, H. Shinagawa, H. Inoue, H. Nakamura, D. Elbaum, D. L. Alkon and T. Yoshioka, *Biophy. J.*, 1998, **74**, 82.
25. M. Suzuki, V. Tseeb, K. Oyama and S. Ishiwata, *Biophy. J.*, 2007, **92**, L46.

26. K. Oyama, M. Takabayashi, Y. Takei, S. Arai, S. Takeoka, S. Ishiwata and M. Suzuki, *Lab Chip*, 2012, **12**, 1591.
27. Y. Takei, S. Arai, A. Murata, M. Takabayashi, K. Oyama, S. Ishiwata, S. Takeoka and M. Suzuki, *ACS Nano*, 2014, **8**, 198.
28. J. V. Houten and R. J. Watts, *J. Am. Chem. Soc.*, 1976, **98**, 4853.
29. L. Yang, H.-S. Peng, H. Ding, F.-T. You, L.-L. Hou and F. Teng, *Microchim. Acta*, 2014, **181**, 743.
30. S. Uchiyama, Y. Matsumura, A. P. de Silva and K. Iwai, *Anal. Chem.*, 2003, **75**, 5926.
31. V. Tseeb, M. Suzuki, K. Oyama, K. Iwai and S. Ishiwata, *HFSP J.*, 2009, **3**, 117.
32. C. Gota, K. Okabe, T. Funatsu, Y. Harada and S. Uchiyama, *J. Am. Chem. Soc.*, 2009, **131**, 2766.
33. S. Uchiyama, K. Kimura, C. Gota, K. Okabe, K. Kawamoto, N. Inada, T. Yoshihara and S. Tobita, *Chem. – Eur. J.*, 2012, **18**, 9552.
34. K. Okabe, N. Inada, C. Gota, Y. Harada, T. Funatsu and S. Uchiyama, *Nat. Commun.*, 2012, **3**, 705.
35. T. Tsuji, S. Yoshida, A. Yoshida and S. Uchiyama, *Anal. Chem.*, 2013, **85**, 9815.
36. Y.-C. Hung, D. M. Bauer, I. Ahmed and L. Fruk, *Methods*, 2014, **67**, 105.
37. G. Ke, C. Wang, Y. Ge, N. Zheng, Z. Zhu and C. J. Yang, *J. Am. Chem. Soc.*, 2012, **134**, 18908.
38. S. Kiyonaka, T. Kajimoto, R. Sakaguchi, D. Shinmi, M. Omatsu-Kanbe, H. Matsuura, H. Imamura, T. Yoshizaki, I. Hamachi, T. Morii and Y. Mori, *Nat. Methods*, 2013, **10**, 1232.
39. J. S. Donner, S. A. Thompson, M. P. Kreuzer, G. Baffou and R. Quidant, *Nano Lett.*, 2012, **12**, 2107.
40. J. S. Donner, S. A. Thompson, C. Alonso-Ortega, J. Morales, L. G. Rico, S. I. C. O. Santos and R. Quidant, *ACS Nano*, 2013, 7, 8666.
41. J.-M. Yang, H. Yang and L. Lin, *Proceedings of 23rd IEEE Micro Electro Mechanical Systems Conference*, 2010, 963.
42. S. Li, K. Zhang, J.-M. Yang, L. Lin and H. Yang, *Nano Lett.*, 2007, 7, 3102.
43. J.-M. Yang, H. Yang and L. Lin, *ACS Nano*, 2011, 5, 5067.
44. A. E. Albers, E. M. Chan, P. M. McBride, C. M. Ajo-Franklin, B. E. Cohen and B. A. Helms, *J. Am. Chem. Soc.*, 2012, **134**, 9565.
45. F. Vetrone, R. Naccache, A. Zamarrón, A. J. de la Fuente, F. Sanz-Rodríguez, L. M. Maestro, E. M. Rodriguez, D. Jaque, J. G. Solé and J. A. Capobianco, *ACS Nano*, 2010, **4**, 3254.
46. L. Shang, F. Stockmar, N. Azadfar and G. U. Nienhaus, *Angew. Chem. Int. Ed.*, 2013, **52**, 11154.
47. G. Kucsko, P. C. Maurer, N. Y. Yao, M. Kubo, H. J. Noh, P. K. Lo, H. Park and M. D. Lukin, *Nature*, 2013, **500**, 54.
48. D. M. Toyli, C. F. de las Casas, D. J. Christle, V. V. Dobrovitski and D. D. Awschalom, *Proc. Natl. Acad. Sci.*, 2013, **110**, 8417.

49. P. Neumann, I. Jakobi, F. Dolde, C. Burk, R. Reuter, G. Waldherr, J. Honert, T. Wolf, A. Brunner and J. H. Shim, *Nano Lett.*, 2013, **13**, 2738.
50. N. Inomata, M. Toda, M. Sato, A. Ishijima and T. Ono, *Appl. Phys. Lett.*, 2012, **100**, 154104.
51. C. Wang, R. Xu, W. Tian, X. Jiang, Z. Cui, M. Wang, H. Sun, K. Fang and N. Gu, *Cell Res.*, 2011, **21**, 1517.
52. L. Gao, L. Wang, C. Li, Y. Liu, H. Ke, C. Zhang and L. V. Wang, *J. Biomed. Optics*, 2013, **18**, 026003.
53. L. Gao, C. Zhang, C. Li and L. V. Wang, *Appl. Phys. Lett.*, 2013, **102**, 193705.
54. K. M. McCabe, E. J. Lacherndo, I. Albino-Flores, E. Sheehan and M. Hernandez, *Appl. Environ. Microbiol.*, 2011, 77, 2863.
55. Y.-F. Chiu, C.-K. Huang and S. Shigeto, *ChemBioChem*, 2013, **14**, 1001.

Thermal Issues in Microelectronics

X. PERPIÑÀ, M. VELLVEHI* AND X. JORDÀ

Institut de Microelectrònica de Barcelona, Centre Nacional de
Microelectrònica (IMB-CNM,CSIC), Campus U.A.B 08193 Cerdanyola del
Vallès, Barcelona, Spain
*Email: miquel.vellvehi@csic.es

13.1 Introduction

Electronic systems are present in our everyday lives. In many applications,
they have complemented or replaced functions performed by other systems
based on mechanic, electromechanic, hydraulic, and pneumatic principles,
as is the case for modern electronic aircraft.[1] This has enhanced innovation
and competitiveness in all sectors of the economy, having a positive impact
on our lives. Another factor responsible for the success of electronic systems
relies on the fact that the interaction between electronic systems and
humans has been enriched by the integration of a generation of smart
low-cost sensors. Apart from smart mobile phones, this integration has
made wearable devices possible, which will give rise to a new commercial
family of products.

These advances have been possible thanks to the microelectronic revolution that occurred in the fifties and sixties, which was searching for more
reliable and rugged electronic systems.[2–4] From then, its growth has been
based on a recipe that combines market demands with technological
development, as stated by Gordon E. Moore.[5] Moore warrants that the
functionality of any digital processing electronic system can be achieved by

RSC Nanoscience & Nanotechnology No. 38
Thermometry at the Nanoscale: Techniques and Selected Applications
Edited by Luís Dias Carlos and Fernando Palacio

Published by the Royal Society of Chemistry, www.rsc.org

increasing the device integration density, which in turn reduces the manufacturing cost. Moreover, it can be substantially improved by increasing the operating frequency of the microelectronic device (clock frequency), making the integration of smaller passive elements, for instance capacitors, also feasible. This scenario has forced a continuous scaling down of element size and increase in operating frequencies of microelectronic devices for digital processing. As a result, the transition from the microelectronic to the nano-electronic world to reach operating frequencies in the GHz range has been achieved. This has given rise to the so-called *More Moore domain*, which is an attempt to develop advanced Complementary Metal Oxide–Semi-conductor (CMOS) technologies. Currently, the technology most suited to developing such digital processing blocks is based on the CMOS process. However, the continuous scaling down has positioned this technology at its limits. Several alternatives are being envisaged to replace CMOS technology as a new platform to obtain novel information processing systems, which are referred to as *Beyond CMOS technologies*. They are based on emerging sub-micronic devices that process information in an alternative way using new state variables (*e.g.*, spin orientation, molecular state, phase state), which exploit other physical phenomena to compute data (*i.e.*, spintronics, molecular, ferromagnetic).[6,7] Despite of this scaling-down career in electronic devices, one of the main achievements of the last decade is to integrate other blocks with different functionalities into electronic systems, along with digital processing sub-systems to create an added value on the final system (*More than Moore domain*). For instance, this is the case when integrating wireless communication, sensors and actuators, power management and lighting in a single electronic system. The integration of several elements belonging to digital and non-digital processing to add more functionality and make the final electronic product more attractive for the final user (*Heterogeneous Integration domain*) can be performed in a given chip, using a System-on-Chip (SoC) approach, or using packaging technologies, *i.e.*, a System-in-Package (SiP) approach.

Electronic systems, as well as their constituting blocks, should meet several requirements regarding their design for manufacturability (optimized fabrication process of all parts), reliability (virtual prototyping), and testability (system assessing), as a consequence of the current nano/microelectronic trends: device miniaturization; operating frequency increase; and heterogeneous integration. Device miniaturization entails an increase in the integration density of elementary functions (data processing, energy management, sensing capability, signal transmission/reception), but it also requires a decrease in the power handled and dissipated within each device as the heat flux increases. However, in the long term, it is observed that the density growth rate exceeds the local dissipation rate, so that there is an unavoidable increase in the density of heat dissipated in the bulk in objects produced by the electronic industry,[8] as can be observed in the More Moore and More than Moore domains. Increasing frequency is a common trend in many electronic systems. Aside from increasing the digital information processing capability, it also allows the final size of electronic systems to decrease, as in the case of power converters. Something similar occurs in wireless communications. In many cases, antenna size fixes the operating frequency, as it is proportional to the operating wavelength, while another constraint comes from a solution to overcrowding of the electromagnetic spectrum in free space (*spectrum crunch*).[8] New frequency spectra are under study to relieve this situation, but they also aim at increasing such operating frequencies. Thus, the thermal effect of such systems will increase much more than their power dissipation, and at the same time it will be difficult to monitor their figures of merit (FOMs). Several solutions at die level require designing multicore processors to facilitate heat extraction; elementary transistors and junctions use shapes, concepts and materials designed to reduce consumption and dissipation.[9] Moreover, substrates themselves include vertical structuring to reduce consumption during transistor switching (*e.g.*, Silicon-on-Insulator, SoI), which sometimes increases their thermal resistance.[9] Finally, heterogeneous integration leads electronic systems into a new scenario, where the heat flux dissipated in a die SoC is increased, and a solution using the SiP approach is claimed to improve the thermal management of new electronic systems.

In this scenario, temperature measurements are crucial for facing challenges related to the thermal management of the design, reliability and testability of the final system. Temperature is a parameter that assists in the design of many micro/nano-electronic systems. Its monitoring is used to assess the thermal management design adopted, to detect failures in the system studied and to determine information about their electrical behaviour (*e.g.*, electrical FOM extraction in high-frequency applications, or monitoring of uneven electric behaviour in power devices). Several approaches have been developed according to the monitoring required. Prior to describing all the techniques used for this purpose, some of the current semi-conductor issues are revised. Following that, every thermal characterization technique used in micro/nano-electronics will be outlined, indicating their advantages and drawbacks. Finally, the applicability of each technique described will be revised. This chapter will conclude with final remarks about all of the challenges foreseen in upcoming electronic products, and by summarizing the performance of all the techniques described.

13.2 Heat Generation and Testability Issues at the Chip Level for Signal Processing

13.2.1 Shrinkage and High-density Integration Effects in Heat Generation at the Chip Level

In the More than Moore domain, tremendous progress in the processing power of integrated circuits (the 'brain' inside electronic devices) has been made, such that complex smart systems have become viable. Downscaling, along with the operating frequency rise, have been the most important and effective ways to achieve high-performance digital processing blocks. However, several factors have slowed the pace of the seventies, eighties and nineties: the tremendous cost increase in lithography, difficulties in developing new technologies, expected large variations in electrical characteristics of the current smaller geometry transistors,[11] and power consumption or heat removal in integrated devices. Device downscaling has been conceived according to the method proposed by Dennard *et al.*,[10] which increases the performance and integration without any increase in power consumption as long as the chip area is kept constant. However, the real scaling trend for the 30 years following 1970 has been much more aggressive than that fixed by Dennard's method: while the gate length, junction depth, and gate oxide thickness decreased by $100\times$, the supply voltage decreased only $10\times$, the chip area increased $10\times$,[11] and the clock frequency increased $1000\times$ (only $10\times$ was expected from the ideal scaling). As a result, power consumption has been scaled by a factor of $100\,000\times$,[12] making it a real concern in several microprocessor technologies during the noughties. This fact has produced several technological efforts in recent years to reduce heat generation in such devices by decreasing the following parameters: power supply voltage, device threshold voltage, gate leakage (use of high-k dielectrics), both series resistance and channel mobility (silicon–germanium strained technology) and junction tunnelling current (short-channel effects, multigate structures).[13] However, device downsizing not only poses technological problems and increases heat flux, it also changes the local thermal conductivity of integrated devices.[14] At the device level, heat dissipation is localized in hot spots formed in the transistor drain region, which may increase the drain series and source injection electrical resistances, degrading their final performance (*e.g.*, timing failures[15]) and reliability.[15] In CMOS technologies close to 10 nm, the introduction of novel materials and non-traditional transistor geometries, including ultrathin body,[13] Fin Field-Effect Transistors (FinFETs) (multigate structures), and nanowire devices, impedes heat conduction decreasing local thermal conductivity.[15,16] This is due to the fact that the classical heat conduction by diffusion becomes ballistic at the nanoscale range and such an exchange mechanism impedes cooling of hot spots through material boundaries.[15] Moreover, the increased surface-to-volume ratio of novel transistor designs also leads to a larger contribution from material boundary thermal resistance.[15]

Concerning the More than Moore domain, sensors and actuators have also been downscaled to exploit ballistic heat transport for harvesting and scavenging applications and they have been used as passive elements to be integrated in analog blocks. Nowadays, sensors and actuators (S&As) are integrated everywhere to sense and monitor parameters and to control actions of importance and interest in any given electronic system. In Mounier[17] the evolution of emerging, under development, mature and obsolete applications of S&As from 2009 to 2012 are analyzed, demonstrating that the S&A market is huge, since a large number and variety of devices are required for performing different functionalities depending on the final application. The majority of today's devices are oriented for stand-alone functionality, and are controlled and supported by vendor electronics. Furthermore, they are taking on more importance due to the trends in wearable devices. Downscaling in power semi-conductor switches has also had an important impact. Such devices are formed by multiple elementary cells designed for increasing their current density capability, operating frequency and temperature. In any integrated circuit, they are used for controlling the powering of each circuit. Obviously, downscaling their size induces several thermal issues, as explained in the case of Integrated Circuits (ICs). As a matter of fact, thermal aspects have been, and are, a hot topic when downsizing is present in whichever microelectronic field.

13.2.2 Heterogeneous Integration in the Low-voltage Scenario

Digital and non-digital blocks can be integrated in the same die (SoC) or stacked in a compact way (3D stack). SoC technology is a pure continuation of transistor size downscaling. However, the scale of the die size is increased, even if the feature size of transistors in a die is decreased. As a result, the length of global interconnects in a die is increased and this becomes one of the major bottlenecks for performance improvement, together with power dissipation along interconnections.[14,18] Another bottleneck is incurred by the number of allowed Input/Output (IO) pads of a chip, which is a trade-off between die size and data bandwidth. Also, the parasitic capacitance and inductance of IO pads are much larger than those within the die, which deteriorates the chip performance. To minimize this, several inter-connection levels are used in current CMOS technologies. Due to down-scaling, there is a material limit in interconnection lines, since grain boundaries of thin film conductors increase their resistivity.[19] As a result, due to high operating frequencies, signal delays or line cross-talk are expected due to parasitic elements (capacitors and inductances) inherent to the materials and dimensions (mainly cross-section) of the interconnections used.[19] Then, the solution adopted is to increase the cross-sections of in-ternal metal (copper) levels to mitigate these effects (double damascene).[20] This requires the use of interconnection vias between metal levels, leading to electromigration issues due to the high current density passing through them. To avoid this, refractory materials (such as tungsten) are used and other materials have been proposed as diffusion barriers to electromigration (*e.g.*, tungsten nitride, titanium nitride, titanium). On the other hand, capacitive effects are reduced by using new low-*k* materials, such as Fluorosilicate Glass (FSG), Carbon-Doped Oxide (CDO) or polyimides. The phenomena stated up to now all have a negative effect on the power con-sumption of the chip, as the resistivity, and consequently the voltage drop, increases along the metal line. A larger number of devices and wires oper-ating at higher clock frequencies means a higher total capacitance, which increases the dynamic losses. However, reducing the power supply voltage decreases the dynamic power, driving up static losses. In the long term, the use of optical communication within the chip is expected to avoid all these problems, however this technology is currently still under investigation.

One of the main successes in current electronic systems is the monolithic or hybrid integration of sensors and S&As onto CMOS platforms.[21] This has been of great importance for reducing the device size and cost, as well as for facilitating new functionalities and better performance of the final product.[21] Hence, S&As on CMOS platforms are important and convincing examples of the More than Moore scenario, as will be analyzed further on, but their packaging technology constitutes one of the bottlenecks in their reliability. Thermal cycling due to night and day cycles could affect their reliability. Even when operating under harsh environment scenarios, they are subject to high-temperature and humid environments. Thus, measuring temperature could be useful to improve the thermomechanical mismatch in accelerated tests, when they are packaged with the other parts of the electronic system.

To obtain more compact systems, three-dimensional (3D) integration technology is an emerging technology for future IC designs,[22–29] which allows the heterogeneous integration that is promising for overcoming barriers in interconnect scaling by different technologies.[30] Blocks of a SoC die can be rearranged in multiple dies that are vertically interconnected as a 3D chip. This results in shorter global interconnects and achieves better performance improvement, reducing power consumption. Furthermore, high bandwidths can be achieved, since IO interconnection densities can be improved significantly. In such structures, thermal issues arise from increasing dynamic power losses, which in turn raise the temperature.[31] Thermal and power constraints are of great concern with 3D ICs, since die stacking can dramatically increase the power density if hot spots overlap each other and additional dies are farther away from the heat sink. Thermally aware floor planning is the key to why the interlayer interconnection plays a more important role than the task of signal transmission or power delivery.[32]

Die stacking can be performed using several technologies. A wafer-to-wafer (WtW) technology is suitable for stacking chips with high production yields such as Dynamic Random Access Memory (DRAM) since the overall yield after stacking rapidly decreases as the number of stacking layers increases. Chip-to-wafer (CtW) technology is suitable for stacking known good dies (KGDs). In addition, chips with different sizes, which are fabricated using different process technologies, can be stacked in the CtW technology. The inherent problem in the CtW technology, however, is the low production throughput. Although wafer-level stacking provides a solution for S&As and CMOS heterogeneous integration, it also raises a number of technological challenges. The compatibility of the process for fabricating a high-density TSVs (Through Silicon Vias) and the technology used to fabricate devices is not obvious. Chip stacks with TSVs are much more sensitive to thermal cycling occurring during fabrication or due to changes in ambient temperature when the device is operating. The coefficient of thermal expansion (CTE) mismatch of the metal via and the chip material can cause non-negligible local stresses and serious reliability problems, with chip cracks occurring during thermal cycling.

However, the adoption of 3D technology is hampered by an insufficient understanding of 3D testing issues and by the lack of testing techniques. Testing techniques for 3D ICs have remained largely unexplored in the research community, even though experts in industry have identified several test challenges related to the lack of probe access for wafers, test access to modules in stacked wafers and dies, thermal concerns, design testability, test economics, and new defects that arise from unique processing steps such as wafer thinning, alignment and bonding. Despite recent advances in architectures, design automation tools and yield enhancement techniques, today's 3D chip designs do not consider test cost and the implications of design decisions on testability, thereby leading to a gap between anticipated (perceptual) benefits and practical value. Among all challenges faced in the heterogeneous integration domain (*e.g.*, IC testability[33]), thermal issues and device testability are the most important ones.

13.2.3 Figure of Merit Monitoring in Radiofrequency/Analog Integrated Circuits

When operating at high-frequency scenarios, several aspects should be addressed to test devices under such conditions. Due to the complexity of any SoC, Built-In Self-Testing (BIST) circuits are integrated with high-frequency analog circuits to monitor their proper operation. This is a common strategy also proposed to increase yield, optimize system performance and reduce the cost of testing.[34–36] It is well known that device parameters differ from their nominal value due to environmental changes (supply voltage, temperature), manufacturing process variation and intradie variability, which tend to increase as technology scales down.[37] Moreover, due to the trend towards device shrinking, the complexity, density and operating frequency of wireless communication circuits continue to increase. This leads to issues such as system optimization, testing time and cost reduction, which are still pending challenges.

The wireless communication circuits are formed from the following blocks: analog, radiofrequency (RF) transceiver and power amplifiers/modulators (PA&PM blocks) as Figure 13.1 shows.[38] In each block, there are key devices or circuits (see Figure 13.1) that fix the sub-system performance: Analog-to-Digital (ADC) and Digital-to-Analog (DAC) converters; serial–deserial (ser–des) processors (in analog circuitry); Low Noise Amplifiers (LNAs); Voltage-Controlled Oscillators (VCOs) (in RF transceiver circuits, driver amplifiers and filters); and Power Amplifiers (PAs) (in PA&PM blocks). In contrast to digital processing devices, the technologies used to implement such blocks depend on many materials, some of which are compatible with

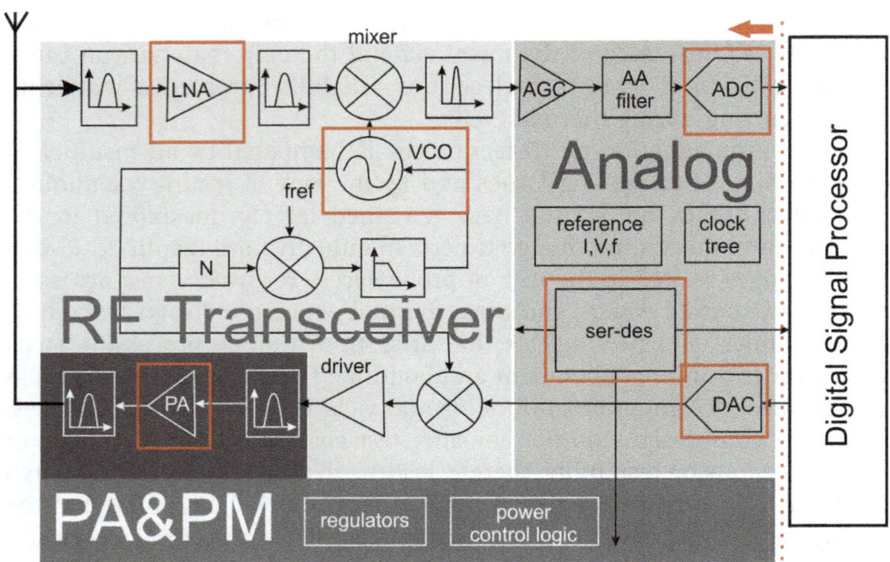

Figure 13.1   System approach for a wireless device to receive and send information coming from a digital signal processor.
Reproduced from ref. 38.

CMOS processing, such as SiGe, whereas, others have not traditionally been compatible with CMOS processing, such as those compound semi-conductors composed of elements from Groups III and V. Because of this, it is important to monitor their FOMs. Most of the built-in monitoring circuits used in RF ICs require electrical contact with the circuit to be character-ized.[35,36] This loading effect has to be taken into account during the design of the circuit under measurement and in some cases, such as in circuits operating at millimetre-wave frequencies, they can significantly degrade the achievable performance. Moreover, if the monitoring device is in contact with a node or path where a high-frequency signal is present, it has to be designed to work at that high frequency, eventually requiring the design of several monitoring circuits to perform the same function but at different operating frequencies. In this sense, temperature measurements could be useful for extracting FOMs from circuits of the wireless communication part, as will be shown later on.

13.3   Heat Generation and Testability Issues at the Chip Level in Energy Processing

13.3.1   Power Semi-conductor Devices

Power electronics is a key technology for managing electrical power efficiently. In this field, semi-conductor devices (power diodes or switches) have been the main contributors to its expansion in recent years. As a consequence of the increasing power demand and reliability requirements on such devices, effective temperature management has become ever more important, representing a central problem for power electronics. Under typical operating conditions (up to 10 kV and 6 kA), these devices dissipate a certain amount of electrical power in all their operation regimes, increasing the average value of their internal temperature. As a result, a complete degradation of the device performance can be observed.

Downscaling of the elementary cells of power semi-conductor switches has also had an important impact. Such devices are formed from multiple elementary cells designed to increase their current density capability, operating frequency and temperature, as well as their ruggedness/reliability under overloading conditions. In this way, the performance and reliability of power electronic systems can be greatly improved, as this allows a simultaneously reduction in volume, weight, and cost.[39,40] However, harsh environmental applications (for instance, spacecraft or well logging) can increase the usual temperature operating conditions beyond the limit of power semi-conductor devices based on silicon technology.[41] Therefore, in both cases, an alternative approach to silicon technology is required. As a solution, wide-bandgap (WBG) semi-conductors, such as silicon carbide (SiC), gallium nitride (GaN) and diamond (D), are currently under study and the first commercial devices based on SiC and GaN technologies are now available.[42] Elementary cell downsizing has been present in the planar-to-trench technology transition (cell pitch decrease) in power Metal Oxide Semi-conductor Field-Effect Transistors (MOSFETs) and Insulated Gate Bipolar Transistors (IGBTs), as well as in wafer thinning to reduce overall losses. Moreover, WBG devices are targeted to extend their use at higher switching frequencies (GaN and SiC) and higher breakdown voltages (SiC and D) than those achieved with silicon.[43] In contrast to GaN, SiC and D have better thermal conductivities than silicon.[43] In many roadmaps and due to its more mature technology than D and GaN, SiC is going to be considered a real choice for industries with breakdown voltages over 1.7 kV and in high-temperature scenarios. In such devices, thermal issues will be of paramount importance, as will be highlighted later on.

Thus, temperature measurement is crucial to maintaining the average of its internal temperature below a maximum value, as it aids and assesses the thermal management design of the power converter. However, with a heat dissipation system, it is not possible to control local temperature rises due to electrothermal coupling. This activates and accelerates several physical degradation mechanisms, varying the electrical, structural, and mechanical properties of power devices. As a consequence, physical changes are induced in the die *e.g.*, burnout failures or die cracking, as well as in its packaging *e.g.*, bond wire fatigue or die attach cracking. To better withstand such high-stress conditions, power devices are being designed with improved local electrothermal behaviour, by performing accurate simulations. Nonetheless, the simulation of their thermal failure phenomena requires transient temperature-field data for verification and improvement of the model parameters used. Therefore, in order to improve thermal management and study electrothermal failure, the development of techniques and systems allowing internal temperature measurements is required as a first step. Moreover, the determination of free-carrier concentration could be complementary to temperature measurement, since it can be used for adjusting simulation parameters, eventually improving the reliability of simulation results.

13.3.2 Light-emission Devices

Solid-state lasers, Light-Emitting Diodes (LEDs) and the emerging Organic Light-Emitting Diodes (OLEDs) are currently the most important light-emission devices. The same requirements as for power semi-conductor devices apply to them, and temperature measurements in such a scenario become a key point, as will be discussed in this section.

Solid-state lasers are leading the field for mass data storage on disk, and are preparing a major offensive in the area of image projection. They are specially arranged to obtain self-stimulated emission and there is a problem of control and uniformity of temperature, especially for narrow band diodes such as Distributed Feedback (DFB) laser diodes.[8] Any temperature fluctuation, even local, shifts the optical gain curve of the semi-conductor stacking. For a simple laser diode with no external filter, subject to a uniform temperature rise, a shift and slight broadening of the emitted line are observed. For a non-uniform temperature distribution, the threshold current of the laser diode increases and its spectral purity is degraded. In some cases, jumps can occur in the guiding mode within the semi-conductor cavity, sometimes accompanied by a splitting of the emitted line. For a laser diode with an external filter (DFB), the gain curve of the semi-conductor can sometimes shift far enough from that of the filter to cause a spectacular shift in the output power or even a disconnection of the laser oscillator. These phenomena can themselves alter the heat dissipation distribution in the diode and sometimes lead to destructive blockages. It is essential when designing optoelectronic systems to optimize the cooling of laser diodes. Such optimization naturally involves reducing the average temperature, but also the temperature gradients, and careful surveillance of changes in the characteristics of the laser which may be influenced by the thermal configuration of the chip. This is especially relevant in cases where the optical load at the laser output is subject to variations, and where a significant part of the power emitted by the laser might be sent back into the cavity.

LEDs present lateral or vertical topologies and can be encapsulated using flip-chip or wire-bonding technologies. Their main application is Solid-State Lighting (SSL). In fact, SSL systems are considered a real alternative to incandescent, high-intensity discharge and compact fluorescent solutions in traditional lighting.[44–46] This is due to LEDs' longer expected lifetime and lower power consumption. Moreover, the removal from the market in many countries, for energy efficiency reasons, of incandescent lamps and the lower efficiency shown by high-intensity discharge bulbs has started a competition between SSL and Compact Fluorescent Lighting (CFL) technologies.[45,47] In this sense, SSL systems promise more eco-friendly technology (*i.e.*, lead-free soldering and lead-free dies) than that used in CFL systems (which contain mercury), while the latter is available at a lower cost.[44] Thus, much effort has been put into developing affordable retrofit LED lamps for the domestic scenario, increasing their lumen output performance and reliability levels according to regulation standards.[48–52] Usually, LED lamps consist of several blue LED dies controlled by a current-based driver and a light dispersion–conversion system, which converts blue light into white and provides an omnidirectional light emission.[44] The dispersion–conversion system enables the design of lamps with the targeted light output distribution, spectrum and colour temperature (*i.e.*, from cold to warm white light). LEDs are individually packaged and connected (serial, parallel or a combination of both) on an LED board to reach the intended radiant flux.[44,53,54] They act as the light source and are externally controlled.[44] LED drivers are usually switched-mode power supplies with a Buck-boost or Fly-back topology and entail several considerations in their design.[55] The scaling down of SSL systems to fit in smaller places (for instance, light spot or traditional light bulb housings) offering higher output lumens according to the market demand,[56] require more compact and smarter drivers, in which the heat generation density might increase considerably and thermal management solutions at lamp engine level should be envisaged.[57] Thus, maintaining the operating temperature of LEDs (LED junction temperature), the light dispersion–conversion system and the driver board components below certain limits is of critical importance for the performance,[58] reliability,[59,60] and life expectancy[61] of an LED lamp.

OLED technology is the youngest of the devices presented here. Based on organic semi-conductors (polymers), OLEDs present better performance than LEDs at a potentially lower cost (they are more thin, flexible and efficient, and have a higher brightness and contrast). They will allow greater scalability and new applications, but are currently still an immature and expensive technology with short lifetimes and a high environmental impact (difficult recycling: high cost and complex techniques). In this sense, thermal management is also a critical factor in the design of OLEDs. High temperatures accelerate degradation in OLEDs,[62,63] can cause thermal runaway[64] and can also lead to surface instabilities in the organic layers.[65] The problem of thermal management is exacerbated in devices with flexible or transparent substrates, which have poor thermal conductivity, as well as in high-power devices or large-area displays.[66,67]

13.3.3 Heterogeneous Integration in Energy Management Systems

Trends in power electronics packaging present some similarities and differences to the low-voltage scenario, as the operating current, frequencies and temperature levels are very different. Power electronics packaging plays an important technical and economic role because it constitutes the interface between the raw semi-conductor device and the circuit application. It electrically interconnects one or several power semi-conductor devices, extracts the dissipated heat through a thermal path separated (or not) from the electric path, isolates the high voltages from other critical parts of the system, and provides a mechanical support to the devices. Power devices can be packaged alone or with other devices in parallel (multichip power modules) to reach the current ratings required by the application.

In the case of power multichip modules, they usually integrate several power switches and also diodes connected in an anti-parallel configuration to implement a current free-wheeling path extensively used in power systems devoted to motor drives (*e.g.*, electrical vehicles or railway traction). Usually, the top-side connection is performed with wire-bonding technology, whereas the die is mounted onto a substrate with a die attach layer. For their development, the use of advanced ceramics (*e.g.*, AlN, SiN) and their adaptation to wide area packaging required for power devices used in high-voltage and current applications (*e.g.*, railway traction) has been crucial. These modules wear out due to the thermomechanical stresses experienced during the operation of power devices, resulting in an increase of the package thermal resistance (base plate solder delamination), an appearance of a non-homogeneous current sharing between devices (wire-bonding lift-off, aluminium reconstruction) or an uneven current distribution in the device itself (die attach delamination). These are the main issues that limit the package life time and it is possible to minimize such effects by exploring other materials (*e.g.*, Ag nanosintering joints, metal matrix composites) or interconnection strategies (*e.g.*, press-packing or bump-bonding). Finally, electronic packaging aims to directly cool down the devices by using the same substrate and integrating more passive elements into the same package, thanks to an increase in the operating frequency. Thus, the weight, cost and volume of the final power converter can be reduced. Moreover, this also enables heterogeneous integration in the power electronics domain, which contemplates integrating several power devices manufactured using different substrates (silicon, SiC, GaN)[68] and directly mounting them in the final application (for instance within a motor).[69]

Wire-bonding technology is the most used interconnection strategy. In spite of its benefits (versatility, low cost, no special wafer processing), its main drawback is to increase the packaging's parasitic effects.[70] Three major solutions have been postulated to overcome this limitation: ribbon bonding, press-packing (spring contact based)[71] and bump-bonding (soldered connection or metal bumps).[70] Ribbon bonding can be used when an application calls for high power and high-frequency packaging. This technique allows for the creation of an interconnect, with a larger cross-section serving to substitute multiple wires to be bonded. Moreover, it has distinct advantages over traditional round wire wedge bonding, coupled with measureable performance, yield and process improvements. The press-packing approach, aside from minimizing the packaging's stray elements, is a fatigue- and explosion-free solution, showing improved thermal behaviour. However, it is suitable only for very high-power applications and requires hermetic sealing, a special cooling scheme and special insulation techniques. On the other hand, bump-bonding technology permits a higher packaging density (*e.g.*, 3D stacking) at low cost, requiring additional wafer processing and presenting the solder alloys fatigue problem.[72] Die attach technology has been based on soldering technology, where the solder paste was based on tin. Now, with the new regulations and the attempts to increase device operating temperatures, new approaches are being investigated for die attach formation. One of the most promising solutions is Ag nanosintering, which allows a die attach layer to be defined at low temperatures (below 200 °C) that could operate at high temperatures (up to 800 °C). In addition, this technology can be used to fix ribbon bonding on the top of devices.

These trends demonstrate the demands on thermal management for such new systems and the requirement for temperature measurements at the chip level to study the suitability of all such solutions, as well as to analyze their wear-out.

13.4 Thermal Characterization Techniques at the Chip Level

Temperature measurements have been required since the early years of microelectronics. They have been used for a long time to perform indirect thermal measurements in devices or systems, to design their thermal management solutions. Currently, thermal measurements also allow other functionalities, such as IC/SoC debugging, locating defects on the die due to catastrophic failures or wafer processing, failure indicator (*i.e.*, thermal performance degradation), extraction of electrical parameters and FOMs in the frequency domain of devices and sub-systems, and control of the technological variability induced by the continuous shrinkage of integrated devices in digital processing.

The first approaches were based on monitoring Temperature-Sensitive Parameters (TSPs). They are based on the inherent temperature dependence of the nominal electrical characteristics of any kind of device, which permit on-chip temperature monitoring. Since then, the integration of other sensors (mainly PN junctions) in more complex systems has been a common approach in many integrated systems. However, its incompatibility with the proper operation of devices, low spatial resolution when using inherent structures (*e.g.*, power electronic devices), limited access to die internal nodes, as well as possible electrical coupling effects, has motivated the development of other non-invasive techniques (off-chip methods) to tackle such challenges. Another important point crucial in their evolution has been the use of punctual measurements using a probe laser beam (punctual techniques) or cameras (multiple sensors), which allow thermal mapping with high spatial resolution. In this sense, the main trend currently is to develop camera-based systems to speed up the acquisition process in off-chip temperature monitoring, due to their recent improvement in signal-to-noise ratio FOMs. Concerning data acquisition, two domains have been used: time and frequency domains. Depending on the purpose of the measurements performed, each solution offers several advantages. All these points are highlighted and revised in this section.

13.4.1   Time *versus* Frequency Domain Thermal Characterization

When monitoring any physical magnitude, measurements can be performed in the time or frequency domain. Characterization in the time domain allows its static (steady-state) and transient monitoring. Static temperature measurements are performed when devices dissipate a constant power, since boundary conditions are time independent, and a thermal steady-state is reached.[73] Such measurements have been traditionally performed for thermal management design (no time dependence). However, with advances in the thermal measurement techniques used, transient characterization has allowed the monitoring of fast phenomena in time. Furthermore, the maximum time resolution inherent to the acquisition system has been overcome by following several approaches (nanosecond or microsecond time scales): single-shot or averaging in time. Single-shot acquisition is performed at low time scales and it can be carried out using appropriate synchronization systems. For instance, this feature can be integrated in modern fast-acquisition intensified cameras. Such cameras allow instant acquisition of short events produced on the nanosecond time scale at an elevated cost. A lower cost solution to reach a good time resolution is the use of boxcar averaging processing. This consists of acquiring multiple readings and averaging them to improve the measurement signal-to-noise ratio.[74,75] With this approach, characterization at the nanosecond or femtosecond time scale has been possible.[76–78] The main drawback of the averaging approach is that only repetitive and non-destructive processes can be studied, whereas single-shot acquisitions can be very interesting for investigating fast and destructive phenomena occurring at the die level. However, time-resolved measurements without averaging bring us the opportunity to observe non-repetitive phenomena, such as destructive mechanisms in power devices. These are single-shot experiments allowing thermal device analysis on the nanosecond[79] or microsecond[80] time scale.

By contrast, when monitoring in the frequency domain (also referred to as *lock-in measurements*), the power dissipated in the studied devices follows a periodic function of time. In such cases, the amplitude and phase of the thermal field at a given frequency can be observed. Such measurements allow hot spots to be located and resolved,[81–85] as well as device or system FOMs in the frequency domain to be extracted. Thus, frequency-domain measurements present the following advantages: robustness to noise;[86–88] information about thermal diffusion in structures (*e.g.*, ref. 81, 86 and 87);

and spatial confinement of thermal fields around the heat source by setting the appropriate frequency.[81,83] Analogously to time-resolved approaches, maximum lock-in frequencies fixed by the acquisition system itself can be overcome by using smart procedures. This is the case for heterodyne detection, which consists of modulating heat[89–91] and light[92,93] sources independently at different and close frequencies and observing the thermal field at its beating frequencies.

This can also be exploited to monitor the real part of the transfer function of any system or device, as it is manifested in the thermal field. However, when doing such characterizations, sometimes we are restricted to a maximum acquisition frequency fixed by the equipment used. Then, to monitor high-frequency electrical behaviour at lower frequencies, heterodyne excitation can be a good choice.[90] The heterodyne excitation approach consists of biasing the sample with a voltage resulting from adding two sinusoidal voltage functions with amplitudes $V_{0,1}$ and $V_{0,2}$ at frequencies $f_{exc,1}$ and $f_{exc,2}$, with a small difference between them ($\Delta f = f_{exc,2} - f_{exc,1}$, beating frequency). When $V(t)$ is applied to the microelectronic system, several heat sources can appear due to the Joule effect, which acts as a mixer. This allows the appearance of electrical information at a lower frequency, Δf, detectable with the system used. Thus, heterodyne excitation allows downconversion of high-frequency electrical phenomena (high-frequency heat source generation) into low-frequency thermal information (heat source modulation).[89–91] This approach also contrasts with that used in modulating light sources, as in thermoreflectance (see later). The illumination and excitation are modulated at $f_{exc,1}$ and $f_{exc,2}$, the Δf component appears as the mixer, which in such cases is the reflection onto the surface of the illumination field used. For instance, this allows an improvement of the frequency response of the thermoreflectance system in frequency,[92,93] as explained later.

13.4.2   Temperature Measurements

The measurement of chip temperature, either at specific locations or as a region mean value, can be undertaken with two main approaches: direct optical or near-field techniques; and electrical methods.[73] Optical techniques (discussed in Section 13.4.2.2) and near-field techniques (discussed in Section 13.4.2.3) require specific equipment and depend on direct access to the semi-conductor. This fact, often translates into special preparation requirements for the Circuits Under Test (CUTs). These restrictions are compensated by the possibility of accessing virtually any possible location, and obtaining complete temperature field distributions, with very good time and space resolutions (depending on each technique). On the other hand, electrical techniques (presented in Section 13.4.2.1) are based on acquiring the temperature from the measurement of an electrical variable or parameter (voltage, current, resistance, time delay, frequency shift, *etc.*) in a given device with a known thermal dependence. Only standard electronic instrumentation is required and the CUT can be analyzed in its final configuration (packaged), but the temperature acquisitions are restricted to the physical locations where the temperature-sensitive devices are placed.

13.4.2.1 Electrical Techniques

As stated before, electrical techniques for temperature measurement are based on the acquisition of electrical variables or parameters in a sensing device that contains the temperature information. In general, this approach also requires additional stable circuitry to bias the sensing device and to process the output signals. The most obvious way to determine the temperature in a microelectronic circuit is to integrate the sensing device at the targeted location. This built-in sensor approach is very convenient as it provides the desired temperature signal decoupled from other functional channels, even though there is a semi-conductor area overhead, and additional power consumption associated with the sensor operation itself. A second thermal measurement approach consists of using a variable or parameter of the microelectronic circuit under analysis to derive its temperature. In this TSP approach, any additional circuitry must be integrated, but previous calibrations are required to infer the temperature from the electrical measurements. The thermal readings are not decoupled from the functional device terminals and the obtained temperature requires a detailed interpretation to understand its real meaning (mean value of the whole chip, local temperature associated with a sub-circuit, *etc.*). In the next sections, we will consider the use of built-in temperature sensors only for signal processing circuits and the use of TSPs only for power processing devices and discrete components. Although this division represents the main practical scenario, it is worth pointing out that in some cases the TSP approach can be used in IC applications[94] and that built-in sensors can be integrated into power devices.[95]

13.4.2.1.1 Integrated Sensors or Built-in Temperature Sensors. From the early days of IC development, thermal stability and protection in critical applications (such as voltage regulators) was a major concern. Sensing the chip temperature was the most efficient way to react against overloading and very soon, solutions based on devices specifically devoted to temperature sensing were proposed.[96,97] These pioneering works relied on the forward voltage drop of the base-emitter diode of bipolar transistors as a temperature-sensitive variable, reaching uncertainties around $\pm 10\,^{\circ}\mathrm{C}$. The strong dependence and good predictability of bipolar device behaviour with temperature[98] and relatively low fabrication process variability has made bipolar transistors and diodes the natural choice for built-in temperature sensors in ICs, even to date. Nevertheless, other sensing devices have also been used, depending on the specific technology or application constraints. In this sense, Schottky diodes have been reported in GaAs Monolithic Microwave Integrated Circuits (MMICs) for microwave applications,[99] SiC Junction Field-Effect Transistors (JFETs) for high-temperature (500 °C) applications,[100] electromechanical resonators with high temperature resolution (8 mK) for CMOS-compatible MEMS technologies,[101] lateral PiN diodes for high-temperature (225 °C) SOI-based ICs,[102] digital delay lines (inverter delay) for advanced CMOS technologies,[103] and resistors (mainly platinum and polysilicon) used in micro hot plates for gas sensors,[104] in thermal test chips[105] or in mobile DRAM memories.[106]

During the eighties, the evolution and development of built-in temperature sensors was clearly driven by the establishment of CMOS as the most used technology to implement VLSI circuits. With technology downscaling, thermal management of ICs (and of microprocessors in particular), became critical, with higher on-chip power densities and temperatures, and the concept of *smart* or *intelligent* temperature sensors soon appeared.[107,108] Basically, these sensing blocks integrate the temperature-sensing device itself, consisting of stable reference circuitry, an ADC block and a means of calibration. This has been performed using the same CMOS process as the functional circuitry, which drastically reduces the overhead costs.[109] This scheme has been gradually improved during the last 20 years with continuous modifications and improvements in terms of accuracy, resolution, silicon area, consumption, speed and calibration requirements. A very detailed and complete description of basic and more spread CMOS-based temperature smart sensor circuits can be found in ref. 110. Basically, two parasitic bipolar transistors of the CMOS technology connected in diode configuration are biased with two different current densities. Under these conditions, the difference between the base-emitter voltages of the transistors is Proportional to the Absolute Temperature (PTAT). A third bipolar transistor is used to generate a constant reference voltage which is used in the ADC (sigma–delta) modulation to digitalize the PTAT voltage. Although this basic configuration is very well established, some authors have proposed alternatives to replace the parasitic bipolar transistors, such as using the MOSFET threshold voltage variation with temperature as a sensing parameter[111] (to avoid bipolar performance degradation in deep sub-micron technologies), designing the smart sensor with MOSFETs operating in their sub-threshold region (to minimize the consumption in ultralow power applications)[112] or using the temperature dependence of inverter delays as a TSP (to reduce area consumption).[103]

The performance evolution of smart CMOS temperature sensors can be quantified when comparing the results of the 1996 reference work[113] with recent implementations described in ref. 103, 114 and 115. All these references show performance tables comparing the characteristics of different state-of-the-art sensors. Each particular realization shows different characteristics and it is difficult to establish a significant comparison among sensors, but looking at the most relevant data, one can conclude that nowadays it is possible to integrate intelligent CMOS temperature sensors with areas around $0.006\,\text{mm}^2$, accuracies of 0.2 °C, resolutions below 30 mK, consumptions well below 1 mA and sampling rates higher than 20 kHz.

The development of high-performing temperature sensors compatible with the most advanced CMOS technologies has also favoured the development of new applications and concepts. In this sense, we will consider here two remarkable examples. First, the introduction of temperature sensors in RF-linked circuits allowed continuous non-contact temperature monitoring in critical locations such as train axles[116] or food monitoring.[117] In these kinds of applications (radiofrequency identification tags) ultralow power consumption is the main design aspect to consider. Secondly, temperature is also a variable that enhances observability and informs us of the state and characteristics of other circuits placed in ICs or complex SoCs, as the silicon surface temperature is directly related to the dissipated power. Temperature sensors are then embedded in the same silicon die with digital or analog circuits under test to observe their status, features and performance. An example of this kind of application is provided in ref. 118, where RF circuit performance characteristics are extracted from the dc output of an on-chip temperature sensor.

13.4.2.1.2 Thermosensitive Parameters. The dependence with temperature of bipolar transistor behaviour has been considered a major research topic since the first microelectronic developments of the fifties. Consequently, the need for experimental semi-conductor temperature measurement methods (also known as *junction temperature*, T_j), grew rapidly. As direct sensing was extremely complex in those early dates, the estimation of T_j from externally accessible signals was the most extensively used approach. A typical implementation is shown in ref. 119. This work presents a circuit for applying power pulses to heat up the Device Under Test (DUT) and then, infer T_j evolution from the open emitter collector current (I_{CO}) measurement, using a previous I_{CO} calibration with temperature. To fix the chip temperature during calibration, the DUT is placed in an oven (or another temperature-controlled environment) and the acquisition of the TSP, I_{CO} in this case, is performed avoiding any possible self-heating effect that could introduce internal temperature differences between the chip and the controlled ambient temperature. The calibration law obtained (often based on polynomial fits) permits direct calculation of T_j from I_{CO} readings.

This basic methodology has been kept unchanged for more than 50 years, not only for bipolar transistors[120] but also for virtually any discrete device by simply selecting the most suitable TSP each time.[121] Ref. 121 also gives a complete overview of the different modifications of the TSP approach for practically any power device in the market. In addition, the relative simplicity of the TSP measurements means they have been used extensively in industry standards related to thermal issues and thermal management of microelectronic circuits, such as the US military series (MIL STD-833, MIL STD-720, *etc.*) or the EIA JESD-51 series. TSP measurements are also applied to the thermal characterization of LEDs as they can be considered as diodes.[57] The only precaution to consider is to account for optical power emitted as light when performing thermal impedance measurements.

A relevant aspect to be considered in temperature measurements through TSPs is the interpretation of the obtained T_j value. It is easy to understand that this single temperature value cannot describe the more or less complex thermal fields present in a modern power device structure.[122] T_j is then a *global* or *average* value, mainly useful for engineering purposes (*i.e.*, the design of cooling systems). When in-depth comprehension of the thermal behaviour of a device is required (for example characterizing local hot spot formation), optical or near-field techniques are recommended.

13.4.2.2   Optical Techniques

Optical methods offer no electrical coupling with inspected devices, granting at the same time off-chip access to the surface or depth of the inspected die. According to this, optical techniques for temperature measurements can be classified into *surface* and *depth-resolved* approaches. In the former case, the dependence of the surface material's optical properties on temperature is used, whereas in the latter, bulk material properties are used as described below.

13.4.2.2.1   Surface Measurement Techniques. In order to optically sense temperature on the die top surface, several physical mechanisms have been exploited. The most commonly used methods benefit from: (1) surface emission in the infrared (IR) range due to heat exchange with the environment by radiation (IR intensity is monitored); (2) deposition of a coating/film with optical properties modulated by temperature (liquid crystal and fluorescence thermographies); (3) modification of the surface reflectance (thermoreflectance) or optical path change (interferometry based thermometry) due to material refractive index dependence on temperature (thermo-optical effect); and (4) spectral thermal modulation of emitted radiation due to bandgap dependence on temperature. All of these are revised and discussed below.

13.4.2.2.1.1   Infrared Thermography. The Infrared Thermography (IRT) technique is the most popular of the thermal imaging methods.[123] It is based on the fact that any object emits IR radiation by the mere fact of being above absolute zero. A black body is a class of object whose wavelength distribution is well known. IRT uses Plank's black body law to determine the absolute temperature of any object. Within the black body assumption, this is made by measuring the radiation intensity at a specific wavelength assuming that the object is in thermal equilibrium with its environment. The spectral radiant emittance of a black body ($W_{bb}(\lambda,T)$) as a function of wavelength (λ) and temperature (T) is given by Plank's law of radiation:[124]

$$W_{bb}(\lambda, T) = \frac{2\pi hc^2}{\lambda^5 \left[\exp\left(\dfrac{hc}{\lambda k_B T}\right) - 1 \right]} \tag{13.1}$$

where h is the Planck's constant, c is the speed of light, k_B is Boltzmann's constant, and λ is the radiation wavelength. In this method, a Charge-Coupled Device (CCD) camera is used with an IR optical system to focus the camera (such as a lens or a microscope) on a surface region of the DUT. However, there are a number of problems, such as the background noise of thermal radiation due to other bodies that are close to the measuring equipment and the knowledge of the emissivity of the DUT surface,[125] that need to be addressed. As such, before performing the measurement it is mandatory to determine the background radiation and thermal emissivity of the material at the measuring wavelength window of the camera used. These two previous steps involve equipment calibration and measurement correction during signal processing, which may be facilitated if the DUT surface is coated with a paint of known emissivity.[126–128] Moreover, a uniform and optimal coating deposition (thickness and roughness) on the die surface is not always easy to achieve, especially in low-dimension components. For this reason, other emissivity correction solutions have been developed.[129] Among them, the usage of off-line strategies, based on image post-processing or equalization, are the most common and interesting correction solutions used in microelectronic applications.[130]

Regarding the experimental setup, the DUT is biased and its surface thermal radiation is acquired by means of an IR CCD camera. The windows of sensitivity of the IR CCD camera are usually placed for wavelengths between 2 and 5 µm (short-wave systems) and between 8 and 13 µm (long-wave systems). These spectral bands are the atmospheric transmission windows for IR radiation. It is a highly standardized method, so the market has commercial systems where the acquisition and analysis of data is done by computer. This speeds up considerably the process of thermal characterization of surface components, obtaining an image showing the surface distribution of temperature. The great difficulty is to establish the equilibrium condition between the device and the environment before acquiring any data, in addition to maintaining controlled environmental conditions throughout the measurement process. The degree of temperature resolution reaches 0.1 K. The spatial resolution is between 5–30 µm. Its maximum time resolution is 100 ms, although specific systems can reach 1 µs.[131] Its range of applications is much broader as it encompasses temperatures between 30–500 °C, but higher temperatures can be achieved if additional filters are used. However, as stated before, the use of lock-in strategies allows detecting temperature modulation amplitudes below 100 µK.[132] Finally, the use of Infrared Lock-in Thermography (IR-LIT) together with a heterodyne excitation technique allows high-frequency capacitive currents due to intradie electrical coupling between microelectronic devices or more complex systems to be detected.[90]

13.4.2.2.1.2 Liquid Crystal Thermography. Liquid Crystal Thermography (LCT) is based on the dependence of the birefringence of a liquid crystal on temperature.[133,134] Isotropic and anisotropic phases of a material can be understood thermodynamically as a change of state. Thus, when a liquid crystal is below a critical temperature, T_c, it presents an ordered optical anisotropic phase (which will cause the phenomenon of birefringence), while above T_c there will be a gradual transition to an optical isotropic phase. This will result in both different light (anisotropic region) and dark (isotropic region) areas on the surface of the crystal distributed according to the divergence of the temperature with respect to T_c. Therefore, with the deposition of a layer of liquid crystal on the device, it is possible to detect surface temperature distributions in a semi-conductor around T_c.[134] Figure 13.2 shows a typical experimental setup for LCT measurements. The DUT is previously prepared and biased with an excitation circuit. A temperature controller fixes a stable ambient temperature during the measurements. In addition, a microscope to focus on the surface of the chip with the CCD camera is used, thus obtaining a high spatial resolution. Using a micropositioner, it is possible to move the sample and make a sweep surface temperature distribution. A PC controller centralizes the control over different parts of the assembly and processes the information obtained with the CCD camera, hence eventually visualizing the screen surface thermal distribution of the DUT. Note that this technique can only measure relative temperature around the critical temperature T_c, thus limiting its application range to between 30 and 110 °C, while the resolution levels range between 50 and 500 mK depending on the sensitivity of the camera used. A spatial resolution of 1 μm can be reached. Depending on the characteristics of the liquid crystal layer deposited, a time resolution from 1 ms to 1 μs can also be obtained.[135,136]

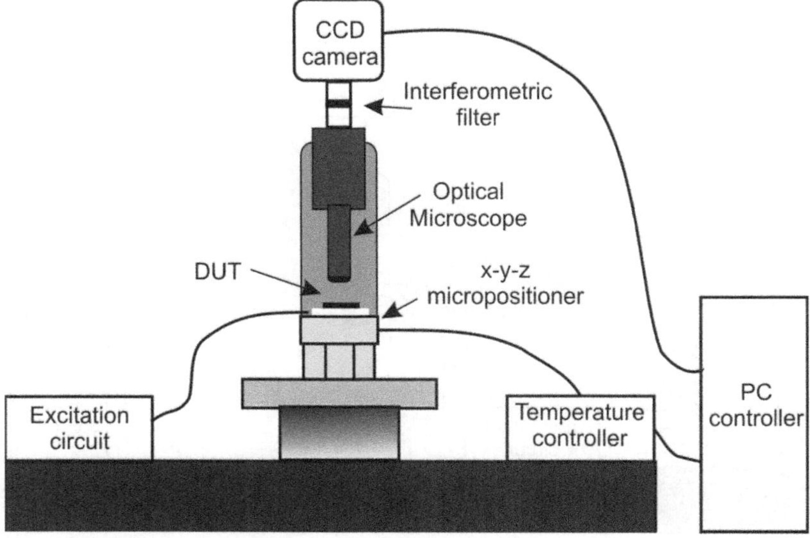

Figure 13.2 Typical setup used for LCT.

13.4.2.2.1.3 Fluorescence Microthermography. Fluorescence Microthermography (FMT) is based on the dependence of the fluorescence of a material with temperature.[137] The basic idea is similar to LCT. A thin layer of a photoluminescent rare earth dye such as Eu^{3+} thenoyltrifluoroacetonate (Eu-TTA) is dissolved in acetone and deposited on the chip surface and lit with an ultraviolet (UV) light (340–380 nm) stimulating the fluorescence of the deposited material (which emits at 612 nm).[138,139] The quantum efficiency of the emission, which is closely related to the fluorescence intensity, is a parameter that depends directly on the temperature. Therefore, once the device is biased by an external circuit, an intensity distribution on the surface of Eu-TTA will be manifested, following the temperature distribution on the semi-conductor surface.

In Figure 13.3, a schematic experimental setup is shown, where the great similarity of this equipment to that used in the LCT can be seen. The only difference between the two setups is the presence of a UV light source, used to excite the Eu-TTA and stimulate its fluorescence. FMT has a temperature resolution of 10 mK, while the spatial resolution reaches 0.3 μm. Regarding temporal resolution, the FMT can distinguish signals on a time scale of at least 200 μs, which corresponds to the response time of the fluorescence of Eu-TTA. Although relative values of temperature on the semi-conductor surface are measured, it is also possible to perform absolute measurements. In this case, a previous calibration of the system is mandatory. In this process, the emissivity calibration coefficient is determined experimentally as a function of temperature when the process of fluorescence occurs. Since the calibration must be repeated for each DUT, the process can become lengthy. However, it is worth pointing out that the handling of the equipment and standardization of the data acquisition and processing is simple. Today, there is equipment commercially available to perform such measurements.

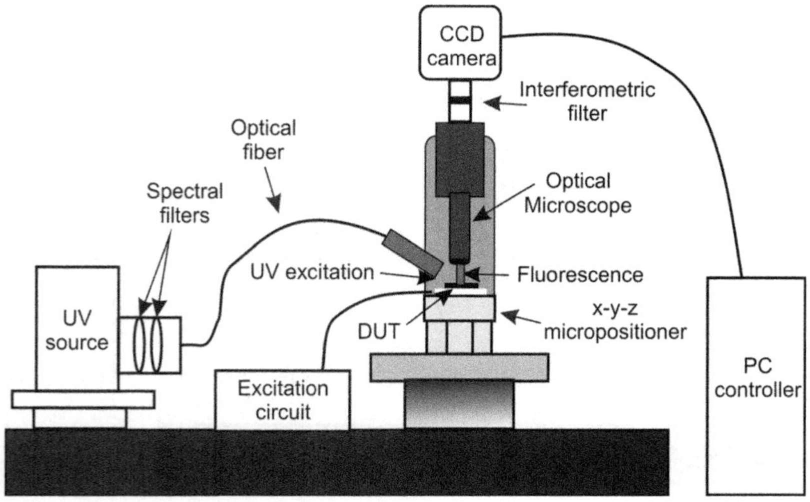

Figure 13.3 Typical setup used for FMT.

Preparation for FMT as well as for LCT, is more complex than that required for the IRT technique, which is a clear disadvantage.

13.4.2.2.1.4 *Thermoreflectance Thermography.*

Thermoreflectance thermometry is based on the temperature dependence of the surface reflection coefficient R of any material (as a consequence of the refractive index change with temperature, *i.e.*, the thermo-optical effect)[140,141] according the following relationship:

$$R(T) = R_0 + \Delta R(T) = R_0(1 + \kappa \Delta T) \tag{13.2}$$

where κ is the thermoreflectance calibration coefficient (typically on the order of 10^{-2}–10^{-5} K^{-1}). To detect $\Delta R(T)$, a probing monochromatic light with intensity I is focused on the inspected surface (punctual or wide area) and the reflected light I' is measured with a single (punctual, single photodetector) or an array of photodetectors (wide area, camera), allowing us to determine the ΔT of the inspected point or area, using this equation:

$$\Delta T = \kappa \frac{I' - I}{I} \tag{13.3}$$

Figure 13.4 shows a typical setup used for thermoreflectance measurements using a photodiode (punctual) or CCD camera (imaging) as the detector. The sample is placed on a translation stage (*xyz* micropositioner). The use of monochromatic non-coherent pulsed light sources is commonly required in CCD thermoreflectance to avoid interference fringes in the thermal image and illumination drifts during the acquisition. For this reason, LED-based illumination systems are used when performing stroboscopic measurements synchronized with the camera acquisition in both time and frequency domains.[74,75,142] Moreover, depending on the

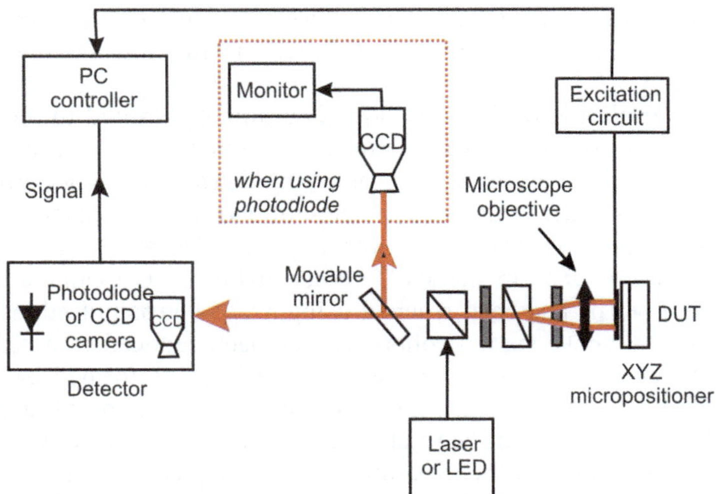

Figure 13.4 Typical setup for thermoreflectance.

wavelength used, temperatures on the top of the passivation layer (UV thermoreflectance), on the semi-conductor or metal (visible thermoreflectance), and through the substrate backside in a flip-chip configuration (NIR thermoreflectance) can be selectively performed.[9] Concerning temperature mapping, it can be carried out by punctual scanning using a laser beam[143] or with a CCD camera,[140,144] which would allow faster surface mapping with a lower signal-to-noise ratio and time resolution. There exists an intermediate solution for surface scanning with a laser using galvanometric mirrors.[145] This solution is a low-cost approach, which improves the time efficiency of moving the sample with a motor stage when punctual thermoreflectance with just one laser beam is performed.

For accurate temperature measurements, the determination of κ is crucial. κ depends on the sample material, the wavelength of the illuminating light,[146] the angle of incidence (and thus, by extension, the surface roughness and Numerical Aperture (NA) of the objective used)[9] and the composition of the sample in the case of multilayer structures (*e.g.*, surface passivation).[140,147] To best perform thermoreflectance calibration, a heating system based on a Peltier stage[9] is used, where the temperature of the sample is known or indirectly calculated by monitoring a TSP parameter. In addition, spectroscopy analysis can be very useful to determine the most suitable wavelength to have the highest sensitivity during the measurements.[146] Another important point is when wide-area measurements using CCD cameras are carried out. In these cases, pixel-by-pixel calibration is challenging when high spatial resolution is required (high-magnification conditions, objectives higher than ×20), as sample dilatations could put it out of focus. Several solutions have been proposed in the literature to refocus the sample, which consist of moving the sample[148] or the objective using a piezoelectric stage.[9] The former solution is more appropriate for frequency-domain measurements, whereas the latter is used in the time domain. In fact, current microscopes add automatic autofocus by hardware (laser autofocus) or by software, moving the sample with a motorized stage.

The time resolution of this technique is ultimately limited to the sensor used, and does not offer a single-shot measurement (boxcar averaging is commonly used). In the case of laser beam thermoreflectance, the time resolution reaches the order of the relaxation times associated with the microscopic scattering processes governing the reflectance, which is typically of the order of ps in metals.[149] When using a CCD camera, a time resolution of 800 ps has been reached using a 300–900 ps pulsed laser as a light source,[150] whose spatial coherence has been reduced to avoid interference fringes in the thermal image. This has been performed by slightly moving the fibre bundle, which allows the light to strike the diffuser from slightly different angles, and the static grain pattern is averaged. With regard to the frequency domain, the maximum lock-in frequency of this technique has been overcome by using heterodyne excitation of illuminating light, as explained before and detailed in ref. 92 and 93. This has enabled the use of

CCD thermoreflectance at higher frequencies (MHz range), improving the limits shown by current CCD cameras.

On the other hand, the spatial resolution is limited by the diffraction of the probe beam or pixel dimensions, being comparable to the beam wavelength (<1 μm) when the pixel dimensions of the CCD camera are smaller. Thermoreflectance temperature resolution is different depending on whether a single photodetector or a CCD camera is used. In the former case, a resolution of 0.6 μm has been reported,[151] whereas in the latter case, it largely depends on the camera's dynamic range (the number of bits of the ADC of the camera). In the literature for CCD-based thermoreflectance, temperature accuracies of 50 mK (14-bit CCD camera)[150] and 10 mK (12-bit CCD camera)[152] have been achieved in the time and frequency domains, respectively. However, its calibration is complex, due to surface non-uniformities and the presence of different materials on the device top, mainly passivation layers, commonly present in all microelectronic systems.[9] This implies the use of several different wavelengths to complete a temperature measurement. Moreover, temperature measurements are difficult to perform in the following two situations. When metal layers are present, they show a low sensitivity that introduces high inaccuracies in the measurements. On the other hand, when steep material change is encountered, different thermoreflectance coefficients are expected. Moreover, when these areas are next to passivation layers, measurements are more complex since a Fabry–Pérot cavity is defined and reflections may occur, in such a way that the reflected light could not show any dependence on temperature.

13.4.2.2.1.5 Thermography by Optical Interferometry. Interferometric thermometry consists of monitoring the temperature by using the interference of a single or multiple laser beams that interact with the die surface. This approach benefits from the dependence on temperature of the refractive index, which in turn modulates the optical path of the laser beam within the device. The interaction can be produced using a single (*e.g.*, Fabry–Pérot interferometer) or multiple paths (*e.g.*, Michelson or Wollaston interferometers), in which the die top metallization acts as a reflecting surface. In the multiple path approach, one of the laser beams is taken as a reference (reference beam) and the other goes through the die top or backside (sensing beam). In Figure 13.5 a schematic of two setup variants using Michelson and Wollaston interferometers is shown.[157] The beam amplitude of a thermally stabilized Super Luminescent Diode (SLD) is modulated by an Acousto-Optical Modulator (AOM). Both stroboscopic detection (boxcar averaging) and single-shot detection can be used. As the refractive index depends on temperature (the thermo-optical effect), the sensing beam finds a variation on its optical path, inducing a phase shift with respect to the reference beam. By superposing both on a photodetector (single or multiple sensors), the phase shift is finally determined. The interferometric methods are divided into *backside multiple path* and *frontside single path* probing techniques. Both techniques can be implemented

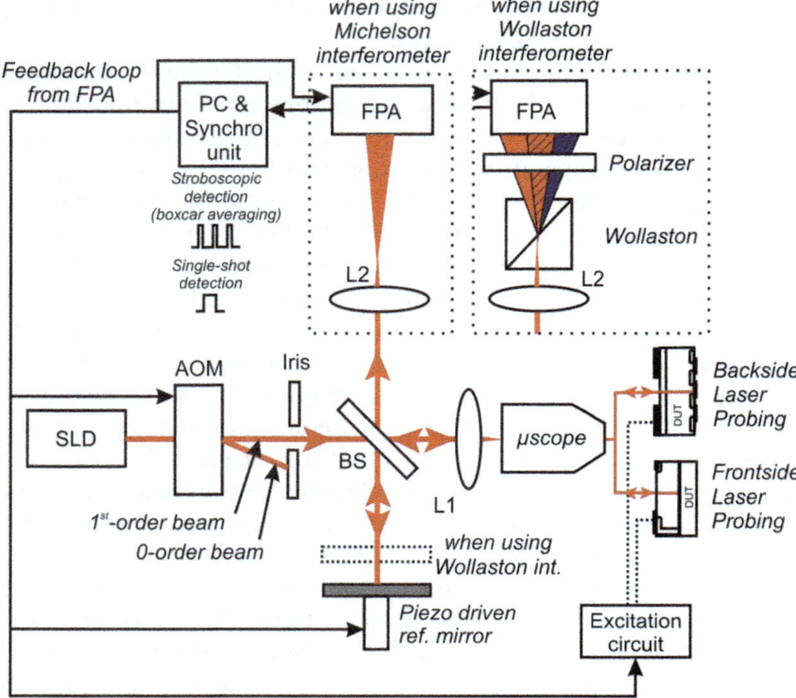

Figure 13.5   Schematic of the stabilized Michelson and Wollaston interferometers for both backside and frontside laser probing techniques.

in Michelson or Wollaston interferometer configurations, as shown in Figure 13.5 where stroboscopic or single-shot detection can be carried out. A detailed explanation of the different setup variations can be found in ref. 157.

13.4.2.2.1.5.1   Backside Multiple Path Probing

Over the years, this technique has evolved from a punctual sensing approach focusing a laser beam onto the die surface, to a multidetector configuration using a camera. At the very beginning, in the nineties, a Michelson configuration was used (sensing and reference beams) with a pulsed laser (minimum time pulse width 5 ns). In Figure 13.6 the experimental setup using a lock-in approach for recovering the phase time evolution ($\Delta\phi(t)$) for backside laser probing is shown. The sensing beam vertically traverses the die from its backside until being reflected by its top metallization layer, while the reference beam is reflected at the rear metallization layer of the DUT.[153,154] The detected quantity was the phase shift $\Delta\phi(t)$ of a vertically propagating probing beam due to modulations in the optical sample thickness:[155,156]

$$\Delta\varphi(t) = 2 \cdot \frac{2\pi}{\lambda} \int_0^L \left[\left(\frac{\partial n_{Si}}{\partial T}\right)_{n,p} \Delta T(x,t) + \left(\frac{\partial n_{Si}}{\partial C}\right)_T \Delta n(x,t) \right] dx \qquad (13.4)$$

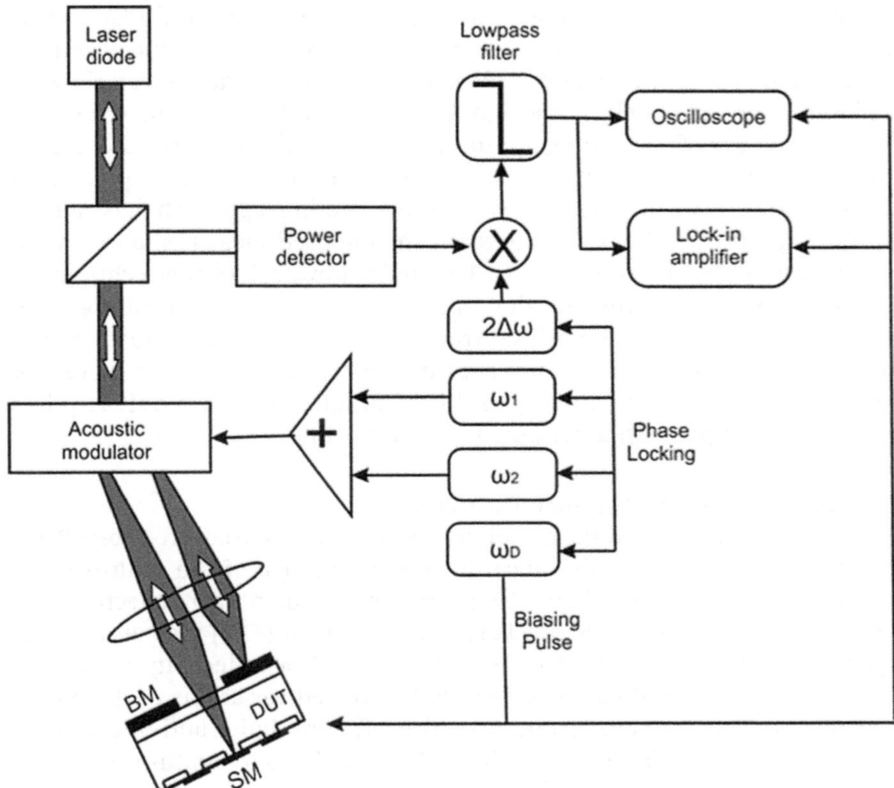

Figure 13.6 Experimental set-up for backside laser probing technique, where SM stands for Surface Metallisation and BM for Backside Metallisation.

In this case, the detected phase shift has two contributions originating from the injection/removal of carriers and the temperature rise during the current pulse. Although they are opposite in sign, they are of the same order of magnitude, provided the power dissipated within the DUT is small. However, for a large temperature rise (*e.g.*, IGBTs in short-circuit operation), the phase modulation by the carriers can be neglected in view of the thermal contribution. To reconstruct an image, the laser scans the area of interest point-by-point, and then the final image is constructed.[157] For each point, the data are acquired by following time averages with a heterodyne approach. This allowed a heat dissipation/temperature map of the internal side of the die top metallization to be extracted with very good time resolution (on the nanosecond time scale).[158,159] Later, with the improvement of fast acquiring cameras, such measurements started to be performed as a single-shot snap-shot, synchronizing the acquisition with a certain delay when a strong thermal signal was present in the measurements (*e.g.*, current filamentation under stressful avalanche conditions).[79] Moreover, a Wollaston interferometer was also implemented to speed up the acquisition process, as two shifted images are obtained and superposed on the Focal Plane Array (FPA) of the camera. However, for low signal-to-noise ratios, time averaging is required when performing measurements (3 mrad uncertainty is achieved). For very small phase signals (<10 mrad, 0.2 mrad uncertainty is reached) the use of a scanning approach is required.[157] The time and lateral resolutions of this technique are within 5 ns (fixed by the laser pulse duration)[157] and 2–3 μm,[79] respectively. The scanning approach has a temperature resolution of 5 °C,[160] whereas camera-based measurements have a poorer resolution, of around 50 °C. The main drawback of the technique is that it requires polishing of the sample backside and depositing anti-reflective coatings to enhance laser transmission through the substrate and internal reflections.[160] Measurements in the frequency domain have not been reported in the literature with this technique, though it only requires biasing the sample with a periodic function of time.

13.4.2.2.1.5.2 Frontside Single Path Probing

Frontside single path probing exploits thermometry based on Fabry–Pérot Interferometry (FPI) by reflectance. It takes advantage of the multiple secondary beams, resulting from the probe beam (main ray) reflections on either opposite sides[161] – Fabry–Pérot Transmission (FPT) – or an internal layer – Fabry–Pérot Reflectivity (FPR) – of the DUT. The reflectivity of a Fabry–Pérot resonator is modulated by thermally induced variations of its optical thickness $n_{Si}(T)L(T)$ where n_{Si} and L are the refractive index and thickness of the semi-conductor layer, respectively. Hence, the temperature evolution within the sample can be detected by monitoring the reflected intensity.[161–163] Representing the reflection coefficients at the front surface and at the rear surface by r_f and r_r, respectively, the reflectivity R_{FP} of a Fabry–Pérot resonator of thickness L becomes:[160]

$$R_{FP} = \frac{(r_f + r_r e^{-\alpha L})^2 - 4 r_f r_r e^{-\alpha L} \sin^2\left(\frac{2\pi}{\lambda} n_{Si} L\right)}{(1 + r_f r_r e^{-\alpha L})^2 - 4 r_f r_r e^{-\alpha L} \sin^2\left(\frac{2\pi}{\lambda} n_{Si} L\right)} \qquad (13.5)$$

As two adjacent maxima of the reflectivity are observed, the optical thickness $n_{Si}(T)L(T)$ is changed by $\lambda/2$. The corresponding temperature difference ΔT is therefore equal to:

$$\Delta T = \frac{\lambda}{2}\left(n_{Si}\frac{\partial L}{\partial T} + L\frac{\partial n_{Si}}{\partial T}\right)^{-1} \qquad (13.6)$$

13.4.2.2.1.6 Thermometries based on Spectral Analysis of Luminescence, Raman Scattering and Photocurrent Emission.

Any phenomena related to radiation or visible light emission in semi-conductors entail a strong dependence on the temperature because of its bandgap. For this reason, spectroscopic measurements, based on luminescence, Raman scattering and photocurrent emission, can be used for thermal monitoring.

13.4.2.2.1.6.1　Luminescence

Luminescence is the emission of light by a substance not resulting from its incandescence. It is thus a form of cold body radiation. We can distinguish between, *chemiluminescence,* caused by a chemical reaction, *sonoluminescence* caused by a sound, *mechanoluminescence* which is caused by a mechanical action, *electroluminescence* caused by a source of electrons and holes injected through a PN junction and *photoluminescence* caused by external optical excitation, like photon absorption. Although, Wiedemann,[164,165] introduced the term *luminescence* in 1888, it was not until in the late 1980s and 1990s that theoretical and experimental works established luminescence to be a highly suitable technique for the thermal characterization of silicon, especially in the context of Photovoltaic (PV) applications.[166–168] Among all the luminescence types mentioned above, electroluminescence[169,170] and photoluminescence[171] have been the most relevant techniques used for measuring temperature in compound semiconductor devices.[172]

Electroluminescence (EL) is an optical and electrical phenomenon in which a material emits light in response to the passage of an electric current or to a strong electric field. EL is the result of a radiative recombination of electrons and holes in a material, usually a semiconductor. The excited electrons release their energy as photons (light). Prior to recombination, electrons and holes are separated either by doping the material to form a PN junction (in semi-conductor electroluminescent devices such as LEDs), or through excitation by the impact of high-energy electrons accelerated by a strong electric field (like in the case of phosphor in electroluminescent displays).

Photoluminescence (PL) describes the light emission from any form of matter after the absorption of electromagnetic radiation; it is, therefore, initiated by photoexcitation. Such excitation typically undergoes various relaxation processes and then photons are reradiated. The period between absorption and emission can be extremely short: it ranges from the femtosecond regime for the emission from, *e.g.,* free-carrier plasma in inorganic semi-conductors,[173] up to milliseconds for phosphorescent processes in molecular systems. Under special circumstances it can also be extended to minutes and even hours. The observation of PL at a certain energy can be seen most-straightforwardly as an indication of the population of the state associated with such a transition energy. While this is generally true in atoms and similar systems, correlations and other more complex phenomena also act as sources for PL in many-body systems such as semi-conductors. PL intensity is proportional to the carrier concentration. This means bright areas generally indicate higher minority carrier lifetime regions, whereas dark areas indicate higher defect concentrations. PL is a contactless technique, which allows it to be applied between the many processing steps within solar cell processing. However, it does require optical filtering to eliminate the reflected illumination wavelength from the PL emissions. A detection range extending from 0.4 to 2.7 μm, sample temperatures of 4 to 300 K, and mapping capabilities with 1 to 2 μm spatial resolution on a Fourier-transform-based system have been achieved with current PL spectroscopy equipment.[174]

13.4.2.2.1.6.2   Raman Thermography

Raman scattering occurs when a photon scatters inelastically from a crystal, with the creation or annihilation of one or more photons. The spectrum of the scattered photon, which will be different to that of the incident photon, is dependent on the crystal temperature as the phonon spectra are temperature dependent. Apart from inelastic scattering, an exchange of energy with lattice vibrations in the material is produced. As the temperature increases, the number of phonons in the excitation mode increases and this will enhance the ratio between the anti-Stokes (upshifted phonons) and Stokes (downshifted phonons) peaks.[175] Using this ratio, the absolute temperature can be calculated. For practical absolute temperature measurement, it is necessary to calibrate the Raman spectrum prior to the experiment. Raman spectroscopy uses short wavelengths from an Ar or He–Ne laser which will allow resolutions of less than 1 μm (limited by the spot size of the laser) and a temperature resolution in the range of 1–2 °C to be obtained.[176] Raman spectroscopy presents a reading velocity and processing that is limited (at least) by the probe movement and by the properties of the material and surface. This technique is well suited to large temperature changes, due to its low sensitivity.

13.4.2.2.1.6.3   Photocurrent Emission

Photocurrent Spectral Analysis (PSA) is a simple and non-invasive evaluation of the temperature of an inner layer, as the channel of a complex system such as a power High-Electron-Mobility Transistor (pHEMT), using photocurrent measurements.[177] By illuminating the gate region at various wavelengths it is possible to collect the photogenerated current spectrum. This spectrum is strongly related to the temperature-dependent absorption characteristics of the different layers forming the device. By analysing the photocurrent spectra at several operating conditions it is thus possible to extract the channel temperature in the DUT without perturbing or modifying the temperature distribution and the dissipation processes.

13.4.2.2.2   Depth-resolved Techniques.   Such techniques are a response to the demand of alternative approaches for thermal sensing in microelectronics. All of the approaches presented up to now have provided a temperature field filtered by the different material layers deposited on the power device (SiO_2, Al). Notice that such measurements can be very useful in the case of monolithically integrated power devices, in which all thermal heat generation is produced close to the surface (lateral topology). In contrast, access to the device's internal structure can be very interesting in the case of discrete power devices (vertical topology) or ICs packaged in 3D stacks. Therefore, effort has been devoted to developing complementary depth-resolved techniques. Basically, they employ an IR laser beam as a probe, which longitudinally passes through the inspected device, whose TSPs are the beam trajectory (Internal Infrared Laser

Deflection thermometry, IIR-LD) and the radiation phase shift (interferometric methods). Both phenomena are due to the modulation of refractive index by temperature (thermo-optical effect). As a result, an average temperature along the device interaction length can be measured. The most relevant techniques are FPI by transmission and IIR-LD.

13.4.2.2.2.1 Fabry–Pérot Interferometry by Transmission. Figure 13.7 shows the principle of FPI, which takes advantage of the multiple secondary beams, resulting from the probe beam (main ray) reflections on the DUT.[160,178] The device behaves as a Fabry–Pérot resonator cavity whose optical path is affected by modulations of the refractive index.[179] This phenomenon can be easily measured by means of a radiation power detector[160,180] or a four-quadrant detector,[181] where the multiple ray interference takes place on its surface. As a result, an oscillating signal pattern evolving in time is obtained. Notice that between two consecutive interference extrema of the detected radiation power, the same temperature increase ($\Delta T_{\text{extrema}}$) is produced. Consequently, by counting the number of extrema as a function of time, the temperature can be finally determined. For deducing $\Delta T_{\text{extrema}}$, the following relationship between n and T is used:[181]

$$n = n_0 \left\{ 1 + \left[L^{-1}\left(\frac{\partial L}{\partial T}\right) + n_0^{-1}\left(\frac{\partial n}{\partial T}\right) \right] T \right\} \tag{13.7}$$

where n_0, $L^{-1}(\partial L/\partial T)$ and $(\partial n/\partial T)$ represent the silicon refractive index at 300 K, and the dilation and thermo-optical coefficients of the material, respectively. By knowing that the phase difference between two consecutive extrema is $\pi/2$, $\Delta T_{\text{extrema}}$ can be determined as:

$$\Delta T_{\text{extrema}} = \frac{\dfrac{\lambda_0}{4L}}{\left(\dfrac{\partial n}{\partial T}\right) + n_0 L^{-1}\left(\dfrac{\partial L}{\partial T}\right)} \approx \frac{\lambda_0}{4L\left(\dfrac{\partial n}{\partial T}\right)} \tag{13.8}$$

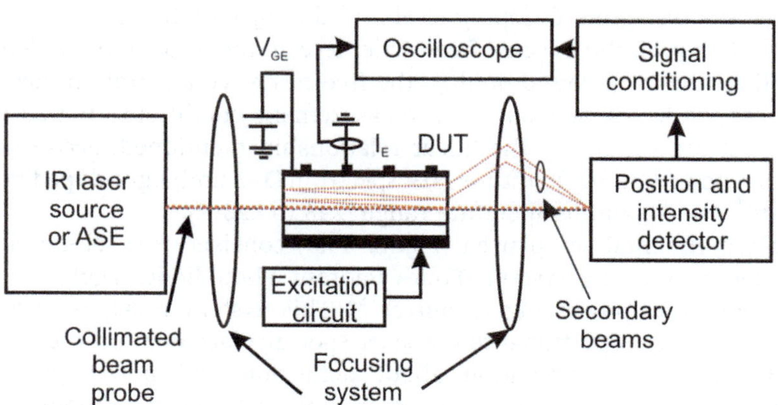

Figure 13.7 Experimental setup for FPI measurements.

where the contribution due to the sample dilation is neglected in comparison to the thermo-optical one, since very high-temperature increments are not expected (less than 300 K).[181] The measurement precision of FPI is related to the resolution power between two consecutive extrema. Apart from having a good time and signal resolution, this technique also requires a power signal waveform leading to extrema that are well defined. This fact is controlled by the polishing quality of the lateral sides of the inspected DUT, which is performed during the DUT preparation stage. Therefore, this issue must be checked, to ensure the suitability of the polishing process followed in FPI thermometry. In contrast, the uncertainty in the initial value determination implies a higher error in temperature measurement than that derived from the sensing system. For the interaction range previously stated (5–10 mm), it corresponds to a maximum resolution in the range of hundreds of mK (325–162 mK). The first time that this technique was experimentally checked using several different techniques is reported in ref. 181.

13.4.2.2.2.2 Internal Infrared Laser Deflection.

In this technique, the refractive index gradient and free-carrier concentration is measured by the deflection and partial power absorption of a probe laser beam passing through the device (IIR-LD).[80,182] A refractive index gradient developed in the DUT alters the propagation direction of a beam (mirage effect). The detection of beam deflection yields information on the average temperature along a laser beam path through the DUT. The average temperature in the device is deduced using independent data for the temperature dependence of the refractive index and carrier-induced absorption.

The technique's viability is based on the linear relationship between the variables involved, the background doping level, and the accomplishment of high-injection-level conditions at the measuring depth.[80] The two last conditions are the most restrictive, since they influence the interaction between the IR radiation and the device. In the first case, the transmission of IR radiation diminishes as the background doping level increases. By contrast, the high-injection-level condition concerns the relative concentration between the background doping level and the injected free-carrier concentration. Namely, if the injected free-carrier level at the measurement depth is lower than the background doping, the free-carrier concentration measurement cannot be carried out. The measurement conditions in this work are within the validity of the linear relationship mentioned, according to studies reported in the literature: NIR spectrum (1–3 μm); light-doped region (10^{14}–10^{16} cm^{-3}); and temperature range (298 to 423 K).

The experimental setup used (Figure 13.8) consists of a self-developed deflection monitoring system (Four-Quadrant Photodiode (FQP), *I/V* converters and post-processing circuitry).[183] This system provides electrical signals that are proportional to the laser spot displacement on the FQP,[183] directly providing information about deflection and power (light absorption). This makes this setup compatible with FPI thermometry, as

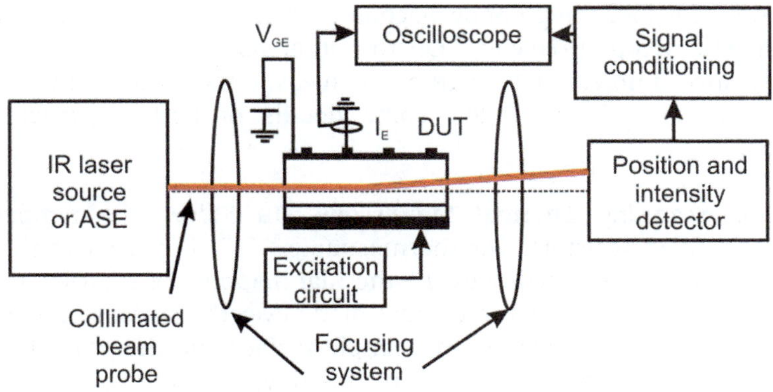

Figure 13.8 Experimental setup for IIR-LD measurements.

presented in ref. 181. Among the problems when implementing IIR-LD, are the arduous sample preparation process (sample lateral wall polishing), which is required to provide optical access to both sides of the device, and the difficulty of performing a precise calibration. In fact, all previous demonstrations have been based on simulation results, the first demonstration of its feasibility in microelectronic devices being provided in ref. 184.

In IIR-LD, the measurement precision and time resolution are limited by the sensing system performance.[183] In all measurements, the working power of the radiation incident onto the FQP is initially set to 2 mW.[181] For this value, the DUT does not experience a noticeable self-heating, and an optimum signal-to-noise ratio is assured at the sensing system output. Subsequently, the beam deflection and power radiation onto the FQP can be precisely measured with up to 1.9 µm and 12.3 µW resolution, respectively. Furthermore, an excellent time resolution (<1 µs) is provided because of its high bandwidth (1.4 MHz). The beam deflection resolution yields, for the above-mentioned interaction length range (5–10 mm), a precision in the temperature gradient measurements ranging from 52 to 25 $\mu K\, \mu m^{-1}$. Thus, a temperature resolution of around 1 mK can be estimated by multiplying the depth-resolution values by the temperature gradient values. This technique has also been demonstrated to be suitable for performing lock-in measurements,[185,186] obtaining an amplitude in the deflection of $2\times10^{-3}\, mK\, mm^{-1}$, which corresponds to a value of 22 µK. One advantage of using this technique in the frequency domain is that the temperature can be easily derived with only one measurement once the thermal gradient is known.[185]

13.4.2.3 Near-field Microscopy Thermography

Thanks to the advances in near-field microscopy,[187] Atomic Force and Near-Field Scanning Optical Microscopies (AFMs and NSOMs, respectively) have allowed the lateral resolution of some of the above-mentioned techniques to be improved. On one hand, AFM allows the placement of a temperature sensor on the inspection point by mechanical contact, the origin of Scanning Thermal Microscopy (SThM).[188] On the other hand, NSOM permits thermoreflectance, fluorescence or IR-based measurements to extract surface temperature maps.[189–191] Below some details of both approaches are provided.

13.4.2.3.1 Scanning Thermal Microscopy. In SThM, the temperature sensor, *e.g.*, a Pt resistor[192] or thermocouple,[193,194] is placed at the AFM tip, which enormously facilitates the thermal mapping of the DUT top surface. When the probe scans in contact mode over the surface of a sample, heat transfer localized between the sample surface and the probe tip leads to a change in the tip temperature. In this way, both the tip–sample heat transfer across the entire surface and the sample topography are obtained simultaneously.[195] However, the roughness of the sample surface can cause variations in the tip–surface thermal contact, leading to noise in the thermal signal. Furthermore, models that take into account such interactions and perform a calibration identifying their parameters are required.[196] This technique is commonly used in the frequency domain and it is used for thermal conductivity extraction[196] or to study the heat exchange between several materials[197] on the nanoscale.

This method shows the highest spatial resolution (around 30–100 nm) with a temperature accuracy of 1 mK.[198] The limitations on temperature extraction are due to the sensor nature, the AFM tip dimensions and the roughness presented by the inspected device surface, related to the calibration of the chosen model. As a matter of fact, its major limitation is the liquid meniscus that forms between the tip and the sample, which is intrinsic to contact measurements, done under ambient atmospheres, and limits the resolution of the technique. Regarding the time response of this technique, the bandwidth is limited to 250 kHz and 1 MHz depending on the capacitive coupling between the tip and the observed sample.[198] The SThM experimental setup is complicated and expensive (shown in Figure 13.9), and data acquisition can be time consuming due to the scanning methods required.[199]

13.4.2.3.2 NSOM-based Thermometry. One way to improve the lateral position of the thermal techniques previously described based on visible light monitoring is the use of NSOM. NSOM can be used for temperature measurements using reflectance thermometry,[189,200] temperature monitoring by fluorescence decay[190] or IRT.[191]

The spatial resolution of a standard optical microscope (far-field) is limited by diffraction to a theoretical spatial resolution of $d = \lambda/\mathrm{NA}$, where λ is the wavelength of light. NSOM has been developed to overcome the diffraction limit. In this technique a tapered optical fibre is coated with an opaque metal film leaving a small aperture of diameter d at the fibre tip. This tip is then brought into close proximity (within 1–10 nm) of a sample surface by controlling the shear or the normal force between the tip and the sample

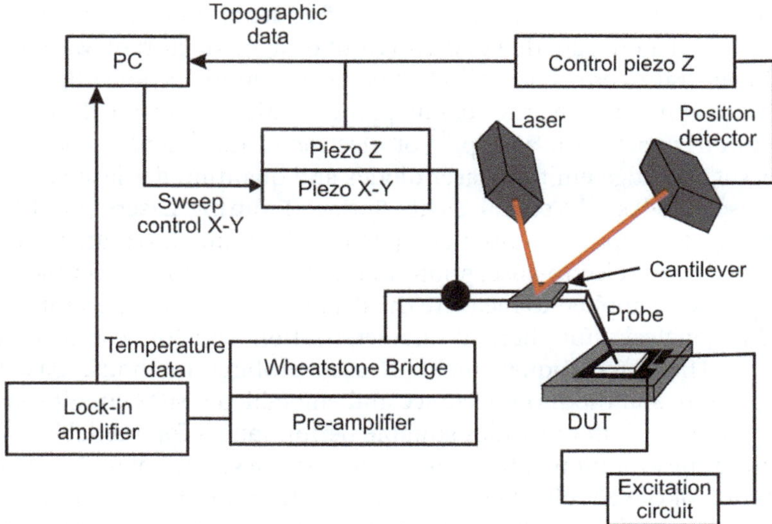

Figure 13.9 Experimental setup for SThM measurements.

surface. By scanning the tip laterally, an optical image can be obtained. Because diffraction is a far-field effect, the close tip–sample proximity allows it to beat the diffraction limit reducing the NA of the microscope used.[190] The spatial resolution of NSOM is limited by the aperture size, which is usually ~50 nm. Efforts to make apertures in the sub-30 nm range have proven to be very difficult, if not impossible. In fact, there exists a physical limit to the aperture because it cannot be made smaller than twice the optical skin depth in the metal coating, which is ~24 nm in Al at a 633 nm wavelength.[201]

13.5 Thermal Studies at the Chip Level in Microelectronics

In this section, some examples of thermal studies applied to microelectronics will be presented. These studies will be focused at the chip level distinguishing two situations: low-power (signal processing electronics) and medium-to-high-power (energy processing electronics) scenarios.

13.5.1 Signal Processing Electronics: Digital Processing and RF/Analog ICs

At the very beginning, temperature measurements were devoted to locally studying thermal phenomena, which was not possible to tackle with average measuring approaches. To this end, Claeys *et al.*[202] initially used thermoreflectance for fault detection in ICs, also allowing electromigration studies.[203] Later, with CCD thermoreflectance, Wang *et al.* demonstrated the characterization of MOS transistors in the frequency domain, showing that they presented a hot spot due to their layout design. All current was crowded to a single drain connection.[204] CCD-based thermoreflectance microscopy can also be applied to the thermal mapping of microelectronic devices such as microrefrigerators on a chip[205] or electronic ICs[206] and optoelectronic devices such as edge-emitting laser diodes and quantum dot lasers,[207] high-power laser diodes,[208] Vertical Cavity Surface Emitting Lasers (VCSELs),[209] Semi-conductor Optical Amplifiers (SOAs), Photonic Integrated Circuits (PICs)[210,211] and electro-absorption modulators.[212] More recently, other techniques were applied to measure the dynamic surface temperature in ICs for failure analysis, for thermal characterization and for testing modern devices.[73] These techniques include laser probing methods, based on interferometry and thermoreflectance and embedded CMOS circuit sensors. Among them, the latter provides suitable performances for a reliable tracing of the dynamic evolution of the temperature on the surface of an IC, which is a projection of the circuit's behaviour and IC structure. The dynamics of the temperature contains information complementary to the traditional voltage and current electrical signals.[73] Furthermore, when such systems (*i.e.*, devices, ICs, or SoCs) are packaged, their electrical characterization, testing and debugging become a challenging task when their operation frequency increases.[213] On one hand, it requires more complex and expensive equipment (*e.g.*, network analyzers). On the other hand, systems are more sensitive to performance degradation due to non-desired couplings at the die level: some parasitic effects can lead to undesired modifications in their nominal frequency behaviour, being very critical to the integration of monitoring circuits that measure current or voltage at critical system nodes.[214,215] In this sense, capacitive coupling is the most common parasitic or defective structure that causes failure or malfunction. It consists of the transfer of energy between IC nodes or embedded blocks through parasitic capacitances.[216,217] By contrast, similar effects can also be observed at the package level: when the package is not properly optimized for the system working frequency, it is possible that not all the power is transferred to the die (package not adapted to the load) due to parasitic inductances. In this case, interconnections or contacts behave as an open circuit when the frequency increases, also inducing delays in the signal transmission. Another particular case is the skin effect and consists of confining the current flow to the material periphery, increasing its effective resistance. This can be observed in solder joints,[218] wire bonds[219] or transmission lines[220] and it can also degrade the expected frequency response of the final encapsulated system. Thus, the study and characterization of such parasitic behaviours is of importance at the die and package levels to have a greater insight into the degradation of the system frequency response during failure analysis or debugging tasks (especially when they are design-related). The electrical frequency response is usually evaluated by non-spatially resolved electrical techniques that perform high-frequency measurements such as capacitance–voltage measurements[221] or Smith charts performed by Vector

Network Analyzers (VNAs), as previously mentioned. These techniques only offer macroscopic measurements, *i.e.*, only provide the averaged electrical behaviour of the packaged microelectronic system without distinguishing the coupling effects coming from inside or outside the die. A solution to perform an accurate characterization and de-embedding is to enhance the system observability by directly accessing the critical internal nodes at the die level. With appropriate fixtures, VNAs also allow local electrical measurements at the die level (100 μm pitch). This technique provides actual information on its frequency behaviour, being very interesting and complementary to other global approaches to account for all mentioned effects during any task concerning electrical characterization, testing and debugging. However, this invasive technique may alter the electrical behaviour of the sample, exempting its use for non-destructive tests. Moreover, its low spatial resolution obstructs the precise location of structural defects causing electric coupling between components or blocks, which can yield hot spot formation. Non-functional RFID pad-free chips have also been analyzed by modulating their powering scheme and non-invasively sensing their surface IR emission with an IR camera following lock-in strategies.[222] This approach is justified by the chip wireless powering strategy and its pad-free design. As a result, latch-up triggering has been identified as the failure mechanism, also showing that electrical FOMs can be extracted non-invasively (*i.e.*, coil coupling frequencies and their bandwidth).

13.5.2 Energy Processing Electronics: Power Devices and Light-emitting Devices

For a long time, thermometry in power devices was based on the use of electrical TSPs.[223] The most used TSP in power electronics is the voltage drop of a PN junction forward biased by a low current level, since all power devices contain inherent PN junctions. Apart from the voltage drop in the PN junction of a power diode, some examples are the voltage drop between the collector–emitter terminals of a bipolar or IGBT in a saturation regime. Other TSPs used in MOS-controlled power devices include the threshold voltage. Despite the threshold voltage being less linear and less sensitive than the forward-biased PN junction, it allows the extraction of a maximum operation temperature much more precisely and accurately, depending on the operation regime of the DUT, *e.g.*, short-circuit for IGBTs.[224] A major advantage of the TSPs used is that temperature measurements can be performed on fully packaged devices with little or no modification, directly measuring the temperature within the device. However, an average value of the device temperature is obtained, because of the calibration experimental conditions.[225,226] This calibration is performed by measuring the desired TSP once the whole device is heated and stabilized to a given steady-state temperature.

In order to improve the low spatial resolution presented by TSP-based techniques, temperature sensors have been placed at DUT inspection points by either monolithic integration or mechanical contact. In the first approach, PN junctions forward biased by a low current,[227] an emitter-based junction of a bipolar transistor,[228] or resistors[229] are employed as temperature sensors. However, sometimes the temperature measurement at the desired inspection point cannot be possible due to technological limitations. Moreover, when multiple sensors are required for measuring different inspection points, the packaging and device design complexity increases, rendering the thermal mapping unviable. Thus, this method is only useful for measuring the maximum temperature rise in operation conditions, as well as for protecting the device when a high temperature rise is produced.[227,228] Pfost *et al.*[228] have calibrated temperature sensors up to 600 °C integrating them in a VDMOS (Vertical Double-Diffused MOS) device.

In recent decades, the dependence of the material optical properties on the temperature has suggested different thermometry techniques applicable to power devices. Basically, they may be categorized as *surface thermal imaging* or *laser probing techniques*. As explained in Section 13.4.2, the former consists of determining the temperature distribution on the DUT top surface by means of a CCD camera. This process may be performed by either directly measuring the radiated heat flux with an IR camera (IRT)[131,230] or previously depositing a thermosensitive film on the inspected surface (LCT)[134] or FMT.[138] As a result, an image containing the superficial thermal mapping of the DUT is recorded with the CCD camera. Although IRT shows lower spatial and temperature resolutions (~ 5 μm and ~ 1 K) than LCT or FMT (~ 1 μm and ~ 50 mK), the commercial availability of IRT equipment (*e.g.*, Flir Systems cameras)[231] has increased, particularly for use *in situ* when high resolution measurement is not required.[231] All these techniques are very useful for hot spot determination especially when lock-in strategies are used.[85,232]

Concerning the laser probing techniques, Abid *et al.*[233] applied thermoreflectance thermometry to power devices. They measured the temperature at several points on the top surface of a thyristor (GTO, concretely), achieving a 20 μm spatial resolution with a minimum measured temperature increase of 10 K. The time resolution of this technique is ultimately limited to the order of the relaxation times associated with the microscopic scattering processes governing the reflectance, which is typically of the order of picoseconds in metals. On the other hand, its spatial resolution is limited by the diffraction of the probe beam, being comparable to the beam wavelength (<1 μm). However, its calibration is complex, due to surface non-uniformities and the presence of different materials on the device top.

On the other hand, interferometric thermometry consists of determining the phase shift between two or more beams. One of them is taken as a reference and the others go through or are reflected inside the DUT (sensing beams). As the refractive index depends on temperature (thermo-optical effect), the sensing beams find a variation in their optical paths, inducing a phase shift with respect to the reference beam. By superposing all the beams on a photodetector, the phase shift is finally determined. The interferometric methods are backside laser probing and FPT.

The backside laser probe was used by Seliger *et al.*[154] and Fürböck *et al.*[234] to determine the temperature and free-carrier concentration at the channel of power MOSFETS and IGBTs, respectively. In order to provide full access through the DUT backside, a window of approximately 70×70 μm^2 must be opened in the rear metallization layer. After that, an anti-reflective coating is deposited to suppress multiple reflections within the substrate.[235] Its spatial resolution is limited by the laser spot size to approximately 2 μm. Although only integral information on the vertical profiles is provided, this technique constitutes a useful characterization method for vertical MOS-gated switches, since it is capable of detecting large temperature rises of 100 K or more. Later, Pogany *et al.*, studied Electrostatic Discharge (ESD) phenomena and the effect of short-circuit conditions on lateral-power MOSFETs (fast destructive events),[236] as well as avalanche phenomena in 90-V VDMOS devices.[237,238] They adapted this scanning method to a single-shot acquisition approach using a camera,[79] demonstrating in lateral-power MOSFETs another approach with the same setup working with an illumination source, which stimulated band-to-band transitions[239] and temperature evolution modulated light absorption in the DUT. The same group also used scanning approaches to study non-destructive events in lateral-power MOSFETs,[155] power superjunction MOSFETs,[240] pHEMTs,[241,242] and GaN Schottky diodes,[243] as well as to assess the proper operation of integrated temperature sensors in VDMOS devices.[228]

Seliger *et al.* performed Fabry–Pérot reflectivity measurements on lateral-power MOSFETs.[244] Although reported works on FPT measurements are not found in the literature, they could be applied to vertical power devices. For FPT, a 5 μm spatial and 10 K temperature resolution are achieved, making it possible to measure temperature increments up to 300 K or more. This technique was appropriately demonstrated by Perpiñà *et al.*[181] They showed the suitability of this technique using a thermal test chip[181] and the temperature of a Punch-Through Insulated Gate Bipolar Transistor (PT-IGBT) was measured.[245] They experimentally obtained the thermo-optical coefficient agreeing with other reported results,[246] but disagreeing with simulation results reported by Seliger *et al.*[244] In addition, they compared FPT transmission results with other laser probing techniques (IIR-LD), obtaining a noticeable agreement.

Using the Raman effect, the temperature between different semiconductor devices, such as MOSFETS, GaAs devices, Si RF devices *etc.* has been measured.[247–250] Raman methods have allowed accurate measurements of GaAs device channel temperatures with high spatial resolution. Significant variations in temperature, not only across a multifinger device, but also inside the drain-gate gap of individual fingers were evident. Thermal resistance as a function of base-plate temperature and finger pitch was determined. Temperature measurements can be employed as a powerful tool not only to predict device lifetime from the device peak temperature, but also to optimize device layout by examining the thermal cross-talk between individual fingers. In this paper, the obtained Raman temperatures were found to be higher than IRT results, which is due to the improved spatial resolution of microRaman spectroscopy. In addition, the integration of Raman and IRT techniques has also been reported.[251] In this work, the authors use both techniques integrated together to analyze the self-heating in AlGaN/GaN Heterostructure Field-Effect Transistor (HFET) devices providing not only an improvement in spatial resolution down to 0.5 μm on the surface, but also unprecedented microscale depth resolution for true 3D thermography. Finally, the thermal resistance of GaAs Metal Semi-conductor Field-Effect Transistors (MESFETs) has also been evaluated using photocurrent spectral analysis in different environmental and temperature-controlled conditions.[252]

13.6 Conclusions

The microelectronic revolution in the fifties and sixties allowed the introduction of microelectronic systems into our daily lives, which become more and more reliable and rugged. However, the continuous scaling down has positioned microelectronic technology at its limits. Several technologies are being envisaged to replace CMOS technology in the different technological domains (More Moore, Beyond CMOS, More Than Moore and Heterogeneous Integration). In this scenario, temperature measurements have been crucial for facing challenges related to final system thermal management design, reliability and testability. In this sense, temperature monitoring is used to assess the thermal management design adopted, to detect failures in the systems studied, and even to determine information about their electrical behaviour.

The beginning of the chapter reviewed the more relevant issues related to heat generation and testability at the chip level, for both signal and energy processing microelectronic devices. Concerning the signal processing, critical issues related to heat generation and testability in the different technological domains were also explained in terms of shrinkage, integration density and analysis of FOMs in RF/Analog ICs. As regards energy processing devices, the chapter has focused on power semi-conductors and light-emission devices and the effect when they are heterogeneously integrated into energy management systems.

A complete review of thermal characterization techniques at the chip level has been presented, differentiating between electrical, optical and near-field microscopy techniques. The electrical techniques include the use of TSPs and built-in temperature sensors to monitor the chip temperature. Optical methods offer no electrical coupling with inspected devices, granting at the same timeoff-chip access to the surface or depth of the inspected die. According to this, optical techniques for temperature measurements can be classified into surface and depth-resolved approaches. In the former case, the dependence of the surface material optical properties on temperature is used (IR, liquid crystal changes, fluorescence, surface reflectance, luminescence, optical path change and spectral thermal modulation) whereas in the latter, bulk material properties are used (thermo-optical effect). Near-field microscopy techniques are used to improve the lateral resolution of some of the above-mentioned techniques such as thermoreflectance, fluorescence and IR-based measurements to extract surface temperature maps.

Finally, some thermal studies at the chip level in microelectronics for signal and energy processing have been presented. These studies have allowed thermal mapping of the chip surface to be obtained, but have also allowed the analysis of failure mechanisms of such devices. The main mechanism being an excess of temperature caused by a local current density increase (current crowding). Such an excess of temperature results in a non-uniform temperature distribution, sometimes concentrated in a hot spot. From the different studies, it has been demonstrated that these structural weak spots can originate in the design process, in the manufacturing process, or be created after specific test campaigns (surge current, power cycling, thermal cycling, ageing tests and others). The thermal measurements are used to determine these hot spots when they operate under real working conditions, eventually contributing to improved circuit design.

List of Acronyms

ADC	Analog-to-Digital Converter
AFM	Atomic Force Microscopy
AlN	Aluminium nitride
AOM	Acousto-Optical Modulator
ASE	Amplifier Spontaneous Emission
BM	Backside Metallisation
BIST	Built-In Self-Testing
BS	Beam Splitter
CCD	Charge-Coupled Device
CDO	Carbon-Doped Oxide
CFL	Compact Fluorescent Lighting
CMOS	Complementary Metal Oxide Semi-conductor
CTE	Coefficient of Thermal Expansion
CtW	Chip-to-Wafer
CUT	Circuit Under Test
D	Diamond
DAC	Digital-to-Analog Converter
DFB	Distributed Feedback
DRAM	Dynamic Random Access Memory
DUT	Device Under Test
EL	Electroluminescence
ESD	Electrostatic Discharge
FinFET	Fin Field-Effect Transistor

FMT	Fluorescence Microthermography
FOM	Figure of Merit
FPA	Focal Plane Array
FPI	Fabry–Pérot Interferometry
FPR	Fabry–Pérot Reflectivity
FPT	Fabry–Pérot Transmission
FQP	Four-Quadrant Photodiode
FSG	Fluorosilicate glass
GaAs	Gallium arsenide
GaN	Gallium nitride
GTO	Gate Turn-Off
HFET	Heterostructure Field-Effect Transistor
HEMT	High-Electron-Mobility Transistor
IC	Integrated Circuit
IGBT	Insulated Gate Bipolar Transistor
IIR-LD	Internal Infrared Laser Deflection
IO	Input/Output
IR	Infrared
IR-LIT	Infrared Lock-in Thermography
IRT	Infrared Thermography
JFET	Junction Field-Effect Transistor
k	dielectric constant
KGD	Known Good Dies
LCT	Liquid Crystal Thermography
LED	Light-Emitting Diode
LNA	Low Noise Amplifier
MEMS	Microelectromechanical Systems
MESFET	Metal Semi-conductor Field-Effect Transistor
MMIC	Monolithic Microwave Integrated Circuit
MOSFET	Metal Oxide Semi-conductor Field-Effect Transistor
MP	Metallization Backside
MS	Metallization Surface
NA	Numerical Aperture
NIR	Near-Infrared
NSOM	Near-Field Scanning Optical Microscopy
OLED	Organic Light-Emitting Diode
PA	Power Amplifier
PIC	Photonic Integrated Circuits
PL	Photoluminescence
PM	Power Modulator
PSA	Photocurrent Spectral Analysis
PTAT	Proportional to the Absolute Temperature
PT-IGBT	Punch-Through Insulated Gate Bipolar Transistor
PV	Photovoltaic
RF	Radiofrequency
RFID	Radiofrequency Identification

S&A	Sensor & Actuator
SiC	Silicon carbide
SiN	Silicon nitride
SiP	System-in-Chip
SLD	Super Luminescent Diode
SM	Surface Metallisation
SOA	Semi-conductor Optical Amplifier
SoC	System-on-Chip
SoI	Silicon-on-Insulator
SSL	Solid-State Lighting
SThM	Scanning Thermal Microscopy
TSP	Temperature-Sensitive Parameter
TSV	Through Silicon Vias
UV	Ultraviolet
VCO	Voltage-Controlled Oscillator
VCSEL	Vertical Cavity Surface Emitting Laser
VDMOS	Vertical Double-Diffused Metal Oxide Semi-conductor
VLSI	Very Large Scale Integration
VNA	Vector Network Analyzers
WBG	Wide Bandgap
WtW	Wafer-to-Wafer

References

1. T. Lhommeau, X. Perpiñà, C. Martin, R. Meuret, M. Mermet-Guyennet and M. Karama, *Microelectron. Reliab.*, 2007, **47**, 1779.
2. J. Bardeen and W. H. Brattain, *Phys. Rev.*, 1949, **75**, 230.
3. J. S. Kilby, *IEEE Trans. Electron Devices*, 1976, **23**, 648.
4. S. E. Thompson and S. Parthasarathy, *Mater. Today*, 2006, **9**, 20.
5. G. E. Moore, *Electronics*, 1965, **38**, 114.
6. D. E. Nikonov and I. A. Young, *Proc. IEEE*, 2013, **101**, 2498.
7. K. Bernstein, R. K. Cavin III, W. Porod, A. Seabaugh and J. Welser, *Proc. IEEE*, 2012, **98**, 2169.
8. C. Brylinski, *Thermal Nanosystems and Nanomaterials*, ed. S. Volz, *Topics in Applied Physics*, Springer Verlag, Heidelberg, Germany, 2009, vol. 118, ch. 12, p. 367.
9. G. Tessier, *Thermal Nanosystems and Nanomaterials*, ed. S. Volz, *Topics in Applied Physics*, Springer Verlag, Heidelberg, Germany, 2009, vol. 118, ch. 13, p. 389.
10. R. H. Dennard, F. H. Gaensslen, H.-N. Yu, V. L. Rideout, E. Bassour and A. R. LeBlanc, *IEEE J. Solid-State Circuits*, 1974, **9**, 256.
11. H. Iwai, *Microelectron. Eng.*, 2009, **86**, 1520.
12. H. Iwai and S. Ohmi, *Microelectron. Reliab.*, 2002, **42**, 1251.
13. S. Takagi, T. Tezuka, T. Irisawa, S. Nakaharai, T. Numata, K. Usuda, N. Sugiyama, M. Shichijo, R. Nakane and S. Sugahara, *Solid-State Electron.*, 2007, **51**, 526.

14. M. Pedrama and S. Nazarian, *Proc. IEEE*, 2006, **94**, 1487.

15. E. Pop, S. Sinha and K. E. Goodson, *Proc. IEEE*, 2006, **94**, 1587.

16. A. Shakouri, Nanoscale Thermal Transport and Microrefrigerators on a Chip, *Proc. IEEE*, 2006, **94**, 1613.

17. E. Mounier, in 2nd Workshop on Design, Control and Software Implementations for distributed MEMS (dMEMS 2012), Besançon (Francia), 2012.

18. T. Simunic, K. Mihic and G. Micheli, *Integrated Circuit and System Design: Power and Timing Modeling, Optimization and Simulation*, ed. V. Paliouras, J. Vounckx and D. Verkest, Springer-Verlag, Heidelberg, Germany, 2005, vol. 3728, p. 237.

19. J. A. Davis, R. Venkatesan, A. Kaloyeros, M. Beylansky, S. J. Souri, K. Banerjee, K. C. Saraswat, A. Rahman, R. Reif and J. D. Meindl, *Proc. IEEE*, 2001, **89**, 305.

20. J. D. Meindl, *Micro IEEE*, 2003, **23**, 28.

21. M. Lantz, C. Hagleitner, M. Despont, P. Vettiger, M. Cortese and B. Vigna, *More than Moore: Creating High Value Micro/Nanoelectronics Systems*, ed. G. Q. Zhang and A. J. van Roosmalen, Springer, 2009, ch. 5, p. 109.

22. Y. Xie, G. H. Loh, B. Black and K. Bersnstein, *ACM J. Emerg. Technol. Comput. Syst.*, 2006, **2**, 65.

23. P. Garrou, C. Bower and P. Ramm, *Handbook of 3D Integration: Technology and Applications of 3D Integrated Circuits*, Wiley-VCH Verlag, Weinheim, 2008.

24. J.-Q. Lu, *Proc. IEEE*, 2009, **97**, 189.

25. M. Koyanagi, T. Fukushima and T. Tanaka, *Proc. IEEE*, 2009, **97**, 49.

26. M. Motoyoshi, *Proc. IEEE*, 2009, **97**, 43.

27. V. F. Pavlidis and E. G. Friedman, *Proc. IEEE*, 2009, **97**, 123.

28. P. Marchal, B. Bougard, G. Katti, M. Stucchi, W. Dehaene, A. Papanikolaou, D. Verkest, B. Swinnen and E. Beyne, 3-D technology assessment: path-finding the technology/design sweet-spot, *Proc. IEEE*, Jan. 2009, **97**(1), 96.

29. T. Zhang, R. Micheloni, G. Zhang, Z. R. Huang and J. J.-Q. Lu, 3-D data storage power delivery, and RF/optical transceiver—case studies of 3-D integration from system design perspectives, *Proc. IEEE*, 2009, **97**, 161.

30. K. Banerjee, S. J. Souri, P. Kapur and K. C. Saraswat, *Proc. IEEE*, 2001, **89**, 602.

31. T. Brunschwiler and B. Michel, *Handbook of 3D Integration: Technology and Applications of 3D Integrated Circuits*, ed. P. Garrou, C. Bower and P. Ramm, Wiley-VCH Verlag, Weinheim, 2008, vol. 2, ch. 33, p. 635.

32. C.-L. Lung, J.-H. Chien, Y.-F. Chou, D.-M. Kwai and S.-C. Chang, *VLSI Design*, ed. Dr E. Tlelo-Cuautle, InTech, ch. 2, p. 19, Available from: http://www.intechopen.com/books/vlsi-design/three-dimensional-integrated-circuits-design-for-thousand-core-processors-from-aspect-of-thermal-man.

33. H.-H. S. Lee and K. Chakrabarty, *IEEE Des. Test Comput.*, 2009, **26**, 26.

34. M. Onabajo, D. Gómez, E. Aldrete-Vidrio, J. Altet, D. Mateo and J. S. Silva-Martínez, *J. Electron. Test.*, 2011, **27**, 225.
35. A. Valdes-Garcia, F. Hussien, J. Silva-Martinez and E. Sánchez-Sinencio, *IEEE J. Solid-State Circuits*, 2006, **41**, 2301.
36. S. Bhattacharya and A. Chatterjee, in *Proceedings of IEEE International Test Conference* (ITC), 2004, p. 801.
37. K. Agarwal, J. Hayes and S. Nassif, *IEEE Trans Semiconductor Manufacturing*, 2008, **21**, 526.
38. H. S. Bennett, in *ITRS Reports*, 2009.
39. S. Tanimoto and H. Ohashi, *Phys. Status Solidi A*, 2009, **206**, 2417.
40. T. Nakamura, M. Miura, N. Kawamoto, Y. Nakano, T. Otsuka, K. Okumura and A. Kamisawa, *Phys. Status Solidi A*, 2009, **206**, 2403.
41. H. S. Chin, K. Y. Cheong and A. B. Ismail, *Metall. Mater. Trans. B*, 2010, **41**, 824.
42. www.cree.com.
43. J. Millán, P. Godignon, X. Perpiñà, A. Pérez-Tomás and J. Rebollo, *IEEE Trans. Power Electron.*, 2014, **29**, 2155.
44. S. Tarashioon, A. Baiano, H. van Zeijl, C. Guo, S. W. Koh, W. D. van Driel and G. Q. Zhang, *Microelectron. Reliab.*, 2012, **52**, 783.
45. J. Y. Tsao, M. E. Coltrin, M. A. Crawford and J. A. Simmons, *Proc. IEEE*, 2010, **98**, 1162.
46. J. Y. Tsao, H. D. Saunders, J. R. Creighton, M. E. Coltrin and J. A. Simmons, *J. Phys. D: Appl. Phys.*, 2010, **43**, 354001.
47. J. Popovic-Gerber, J. A. Oliver, N. Cordero, T. Harder, J. A. Cobos, M. Hayes, S. C. O'Mathuna and E. Prem, *IEEE Trans. Power Electron.*, 2012, **27**, 2338.
48. J. Jakovenko, J. Formánek, X. Perpiñà, X. Jordà, M. Vellvehi, R. J. Werkhoven, M. Husáka, J. M. G. Kunen, P. Bancken, P. J. Bolt and A. Gasse, *Microelectron. Reliab.*, 2013, **53**, 1076.
49. ASSIST Recommendation. *LED Life for General Lighting: Definition of Life*, vol. 1, issue no. 1, February 2005.
50. *IESNA LM-80-08 Standard*. IES approved method for lumen maintenance of LED light sources; 2008.
51. *IESNA LM-79-08 Standard*. IES approved method for the electrical and photometric measurements of solid state lighting products; 2008.
52. *IEEE Std 1413.1-2002*. IEEE guide for selecting and using reliability prediction based on IEEE 1413. IEEE standard coordinating committee 37.
53. M. Ha and S. Graham, *Microelectron. Reliab.*, 2012, **52**, 836.
54. E. Juntunen, A. Sitomaniemi, O. Tapaninen, R. Persons, M. Challingsworth and V. Heikkinen, *IEEE Trans. Compon., Packag., Manuf. Technol.*, 2012, **12**, 1957.
55. J. M. Alonso, J. Vina, D. G. Vaquero, G. Martínez and R. Osorio, *IEEE Trans. Ind. Electron.*, 2012, **59**, 1689.
56. M. Harris, *IET Magazine*, 2009, **4**, 18.

57. X. Jordà, X. Perpiñà, M. Vellvehi, W. Hertog, M. Perálvarez and J. Carreras, in *Proceedings of Therminic*, 2013, Berlin, Germany, p. 194.

58. L. Liu, D. Yang, G. Q. Zhang, Z. You, F. Hou and D. Liu, in *Proceedings of 12th EuroSimE conference*, 2011, Linz, Austria.

59. W. D. van Driel, C. A. Yuan, S. Koh and G. Q. Zhang, in *Proceedings of 12th EuroSimE conference*, Linz, Austria.

60. M.–H. Chang, D. Das, P. V. Varde and M. Pecht, *Microelectron. Reliab.*, 2012, **52**, 762.

61. G. Meneghesso, S. Levada, E. Zanoni, G. Scamarcio, G. Mura, S. Podda, M. Vanzi, S. Du and I. Eliashevich, *Microelectron. Reliab.*, 2003, **43**, 1737.

62. M. Ishii and Y. Taga, *Appl. Phys. Lett.*, 2002, **80**, 3430.

63. G. Nenna, G. Flaminio, T. Fasolino, C. Minarini, R. Miscioscia, D. Palumbo and M. Pellegrino, *Macromol. Symp.*, 2007, **247**, 326.

64. N. Tessler, Y. Preezant, N. Rappaport and Y. Roichman, *Adv. Mater.*, 2009, **21**, 2741.

65. W. O. Akande, O. Akogwu, T. Tong and W. Soboyejo, *J. Appl. Phys.*, 2010, **108**, 1.

66. P. Schwamb, T. C. Reusch and C. J. Brabec, *Org. Electron.*, 2013, **14**, 1939.

67. J. C. Sturm, W. Wilson and H. Iodice, *IEEE J. Sel. Top. Quantum Electron.*, 1998, **4**, 75.

68. H. Ohashii, in *Proceedings of ISPSD*, 2012, Bruges, Belgium.

69. J. G. Kassakian and T. M. Jahns, *IEEE J. Emerg. Sel. Top. Power Electron.*, 2013, **1**, 47.

70. S. Mandray, J.-M. Guichon, J.-L. Schanen, M. M. Guyennet and J.-M. Dienot, *IEEE Trans. Ind. Appl.*, 2009, **45**, 871.

71. S. Eicher, M. Rahimo, E. Tsyplakov and D. Schneider, *Proc. IAS Ann. Meet.*, 2004, 1534.

72. P. Solomalala, J. Saiz, M. Mermet-Guyennet, A. Castellazzi, M. Ciappa, X. Chauffleur and J. P. Fradin, *Microelectron. Reliab.*, 2007, **47**, 1343.

73. J. Altet, W. Claeys, S. Dilhaire and A. Rubio, *Proc. IEEE*, 2006, **94**, 1519.

74. Y. Ezzahri, J. Christofferson, G. Zeng and A. Shakouri, *J. Appl. Phys.*, 2009, **106**, 114503.

75. K. Maize, J. Christofferson, and A. Shakouri, in Proceedings of Semitherm, San Jose, USA, 2008, p. 55.

76. A. Vertikov, M. Kuball, A. V. Nurmikko and H. J. Maris, *Appl. Phys. Lett.*, 1996, **69**, 2465.

77. B. Vermeersch, J.-H. Bahk, J. Christofferson and A. Shakouri, *J. Appl. Phys.*, 2013, **113**, 104502.

78. J. Christofferson, K. Yazawa and A. Shakouri, in *Proceedings of* IHTC, Washington, USA, 2010.

79. D. Pogany, V. Dubec, S. Bychikhin, C. Fürböck, M. Litzenberger, G. Groos, M. Stecher and E. Gornik, *IEEE Electron. Device Lett.*, 2002, **23**, 606.

80. X. Perpiñà, X. Jordà, N. Mestres, M. Vellvehi and P. Godignon, *Meas. Sci. Technol.*, 2004, **15**, 1011.
81. J. Altet, J. M. Rampnoux, J. C. Batsale, S. Dilhaire, A. Rubio, W. Claeys and S. Grauby, *Microelectron. Reliab.*, 2004, **44**, 95.
82. S. Dilhaire, E. Schaub, W. Claeys, J. Altet and A. Rubio, *Int. J. Therm. Sci.*, 2000, **39**, 544.
83. S. Dilhaire, J. Altet, S. Jorez, E. Schaub, A. Rubio and W. Claeys, *Microelectron. Reliab.*, 1999, **39**, 919.
84. J. Altet, M. A. Salhi, S. Dilhaire and A. Ivanov, *Electron. Lett.*, 2004, **40**, 241–242.
85. J. León, M. Berthou, X. Perpiñà, V. Banu, J. Montserrat, M. Vellvehi, P. Godignon and X. Jordà, *J. Phys. D: Appl. Phys.*, 2014, **47**, 055102.
86. O. Breitenstein, M. Langenkamp, F. Altmann, D. Katzer, A. Lindner and H. Eggers, *Rev. Sci. Instrum.*, 2000, 7, 4155.
87. J. Altet, S. Dilhaire, S. Volz, J. M. Rampnoux, A. Rubio, S. Grauby, L. D. Patino-López, W. Claeys and J. B. Saulnier, *Microelectron. J.*, 2002, **33**, 689.
88. J. Altet, M. A. Salhi, S. Dilhaire, A. Syal and A. Ivanov, *Electron. Lett.*, 2003, **39**, 1140.
89. J. Altet, E. Aldrete, D. Mateo, X. Perpiñà and X. Jordà, *Meas. Sci. Technol.*, 2008, **19**, 115704.
90. J. León, X. Perpiñà, J. Altet, M. Vellvehi and X. Jordà, *Appl. Phys. Lett.*, 2013, **102**, 054103.
91. J. León, X. Perpiñà, J. Altet, M. Vellvehi and X. Jordà, *J. Phys. D: Appl. Phys.*, 2013, **46**, 445501.
92. B. C. Forget, S. Grauby, D. Fournier, P. Gleyzes and A. C. Boccara, *Electron. Lett.*, 1997, **33**, 1688.
93. S. Grauby, B. C. Forget, S. Holé and D. Fournier, *Rev. Sci. Instrum.*, 1999, **70**, 3603.
94. S. Lopez-Buedo, J. Garrido and E. Boemo, *IEEE Des. Test Comput.*, 2000, **17**, 84.
95. E. R. Motto and J. F. Donlon, in Proceedings of *Applied Power Electronics Conference and Exposition* (APEC), 2012, p. 176.
96. D. F. Hilbiber, *ISSCC Digest Tech. Pap.*, 1964, 7, 32.
97. R. J. Widlar, *IEEE J. Solid-State Circuits*, 1971, **6**, 2.
98. Y. P. Tsividis, *IEEE J. Solid-State Circuits*, 1980, **15**, 1076.
99. F. Mascart, J. Vindevoghel, E. Constant, G. Blondel and J. Magarschack, *Electron. Lett.*, 1986, **22**, 122.
100. J. B. Casady, W. C. Dillard, R. W. Johnson and U. Rao, *IEEE Trans. Compon., Packag., Manuf. Technol., Part A*, 1996, **19**, 416.
101. C. M. Jha, G. Bahl, R. Melamud, S. A. Chandorkar, M. A. Hopcroft, B. Kim, M. Agarwal, J. Salvia, H. Mehta and T. W. Kenny, in *Proceedings of Solid-State Sensors, Actuators and Microsystems Conference*, 2007, p. 229.
102. B. Rue and D. Flandre, in *Proceedings of IEEE International SOI Conference*, 2007, p. 111.

103. C.-C. Chen and H.-W. Chen, *IEEE Sens. J.*, 2014, **14**, 278.
104. D. Barrettino, M. Graf, K. Kirstein, A. Hierlemann and H. Baltes, in *Proceedings of International Symposium on Circuits and Systems* ISCAS, 2004, vol. 4, p. 888.
105. X. Jordà, X. Perpiñà, M. Vellvehi, F. Madrid, D. Flores, S. Hidalgo and J. Millán, *Appl. Therm. Eng.*, 2011, **31**, 1664.
106. C.-K. Kim, B.-S. Kong, C.-G. Lee and Y. H. Jun, in *Proceedings of IEEE International Symposium on Circuits and Systems ISCAS*, 2008, p. 3094.
107. P. Krummenacher and H. Oguey, *Sens. Actuators, A*, 1990, **22**, 636.
108. V. Szekely, C. Marta, Z. Kohari and M. Rencz, *IEEE Trans. Very Large Scale Integr. (VLSI) Syst.*, 1997, **5**, 270.
109. A. Bakker, *Proc. IEEE*, 2002, **2**, 1423.
110. M. A. P. Pertijs, J. H. Huijsing, *Precision Temperature Sensors In CMOS Technology*, Springer, 2006.
111. Y. Jeong and F. Ayazi, in *Proceedings of IEEE International Symposium on Circuits and Systems* (ISCAS), 2012, p. 3118.
112. K. Ueno, T. Hirose, T. Asai and Y. Amemiya, in *Proceedings of International Symposium on Intelligent Signal Processing and Communications*, ISPACS, 2006, p. 546.
113. A. Bakker and J. H. Huijsing, *IEEE J. Solid-State Circuits*, 1996, **31**, 933.
114. F. Sebastiano, L. J. Breems, K. A. A. Makinwa, S. Drago, D. M. W. Leenaerts and B. Nauta, *IEEE J. Solid-State Circuits*, 2010, **45**, 2591.
115. J. S. Shor and K. Luria, *IEEE J. Solid-State Circuits*, 2013, **48**, 2860.
116. J. Qian, J. Chen, C. Zhang and L. Wu, in *Proceedings of 10th IEEE International Conference on Solid-State and Integrated Circuit Technology* ICSICT, 2010, p. 1404.
117. M. K. Law, A. Bermak and H. C. Luong, *IEEE J. Solid-State Circuits*, 2010, **45**, 1246.
118. M. Onabajo, J. Altet, E. Aldrete-Vidrio, D. Mateo and J. Silva-Martinez, *IEEE Trans. Circuits Syst.*, 2011, **58**, 458.
119. K. E. Mortenson, *Proc. IRE*, 1957, 504.
120. D. L. Blackburn, in *Proceedings of Fourth Annual IEEE Semiconductor Thermal and Temperature Measurement Symposium*, SEMITHERM, 1988, p. 1.
121. Y. Avenas, L. Dupont and Z. Khatir, *IEEE Trans. Power Electron.*, 2012, **27**, 3081.
122. R. Schmidt and U. Scheuermann, in *Proceedings of 13th European Conference on Power Electronics and Applications*, EPE, 2009, p. 1.
123. G. Gaussorgues, *Infrared Thermography*, Chapman & Hall, Cambridge, 1994.
124. G. Rybicki and A. P. Lightman, *Radiative Processes in Astrophysics*, Wiley-Interscience, New York, 1979.
125. W. Minkina and S. Dudzik, *Infrared Thermography. Errors and Uncertainties*, Wiley, 2009.

126. R. Brandt, C. Bird and G. Neuer, *Measurement*, 2008, **41**, 731.
127. M. R. Dury, T. Theocharous, N. Harrison, N. Fox and M. Hilton, *Opt. Commun.*, 2007, **270**, 262.
128. M. Gradhand and O. Breitenstein, *Rev. Sci. Instrum.*, 2005, **76**, 053702.
129. G. A. Bennett and S. D. Briles, *IEEE Trans. Compon., Hybrids, Manuf. Technol.*, 1989, **12**, 690.
130. M. Vellvehi, X. Perpiñà, G. L. Lauro, F. Perillo and X. Jordà, *Rev. Sci. Instrum.*, 2011, **82**, 114901.
131. H. Köck, V. Košel, C. Djelassi, M. Glacanovics and D. Pogany, *Microelectron. Reliab.*, 2009, **49**, 1132.
132. O. Breitenstein, W. Warta and M. Langenkamp, *Lock-in Thermography: Basics and Use for Evaluating Electronic Devices and Materials*, Springer, 2nd edn, 2010.
133. K. Azar, J. R. Benson and V. P. Manno, in *IEEE Semiconductor Thermal Measurement and Management Symposium*, Phoenix, USA, 1991, p. 23.
134. A. Csendes, V. Székely and M. Rencz, Thermal mapping with liquid crystal method, *Microeletron. Eng.*, 1996, **31**, 281290.
135. H. Lin, M. Khan and T. Giao, in *Proceedings of the 20th International Symposium for Testing and Failure Analysis* ISTFA, Los Angeles, USA, 1994, p. 81.
136. J. Hiatt, *Proceedings of the IEEE International Reliability Physics Symposium IRPS*, Orlando, USA, 1981, p. 130.
137. E. J. Bowen and J. Sahu, *J. Phys. Chem.*, 1959, **63**, 4.
138. D. L. Barton and P. Tangyunyong, *Microeletron. Eng.*, 1996, **31**, 271.
139. P. Kolodner and J. A. Tyson, *Appl. Phys. Lett.*, 1982, **40**, 782.
140. M. Farzaneh, K. Maize, D. Lüerßen, J. A. Summers, P. M. Mayer, P. E. Raad, K. P. Pipe, A. Shakouri, R. J. Ram and J. A. Hudgings, *J. Phys. D: Appl. Phys.*, 2009, **42**, 143001.
141. V. Quintard, G. Deboy, S. Dilahire, D. Lewis, T. Plan and W. Claeys, *Microelectron. Eng.*, 1993, **31**, 303.
142. S. Grauby, B. C. Forget, S. Holé and D. Fournier, *Rev. Sci. Instrum.*, 1999, **70**, 3603.
143. W. Claeys, S. Dilhaire, V. Quintard, J. P. Dom and Y. Danto, *Qual. Reliab. Eng. Int.*, 1993, **9**, 303.
144. G. Tessier, M.-L. Polignano, S. Pavageau, C. Filloy, D. Fournier, F. Cerutti and I. Mica, *J. Phys. D: Appl. Phys.*, 2006, **39**, 4159.
145. S. Grauby, A. Salhi, J.-M. Rampnoux, H. Michel, W. Claeys and S. Dilhaire, *Rev. Sci. Instrum.*, 2007, **78**, 074902.
146. G. Tessier, D. Hole and D. Fournier, *Appl. Phys. Lett.*, 2001, **78**, 2267.
147. J. A. Summers, T. Yang, M. T. Tuominen and J. A. Hudgings, *Rev. Sci. Instrum.*, 2010, **81**, 014902.
148. S. Dilhaire, S. Grauby and W. Claeys, *Appl. Phys. Lett.*, 2004, **84**, 822.
149. S. Dilhaire, D. Fournier, and G. Tessier, in *Microscale and Nanoscale Heat Transfer*, ed. S. Volz, Springer-Verlag, Heidelberg, Germany, 2007, vol. 107, p. 239.

150. J. Christofferson, K. Yazawa and A. Shakouri, in *Proceedings of International Heat Transfer Conference* IHTC, Washington DC, USA, 2010.
151. S. A. Thorne, S. B. Ippolito, M. S. Ünlü and B. B. Goldberg, *Mater. Res. Soc. Proc.*, 2003, **738**, G12.9.1.
152. P. M. Mayer, D. Lüerssen, R. J. Ram and J. A. Hudgings, *J. Opt. Soc. Am. A*, 2007, **24**, 1156.
153. M. Goldstein, G. Sölkner and E. Görnik, *Microelectron. Eng.*, 1994, **31**, 87.
154. N. Seliger, P. Habaš and E. Gornik, *Microelectron. Eng.*, 1996, **24**, 431.
155. D. Pogany, S. Bychikhin, C. Furbock, M. Litzenberger, E. Gornik, G. Groos, K. Esmark and M. Stecher, *IEEE Trans. Electron. Devices*, 2002, **49**, 2070.
156. R. Thalhammer, C. Fürböck, N. Seliger, G. Deboy, E. Gornik, and G. Wachutka, in *Proceedings of International Symposium of Power Semiconductor Devices and ICs ISPSD*, Kyoto, Japan, 1998, p. 199.
157. V. Dubec, S. Bychikhin, D. Pogany, E. Gornik, T. Brodbeck and W. Stadler, *Microelectron. Reliab.*, 2007, **47**, 1539.
158. D. Pogany, S. Bychikhin, M. Litzenberger, E. Gornik, G. Groos and M. Stecher, *Appl. Phy. Lett.*, 2002, **81**, 2881.
159. M. Goldstein, G. Sölkner and E. Görnik, *Rev. Sci. Instrum.*, 1993, **64**, 3009.
160. R. Thalhammer, *Advances in Imaging and Electron Physics*, ed. P. W. Hawkes, Elsevier, 2005, vol. 135, p. 225.
161. D. Pogany, N. Seliger, E. Gornik, M. Stoisiek and T. Lalinsky, *J. Appl. Phys.*, 1998, **84**, 4495.
162. N. Seliger, E. Gornik, C. Fürböck, D. Pogany, P. Habǎs, R. Thalhammer and M. Stoisiek, *Elektrotechnik Informationstechnik*, 1998, **115**, 403.
163. N. Seliger, D. Pogany, C. Fürböck, P. Habǎs, E. Gornik and M. Stoisiek, *Microelectron. Reliab.*, 1997, **37**, 1727.
164. E. Wiedemann, *Ann. Phys.*, 1888, **34**, 446.
165. B. Valeur and M. N. Berberan-Santos, *J. Chem. Educ.*, 2011, **88**, 731.
166. P. Würfel, S. Finkbeiner and E. Daub, *Appl. Phys. A: Mater. Sci. Process.*, 1995, **A60**, 67.
167. K. Schick, E. Daub, S. Finkbeiner and P. Würfel, *Appl. Phys. A: Solids Surf.*, 1992, **A54**, 09.
168. P. Würfel, *Physik der Solarzellen*, Springer Spektrum, Berlin, 2000.
169. T. Fuyuki, H. Kondo, T. Yamazaki, Y. Takahashi and Y. Uraoka, *Appl. Phys. Lett.*, 2005, **86**, 262108.
170. R. Kane and H. Sell, *Revolution in Lamps: a Chronicle of 50 Years of Progress*, The Fairmont Press, Inc, 2nd edn, 2001.
171. T. Trupke, R. A. Bardos, M. C. Schubert and W. Warta, *Appl. Phys. Lett.*, 2006, **89**, 44107.
172. S. M. He, X. D. Luo, B. Zhang, L. Fu, L. W. Cheng, J. B. Wang and W. Lu, *Rev. Sci. Instrum.*, 2011, **82**, 123101.

173. G. R. Hayes and B. Deveaud, *Phys. Status Solidi A*, 2002, **190**, 637.
174. http://www.nrel.gov/.
175. J. Christofferson, K. Maize, Y. Ezzari, J. Shabani, X. Wang and A. Shakouri, *J. Electron. Packag.*, 2008, **130**, 041101.
176. F. LaPlant, G. Laurence and D. Ben-Amotz, *Appl. Spectrosc.*, 1996, **50**, 1034.
177. D. Savian, A. Di Carlo, P. Lugli, M. Peroni, C. Cetronio, C. Lanzieri, G. Meneghesso and E. Zanoni, *Electron. Lett.*, 2003, **39**, 247.
178. R. Thalhammer, *J. Appl. Phys.*, 2005, **97**, 023102.
179. M. Born and E. Wolf, *Principles of Optics*, Pergamon Press, Exeter, 6th edn, 1989.
180. R. Thalhammer and G. Wachutka, *J. Opt. Soc. Am. A*, 2003, **20**, 698.
181. X. Perpiñà, X. Jordà, M. Vellvehi, J. Millán and N. Mestres, *Semicond. Sci. Technol.*, 2006, **21**, 1537.
182. G. Deboy, G. Sölkner, E. Wolfgang and W. Claeys, *Microeletron. Eng.*, 1996, **31**, 299.
183. X. Perpiñà, X. Jordà, M. Vellvehi, J. Millán and N. Mestres, *Rev. Sci. Instrum.*, 2005, **76**, 025106.
184. X. Perpiñà, X. Jordà, F. Madrid, S. Hidalgo, D. Flores, M. Vellvehi and N. Mestres, *Rev. Sci. Instrum.*, 2005, **76**, 094905.
185. X. Perpiñà, X. Jordà, M. Vellvehi, J. Altet and N. Mestres, *J. Phys. D: Appl. Phys.*, 2008, **41**, 155508.
186. X. Perpiñà, X. Jordà, M. Vellvehi and J. Altet, *Appl. Phys. Lett.*, 2011, **98**, 164104.
187. B. T. Rosner and D. W. van der Weide, *Rev. Sci. Instrum.*, 2002, **73**, 2505.
188. G. B. M. Fiege, F. J. Niedernostheide, H. J. Schulze, R. Barthelmess and L. J. Balk, *Microelectron. Reliab.*, 1999, **39**, 1149.
189. A. Vertikov, M. Kuball, A. V. Nurmikko and H. J. Maris, *Appl. Phys. Lett.*, 1996, **69**, 2465.
190. T. Fujii, Y. Taguchi, T. Saiki and Y. Nagasaka, *Rev. Sci. Instrum.*, 2012, **83**, 124901.
191. C. Feng, M. S. Unlu, B. B. Goldberg and W. D. Herzog, in *Lasers and Electro-Optics Society Annual Meeting LEOS*, 1996, p. 249.
192. G. B. M. Fiege, A. Altes, R. Heiderhoff and L. J. Balk, *J. Phys. D: Appl. Phys.*, 1999, **32**, 13.
193. A. Mujamdar, J. P. Carrejo and J. Lai, *Appl. Phys. Lett.*, 1993, **62**, 2501.
194. A. Bontempi, L. Thiery, D. Teyssieux, D. Briand and P. Vairac, *Rev. Sci. Instrum.*, 2013, **84**, 103703.
195. W. Liu and B. Yang, *Sensor Rev.*, 2007, **27**, 298.
196. E. Puyoo, S. Grauby, J.-M. Rampnoux, E. Rouvière and S. Dilhaire, *J. Appl. Phys.*, 2011, **109**, 024302.
197. L. Shi, J. Zhou, P. Kim, A. Bachtold, A. Majumdar and P. L. McEuen, *J. Appl. Phys.*, 2009, **105**, 104306.
198. J. Kölzer, E. Oersterschulze and G. Deboy, *Microeletron. Eng.*, 1996, **31**, 251.

199. J. Christofferson, K. Maize, Y. Ezzahri, J. Shabani, X. Wang and A. Shakouri, *J Electron. Packag.*, 2008, **130**, 041101.
200. A. Vertikov, M. Kuball, A. V. Nurmikko and H. J. Maris, *Appl. Phys. Lett.*, 1996, **69**, 2465.
201. L. Shi and A. Majumdar, Applied Scanning Probe Methods, *NanoScience and Technology*, Springer, 2004, p. 327.
202. W. Claeys, S. Dilhaire, V. Quintard, J. P. Dom and Y. Danto, *Qual. Rel. Eng. Int.*, 1993, **9**, 303.
203. S. Dilhaire, T. Phan, E. Schaub, and W. Claeys, in *Proceedings of the 6th International Symposium on the Physical and Failure Analysis of Integrated Circuits IPFA*, Singapore, 1997, p. 62.
204. X. Wang, Y. Ezzahri, J. Christofferson and A. Shakouri, *J Phys. D: App. Phys.*, 2009, **42**, 075101.
205. J. Christofferson, K. Maize, Y. Ezzahri, J. Shabani, X. Wang and A. Shakouri, *J. Electron. Packag.*, 2008, **130**, 041101.
206. P. E. Raad, P. L. Komarov and M. G. Burzo, in *Proceedings of 23rd IEEE SEMITHERM Symposium*, San Jose, USA, 2007.
207. P. K. L. Chan, K. P. Pipe, Z. Mi, J. Yang, P. Bhattacharya and D. Lüerßen, *Appl. Phys. Lett.*, 2006, **89**, 011110.
208. P. K. L. Chan, K. P. Pipe, J. J. Plant, R. B. Swint and P. W. Juodawlkis, *Appl. Phys. Lett.*, 2006, **89**, 201110.
209. C. Degen, I. Fischer and W. Elsasser, *Appl. Phys. Lett.*, 2000, **76**, 3352.
210. E. W. Kapusta, D. Lüerßen and J. A. Hudgings, *IEEE Photon. Technol. Lett.*, 2006, **18**, 310.
211. D. Lüerßen, R. J. Ram, A. Hohl-AbiChedid, E. Clausen Jr. and J. A. Hudgings, *IEEE Photon. Technol. Lett.*, 2004, **16**, 1625.
212. Z. Bian, J. Christofferson, A. Shakouri and P. Kozodoy, *Appl. Phys. Lett.*, 2003, **83**, 3605.
213. P. J. Petersan and S. M. Anlage, *J. Appl. Phys.*, 1992, **84**, 3392.
214. J. Altet, A. Rubio, J. L. Rossello and J. Segura, *IEEE Commun. Mag.*, 2003, **41**, 98.
215. J. Altet, D. Gomez, X. Perpiñà, D. Mateo, J. L. González, M. Vellvehi and X. Jordà, *Sens. Actuators, A*, 2013, **192**, 49.
216. Y. I. Bontzios, *IEEE Trans. Electron. Devices*, 2010, **57**, 1751.
217. J. Y. Cheng, C. T. Huang and J. G. Hwu, *J. Appl. Phys.*, 2009, **106**, 074507.
218. Y. Wei, B. Cemal, in *Proceedings of 13th IEEE Intersociety Conference on Thermal and Thermomechanical Phenomena in Electronic Systems (ITherm)*, 2012, San Diego, USA, p. 408.
219. J. H. Ki and M. Swaminathan, *IEEE Trans. Comput.-Aid. Des. Integr. Circuits Syst.*, 2009, **28**, 846.
220. D. Kim, H. Kim and Y. Eo, *Electron. Lett.*, 2013, **49**, 1084.
221. L. Boyer, P. Notingher and S. Agnel, in *Proceedings of 24th Annual IEEE Applied Power Electronics Conference and Exposition (APEC)*, Washington, USA, 2009, p. 2055.

222. J. León, X. Perpiñà, M. Vellvehi, A. Baldi, J. Sacristán and X. Jordà, *Appl. Phys. Lett.*, 2013, **102**, 084106.

223. J. W. Sofia, *Electron. Cooling*, 1997, **3**, 22.

224. A. Ammous, B. Allard and H. Morel, *IEEE Trans. Power Electron.*, 1998, **13**, 12.

225. F. Oettinger, D. L. Blackburn and E. Rubin, *IEEE Trans. Electron. Devices*, 1976, **23**, 831.

226. A. Hamidi and G. Coquery, *Microelectron. Reliab.*, 1997, **37**, 1755.

227. T. Kajiwara, A. Yamaguchi, Y. Hoshi, and K. Sakurai, in *Proceedings of International Symposium of Power Semiconductor Devices and ICs ISPSD*, Kyoto, Japan, 1998, p. 281.

228. M. Pfost, D. Costachescu, A. Mayerhofer, M. Stecher, S. Bychikhin, D. Pogany and E. Gornik, *IEEE Trans. Semicond. Manuf.*, 2012, **25**, 294.

229. Y. K. Leung, S. C. Kuehne, V. S. K. Huang, C. T. Nguyen, A. K. Paul, J. D. Plummer and S. S. Wong, *IEEE Electron. Device Lett.*, 1997, **18**, 13.

230. M. Grecki, J. Pacholik, B. Wiecek and A. Napieralski, *Microelectron. J.*, 1997, **28**, 337.

231. www.flir.com.

232. G. Breglio, A. Irace, E. Napoli, M. Riccio, P. Spirito, K. Hamada, T. Nishijima and T. Ueta, *Microlectron. Reliab.*, 2008, **48**, 1432.

233. R. Abid, F. Miserey and F.-Z. Mezroua, *J. Phys. III*, 1996, **6**, 279.

234. C. Fürböck, R. Thalhammer, M. Litzenberger, N. Seliger, D. Pogany, E. Gornik and G. Wachutka, in *Proceedings of International Symposium of Power Semiconductor Devices and ICs ISPSD*, Toronto, Canada, 1999, p. 193.

235. R. Thalhammer, *Internal Laser Probing Techniques for Power Devices: Analysis, Modeling and Simulation*, Doctoral dissertation, Technishen Universität München, Fakultät für Elektrotechnik und Informationstechnik, 2000.

236. M. Heer, P. Grombach, A. Heid and D. Pogany, *Microelectron. Reliab.*, 2008, **48**, 1525.

237. M. Denison, M. Blaho, P. Rodin, V. Dubec, D. Pogany, D. Silber, E. Gornik and M. Stecher, *IEEE Trans. Electron. Devices*, 2004, **51**, 1695.

238. G. Haberfehlner, S. Bychikhin, V. Dubec, M. Heer, A. Podgaynaya, M. Pfost, M. Stecher, E. Gornik and D. Pogany, *Microelectron. Reliab.*, 2009, **49**, 1346.

239. D. Pogany, V. Dubec, S. Bychikhin, C. Furbock, M. Litzenberger, S. Naumov, G. Groos, M. Stecher and E. Gornik, *IEEE Trans. Device Mater. Reliab.*, 2003, **3**, 85.

240. J. Rhayem, B. Besbes, R. Blečić, S. Bychikhin, G. Haberfehlner, D. Pogany, B. Desoete, R. Gillon, A. Wieers and M. Tack, *Microelectron. J.*, 2012, **43**, 618.

241. J. Kuzmik, S. Bychikhin, M. Neuburger, A. Dadgar, A. Krost, E. Kohn and D. Pogany, *IEEE Trans. Electron. Devices*, 2005, **52**, 1698.

242. J. Kuzmik, S. Bychikhin, E. Pichonat, C. Gaquière, E. Morwan, E. Kohn, J.-P. Teyssier and D. Pogany, *Int. J. Microwave Wireless Technol.*, 2009, **1**, 153.
243. C. Shih-Hung, A. Griffoni, P. Srivastava, D. Linten, S. Thijs, M. Scholz, M. Denis, A. Gallerano, D. A. Lafonteese, A. Concannon, V. A. Vashchenko, P. Hopper, S. Bychikhin, D. Pogany, M. Van Hove, S. Decoutere and G. Groeseneken, *IEEE Trans. Device Mater. Reliab.*, 2012, **12**, 589.
244. N. Seliger, D. Pogany, C. Fürböck, P. Habas, E. Gornik and M. Stoisiek, in *Proceedings of European Solid-State Device Conference ESSDERC*, Stuttgart, Germany, 1997, p. 512.
245. X. Perpiñà, X. Jordà, M. Vellvehi, M. Mermet-Guyennet, J. Millán and N. Mestres, in *Proceedings of 19th International Symposium on Power Semiconductor Devices and ICs ISPSD*, Jeju, Korea, 2007.
246. A. N. Magunov, *Opt. Spectrosc.*, 1992, **76**, 205.
247. R. M. B. Agaiby, S. H. Olsen, G. Eneman, E. Simoen, E. Augendre and A. G. O'Neill, *Electron. Device Lett.*, 2010, **31**, 4219.
248. B. Ketterer, Raman Spectroscopy of GaAs Nanowires: Doping Mechanisms and Fundamental Properties, Thèse. École Polytechnique Fédérale de Lausanne, 2011.
249. Z. Guangchen, F. Shiwei, L. Jingwan, Z. Yan and G. Chunsheng, *J. Semicond.*, 2012, **33**, 044003.
250. A. Sarua, A. Bullen, M. Haynes and M. Kuball, *IEEE Trans. Electron. Device*, 2007, **54**, 1838.
251. A. Sarua, H. H. Wills, J. Pomeroy, M. Kuball and A. Falk, in *Proceedings of the 15th International Symposium on the Physical and Failure Analysis of Integrated Circuits IPFA*, Singapore, 2008.
252. P. Regoliosi, A. Di Carlo, A. Reale, P. Lugli, M. Peroni, C. Lanzieri and A. Cetronio, *Semicond. Sci. Technol.*, 2005, **20**, 135.

Heat Transport in Nanofluids

EFSTATHIOS E. MICHAELIDES

Department of Engineering, Texas Christian University, Fort Worth
TX 76129, USA
Email: e.michaelides@tcu.edu

14.1 Thermal Conductivity Enhancement

14.1.1 Experimental Data

Early experimental work on nanofluids[1] reported that the addition of a small fraction of carbon nanotubes (CNTs) to engine oil increased the thermal conductivity of the base fluid by a factor of two to three. Such conductivity enhancements represent significant increases in the convective heat-transfer coefficients of engineering systems, and hold promise for making nanofluids the heat-transfer media of the future. The magnitude of thermal conductivity enhancement was initially characterized as 'anomalous' and sparked immense scientific interest during the first decade of the 21st Century. Several laboratories throughout the globe experimented with different types of fluids and nanoparticles. Among the most common base fluids that were investigated were: water; engine oil and other oils; and ethylene glycol. Among the nanoparticles were: aluminium oxide (Al_2O_3); copper oxides (both CuO and Cu_2O); single-walled and multiwalled CNTs; copper (Cu); and gold (Au).

More recent experiments on thermal conductivity show that several other types of nanofluids enhance the thermal conductivity of the base fluid only moderately. Typical enhancements with metal oxide nanofluids are in the range 5–50%.[2,3] However, a few studies, typically with metal/metal alloys

RSC Nanoscience & Nanotechnology No. 38
Thermometry at the Nanoscale: Techniques and Selected Applications
Edited by Luís Dias Carlos and Fernando Palacio
© The Royal Society of Chemistry 2016
Published by the Royal Society of Chemistry, www.rsc.org

nanoparticles or CNTs have reported thermal conductivity enhancements close to 100% or above.[4,5] From the several experimental studies it appears that the most important parameters of conductivity enhancement in nanofluids are:

1. The type and properties of the nanoparticles;
2. The type of base fluid;
3. The volumetric fraction of the nanoparticles; and
4. The pH and type of surfactant used to stabilize the nanofluid.

The highest increases of thermal conductivity have been observed with suspensions of CNTs and in metallic nanoparticle suspensions. The lowest enhancements were observed with metal oxide nanoparticles, such as TiO_2 and Al_2O_3. Figure 14.1 depicts the ratio of the thermal conductivities of the nanofluid to that of the base fluid, k_e/k_f, – this ratio is often referred to as the *relative conductivity* – as a function of the volumetric fraction of solids. The figure is a compilation of several sets of experimental data from the literature with the most typical nanoparticles.[4–11] The two lines represent trends with a metallic nanofluid and a metal oxide nanofluid. Most of the experimental measurements were accomplished with hot wire instruments, which are considered relatively accurate for suspensions of nanoparticles. It is apparent in this figure that CNTs and metallic nanoparticles are associated with higher values of the relative conductivity of the suspension.

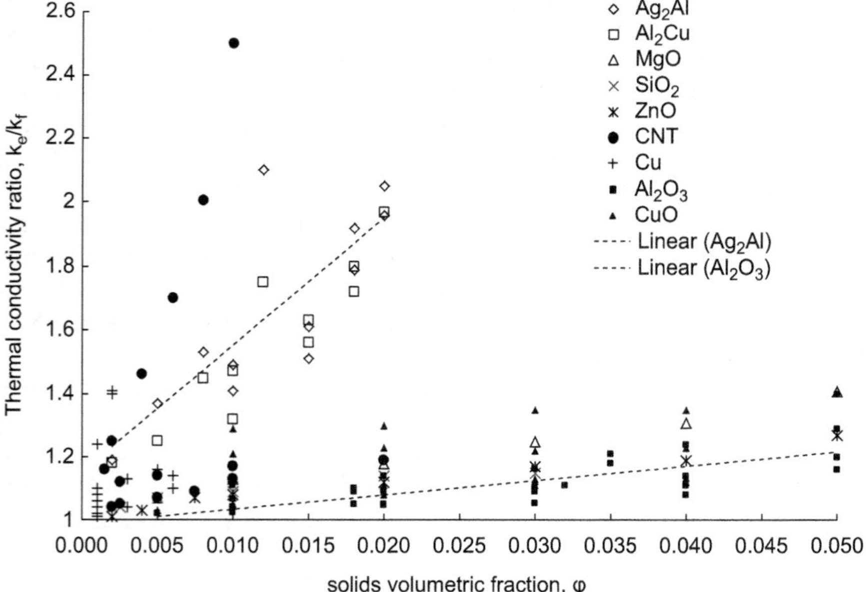

Figure 14.1   The effective thermal conductivity enhancement of several nanofluid systems.

The following conclusions may be drawn from Figure 14.1:

1. The percentage conductivity enhancement is significantly higher than the percentage concentration of the particles. A very low concentration of particles (less than 1%) often causes a 20–40% enhancement of the conductivity;
2. The thermal conductivity enhancement depends primarily on the type of particles: nanofluids with particles, such as CNTs, Ag_2Al and Al_2Cu exhibit significantly higher effective conductivities, while the effective conductivities of nanofluids with Al_2O_3 nanoparticles are lower. In some experiments with Al_2O_3 nanoparticles the thermal conductivity enhancement was very low, almost insignificant, even at volumetric fractions close to 5%; and
3. It appears that there are diminishing returns to adding more particles beyond a concentration of approximately 1%, as the effective conductivity increase is not simply proportional to the volumetric ratio.

The thermal conductivity enhancement of the same nanofluid system (same base fluid and particles) is not consistent among all the data sets. This is apparent in the several sets of data for Al_2O_3 and Cu nanofluids, which differ significantly between different studies. Figure 14.2, which shows the relative conductivity of CNT nanofluids,[1,12–15] demonstrates this point. The implication of the discrepancy among different data sets is that, while the type and concentration of nanoparticles is a major determinant of the conductivity enhancement, variables other than these two play an important role. The identification and quantification of the effect of such variables is

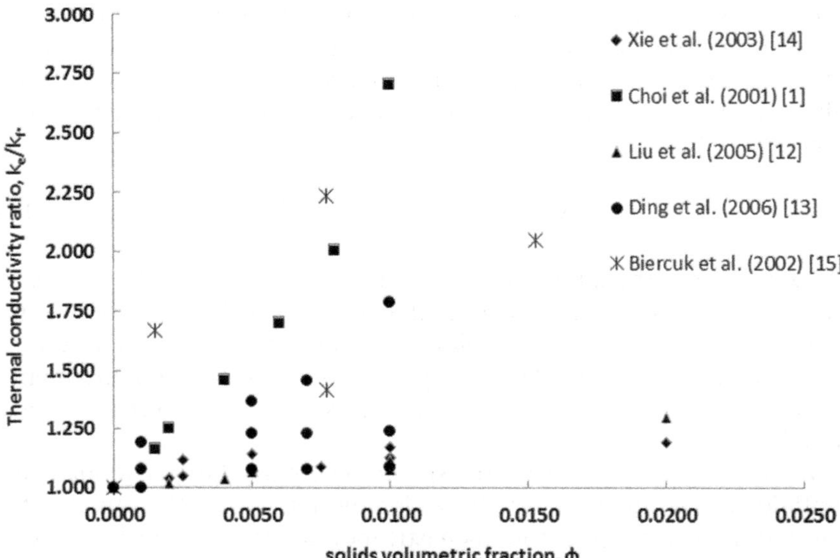

Figure 14.2   Thermal conductivity data for several CNT nanofluid systems.

very important in engineering studies for the design of reliable and effective heat-transfer media using nanofluids.

14.1.2 Analytical Expressions and Correlations

Maxwell[16] derived an expression for the effective electrical conductivity of a heterogeneous mixture with spherical particles. Because of the analogy in the thermal and electrical conduction processes, his expression is also valid for the thermal conductivity of mixtures. When written in terms of the conductivities of the fluid and solid spheres, Maxwell's equation becomes:

$$k_e = k_f \left[1 + \frac{3(k_s - k_f)\varphi}{(k_s + 2k_f) - (k_s - k_f)\varphi} \right], \qquad (14.1)$$

where k_f is the conductivity of the base fluid, k_e the effective conductivity of the suspension, k_s the conductivity of the solid particles, and φ the volumetric fraction of the nanoparticles. Later, Bruggeman,[17] and Hamilton and Crosser[18] extended this expression to short cylinders and irregular particles. It must be pointed out that at the limit of solid conductivity $(k_s \gg k_f)$ eqn (14.1) and the similarly obtained analytical expressions yield the following asymptotic expression for the thermal conductivity of the suspension:

$$k_e = k_f \left[1 + \frac{3\varphi}{1 - \varphi} \right]. \qquad (14.2)$$

It is apparent that the last expression does not include the effect of the very high conductivity of the solid particles $(k_s \gg k_f)$. Of particular interest in the study of the thermal conductivity of nanofluids, and especially nanofluids with CNTs, is the analytical study by Nan *et al.*[19] who derived an expression for the conductivity of composites with high-aspect-ratio fibres. They later derived a simplified expression, applicable to nanofluids with high-aspect-ratio CNTs, at low solid concentrations $(\varphi \ll 1)$, which is as follows:[20]

$$k_e = \frac{3k_f + \varphi k_s}{3 - 2\varphi} \approx k_f + \frac{\varphi k_s}{3} \qquad (14.3)$$

Because $k_s \gg k_f$, the addition of very small amounts of CNTs – materials that have thermal conductivities in the range 2000–3000 W mK^{-1} – may triple or quadruple the conductivity of the suspension even at volumetric concentrations of 1%. This largely explains the significant enhancement of the thermal conductivity of CNT nanofluids that was observed by early researchers.[1,15]

Several researchers have tried to explain the observed high conductivity of nanofluids using models for the Brownian motion of particles. An analytical study[21] used the concept of the stochastic movement of the particles and the resulting Langevin equation to describe the fluctuations of nanoparticle suspensions. The authors modelled the *heat transfer chain* between nanoparticles and base fluid and, superimposed the effects of the Brownian motion on the effective conductivity predicted by Maxwell. The authors derived the following expression for the effective thermal conductivity of the suspension:

$$k_e = k_f \left[1 + \frac{3(k_s - k_f)\varphi}{(k_s + 2k_f) - (k_s - k_f)\varphi} \right] + \frac{9\varphi h_c k_B T}{8\pi\rho_s\alpha^4} \qquad (14.4)$$

where k_B is the Boltzmann constant and h_c is the convective heat-transfer coefficient between the fluid and the particles, which also includes the interfacial resistance to heat transfer, and ρ_s is the density of the nanoparticles.

Among the other expressions, the correlation of several sets of experimental data for aqueous nanofluids resulted in the following expression:[22]

$$\frac{k_e}{k_f} = 1 + 1.0112\varphi + 2.4375\varphi\frac{47}{d_s} - 0.0248\varphi\frac{k_s}{k_w}, \qquad (14.5)$$

where d_s is the particle size in nm and k_w is the thermal conductivity of water, where $k_w = 0.613$ W mK^{-1}.

14.2 Mechanisms of Thermal Conductivity Enhancement

The plethora of experimental data collected on the enhanced conductivity of nanofluids has also resulted in several analytical studies, through which the mechanisms for the enhanced conductivities were studied. The following mechanisms have been suggested for the enhanced conductivity of nanofluids:

1. The higher heat conductivity of the particles;
2. The shape (spherical, cylindrical, irregular) of the particles;
3. The Brownian motion of the particles, which also includes the fluid that follows the motion of the particles;
4. The formation of aggregates and chains that form highly conductive paths in the nanofluid or contribute to the development of such paths;
5. Significant changes in the thermodynamic properties of the fluid at the solid–fluid interface and the formation of a *solid layer*.
6. The electric charge on the surface of the particles;
7. Transient local heat-transfer effects; and
8. Preparation and surfactants.

These mechanisms will be examined in more detail in the following sections.

14.2.1 Particle Conductivity

Typically, the conductivity of the solid nanoparticles is three orders of magnitude higher than the conductivity of the base fluid, as may be seen in

Table 14.1 Thermal conductivities of materials in common nanofluids.

Solid	k_s/W mK^{-1}	Liquid	k_f/W mK^{-1}
Silver (Ag)	427	Water	0.613
Copper (Cu)	395	Ethylene glycol	0.253
Aluminum (Al)	237	Engine oil	0.145
Carbon nanotubes	3200–3500	Alcohol	0.115
Brass	120	Glycerol	0.285
Nickel	91		
Quartz (single crystal)	7–12		
Alumina (Al$_2$O$_3$)	39		

Table 14.1, which lists the thermal conductivities of several materials commonly used as base fluids and nanoparticles. Because of the significantly higher conductivities of the solids, it is expected that the addition of even a small fraction of particles ($\varphi \sim 1\%$) in the base fluid would result in a substantial increase of the thermal conductivity of the liquid–solid mixture.

Effective conductivity enhancements have been observed in almost all experimental and analytical studies, regardless of the type, shape and volume fraction of the particles used. A similar enhancement has also been observed in composite materials, where there is no apparent flow.

The very high values of thermal conductivity, which were observed in the early experiments with nanofluids, have led a few authors to characterize the enhancement of the thermal conductivity of nanofluids as 'anomalous'. These early conductivity enhancements were of the order of 150% and were most commonly observed with CNT nanofluids. CNTs have very high aspect ratios (100 to 1000) and their thermal conductivities are of the order of 1000 times higher than the conductivity of water. One of the characteristics of long spheroids and long fibres is that the conduction within a heterogeneous mixture is significantly higher in the directions of their long axes and the effective conductivity is more accurately modelled by eqn (14.3). It is rather unfortunate that the results of the earlier studies with fibrous solid particles (CNTs) were compared to analytical models of spheres and short cylinders[16–18] similar to eqn (14.1), which yield significantly lower values for the thermal conductivity. This discrepancy led to the characterization of the thermal conductivity of nanofluids as 'anomalous'. A benchmark study,[23] which was conducted with the participation of 34 laboratories worldwide, used several sets of well-characterized experimental data and appropriate analytical expressions and concluded that the observed heat-transfer enhancement with nanofluids is not anomalous, and that 86% of the available experimental data are explained by one of the pertinent equations or their approximations, *e.g.*, eqn (14.1)–(14.3).

14.2.2 Particle Shape and Orientation

It is apparent from eqn (14.1) and (14.3) that the shape of the particles plays a significant role in the conductivity of the nanofluid. All the analytical and experimental studies indicate that the effective conductivity of heterogeneous mixtures is significantly higher with elongated rather than with spherical particles.[19,24] Since elongated particles, such as spheroids and fibres, have a longer dimension, where heat is conducted at a higher rate, the aspect ratio and the orientation of the elongated particles is important in the calculation of the thermal conductivity of the heterogeneous mixtures.

The importance of particle orientation in relation to the direction of heat flow is easily demonstrated by the following example:[24] let us consider the heat transfer between two parallel plates in a liquid–solid heterogeneous mixture and assume that all the solid particles are packed together (have fused) in a single layer. This particle configuration will yield two extreme orientation cases that correspond to the upper and lower limits of heat conduction:

1. The fused particle layer is parallel to the direction of heat transfer; or
2. The fused particle layer is perpendicular to the direction of heat transfer.

The two limit cases are depicted in Figure 14.3, where the *fused* particle layer is portrayed in grey. The heat flux direction is the same as the direction of the temperature difference, ΔT. The volumetric fraction of the solids is φ. Hence, the dimensionless thickness of the particle layer is φ, and that of the liquid layer is $1 - \varphi$.

In the first case (A) where the orientation of the particle layer is perpendicular to the temperature gradient, the induced heat flux, q, is conducted from one plate to the other and causes temperature differences ΔT_1, ΔT_2 and ΔT_3 between the plates as shown in the figure. From elementary considerations of heat conduction between the two plates, we obtain the following expression for the temperature differences in case (A):

$$\Delta T_1 + \Delta T_3 = \frac{qL(1 - \varphi)}{k_\mathrm{f}} \quad \text{and} \quad \Delta T_2 = \frac{qL\varphi}{k_\mathrm{s}}. \tag{14.6}$$

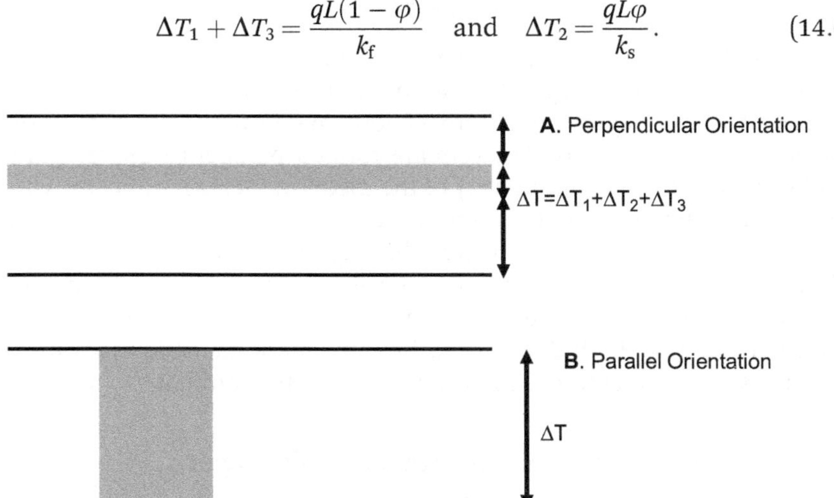

Figure 14.3 The two limiting cases of particle orientation in a liquid.

The sum of the three temperature differences is equal to the total temperature difference, ΔT, as shown in the figure. Hence, we have:

$$\Delta T = q \left[\frac{L(1 - \varphi)}{k_f} + \frac{L\varphi}{k_s} \right].$$

(14.7)

Given that $q = k_e \Delta T$, eqn (14.7) yields the following expression for the effective conductivity of the medium between the two conducting plates:

$$\frac{1}{k_e} = \frac{1 - \varphi}{k_f} + \frac{\varphi}{k_s}.$$

(14.8)

Nanofluids typically have very low volumetric fractions of solids ($\varphi \ll 1$) and the conductivity of the nanoparticles is much higher than the conductivity of the fluid $k_s \gg k_f$. Under these conditions the asymptotic limit of the last expression is:

$$(k_e)_{\varphi \ll 1} = k_f(1 + \varphi)$$

(14.9)

In the second case, (B) where the orientation of the solid particle layer is parallel to the temperature gradient, the total heat flux is equal to the sum of the heat fluxes in the two layers:

$$q = [k_f(1 - \varphi) + k_s\varphi] \frac{\Delta T}{L},$$

(14.10)

This expression yields the following equivalent conductivity for the mixture:

$$k_e = k_f(1 - \varphi) + k_s\varphi.$$

(14.11)

For typical nanofluids, where $k_s \gg k_f$, and $\varphi \ll 1$, the thermal conductivity of this limiting case is:

$$(k_e)_{\varphi \ll 1} = k_f + \varphi k_s.$$

(14.12)

Eqn (14.9) and (14.12) prove that when particle layers are formed within the fluid, which are perpendicular to the direction of the heat flux, the effective conductivity of the heterogeneous medium does not increase significantly. When the particle layers are formed parallel to the temperature gradient, the effective conductivity of the medium increases linearly with the volumetric fraction and the thermal conductivity of the fibres. A recent study[25] showed that most of the available experimental data fall within these 'classical' limits. Such observations have significant implications in the possible 'design' of nanofluids in the future, where the highly conducting nanoparticles may be induced (*e.g.*, by a magnetic field) to align in a direction that is parallel to the imposed temperature gradient.

14.2.3   Formation of an Interfacial Solid Layer

Early research on nanofluids[26,27] suggested that a 'solidified' liquid layer composed of a few molecular layers of the base fluid is formed on the surface of the solid particles. With this layer, the nanoparticle is a composite material consisting of the nanoparticle proper, which has conductivity k_s, and the solidified layer, whose conductivity is approximately the same as the conductivity of the solid phase of the base fluid k_{sf}. When $k_{sf} > k_f$, the formation of the solid layer enhances the conductivity obtained by the particle alone under all expressions emanating from the effective medium model. However, for this enhancement to be significant, the ratio of the conductivity of the formed 'solid' to the nanoparticle conductivity, k_{sf}/k_s, must be higher than 0.02 and the relative thickness of the interfacial layer, $\delta\alpha/\alpha$, must be less than 0.05. In the case of aqueous nanofluids, for which the conductivity of ice is $k_{sf} = 2.2$ W mK^{-1}, the first condition is not satisfied with most of the nanoparticles listed in Table 14.1. Similar considerations with the solids of other nanofluids, lead to the conclusion that the formation of the solid layer alone does not explain the vast majority of the experimental observations.

A more recent NMR study measured the thickness of the "solid nanolayer" around magnetic alumina nanoparticles of 43 nm size in water. The thickness of the "nanolayer" was only 1.4 nm.[28] This layer of relatively immobile water molecules is equivalent to five water molecule layers formed on the surface of the nanoparticles. However, the measurement of the thermal conductivity of this nanofluid revealed that the formation of this interfacial layer had no effect on the measured effective conductivity of the nanofluid. This may imply that the interfacial layer of the water molecules did not have the properties of ice. Molecular Dynamics (MD) simulations, which probe the structure and properties of this molecular layer, also do not support this theory, especially in the case of the hydrophobic carbon atoms that form CNTs.[24,29,30]

14.2.4 Electric Surface Charge and pH

It is well known that electric charges and the value of the liquid's pH do not have an appreciable effect on the transport properties of pure fluids, including the base fluids used in all nanofluids. Any effect of these variables on the conductivity of nanofluids must take place *via* mechanisms that act through the structure of the particles, primarily clustering, aggregation and the formation of structures that facilitate the transfer of energy. An extensive experimental investigation on the influence of surface charges of the nanoparticles on the effective thermal conductivity of suspensions concluded that surface charges significantly influence the effective conductivity through the structure of the nanofluid. Departures from electric neutrality for the nanoparticles, which appear as high zeta-potential measurements, result in the higher stability of the nanoparticle clusters, higher stability of the suspension overall and better dispersion of nanoparticles within the base fluid. This has an enhancing effect on the thermal conductivity of the suspensions.[31]

The structure of the suspension and the zeta-potential effect may partly explain the disparities between some sets of experimental data, where researchers used surfactants that alter the electric surface charge of the nanoparticles and the internal structure of the suspensions.[24,32] Two independent studies that examined the effect of the pH of the base fluid, found significant changes in the effective conductivity of the suspension, which were attributed to the altered shapes of the nanoparticle aggregates and the altered structure of the nanofluid.[33,34]

14.2.5 Brownian Movement

The Brownian movement of nanoparticles in the base fluid is the primary mechanism for the *microconvection* and mechanical agitation in nanofluids – a mechanism that always contributes to higher heat and mass transfer. For this reason, several studies have been undertaken to quantify the effect of Brownian movement on the thermal conductivity of nanofluids. Since the Brownian movement and the resulting microconvection are difficult to observe experimentally, most of the studies on this mechanism are analytical and numerical. However, not all of the authors arrive at the same conclusions and several disagree significantly on the importance of Brownian movement to nanofluid conductivity. A recent review on the subject and a monograph[3,24] provide a good exposition of all the important studies on the effect of the Brownian movement of particles.

At first, one must consider that the two effects of higher particle conductivity and particle shape (Sections 14.2.1 and 14.2.2 above) account for a large part of the observed higher effective conductivity of nanofluids: 86% of the experimental data in the international benchmark study[23] are explained to within a 15% experimental error caused by these two effects.[23] Therefore, the Brownian movement and other effects may explain a smaller portion of the data and may help reduce the uncertainty of the experimental measurements.

Secondly, it is known that the Brownian movement causes 'agitation' in the fluid and that all forms of such agitation facilitate the transport of energy and mass. For example, agitation by turbulence and the turbulent motion of larger particles is a major contributor to heat and mass transfer of particulate systems.[35,36] Of course, Brownian-induced agitation in the fluid is of lesser magnitude than turbulent agitation and is expected to have a lesser (but not vanishing) effect. The benchmark study[23] leaves open the possibility of lesser contributions by other mechanisms and the Brownian movement is very likely the principal mechanism among them.

Thirdly, the effect of the Brownian movement of particles gives an excellent explanation for the very high dependence of the conductivity of nanofluids on temperature.[37,38] The increased Brownian movement at higher temperatures and the associated microconvection is touted by most researchers as the main mechanism for the significant temperature effect. The acceptance of Brownian movement as the primary mechanism for the high dependence of the thermal conductivity on temperature by the vast majority of researchers is in itself an indication that the Brownian movement is a contributing mechanism to the enhanced effective thermal conductivity of nanofluids, but most likely, a mechanism of lesser importance than particle concentration, shape and conductivity.

14.2.6 Transient Motion of Particles

The transient effects of particulate motion are generated by the Brownian movement and are related to the following:[24,39,40]

1. The movement of the *virtual* or *added mass* of the base fluid that accompanies the motion of the particles;
2. The mass of the fluid that rushes in to replace the volume of the moving particles and the fluid volume of the added mass; and
3. The contribution of the history term to the heat transfer from the particles.

All three effects contribute to the enhancement of the microconvection or fluid agitation, which leads to higher heat transfer. It must be noted that the Brownian movement of the particles is caused by the motion of the fluid molecules (thermal motion), but manifests itself through motion of the particles, which have a significantly higher inertia than that of the fluid molecules and significantly higher characteristic time scales. It is most likely that the effects of the transient motion would materialize at time scales that span the range of the particle characteristic time scales. For 10 nm copper nanoparticles in water, these characteristic time scales are approximately 2.0×10^{-10} s, while the time scales of the fluid molecules are significantly shorter (of the order of 10^{-14} s). Hence, the Brownian movement of particles and the associated transient particle and fluid movement would not be expected to play a major role on the effective thermal conductivity of the suspension, which is associated with the molecular momentum and energy transfer at molecular time scales. The Brownian movement and the transient effects are expected to have a more significant influence on the convective heat-transfer coefficient of the suspensions, a subject that is treated in Section 14.3.

14.2.7 Particle Distribution and Aggregation

Single particles and aggregates form structures and layers that may significantly enhance the conductivity of the suspension, even when the volumetric concentration does not change. For example, when two or more spherical particles aggregate, the resulting composite particle is longer. A glance at the theoretical models[16–20] proves that the formation of longer particles, and especially of linear chains of particles, brings the effective conductivity of the suspension closer to that of fibrous nanoparticles, whose thermal conductivity is given by eqn (14.3) rather than eqn (14.1). This occurs because a fraction of the elongated particles is always oriented in the direction of the temperature gradient and, thus, creates conducting paths.

Some of the experimental and numerical studies on particle aggregation that have observed the formation of highly conducting *paths* or *bridges*, support this mechanism of effective conductivity enhancement.[41–44]

14.2.8   Preparation and Surfactants

The method of preparation of nanofluids and the specific chemicals used as surfactants significantly influence the particle size, the formation of clusters and aggregates, and the particle distribution. Every laboratory that chemically manufactures nanoparticles, uses their own methods for the stabilization of their nanofluids and these are typically not reported in the literature. Mechanical methods, such as milling and shearing the nanofluid, have been used extensively to produce stable nanofluids. Electromechanical methods, such as the application of ultrasonic waves (sonification or ultrasonification), have also been used extensively to disperse the nano-particles and keep them from aggregating and forming sediments. Chemical methods are routinely used with a variety of surfactants, including: sodium laurate; sodium dodecyl benzene sulfonate; sodium dodecylsulfate; and gum arabic. A review of the stability and properties of nanofluids[45] provides an extensive list of surfactants, and stipulates that "…choosing the right surfactant is the most important part of the [preparation] procedure." It must be noted that the choice of chemical surfactant (for the optimum chemical compound and concentration) is typically accomplished by a trial-and-error method.

Oftentimes, the amount or concentration of the surfactant is close to the concentration of the nanoparticles. This implies that the surfactant must be considered a component that affects the thermodynamic and transport properties (including thermal conductivity) of the mixture.[24] When the actual concentration of the surfactants in the heterogeneous mixture is not high enough to have its own influence, the effect of the surfactant and the method of preparation of nanofluids on the effective conductivity is realized *via* the size and shape mechanism that was expounded in Section 14.2.2.

14.2.9   Other Mechanisms

From the large number of experimental investigations it follows that a few other mechanisms or parameters have a rather minor influence on the thermal conductivity of nanofluids.[24] Among these are:

1. Magnetic fields and magnetic nanoparticles, which most likely work *via* the particle aggregation/distribution mechanism;[46,47]
2. The amount of fluid 'trapped' within the volume defined by the boundary of irregular particles;[48,49] and
3. Velocity discontinuity (*slip*) and thermal discontinuity at the fluid–particle interface.[50,51]

A purely statistical study[52] surveyed the results of 141 nanofluid experimental studies and performed a statistical analysis on heat-transfer enhancement mechanisms. Among the statistical observations of this study is that the Brownian movement of the particles is reported as one of the mechanisms of heat-transfer enhancement by approximately 70% of the researchers. Thirty-three percent of the studies stated that Brownian movement is the only mechanism for the observed enhancement. The four leading mechanisms or combinations of mechanisms promulgated by the researchers were:

1. Brownian motion (solely), 33%;
2. Interfacial layer formation, 22%;
3. Brownian motion with diffusion and aggregation, 11%; and
4. Brownian motion with thermophoresis, 9%.

Regarding the last item it must be noted, however, that the cause of thermophoresis is the Brownian movement of particles in a temperature gradient.[24,53] Thermophoresis and Brownian movement are not independent phenomena.

14.3 Additional Augmentation of the Convective Heat-transfer Coefficients

Nanofluids as heat-transfer media could be used in microchannels to remove heat by convection from a hotter surface (*e.g.*, from an electronic component). The primary parameter of interest in convection heat-transfer applications is the value of the heat-transfer coefficient, h_c, which is typically given by a closure equation in terms of the Nusselt number:

$$\dot{Q} = h_c A \Delta T \quad \text{with} \quad h_c = Nu \frac{k_e}{L_{char}}, \tag{14.13}$$

where L_{char} is the characteristic dimension of the heat-transfer system. The Nusselt number (Nu) is typically a function of the Reynolds (Re) and Prandl numbers and is given by an empirical equation. It is apparent from eqn (14.13) that if the thermal conductivity of a heat-transfer medium increases, the convective coefficient increases proportionately.

Several experiments that independently measured the thermal conductivity and the convective coefficients of nanofluids in laminar flows concluded unequivocally that the increase of the convective heat-transfer coefficient is significantly above what the simple relationship of eqn (14.13) predicts.[54,55] That is, the experiments concluded that there is an augmentation[†] of the convective coefficient, h_c, in addition to what the

---

[†]In order to differentiate between the two heat-transfer mechanisms, the term *augmentation* is used for the observed increase of the heat-transfer coefficient, h_c, and the term *enhancement* for the increase in the effective thermal conductivity, k_e.

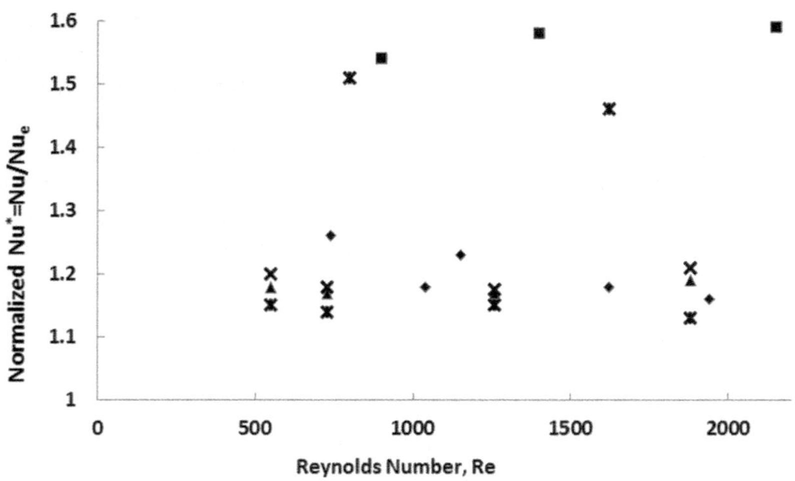

Figure 14.4   Normalized Nusselt numbers for several types of nanofluid.[53–56]

effective thermal conductivity of the fluid signifies. It appears from most studies that the additional augmentation increases with the volumetric fraction, φ, but the relationship is not linear.

Figure 14.4 shows representative data[54–57] of this heat-transfer augmentation as the ratio Nu_{obs}/Nu_e *vs. Re* for nanofluids composed of Al_2O_3 nanoparticles in water. Nu_{obs} is the measured Nusselt number in the heat-transfer experiments and Nu_e is the Nusselt number calculated analytically using the effective conductivity of the nanofluid $(Nu = h_c L_{char}/k_e)$ and the flow parameters. It can be observed in this figure that, even though the trend of the ratio Nu_{obs}/Nu_e *vs. Re* is not very clear, the data consistently demonstrate that this ratio is greater than one, sometimes by as much as 60%. This implies that, other than the increased conductivity of the nanofluids, there is an additional increase of h_c, which needs to be accounted for. The most likely mechanism for this additional heat-transfer augmentation is micro-advection within the nanofluid, which is caused by the Brownian movement of nanoparticles and of the fluid that follows the transient nanoparticle motion.

The vast majority of the experimental studies on the convective heat-transfer coefficient for laminar flows prove that there is such an additional heat-transfer augmentation. In contrast, the experimental evidence with turbulent flow measurements is not clear: a few experimental studies observed some measure of an augmentation in the range 10–20%,[58,59] while most of the others did not observe any augmentation and concluded that the ratio Nu_{obs}/Nu_e is approximately equal to one.[60,61]

The most likely explanation for the observed augmentation of the convective heat-transfer coefficient, above the values allowed by the enhanced thermal conductivity of the nanofluid, is fluid microconvection, which is induced by the Brownian movement of the particles. This microconvection includes the movement of the added mass that follows the particles, as well as the masses of fluid that rush in to fill the void left by the particles in their original positions, and rush out to allow the particles to move into their final positions. The absence of such augmentation in the turbulent heat-transfer regime corroborates this mechanism: the magnitude of turbulent velocity fluctuations and the agitation of the fluid are much higher than that of the microconvection process. Therefore, in turbulent flows, the contribution of the microconvection to the value of the convective heat-transfer coefficient would be negligible and almost vanishing. In contrast, any systematic velocity fluctuations in laminar flows, where there is virtually no other mechanism for fluid agitation, contribute significantly to the overall heat-transfer coefficient.

14.4 Natural Convection

The addition of nanoparticles to a base fluid has the following effects on the process of natural (free) convection.[24] Some of the effects promote, and some inhibit, the fluid instabilities that start and maintain the natural convection process:

1. Increase of the fluid viscosity (inhibiting);
2. Local instability damping on the surface of the particles (inhibiting);
3. Interactions with the heating surface. Depending on the nature of particles and the interactions, this may have either a promoting or an inhibiting effect;
4. Particle sedimentation that induces fluid currents and promotes instabilities (promoting);
5. Increase of the fluid conductivity (promoting); and
6. Chemically active nanoparticles, which are sources or sinks of energy may cause local thermal and flow instabilities around the particles (promoting).

It appears from these observations that there is not a clear *a priori* reason for the enhancement or reduction of the heat-transfer coefficients in a natural convection process when nanoparticles are added to the base fluid, and this is reflected in the sets of experimental data, which are divided, with some showing enhancements and others reduction in the Nusselt numbers and the heat-transfer coefficients. For that matter, the picture of natural convection in nanofluids is not clear: on one hand, almost all of the analytical studies[62,63] and a few experimental studies[64] advocate that there is an enhancement of the natural convection heat-transfer coefficients, when nanoparticles are added to the base fluid. On the other hand, several experimental studies[65,66] show a clear and systematic reduction of the natural heat-transfer coefficients. More accurate and better documented experiments, where the size and distribution of nanoparticles are also measured, are needed to improve our understanding of the effect of nanoparticle addition on natural convection within nanofluids.

There is a significant implication, of a possible earlier onset of natural convection within nanofluids, for the popular method of thermal conductivity measurement using the Transient Hot Wire (THW) method. One of the fundamental assumptions of the operation of the THW instrument is that conduction only occurs during the time of the measurements and that the onset of natural convection takes place a long time after the completion of the measurements. An earlier onset of natural convection in a THW instrument would result in a significantly higher and erroneous value of the measured thermal conductivity.[24]

14.5   Boiling with Nanofluids

14.5.1   Pool Boiling

The addition of nanoparticles to base fluids, which are used for boiling applications, has the following promoting and inhibiting effects:[24]

1. Nanoparticles provide additional nucleation sites;
2. Nanoparticles, especially at higher concentrations, may deposit, cover and deactivate a fraction of the nucleation sites;
3. Nanoparticles left behind in the evaporation process deposit on the heating surface, thus altering the characteristics of the surface;
4. The surfactants added for the stability of nanoparticles in suspension alter the surface tension of the base fluid;
5. Nanoparticles enhance the microlayer evaporation on heated surfaces; and
6. Nanoparticles alter, sometimes significantly, the thermal and transport properties of the base fluid.

Most of the experimental data on nanofluid nucleate boiling indicate that the presence of nanoparticles decreases the pool boiling heat-transfer coefficient of base fluids. Typical pool boiling experimental results with the addition of metal oxide nanoparticles are shown in Figure 14.5. The Nukiyama curve,[67] which characterizes the boiling of a homogeneous fluid, is also shown in the figure.

Most experimental studies indicate that the addition of metal oxide nanoparticles to a base fluid causes a reduction of the pool boiling heat-transfer coefficient.[68–70] However, most of the experimental studies with metallic nanoparticles and CNTs indicate an increase of the pool boiling heat-transfer coefficient.[71,72] While more studies on pool boiling coefficients may be needed, a tentative conclusion is that the addition of metal oxide nanoparticles decreases the pool boiling heat-transfer coefficient, whereas

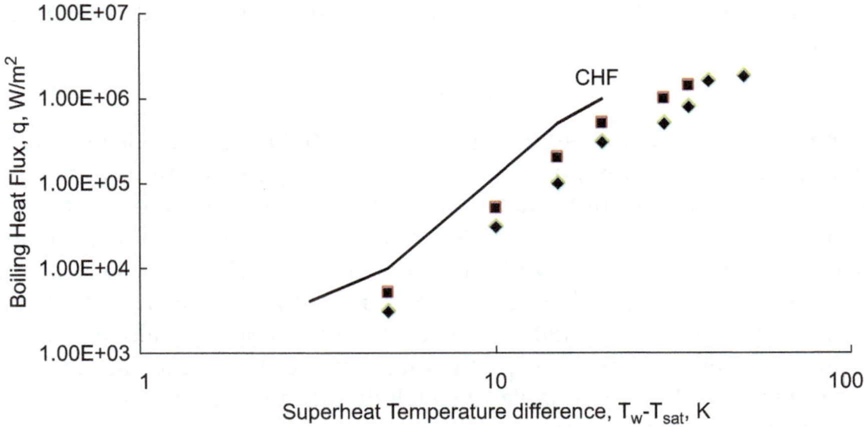

Figure 14.5   Typical experimental data for pool boiling with nanofluids: the pool boiling heat transfer coefficient is the ratio $q/(T_w - T_{sat})$ and is lower than that obtained from the Nukiyama curve. The CHF is enhanced significantly.

the addition of metallic nanoparticles and CNTs enhances the pool boiling heat-transfer coefficient.

14.5.2   Convective Boiling

Convective or *forced boiling* occurs in channels, where the nanofluid is forced by a pump. The main differences between pool boiling and convective boiling are:

1. The vapour is advected downstream and does not stay at the nucleation site;
2. Because the volume and cross-sectional area of the channel are restricted, the liquid and vapour phases arrange themselves in one of the flow regimes that characterize the vapour–liquid (or gas–liquid) two-phase flow; and
3. There is frictional pressure drop along the channel that affects the saturation temperature. In most applications though, the pressure drop is very low and has a minimal effect on the heat transfer in the channel.

The formation of known vapour–liquid flow regimes (*e.g.*, annular, churn, bubbly, *etc.*)[73] in convective boiling determines the conditions at the boundaries of the channel and, hence, the friction factor and the convective heat-transfer coefficients. The mass and volume of the vapour phase increase along the flow direction, because of heat influx, evaporation and (by a small amount) pressure reduction. There is not a general expression that applies with high accuracy for this type of boiling, but a working and fairly accurate suggestion[74] is that the heat flux for convective boiling is equal to the sum of the single-phase boiling and the pool boiling heat flux for the liquid:

$$q_{CB} = q_{NB} + q_1. \qquad (14.14)$$

The subscripts CB and NB refer to convective boiling and nucleate boiling respectively. Because for most fluids, $q_{NB} \gg q_1$, the approximation $q_{CB} \approx q_{NB}$ is often used. A recent study that was conducted for nanofluids[75] presents several of the more recently developed correlations for the heat-transfer coefficients for the convective boiling of liquids in channels. One of the prominent observations from the data of this study is the very high disagreement among the available correlations for liquid flow alone: the uncertainty associated with the experimental correlations for the base fluids is more than 50%. It is reasonable to conclude that the uncertainty for nanofluids would be even higher.

Unlike pool boiling, the majority of experimental evidence for the heat-transfer coefficients of convective heat-transfer points to the enhancement of the convective heat-transfer coefficient with the addition of nanoparticles. There is no experiment that observed a decrease in the convective boiling heat-transfer coefficient without a clear explanation of the reasons that led to the decrease. Most of the experimental studies indicate a significant enhancement, or at least no deterioration of the convective boiling heat-transfer coefficient.[24,76,77] Some of the experimental studies observed significant enhancements of more than 50%.[78] The main reasons for the enhancement of the convective boiling heat-transfer coefficients in channels are:

1. Channel boundary modification with an increased number of nucleation sites;
2. Increase of bubble–surface contact angles and inhibition of the formation of long-lasting dry vapour patches by higher concentrations of deposited particles at the bubble–surface interface;
3. Increased frequency of bubble departure from the nucleation sites; and
4. The formation of larger bubbles.

Because of the very high heat-transfer coefficients associated with boiling, nanofluids have been touted as the heat-transfer media with convective boiling in microchannels. A cause of concern with this application is the long-term reliability of such cooling systems: as the base fluid evaporates, nanoparticles deposit close to the exit of the channel and cause channel blocking, something that has been observed in one of the failed experiments in this area.[79] One may guard against partial or total blocking of the cooling channels by not allowing the complete evaporation of the base fluid.

Since one of the important nanofluid applications is the cooling of tiny electronic components – a very important technological area – well-coordinated experiments are needed in the future that will determine under what conditions (types of nanoparticles, base fluid, volumetric fraction, flow conditions, shape of channels, *etc.*) there is significant enhancement, and what are the optimum conditions for consistent and reliable heat transfer removal with boiling in microchannels.

14.5.3 Critical Heat Flux

The onset of Critical Heat Flux (CHF) occurs with the expansion of dry vapour patches on heated surfaces, including heated microchannels. The vapour expansion process spreads the vapour patches sideways, joins two or more vapour patches and creates thin vapour layers over large parts of the heating surface. Because particles do not evaporate with the fluid, they are trapped within the liquid layer at the heated surface. This creates higher particle concentrations adjacent to the dry patches. The increased concentration of particles has three effects on the vapour phase:

1. Restricts the sideways expansion of the vapour patch by forming a solid residue that impedes the spreading of vapour;
2. Maintains, through the interfacial surface tension, thin interstitial liquid layers between the individual particles and between the particles and the heating surface; and
3. The presence of particles close to the heated surface disturbs the shape of the meniscus at the vapour–liquid interface. This creates a "structural disjoining pressure" which increases the wettability and inhibits the spreading of dry patches.[80]

The three effects are adverse to the spreading of vapour pockets and bubbles. Therefore, one would expect *a priori* that the addition of particles, including nanoparticles, to a base fluid would enhance the CHF by modifying the structure of the heated surface and the concentration close to this surface. This is corroborated by all the experimental studies on CHF with nanofluids.

Unlike the experimental results for the boiling heat-transfer coefficients, all the experimental data on CHF, with no exceptions, indicate that the addition of nanoparticles – in both pool boiling and convective boiling – increases the CHF of base fluids, sometimes by a factor of two or three.[81,82] This trend is also illustrated in Figure 14.5, where the CHF increases from approximately $1\,000\,000$ W m^{-2} to $1\,800\,000$ W m^{-2} (this is not entirely obvious in the logarithmic scale of the figure). Significant increases in the CHF have been observed even when the solid's volumetric ratio is very low, *e.g.*, in the range $10^{-6} < \varphi < 10^{-5}$. Significant CHF enhancements were observed with all types of nanoparticles (metallic, metal oxides, CNTs, *etc.*) and with several types of heating surface (copper plates, aluminium, stainless steel, copper wires, steel, NiCr alloys). General observations and conclusions from the experimental studies on CHF are:[24]

1. CNTs cause the highest CHF enhancement, followed by metal nanofluids and metal oxide nanofluids;

2. CHF enhancement in convective boiling is, in general, lower than that of pool boiling;
3. The vapour bubbles departing from the surface are significantly larger in nanofluid experiments than in pure water experiments. This implies that the nanoparticles or, most likely, the surfactants used for their stabilization, modify the surface tension of the water;
4. The heating surfaces are modified by the deposition of a layer of particles, which sometimes form a thin, solid porous layer.

Surface heat fluxes achieved with boiling media (typically of the order of 10^6 W m^{-2}) are significantly higher than heat fluxes for single-phase fluids. For this reason, boiling heat-transfer media, and in particular boiling nanofluids, are touted as the cooling media that may achieve surface fluxes close to 10^7 W m^{-2}, and could be used with the next generation of microelectronic devices.[79] An important consideration for this application is that the surface temperature is kept under 130 °C at all times including during fast transient operation. The high CHF of nanofluids provides a significant safety margin, which would ensure that the temperature of the heated surface does not exceed the design value.

A glance at Figure 14.5 illustrates this point and proves the promise of nanofluids as reliable heat-transfer media in electronic devices: the base fluid boiling curve (solid line) shows a CHF of 1 000 000 W m^{-2} with superheating at 20 °C. The experimental data (diamonds) extend to a CHF of 1 800 000 W m^{-2} with a higher superheating temperature of 50 °C. The difference in the superheating temperatures (30 °C) is not prohibitive for a thermal system with higher reliability. One may actually design a cooling system using this nanofluid to operate at a heat flux of 1 400 000 W m^{-2} with superheating at 40 °C. This system will have an adequate safety margin (400 000 W m^{-2}) before the CHF is exceeded and the electronic component fails.

14.6 Conclusions and General Observations

The addition of nanoparticles to base fluids significantly modifies the transport properties of these fluids. In particular, the addition of nanoparticles increases the thermal conductivity of the fluids, sometimes by a factor of two or three. Higher thermal conductivity enhancements were observed with fibrous carbon nanotubes and metallic nanofluids, which also tend to use the most expensive nanoparticles. Most of the experimental results on metal oxide nanofluids, which utilize less expensive nanoparticles, indicate that the thermal conductivity enhancement is modest, in the range 5–50%. The primary mechanism for the enhancement of the thermal conductivity is the significantly higher conductivity of the solid nanoparticles and in particular of elongated fibres. Electric surface charges, particle aggregation, and surfactant choice also influence the thermal conductivity of the suspension by affecting the shape and distribution of the nanoparticles.

The Brownian movement of the nanoparticles has a rather minor effect on the thermal conductivity.

In addition to the observed thermal conductivity enhancement, nanofluids also exhibit an augmentation of their convective heat-transfer coefficient, above the values that are expected from the observed thermal conductivity enhancement. This augmentation occurs in laminar flows only and is most likely due to microconvection and transient effects caused by the Brownian movement of the nanoparticles.

Regarding boiling with nanofluids, the majority of experimental evidence is that the pool boiling and convective boiling coefficients are reduced with the addition of all types of nanoparticles. However, the critical heat flux of the suspension increases, often by a factor of two or three. Despite the lower boiling heat-transfer coefficients, this critical heat flux enhancement is very significant for the design of reliable thermal systems with boiling nanofluids, because it provides a safety factor that is crucial to the design of electronic cooling systems.

It must be noted that all the reliable experimental observations on heat transport using nanofluids indicate that there is nothing 'anomalous' about these heat-transfer media. The high transport coefficients, including thermal conductivity, are readily explained by known mechanisms of heterogeneous suspensions, of which the most important are the type, shape and the distribution/configuration of nanoparticles. The successful manipulation of the shapes of particle aggregates, the formation of particle structures and the optimized distribution of nanoparticles in base fluids will create heat-transfer media in the future that are far superior to the base fluids.

Acknowledgements

This work was partly supported by a grant from NSF to UTSA (TCU subcontract), HRD-1137764, and by a DOE grant through the National Energy Technology Laboratory to UTSA (DE-NT0008064 and DE-FE0011453). The author is thankful for this support.

References

1. S. U. S. Choi, Z. G. Zhang, W. Yu, F. E. Lockwood and E. A. Grulke, *Appl. Phys. Lett.*, 2001, **79**, 2252.
2. S. Kakaç and A. Pramuanjaroenkij, *Int. J. Heat Mass Transfer*, 2009, **52**, 3187.
3. E. E. Michaelides, *J. Non-Equilib. Thermodynamics*, 2013, **38**, 1.
4. M. Chopkar, S. Sudarshan, P. K. Das and I. Manna, *Metall. Mater. Trans. A*, 2008, **39**(7), 1535.
5. M. Chopkar, P. K. Das and I. Manna, *Scr. Mater.*, 2006, **55**, 549.
6. M. S. Liu, M. C. C. Lin, C. Y. Tsai and C. C. Wang, *Int. J. Heat Mass Transfer*, 2006, **49**, 3028.
7. A. Turanov and Y. Tolmachev, *Heat Mass Transfer*, 2009, **45**, 1583.

8. H. Xie, W. Yu, Y. Li and L. Chen, *Nanoscale Res. Lett.*, 2011, **6**, 124.

9. J. A. Eastman, S. U. S. Choi, S. Li, W. Yu and L. J. Thompson, *Appl. Phys. Lett.*, 2001, **78**(6), 718.

10. H. Xie, J. Wang, T. Xi, Y. Liu, F. Ai and Q. Wu, *J. Appl. Phys.*, 2002, **91**(7), 4568.

11. S. K. Das, P. Nandy, P. Thiesen and W. Roetzel, *J. Heat Transfer*, 2003, **125**, 567.

12. M. S. Liu, Lin MC-C, I. T. Huang and C.-C. Wang, *Int. Commun. Heat Mass Transfer*, 2005, **32**, 1202.

13. Y. Ding, H. Alias, D. Wen and R. A. Williams, *Int. J. Heat Mass Transfer*, 2006, **49**, 240.

14. H. Xie, H. Lee, W. Youn and M. Choi, *J. Appl. Phys.*, 2003, **94**(8), 4967.

15. M. Biercuk, M. Llaguno, M. Radosavljevic, J. Hyun, A. Johnson and J. Fischer, *Appl. Phys. Lett.*, 2002, **80**(15), 2767.

16. J. C. Maxwell, *A Treatise on Electricity and Magnetism*, Clarendon Press, Oxford, 2nd edn, 1881.

17. D. A. G. Bruggeman, *Ann. Phys.*, 1935, **24**, 636.

18. R. L. Hamilton and O. K. Crosser, *Ind. Eng. Chem. Fundam.*, 1962, **1**, 187.

19. C. W. Nan, R. Birringer, D. R. Clarke and H. Gleiter, *J. Appl. Phys.*, 1997, **81**, 6692.

20. C. W. Nan, Z. Shi and Y. Lin, *Chem. Phys. Lett.*, 2003, **375**, 666.

21. Y. Xuan, Q. Li, X. Zhang and M. Fujii, *J. Appl. Phys.*, 2006, **100**, 043507.

22. K. Khanafer and K. Vafai, *Int. J. Heat Mass Transfer*, 2011, **54**, 4410.

23. J. Buongiorno, D. C. Venerus, N. Prabhat, T. McKrell, J. Townsend, R. Christianson, Y. V. Tolmachev, P. Keblinski, L. Hu, J. L. Alvarado, I. C. Bang, S. W. Bishnoi, M. Bonetti, F. Botz, A. Cecere, Y. Chang, G. Chen, H. Chen, S. J. Chung, M. K. Chyu, S. K. Das, R. Di Paola, Y. Ding, F. Dubois, G. Dzido, J. Eapen, W. Escher, D. Funfschilling, Q. Galand, J. Gao, P. E. Gharagozloo, K. E. Goodson, J. G. Gutierrez, H. Hong, M. Horton, K. S. Hwang, C. S. Iorio, S. P. Jang, A. B. Jarzebski, Y. Jiang, L. Jin, S. Kabelac, A. Kamath, M. A. Kedzierski, L. G. Kieng, C. Kim, J. H. Kim, S. Kim, S. H. Lee, K. C. Leong, I. Manna, B. Michel, R. Ni, H. E. Patel, J. Philip, D. Poulikakos, C. Reynaud, R. Savino, P. K. Singh, P. Song, T. Sundararajan, E. Timofeeva, T. Tritcak, A. N. Turanov, S. V. Vaerenbergh, D. Wen, S. Witharana, C. Yang, W. H. Yeh, X. Z. Zhao and S. Q. Zhou, *J. Appl. Phys.*, 2009, **106**, 094312.

24. E. E. Michaelides, *Nanofluids – Their Thermodynamic and Transport Properties*, Springer, New York, 2014.

25. J. Eapen, R. Rusconi, R. Piazza and S. Yip, *J. Heat Transfer*, 2010, **132**, 102402.

26. P. Keblinski, S. R. Phillpot, S. U. S. Choi and J. A. Eastman, *Int. J. Heat Mass Transfer*, 2002, **45**, 855.

27. W. Yu and S. U. S. Choi, *J. Nanopart. Res.*, 2003, **5**, 167.

28. C. Gerardi, D. Cory, J. Buongiorno, L. W. Hu and T. McKrell, *Appl. Phys. Lett.*, 2009, **95**, 253104.

29. X. D. Din and E. E. Michaelides, *AIChE J.*, 1998, **44**, 35.
30. Q. Z. Xue, P. Keblinski, S. R. Philpot, S. U. S. Choi and J. A. Eastman, *Int. J. Heat Mass Transfer*, 2004, **47**, 4277.
31. D. Lee, J. W. Kim and B. G. Kim, *J. Phys. Chem. B*, 2006, **110**, 4323.
32. J. Y. Jung and J. Y. Yoo, *Int. J. Heat Mass Transfer*, 2009, **52**, 525.
33. C. T. Wamkam, M. K. Opoku, H. Hong and P. Smith, *J. Appl. Phys.*, 2011, **109**, 024305.
34. J. Fan and L. Wang, *Int. J. Heat Mass Transfer*, 2011, **54**, 4349.
35. R. Pfeffer, S. Rosetti, S. Licklein, NASA report TN D-3603, 1966.
36. E. E. Michaelides, *Int. J. Heat Mass Transfer*, 1986, **29**, 265.
37. S. K. Das, P. Nandy, P. Thiesen and W. Roetzel, *J. Heat Transfer*, 2003, **125**, 567.
38. C. H. Li and G. P. Peterson, *J. Appl. Phys.*, 2006, **99**, 084314.
39. M. R. Maxey and J. J. Riley, *Phys. Fluids*, 1983, **26**, 883.
40. E. E. Michaelides, *Particles, Bubbles and Drops – Their Motion, Heat and Mass Transfer*, World Scientific Publ., New Jersey, 2006.
41. R. Prasher, W. Evans, P. Meaking, J. Fish, P. Phelan and P. Keblinski, *Appl. Phys. Lett.*, 2006, **89**, 143119.
42. E. V. Timofeeva, A. N. Gavrilov, J. M. McCloskey, W. V. Tolmachev, S. Sprunt, L. M. Lopatina and J. V. Sellinger, *Phys. Rev. E: Stat., Nonlinear, Soft Matter Phys.*, 2007, **76**, 061203.
43. J. Philip, P. D. Sharma and B. Raj, *Nanotechnology*, 2008, **19**, 305706.
44. S. Kondaraju, E. K. Jin and J. S. Lee, *Int. J. Heat Mass Transfer*, 2010, **53**, 862.
45. A. Ghadimi, R. Saidur and H. S. C. Metselaar, *Int. J. Heat Mass Transfer*, 2011, **54**, 4051.
46. J. Philip, J. M. Laskar and B. Raj, *Appl. Phys. Lett.*, 2008, **92**, 22191.
47. H. Younes, G. Christensen, X. Luan, H. P. Hong and P. Smith, *J. Appl. Phys.*, 2012, **111**, 064308.
48. A. S. Cherkasova and J. W. Shan, *J. Heat Transfer*, 2010, **132**, 082402.
49. R. Fantoni, A. Giacometti, F. Sciortino and G. Pastore, The Royal Society of Chemistry, *Soft Matter*, 2010, **7**, 2419.
50. Z. G. Feng, E. E. Michaelides and S. L. Mao, *Fluid Dyn. Res.*, 2012, **44**, 025502.
51. Z. G. Feng and E. E. Michaelides, *Int. J. Heat Mass Transfer*, 2012, **55**, 6491.
52. A. Sergis and Y. Hardalupas, *Nanoscale Res. Lett.*, 2011, **6**, 391.
53. E. E. Michaelides, *Int. J. Heat Mass Transfer*, 2015, **81**, 179.
54. Q. Li and Y. Xuan, *Sci. China, Ser. E: Technol. Sci.*, 2002, **45**, 408.
55. D. Wen and Y. Ding, *Int. J. Heat Mass Transfer*, 2004, **47**, 5181.
56. Y. Ding, H. Alias, D. Wen and R. A. Williams, *Int. J. Heat Mass Transfer*, 2006, **49**, 240.
57. M. Hojjat, S. G. Etemad and R. Bagheri, *Korean J. Chem. Eng.*, 2010, **27**, 1391.
58. Y. Xuan and Q. Li, *J. Heat Transfer*, 2003, **125**, 151.
59. S. Torii and W. J. Yang, *J. Heat Transfer*, 2009, **131**, 043203.

60. W. Williams, J. Buongiorno and L. W. Hu, *J. Heat Transfer*, 2008, **130**, 042412.
61. N. Prabhat, J. Buongiorno and L. W. Hu, *J. Nanofluids*, 2012, **1**, 55.
62. D. Y. Tzou, *J. Heat Transfer*, 2008, **130**, 072401.
63. R. Shukla and V. Dhir, *J. Heat Transfer*, 2008, **130**, 042406.
64. M. Okada and T. Suzuki, *Int. J. Heat Mass Transfer*, 1997, **40**, 3201.
65. N. Putra, W. Roetzel and S. K. Das, *Heat Mass Transfer*, 2003, **39**, 775.
66. D. Wen and Y. Ding, *Int. J. Heat Fluid Flow*, 2005, **26**, 855.
67. S. Nukiyama, *J. Japan Soc. Mech. Eng.*, 1934, **37**, 367, (English Translation, *Int. J. Heat Mass Transfer*, 1966, **9**, 1419).
68. S. K. Das, S. K. Putra and W. Roetzel, *Int. J. Heat Mass Transfer*, 2003, **46**, 851.
69. I. C. Bang and S. H. Chang, *Int. J. Heat Mass Transfer*, 2005, **48**, 2407.
70. S. J. Kim, I. C. Bang, J. Buongiorno and L. W. Hu, *Int. J. Heat Mass Transfer*, 2007, **50**, 4105.
71. S. Witharana, *Boiling of refrigerants on enhanced surfaces and boiling of nanofluids*, Lic. Thesis, Royal Inst. of Technology, Stockholm, Sweden, 2003.
72. K. J. Park and D. Jung, *Int. J. Heat Mass Transfer*, 2007, **5**, 4499.
73. A. E. Bergles, J. G. Collier, J. M. Delhaye, G. F. Hewitt, F. Mayinger, *Two-phase Flow and Heat Transfer in the Power and Process Industries*, Hemisphere, New York, 1981.
74. P. B. Whalley, *Boiling, Condensation and Gas-liquid Flow*, Clarendon Press, Oxford, 1987.
75. A. A. Chehade, H. L. Gualous, S. L. Masson, F. Fardoun and A. Besq, *Nanoscale Res. Lett.*, 2013, **8**, 130.
76. H. Peng, G. Ding, W. Jiang, H. Hu and Y. Gao, *Int. J. Refrig.*, 2009, **32**(6), 1259.
77. K. B. Rana, A. K. Rajvanshi and G. D. Agrawal, *J. Visualization*, 2013, **16**, 133.
78. M. Boudouh, H. L. Gualous and M. De Labachelerie, *Appl. Therm. Eng.*, 2010, **30**(17–18), 2619.
79. D. Faulkner, M. Khotan, R. Shekarriz, *19th Annual IEEE Semiconductor Thermal Measurement and Management Symposium*, 2003, 223.
80. D. Wen, *Int. J. Heat Mass Transfer*, 2008, **51**, 4958.
81. D. Milanova and R. Kumar, *J. Heat Transfer*, 2008, **130**, 042401.
82. K. J. Park, D. Jung and S. E. Shim, *Int. J. Multiphase Flow*, 2009, **35**, 525.

Thermometry in Micro and Nanofluidics

C. BERGAUD[a,b]

[a] CNRS, LAAS, 7 Avenue du Colonel Roche, F-31400 Toulouse, France;
[b] Université de Toulouse, LAAS, F-31400 Toulouse, France
Email: bergaud@laas.fr

15.1 Thermal Issues in Micro and Nanofluidics: General Context

Measuring and controlling temperature on a reduced scale is an increasingly active area in micro and nanofluidics, more specifically when it concerns chemical and biological processes. The main motivation relies on the increased capability to obtain localized heating, strong thermal gradients and fast temperature cycling with an active control of temperature. Indeed, the design optimization of microfluidic systems takes advantage of parallelization and miniaturization where smaller volumes give rise to lower thermal mass and heat capacitance. For example, in the context of lab-on-a-chip, microfabricated polymerase chain reaction (PCR) devices[1,2] and microfluidic calorimeters[3–8] have benefited from both recent technological advances and improved methods for measuring and monitoring local temperature changes.

On the other hand, from a thermal point of view, the reduction in size can generate unexpected localized temperature variations in microfluidic devices since Joule heating is likely to occur when, for example, an electric field is applied across a channel containing a conducting medium.[9] In that case,

RSC Nanoscience & Nanotechnology No. 38
Thermometry at the Nanoscale: Techniques and Selected Applications
Edited by Luís Dias Carlos and Fernando Palacio
© The Royal Society of Chemistry 2016
Published by the Royal Society of Chemistry, www.rsc.org

the temperature is not the physical parameter used for inducing the desired phenomenon and an unwanted increase in temperature can dramatically impact the overall behaviour of the device and/or the surrounding liquid and consequently the sample under study by changing some of its physical properties.

A huge variety of physical phenomena are to be considered in thermometry. Most of them are described in the previous chapters of this book. Here, rather than exploring these phenomena in the context of micro and nanofluidics, we will draw a general and non-exhaustive picture of recent achievements concerning thermometry in micro and nanofluidics. The selected examples concern thermally induced phenomena in confined volumes when heat is intentionally generated for a targeted application or when an external physical perturbation causes an unwanted temperature increase.

15.1.1 Inducing Heat for Chemical and Biological Applications

In addition to PCR devices and calorimeters, local heat sources and steep thermal gradients can be used in microfluidic systems to sort and concentrate molecules or to monitor biochemical reactions. Integrated resistive heaters, laser-based heating,[10–12] Joule heating under electrophoretic flow,[9,13,14] dielectrophoresis[15] or radiofrequency,[16–18] are now commonly used to obtain the desired temperature variations in confined volumes.

In the field of separation methods, in temperature gradient focusing (TGF), a separation method for ionic species that can be compared to electric field gradient focusing (EFGF), the electric field gradient is obtained by applying a temperature gradient in a separation channel filled with a buffer exhibiting a temperature-dependent ionic strength.[19–21] Isoelectric focusing (IEF) has been improved through the use of thermally induced pH gradients in small volumes.[22,23] Thermal gradients of the order of 1 K μm^{-1} were obtained by focusing an ionic current through a microhole and used to determine the melting profile of DNA duplexes.[14] As illustrated in Figure 15.1, the advantage of using spatial temperature gradients is that a wide range of temperatures can be explored simultaneously.

Thermophoresis which refers to the motion of molecules or particles induced by a thermal gradient is also a powerful approach for sorting and trapping molecules or particles[24–26] and analysing interactions of small molecules or proteins in biological media[27] as shown in Figure 15.2.

The use of thermally induced perturbations in microfluidic devices in a wide range of medium conditions and temperatures has also emerged as a rapid and efficient way of determining protein conformation (see Figure 15.3),[28–30] mixing solutions[31–33] or moving droplets.[34,35]

Concerning chemical reactions, it has been exploited for obtaining activation energies from catalytic reactions or phase transitions from phospholipid membranes.[36] Small temperature modulations have been

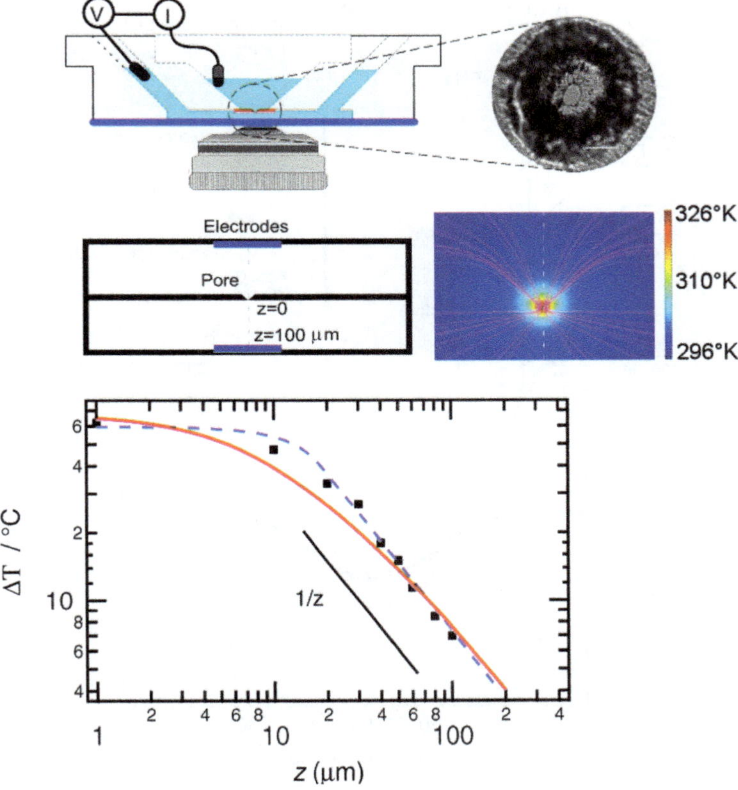

Figure 15.1 (Top) Schematic representation of a fluidic cell composed of two chambers separated by a 50 μm thick Teflon sheet in which a microhole has been drilled. (Bottom) Local temperature can be increased up to 100 °C when an ionic current is focused in the microhole; temperature variation along the vertical axis below the microhole. The diameter of the microhole is 7.5 μm and the ionic current is 1.81 mA. An average thermal gradient of 1 °C μm^{-1} has been measured over a depth of 40 μm. The temperature was derived from the fluorescence intensity of Tetramethylrhodamine (TMR) grafted to DNA oligomers.
Reproduced with permission from ref. 14. © AIP Publishing 2010.

employed to obtain chemical dynamics.[37] This elegant approach was applied for selective titration in complex mixtures.[38] A strong interest in the field of thermoplasmonics is also growing fast for potential applications in nano-chemistry, drug delivery and photothermal therapy.[39–44]

The field of thermoelectrochemistry is focused on the development of new ways to vary the temperature while performing electrochemical measurements.[45] Based on the use of Joule-heated working electrodes, laser heating, as well as electric heaters for direct and indirect heating have been considered.[46] More recently, modern methods based on the use of non-isothermal electrolysis cells have been reported for exploring fast thermal transient regimes.[45]

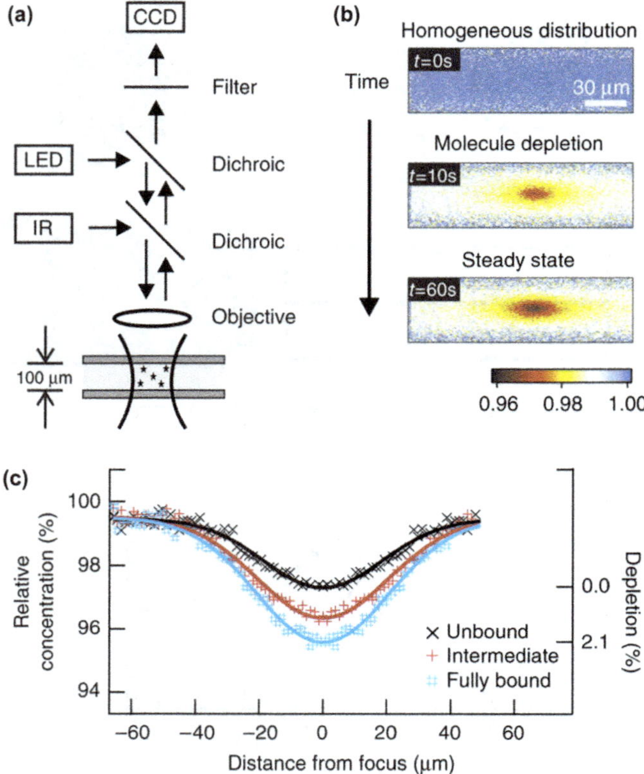

Figure 15.2   (a) Experimental setup consisting of a heating infrared laser focused into a silica capillary with a diameter of 100 μm; (b) fluorescence images of the molecules labelled with a fluorescent dye experiencing the thermophoretic force within the thermal gradient created by the laser; (c) steady-state profiles of human interferon gamma (hIFN-γ) interacting at different concentrations with a specific anti-body within the thermal gradient.
Reproduced with permission from ref. 27. © Macmillan Publishers 2010.

As shown in Figure 15.4, another interesting example concerns a thermal sensor made of a gold nanowire that has been used for ultrasensitive molecular detection based on the thermal conductance change at a solid–liquid interface.[47] A Joule-heated nanowire array was also proposed to selectively control a chemical reaction in a confined volume.[48]

15.1.2   Unexpected Heat in Micro and Nanofluidics

Temperature is a parameter that can strongly impact the overall properties of a sample under study. Indeed, temperature variations induced by local electrical fields or optical excitations can become critical with the reduction in size of micro and nanofluidic devices.[49,50] Laser-induced heating in

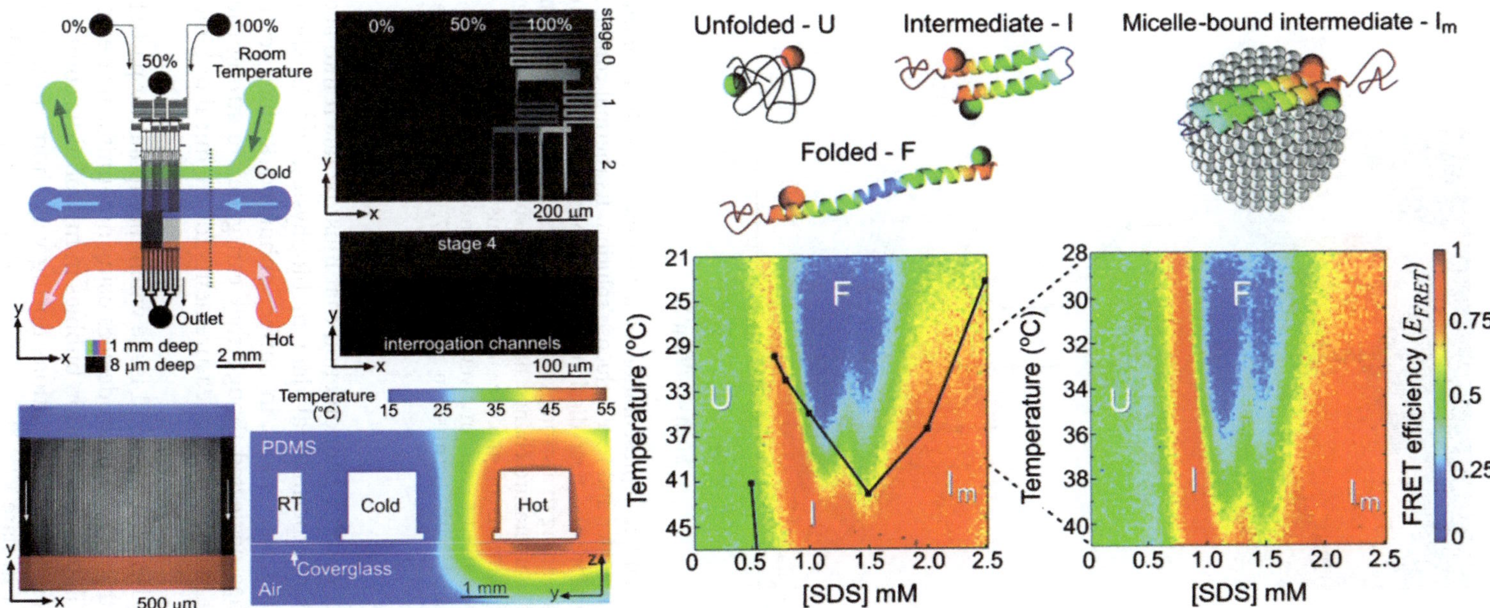

Figure 15.3 (Left) Schematic representation of a microfluidic device composed of three channels at different temperatures (room temperature, cold and hot water). Three inlets are used to study the relative concentrations of the solutions at 0%, 50% and 100%. (Right) This device generates orthogonal gradients of concentration and temperature and was used to obtain a diagram of protein conformation as a function of temperature and medium conditions. Reproduced with permission from ref. 28. © American Chemical Society 2009.

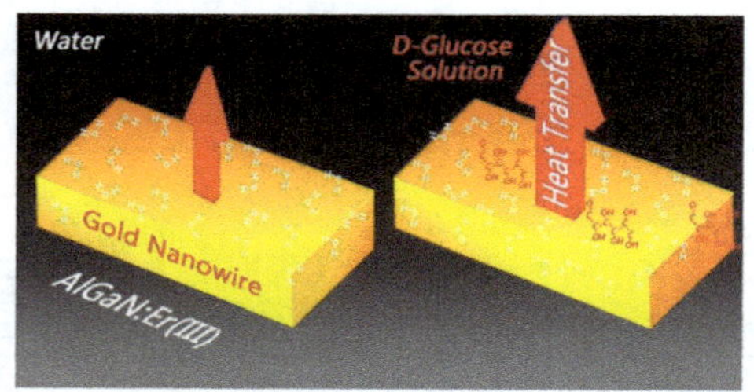

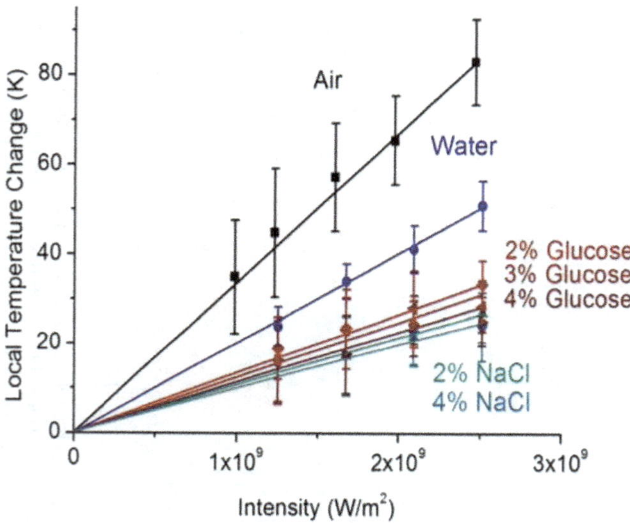

Figure 15.4   (Top) Schematic representation of a laser-heated gold nanowire immersed in water and in a D-glucose solution. Its thermal conductance, and therefore the heat dissipation, is impacted by the surrounding medium leading to a strong variation of the temperature. (Bottom) The induced temperature change has been measured in air, water and for various solutions of glucose and NaCl. This variation in heat dissipation was used to design a very sensitive molecular sensor. Reproduced with permission from ref. 47. © American Chemical Society 2013.

optical traps is a well-known phenomenon that must be taken into account for obtaining reliable experimental data in biophysics.[51,52] As shown in Figure 15.5, photothermal effects in surface-enhanced Raman scattering (SERS) are sometimes neglected though they can generate strong thermal gradients and result in misleading measurements for chemical or biological sensing.[53]

Joule heating can become a critical issue when manipulating particles or living cells in dielectrophoresis microdevices using non-uniform ac electrical fields.[15,54–57] The collapse of cavitation bubbles in microfluidic confinement gives rise to huge temperature variations during very short duration times that are not easy to estimate.[43,58–61] As illustrated in Figure 15.6, it has also

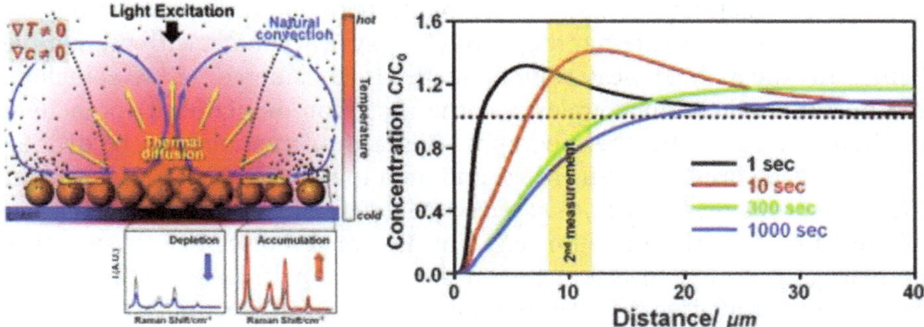

Figure 15.5   (Left) Schematic representation of the thermophoretic effect in SERS measurements. The SERS intensity is directly impacted by the depletion or accumulation phenomena induced by thermophoresis. (Right) Simulation results for the concentration distribution induced by laser heating for various time durations.
Reproduced with permission from ref. 53. © Wiley-VCH 2010.

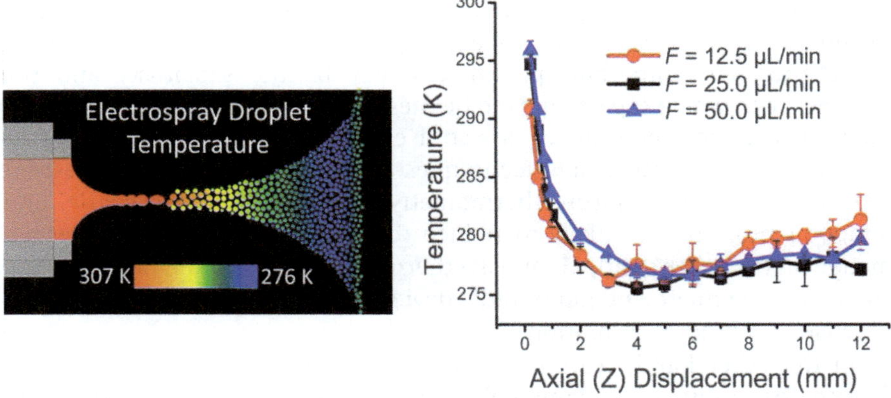

Figure 15.6   (Left) Schematic representation of droplet temperature changes in an electrospray plume. (Right) Axial temperature profiles along the spray axis for various solvent flow rates. Temperature variations up to 30 K have been measured.
Reproduced with permission from ref. 63. © American Chemical Society 2014.

been shown that non-negligible droplet temperature changes due to friction forces inside the injector take place during spray experiments and can impact chemical equilibria.[62,63]

These few examples show that temperature can play a key role as a perturbative parameter in the context of micro and nanofluidics whether in a positive or negative way. Controlling temperature is therefore often necessary and relies on measuring temperature distribution in the reacting medium on a reduced scale.

The main distinction that can be made between various thermometry techniques concerns whether they involve a physical contact with the sample being investigated. Typical contact methods considered *invasive* include thermistors and thermocouples and the most commonly used contactless method is infrared (IR) thermography. Contactless methods can be divided into two categories: semi-invasive and non-invasive.[64] The measurement method is considered as *semi-invasive* when it implies a slight modification of the sample or the medium (*e.g.*, the use of fluorescent dyes or nanoparticles); it is *non-invasive* when it is based on intrinsic temperature-dependent properties of the medium (*e.g.*, refractive index, viscosity, absorption or emission of light). We will see below that invasive techniques are not compatible with measurements in liquid, while among semi-invasive methods, fluorescent thermometry has emerged as a powerful approach in the context of micro and nanofluidics. Non-invasive techniques such as Raman spectroscopy or interferometric measurements often suffer from poor temporal resolution.

One should keep in mind that there is no universal thermometer, and that specific features must be considered for a given application. Depending on the method used, temperature-dependent values that can be measured include electrical resistance, voltage, optical intensity, refractive index, absorbance, *etc*. Important parameters may be the sensitivity and the accuracy with which the signal is to be measured. The temperature dynamic range of the measured signal is also a critical aspect that can render irrelevant certain measurement techniques or the choice of some fluorescent probes for high-temperature thermometry. The need for high spatial and temporal resolution is likely to reduce the choice to fluorescent thermometry. On the other hand, 2D and 3D thermal imaging is often needed when characterizing micro and nanofluidic devices but is generally achieved at the expense of temporal resolution.

On the other hand, real-time temperature mapping is obviously a key feature for studying fundamental chemical and biological reactions. In addition, thermodynamic or kinetic molecular probes may be used for steady-state or transient measurements but strong differences in their respective thermal relaxation times can become an important criterion when performing fast transient thermal characterization.[65,66]

Typical characteristics of the main thermometric methods used at the nanoscale are reviewed in the excellent article written by Brites *et al.*[67] The measurement of droplet temperature by optical techniques has been reviewed by Lemoine and Castanet.[62] Concerning the use of molecular thermal probes in the context of micro and nanofluidics, the reader is referred to the book chapter written by Gosse *et al.*[65]

Here, the emphasis is put on presenting the main thermometric methods usable in the field of micro and nanofluidics. Contact methods will be discussed briefly, this chapter being mainly devoted to contactless techniques with selected illustrative examples.

15.2   Contact Methods

External temperature sensors such as resistive sensors (thermistors) or thermocouples are hardly used in micro and nanofluidics. A few examples concern thermocouples or Pt microelectrodes that have been developed for measuring the temperature gradient using a scanning electrochemical microscope (SECM, see Figure 15.7).[68,69] A thermocouple probe has been described for determining the intracellular temperature in a single cell with a temperature resolution below 0.1 °C.[70] An improved version of scanning thermal microscopy (SThM) has recently been illustrated for measurements in a liquid environment.[71] This technique, called liquid-immersion SThM (iSThM), was applied in a dodecane environment with a spatial resolution of about 30 nm mainly impacted by heat transfer through the liquid rather than through the solid–solid contact.

The common drawback of these approaches is that it remains very difficult to determine the temperature without a comprehensive modelling of the heat transfer between the temperature sensor and the sample immersed in the liquid. Indeed, for most configurations, the thermal mass of the sensor, if comparable to the value of the sample under test, will undoubtedly impact the temperature measurement through important heat losses. Moreover, this approach is only relevant for open fluidics. The insertion of an external thermal sensor within a channel is in many cases not achievable, and therefore only an estimated value of the temperature is obtainable from an external contact point.

The integration within a microfluidic device of thermistors and thermocouples, made of a single conductive material or a combination of two conductive materials respectively, can constitute an alternative for *in situ* thermal measurement with microscale resolution and response times

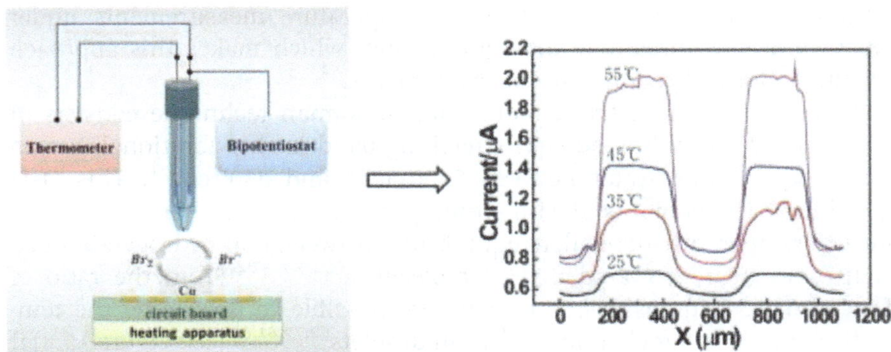

Figure 15.7   (Left) Graphic representation of a thermocouple microelectrode designed for local temperature measurement and SECM. (Right) Lateral scanning current curves measured at various temperatures on a circuit board with parallel copper strips.
Reproduced with permission from ref. 69. © Elsevier 2014.

down to the microsecond.[65,72–76] Nonetheless, only surface temperature measurements are achievable and 2D thermal mapping is not possible except by using arrays of resistive sensors or thermocouples, but at the expense of the spatial resolution and with an increased complexity of the fabrication process.[3] In addition, these sensors are also sensitive to electromagnetic fields.

15.3   Contactless Methods

Within contactless methods, one may distinguish between semi-invasive and non-invasive methods or whether they rely on luminescence or not. Among non-invasive techniques, IR thermometry based on measuring the thermal radiation in the IR spectrum has been widely used and optimized in the field of microelectronics for thermal imaging, but it is limited in terms of spatial resolution (a few microns). Moreover, the strong absorption of water for wavelengths above 1 μm makes this technique incompatible with thermal characterization in micro and nanofluidics. In the following, we will illustrate different thermometric approaches and categorize them on the basis of non-luminescent and luminescent measurement techniques.

15.3.1   Non-luminescent Measurements

15.3.1.1   *Raman Spectroscopy*

Raman spectroscopy relies on the inelastic scattering of monochromatic light by molecules or atoms. The intensity ratio of the Stokes and anti-Stokes signals from the same spectral line can be used for thermometry at temperatures up to 2230 °C, but as the anti-Stokes signal is weak at room temperature, it is not really effective for temperature measurements under 100 °C and it requires long integration times which makes this approach unsuitable for fast temperature measurements.[64]

For liquid thermometry, a more familiar Raman technique consists of using the strong and broad O–H stretching band in the vibrational Raman spectrum of liquid water between 2800 cm^{-1} and 3800 cm^{-1}. This O–H stretching band is bi-modal with strong peaks at 3250 cm^{-1} and 3450 cm^{-1} due to a temperature-dependent equilibrium between non-hydrogen-bonded (NHB) and hydrogen-bonded (HB) molecules.[62,65,77–79] From the ratio of these two peak intensities, it is therefore possible to determine the temperature inside microchannels[79,80] or in droplets[62,77,81,82] (see Figure 15.8) if a temperature calibration of the Raman intensity of the O–H stretching band has been previously done, using *e.g.*, liquid crystal thermometry[79] or a small calibrated thermocouple.[81] Near-surface temperature measurements based on the O–H stretching band using cavity-enhanced Raman scattering (CERS) induced by whispering gallery mode propagation have been also proposed and combined with laser-induced fluorescence (LIF).[83] This non-invasive

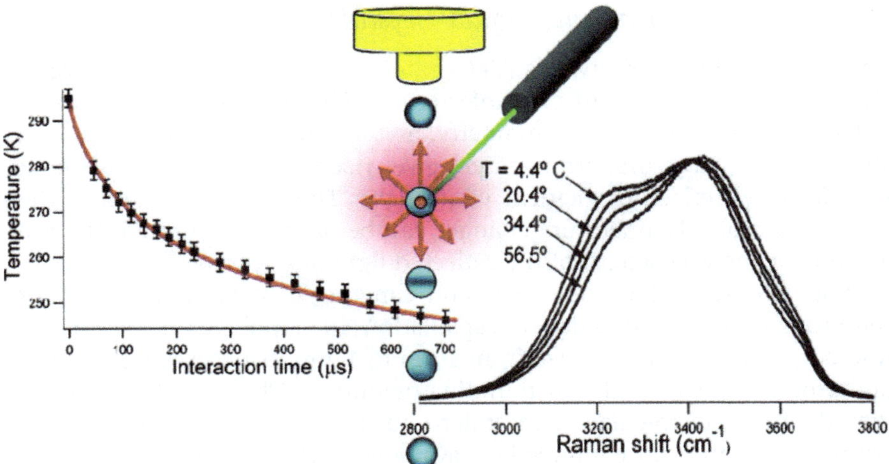

Figure 15.8 (Left) Water droplet temperature as a function of vacuum interaction time to study the evaporative cooling rate. (Right) The liquid droplet temperature was determined using the Raman spectra of the O–H stretching band over the range of 0–60 °C.
Reproduced with permission from ref. 81. © American Chemical Society 2006.

technique that does not involve any perturbation of the fluid in the channel is nonetheless a time-consuming technique with a rather slow image point analysis.[84] A thermal dependence of 0.45 °C/% has been demonstrated with a temporal resolution of 0.3 s and a spatial resolution in the micrometre range.[85]

15.3.1.2   Magnetic Resonance Imaging

In magnetic resonance imaging (MRI), many parameters such as the diffusion coefficient, the relaxation time, the magnetization transfer, the proton density and the proton resonance frequency are temperature dependent.[86] Among them, proton resonance frequency (PRF) has become the method of choice for temperature mapping in the field of thermal therapies, as it exhibits an excellent linearity with temperature over a large physiological temperature range.[87] The measurement principle is based on the increase with temperature of the nuclear shielding effect of electrons in water molecules, which leads to a reduction of the local magnetic field compared to the external macroscopic magnetic field. As a consequence, the PRF decreases with temperature. Both spectroscopic and phase imaging can be implemented for PRF-based thermometry. This measurement technique has been used to estimate Joule heating and monitor the electrolyte temperature in capillary electrophoresis and electrochromatography columns.[88] Lacey *et al.* reported an accuracy of 0.2 °C with a spatial resolution of about 1 mm and a temporal resolution under 1 s.[89,90]

15.3.1.3 *Thermochromic Liquid Crystals*

Thermochromic liquid crystals (TLCs) are made of optically active organic chemicals. They are based on cholesteric or chiral nematic compounds.[65] TLCs are easy to use for thermometry, as a thermally induced phase transition of their structural arrangement gives rise to a visible spectral change by selectively reflecting incident white light. This transformation is thermally reversible. Temperature changes can be quantified through the average hue or the wavelength of the reflected light after proper calibration[91,92] with an accuracy of 0.1 to 0.5 K and time scale of around 0.1 s.[65] For reproducibility and stability, encapsulation in a polymer shell to form microcapsules, ranging in size from 1 μm to 1 mm, is necessary to stabilize and protect TLCs from the surrounding medium.[93] The spatial resolution is directly linked to the size of the microcapsules or the laser spot size. The temperature range for commercially available TLCs goes from −30 °C up to +120 °C. In micro and nanofluidics, TLCs have been mainly used for temperature measurement in PCR microdevices (see Figure 15.9).[91–95]

15.3.1.4 *Particle Image Velocimetry*

Cross-correlation analysis using particle image velocimetry (PIV) can yield both velocity and temperature information in low-speed flows ($<$10 μm s^{-1}).[96] This technique is based on observing the Brownian motion of small particles. Since the diffusion coefficient depends on the absolute temperature and the dynamic viscosity, the temperature can be inferred by measuring the change in viscosity of a liquid medium, which is strongly temperature dependent. Based on this measurement principle, mean square displacement (MSD) based thermometry was reported by Cheng *et al.*[97] to

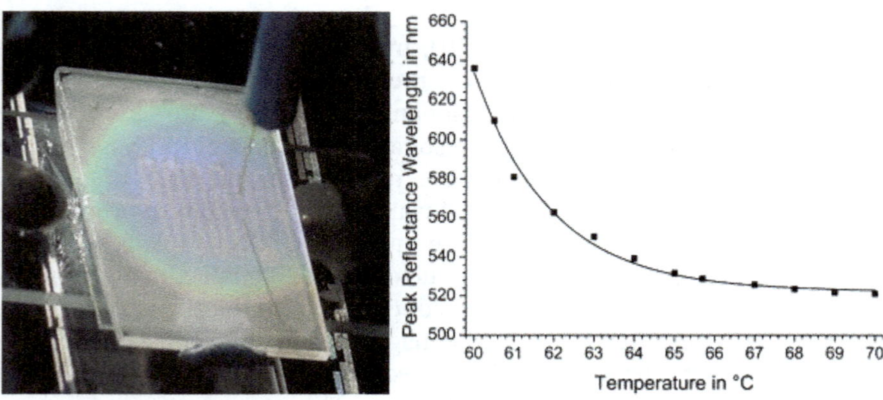

Figure 15.9   (Left) Thermal imaging of a microfluidic reactor chip filled with TLCs. (Right) Variation of the maximum reflectance wavelength of the TLCs as a function of temperature.
Reproduced from ref. 93 with permission from the Royal Society of Chemistry.

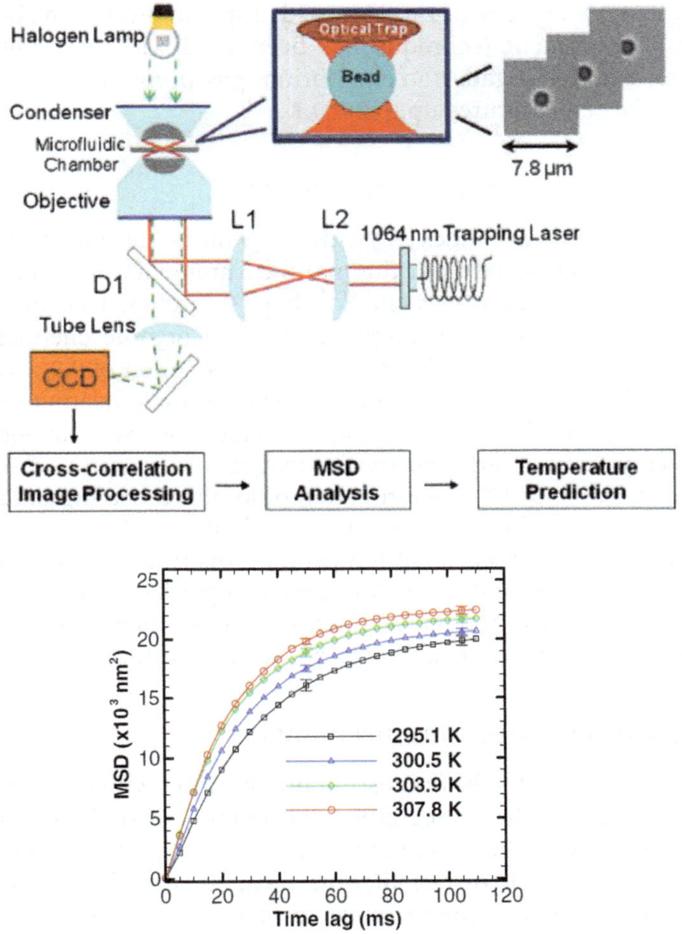

Figure 15.10 (Top) Schematic representation of MSD-based thermometry for fluid temperature measurements. (Bottom) MSD of an optically trapped silica bead for temperatures ranging from 22 to 35 °C.
Reproduced with permission from ref. 97. © The Japan Society of Applied Physics 2012.

determine the temperature in a microfluidic flow cell using trapped silica beads (see Figure 15.10). Abbondanzieri *et al.*[98] used power spectral measurements that were fit to a Lorentzian form to determine the temperature and study the influence of temperature on RNA polymerase. Three-dimensional *in situ* temperature measurements exploiting the Brownian motion of nanoparticles have been performed using a videomicroscopy setup.[99] In this study, suspensions of carboxylate-modified polystyrene particles in deionized water or in cell medium were used in a poly-(dimethylsiloxane) (PDMS) microfluidic system made of three independent channels where thermal gradients could be generated in the horizontal or vertical directions. An overall temperature accuracy of 1 °C was established within a temperature range of 1 to 50 °C with a spatial resolution about 1 μm. A similar measurement technique has been used with optically trapped silica spheres to investigate non-equilibrium gas properties in two spatial dimensions for temperatures up to 2000 K.[100]

15.3.1.5 *Interferometric Measurements*

The variation of refractive index (RI) with temperature for most fluids can be advantageously used to perform thermal measurements in small volumes. Based on this principle, Swinney and Bornhop[101] developed a simple on-chip RI detector to determine the temperature of an organic chemical buffer solution in a microchannel with a temperature resolution of 9.9×10^{-4} °C. It was also employed by the same group to evaluate Joule heating in on-chip capillary electrophoresis.[102] Other examples include the use of an optical fibre extrinsic Fabry–Pérot interferometry for temperature control of enzymatic reactions in microchips.[103] The estimation of localized heating induced by a laser optical trap was done by exploiting the change of the RI of water with temperature.[51] Three-dimensional temperature distributions of a resistively heated metal microwire and optically heated nanoparticle arrays immersed in water were obtained using an optical microscopy technique known as thermal imaging using quadriwave shearing (TIQSI, see Figure 15.11).[43,104,105]

15.3.2 Luminescence Measurements

The luminescence emitted by fluorescent probes varies with the surrounding temperature. Indeed, the temperature dependence of their photophysical properties gives rise to changes in their fluorescence spectrum, fluorescence quantum yield and excited-state lifetime through several processes: intersystem crossing, delayed fluorescence, protonation, excimer or exciplex formation, *etc.*[65,106] The most common measurements consist of measuring in 2D or 3D the variation of the fluorescence intensity or lifetime with temperature. For most cases, they decrease with an increase in temperature. The origin of this behaviour is mainly due to an increase in thermal nonradiative processes (*e.g.*, collisions with other molecules, intramolecular vibrations and rotations, *etc.*). In liquid environments, luminescence can be strongly influenced by oxygen quenching, pH variations or a change in probe concentration.

15.3.2.1 *One- or Two-colour Intensity Based Measurements*

The easiest method for fluorescence analysis is to measure the integrated intensity of the light emitted by a luminescent probe either within a certain band of the spectrum, typically in the vicinity of the emission peak, or throughout the entire emission spectrum. This approach, generally referred to as LIF, has been widely used to obtain, in 2D or 3D, spatial and temporal temperature mapping in microfluidic systems within a range between

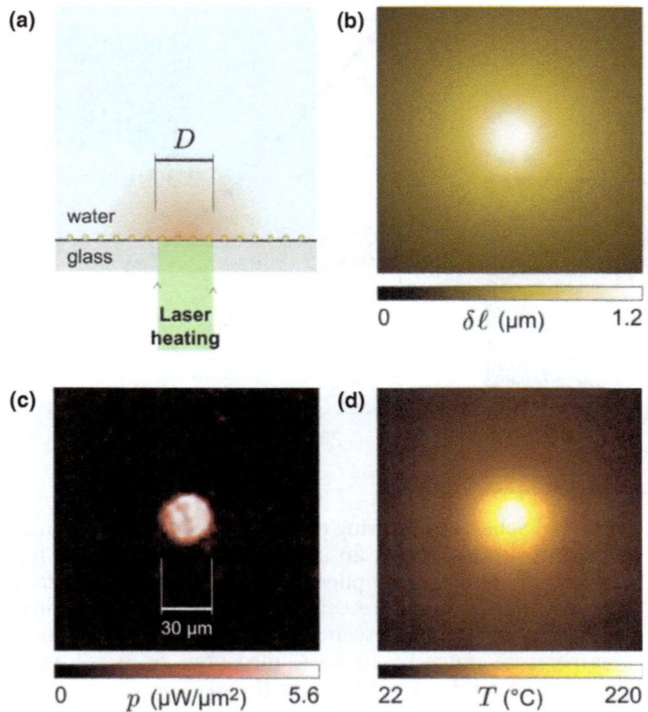

Figure 15.11   (a) Schematic drawing of the experiment; (b) optical patch difference induced by laser heating and measured by TIQSI; (c) corresponding heat source density; (d) corresponding temperature mapping. Reproduced with permission from ref. 43. © American Chemical Society 2014.

20 and 70 °C with a precision better than 1 °C and with millisecond response times. The luminescent probes used are either organic (*e.g.*, Rhodamine B,[11,13,15,76,83,107,108] SYBR Green I dye[109] or fluorescein[110]), organometallic (*e.g.*, complexes of $[Ru(bpy)_3]^{2+}$)[111,112] or inorganic (*e.g.*, CdSe nanocrystals),[36] to name a few (see Figure 15.12).

The main drawback of intensity based measurements is that they generally suffer from losses in accuracy. These are mainly due to factors such as drift of the optical excitation and/or variations in the optical properties of the sample due to photobleaching or changes in concentration due to thermal gradients or adsorption of the fluorescent probes onto surfaces. A strategy to get rid of oxygen quenching or pH variation relies on the use of a passivation layer acting as a shield that protects the fluorescent probes. For example, encapsulation of fullerene-containing polymer nanoparticles within an oxygen-impermeable polymer enables measurements to be made in the presence of oxygen.[113] PDMS or SU8 thin films containing Rhodamine B have been employed to avoid measurement errors arising from a change in concentration due to adsorption (or absorption) onto the channel surfaces of microfluidic devices.[114,115] Three-dimensional mapping inside a

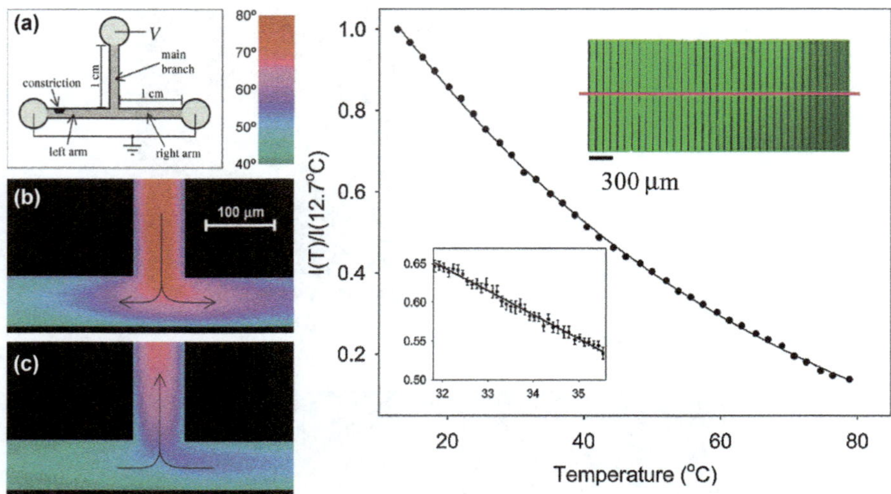

Figure 15.12 (Left) (a) Schematic drawing of the T-shaped microchannel; (b) temperature measurement with an applied voltage of 2500 V; (c) temperature measurement with an applied voltage of −2500 V. The temperature was derived from the fluorescence intensity of Rhodamine B.[13] (Right) Variation of the fluorescence intensity of CdSe nanocrystals over a temperature range of 10–80 °C measured from a linear temperature gradient in a microfluidic system (inlet).
Reproduced with permission from ref. 36. © American Chemical Society 2002.

microfluidic device has been obtained using a similar approach.[116] The main drawback is that the measured temperature within the passivation layer may greatly differ from that within the fluid. Moreover, the thermal conductivity and heat capacity of the passivation layer may impact the overall thermal properties of the microfluidic device and its operating features. Another way to improve temperature measurements obtained from fluorescence spectra consists of optimizing the parameter extractions. For that purpose, neural network recognition (NNR) has been proposed to determine the sample temperature from the magnitude and the shape of measured fluorescence spectra. Temperature oscillations were induced through photothermal excitation in glycerol containing Rhodamine B and $CuCl_2$ and measured with an accuracy of 4.2 mK Hz$^{1/2}$.[117] In any case, for one-color fluorescence thermometry, photobleaching remains a critical issue even if it can be used for improving temperature measurements as its rate greatly depends on temperature.[118,119]

Indeed, when considering luminescence intensity measurements, ratiometric thermometry using dual-emitting temperature probes should be the preferred method to get rid of local environmental perturbations that alter absolute luminescence intensities. Concerning dual-emitting probes and ratiometric measurements, the reader should refer to the excellent review articles written by McLaurin *et al.*[120] and Brites *et al.*[67] In the following, we will just give a few representative examples focused on the field of micro and nanofluidics. The different scenarios presented here rely on the use of fluorescent thermodynamics or kinetics probes, the properties of which are described by Gosse *et al.*[65]

Dual emission from independent fluorescent probes is the simplest method for obtaining reliable luminescence data as a function of temperature. In this case, a fluorophore, whose fluorescence is strongly temperature dependent, is used with a second one, which should be less sensitive to temperature. The accuracy in determining the temperature is directly correlated with the difference of the temperature sensitivities of the two fluorophores: the higher the difference, the more accurate the temperature determination. The temperature dependence of the fluorescence emission of Rhodamine B is due to the formation of a twisted intramolecular charge transfer (TICT) state.[65] It has been used with fluorescein or carboxyfluorescein to characterize the temperature gradient generated by internal Joule heating in a microchannel.[19,110] It has also been employed with Rhodamine 110[63,121] or Sulforhodamine 101[122] to study thermal transport in microfluidic systems. Thermal sensors based on the use of droplets of an ionic liquid containing Rhodamine B and Rhodamine 110 encapsulated with Parylene have also been described by Kan *et al.*[123] Based on the fluorescence intensity ratio of Rhodamine B and Rhodamine 6G, an optofluidic temperature probe has been proposed to study the thermal activation of ion channels.[124] The use of single quantum dot (QD) intensity measurements has also been reported with Rhodamine 110 as a reference dye to simultaneously map the fluid temperature and velocity in microflows using a total internal reflection fluorescence configuration (see Figure 15.13).[125]

Temperature measurements using two independent fluorophores may be limited in terms of accuracy and reproducibility if the two fluorophores do not respond in the same way to external perturbations like, for example, oxygen quenching or pH variations. These limitations may be overcome by performing two-colour single-dye fluorescence measurements. Indeed, emission spectra of fluorescent probes are often not uniformly impacted by temperature and therefore ratiometric intensity measurements can be performed using two different spectral bands within the fluorescence spectrum (see Figure 15.14).[126,127] This technique was first used to study the evaporation and combustion of ethanol droplets containing Rhodamine B.[128] It has been further improved for simultaneously measuring droplet temperature and velocity.[62]

Temperature measurements based on energy transfer between two luminescent probes through Förster resonance energy transfer (FRET) is another way to perform two-colour thermometry. For example, the thermal characterization of self-assembled hemi-spherical droplets, made of epoxy resin and containing complexes of lanthanide-doped upconversion nanoparticles with Rhodamine 6G dye molecules, was obtained with a temperature sensitivity of about 1% per °C for temperatures between 25–45 °C, and 2.7% per °C for temperatures between 47–57 °C.[129] In the context of

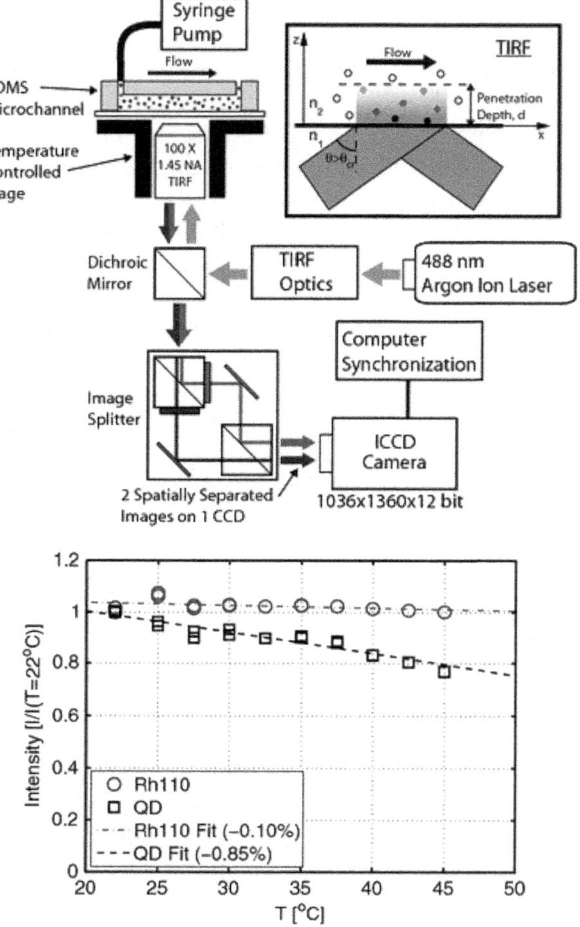

Figure 15.13   (Top) Schematic representation of the experimental setup for simultaneously measuring the fluorescence emission of QDs and Rhodamine 110 in a PDMS microchannel using a total internal reflection fluorescence configuration. (Bottom) Temperature variation of the fluorescence intensity of the QDs and Rhodamine 110, used here as a reference dye.
Reproduced with permission from ref. 125. © Springer 2008.

microfluidic devices, the use of two-colour molecular beacons (MBs) is an interesting alternative for dual fluorescence emission using FRET, provided two fluorophores, a donor and an acceptor, are located at the extremities of the oligonucleotide probe. In that case, the conformational change of the MB upon temperature change between the closed and open states will induce a modification of the emission spectra of both fluorophores. As illustrated in Figure 15.15, this approach has been developed by Barilero *et al.*,[66] using fluorescein and Texas Red as the donor and acceptor, respectively. A two-colour mapping of a microfluidic device heated by Joule

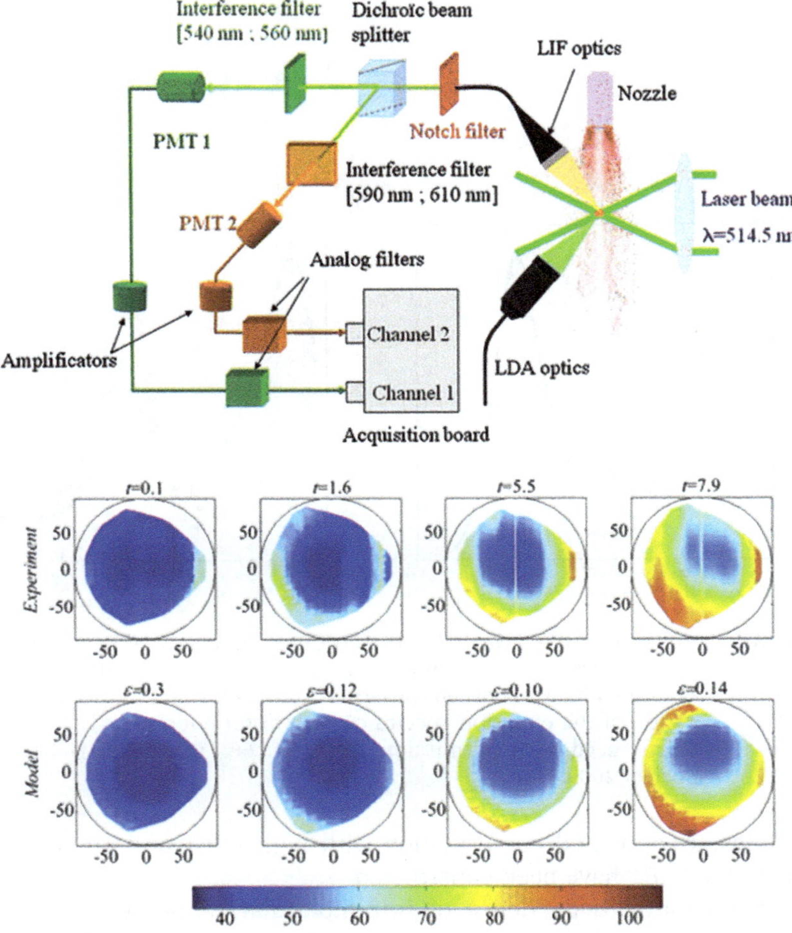

Figure 15.14 (Top) Schematic representation of the experimental setup for two-colour LIF measurements of droplet temperature in a hydrocarbon spray. Reproduced with permission from ref. 126. © Springer 2011. (Bottom) Comparison between theoretical and experimental temperature field of droplets made of *n*-decane. Reproduced with permission from ref. 127. © Elsevier 2011.

effect with a resistive strip has been obtained with a working temperature range between 5 and 35 °C, a relative sensitivity of about 10^{-3} K^{-1} and a response time of a few milliseconds. These features can be easily modified by optimizing the design of the oligonucleotide probe in terms of length and sequence composition. The use of L-DNA MBs as nanothermometers has been described for temperature sensing in living cells.[130] More sophisticated DNA switches with folding kinetics controlled by the rate of temperature variation have been designed by Viasnoff *et al.*[131] Under fixed chemical conditions, out-of-equilibrium and equilibrium states can be reached through fast and slow cooling rates, respectively. For distinguishing between

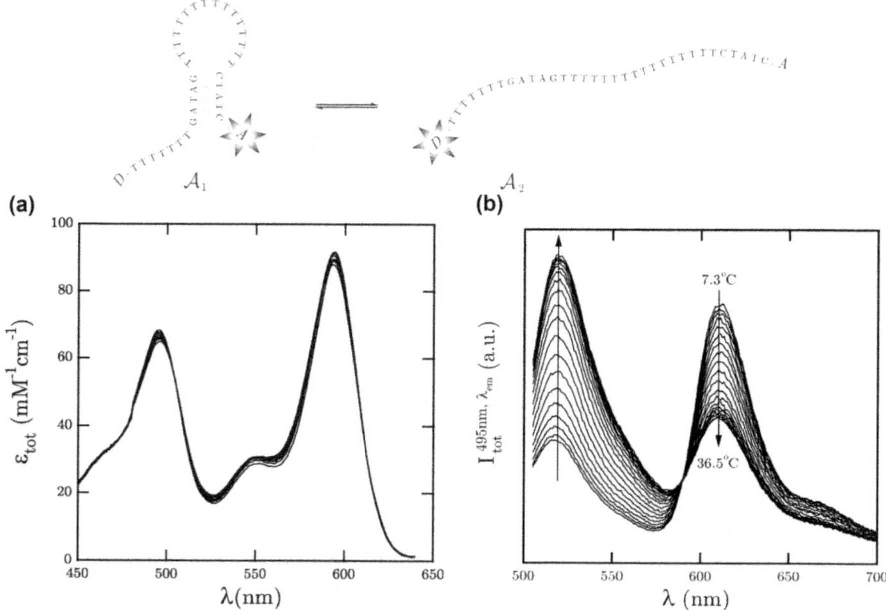

Figure 15.15   (a) Design of a DNA MB with two fluorophores, used as a molecular thermometer; (b) the conformational change of the MB upon temperature change between the closed and open states induces a modification of the emission spectra of both fluorophores.
Reproduced with permission from ref. 66. © American Chemical Society 2009.

both states, a contracted conformation with high FRET and an elongated one with low FRET have been constructed.

The temperature dependence of intramolecular monomer/excimer (or exciplex) interconversion is a well-documented phenomenon.[132] Ratiometric intensity or decay time measurements of the flurorescence emitted by a monomer and excimer/exciplex have been reported for monitoring the temperature in organic solvents over a temperature range of 20–100 °C[133] and in hydrocarbon liquids at temperatures up to 400 °C.[134] The calibration of the temperature distribution in a microchannel chip for a continuous-flow PCR has been achieved using an aqueous solution of 1-pyresulfonic acid sodium salt (PS-Na).[135]

Another approach for dual-emission intensity measurements relies on the temperature dependence of a fluorescent probe sensitive to protonation. For example, as illustrated in Figure 15.16, the temperature-dependent fluorescence of PYMPON acidic and basic states was used to perform thermal imaging of a solution inside a microfluidic channel heated by the Joule effect. Under the same experimental conditions, the thermal response time (on the microsecond time scale) for dynamic characterization was greatly improved compared to the one obtained when using a

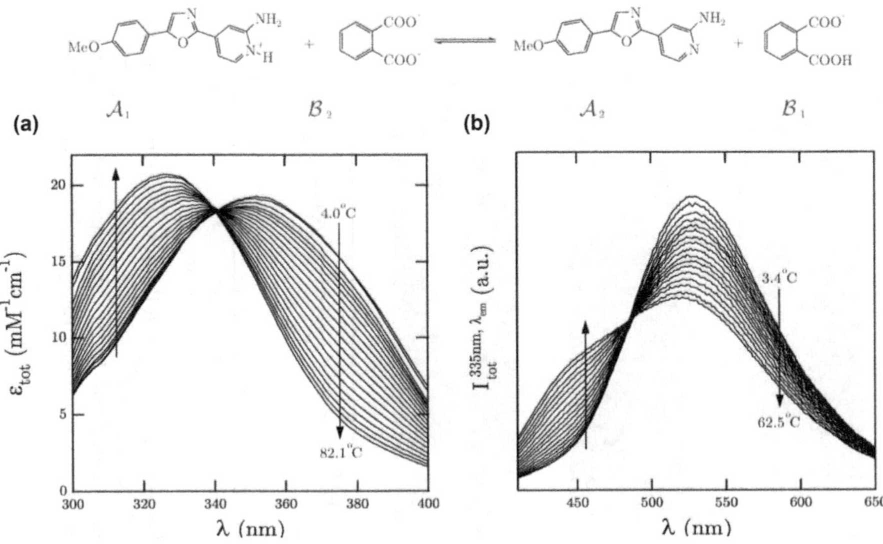

Figure 15.16   (a) The fluorescent pH probe PYMPON used as a molecular thermometer; (b) upon protonation and deprotonation with temperature, the
A_1 acidic state and the A_2 basic state will emit fluorescence at different
wavelengths.
Reproduced with permission from ref. 66. © American Chemical
Society 2009.

temperature-induced conformational change of MBs (on the millisecond
time scale).[66]

The use of excited states in thermal equilibrium constitutes a very
powerful technique for obtaining reliable temperature measurements.[136]
Many lanthanide ions exhibit luminescence from multiple excited states and
have been extensively used for ratiometric thermometry in liquid or biological environments.[47,137–139] For example, Aigouy et al.[140] have developed a
scanning thermal probe microscope based on the use of a fluorescent
nanoparticle glued at the extremity of a tungsten tip. The nanoparticle is
made of a fluoride glass co-doped with Er^{3+} and Yb^{3+} ions. After direct near-
IR optical excitation ($\lambda = 975$ nm) of the Yb^{3+} ions through a two-photon
upconversion process, the Er^{3+} ions emit visible light after energy transfer
from the excited Yb^{3+} ions. The temperature is then obtained from the
analysis of the relative intensities of two thermally coupled levels in the Er^{3+}
ions located at $\lambda \sim 520$ nm and $\lambda \sim 550$ nm. As shown in Figure 15.17, both
topographical and thermal images of a Joule-heated microheater immersed
in a water/glycerol solution were obtained using this approach. The same
temperature measurement mechanism was proposed by Dong et al.[141]
with magnetically heated mesoporous silica nanoparticles containing
superparamagnetic nanocrystals. These nanoparticles were used both as
the heater and the thermometer to quantify the temperature increase of a
surrounding fluid during exposure to a magnetic field.

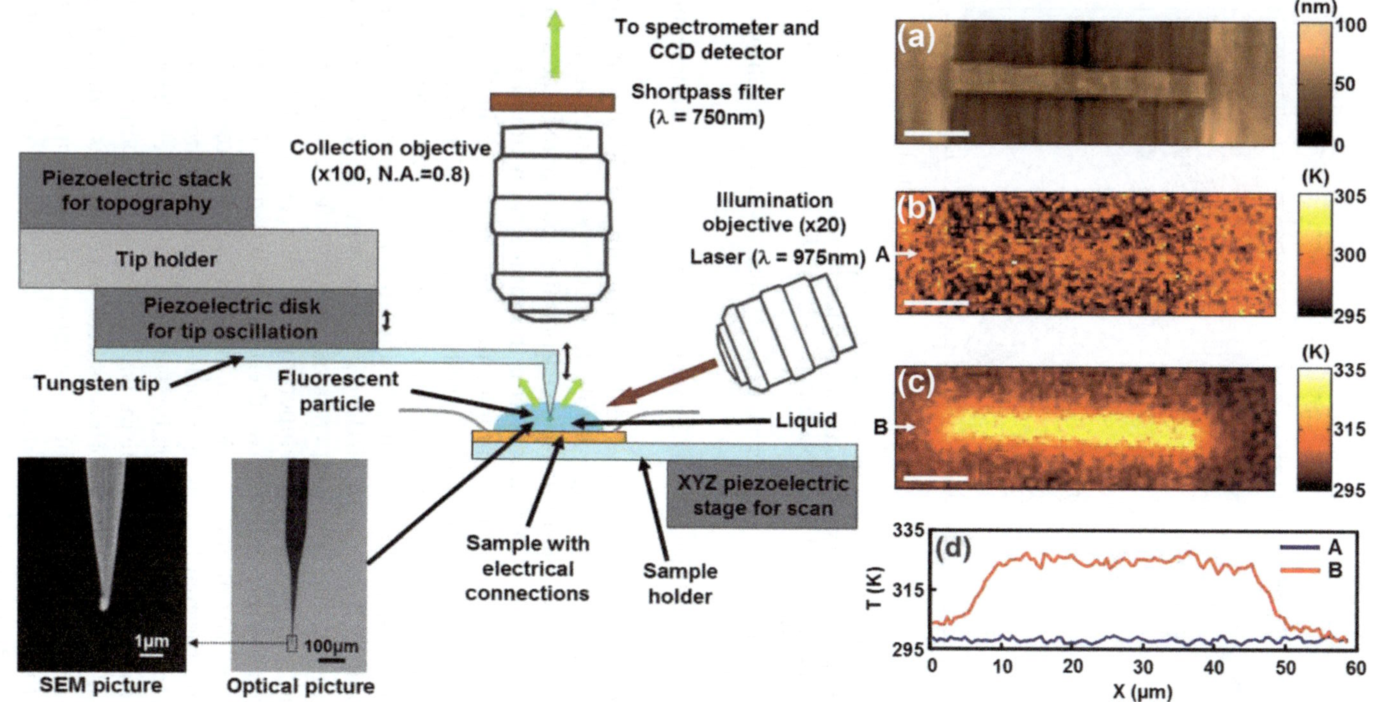

Figure 15.17 (Left) Schematic representation of the experimental setup of a scanning thermal microscope. A fluorescent particle is glued at the end of a sharp tungsten tip. The particle is made of fluoride glass co-doped with Er^{3+} and Yb^{3+} ions. The temperature is obtained from the analysis of the relative intensities of two thermally coupled levels of the Er^{3+} ions located at $\lambda \sim 520$ nm and $\lambda \sim 550$ nm. (Right) Both topographical and thermal mappings have been obtained for a Joule-heated microwire immersed in a water/glycerol droplet.
Reproduced with permission from ref. 140. © AIP Publishing 2011.

15.3.2.2 Fluorescence Lifetime Imaging

Upon excitation, the lifetime during which a fluorophore remains in an excited state is to a large extent dependent on the environment. In the fluorescence lifetime imaging (FLIM) technique, this is exploited by determining the dependence between the lifetime and a specific environmental parameter. Temperature in particular tends to affect the lifetime of the excited state. The measured lifetime data are generally fitted to a single exponential decay and single-point measurements as well as 2D or 3D imaging can be performed. Both time-domain and frequency-domain measurements can be used, depending on the time scale of the luminescence lifetime.[106,142] As compared to intensity based measurements, FLIM requires a relatively complex experimental setup when implemented in the time domain. If the time scale of fluorescence lifetimes is typically on the order of nanoseconds, femtosecond pulsed lasers must be employed to precisely control the excitation process. Furthermore, the acquisition of the fluorescence is generally achieved using lock-in techniques in order to precisely select the time window during which fluorescence should be captured. One important feature is that, since FLIM is a ratiometric measurement technique, it is independent of external parameters such as the dye concentration and the excitation/detection efficiency, thereby facilitating quantitative temperature measurements. On the other hand, the method typically requires the acquisition of several data series in order to reach a satisfactory signal-to-noise ratio. This in turn leads to image capture times that are generally longer than those for intensity based measurements. As a result, the temporal resolution of FLIM is rather low, but it has been shown that it could be improved by several orders of magnitude, up to the microsecond scale, using a statistical implementation over a large number of droplets.[143,144]

This technique was described for the first time by Benninger *et al.*[145] to obtain 3D temperature distributions in a fluid within a microchannel. Fluorescence lifetime values of Rhodamine B in methanolic solutions were exploited in a temperature range of 25–95 °C. A two-photon excitation was used with a spatial resolution of 1 µm and a precision of 1 °C. For 3D measurements, an automatic stage was moved up and down to vary the imaging depth. Other studies include the use of a Rhodamine-labelled DNA oligomer for temperature measurements in microlitre-sized volumes within a temperature range of 15–35 °C.[146] The background signals originating from fluorescence thermometry measurements in PDMS microchannels were removed by discriminating the fluorescence lifetime of Rhodamine B when absorbed in PDMS or when dissolved in water.[147] Mendels *et al.*[148] have employed Kiton Red, a water-soluble, sulfonated derivative of Rhodamine to overcome the low solubility of Rhodamine in water. Quantitative FLIM measurements along with computational fluid dynamics have been performed in aqueous solutions to study thermal and solutal transport in a mixing T-junction. Optically trapped aqueous droplets of Kiton Red in an oil environment were also used for quantitative temperature

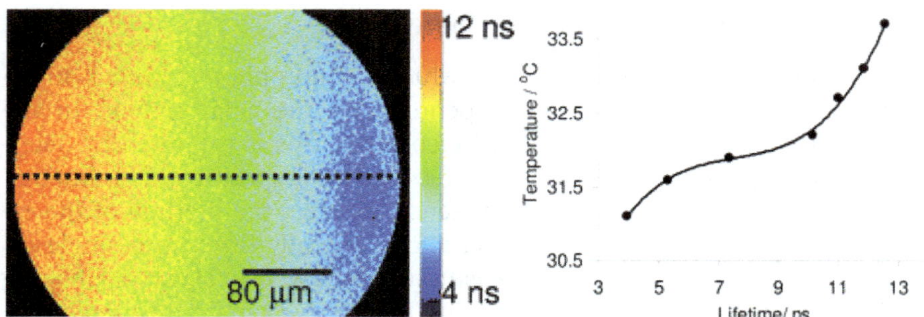

Figure 15.18   (Left) FLIM image of a microfluidic chamber obtained with a sub-degree temperature resolution and a microscale spatial resolution. (Right) Temperature calibration curve as a function of the average lifetime of a thermosensitive polymer within the phase transition range of 30 to 34 °C.
Reproduced from ref. 150 with permission from the Royal Society of Chemistry.

measurements using FLIM.[149] The thermal resolution has been improved to 0.1 °C using a water-soluble, thermoresponsive polymer poly(N-alkylacrylamide) labelled with a benzofurazan fluorophore. As can be seen in Figure 15.18, around the polymer phase transition between 30 and 34 °C, an abrupt temperature-dependent change is observed, leading to an increase of the average fluorescence lifetime in aqueous solution on temperature change.[150]

The detection of the blinking of fluorescent probes induced by protonation and deprotonation may be also exploited for nanothermometry.[151,152] Here, the presence of a hydroxyl group on the fluorophore makes the fluorescent molecule switch from a protonated state to a deprotonated state. The relaxation time of the fluorescence signal fluctuations associated with the two states is temperature dependent, and can be measured using fluorescence correlation spectroscopy (FCS) which is also a ratiometric measurement technique. As illustrated in Figure 15.19, the variation of the relaxation times of the enhanced green fluorescent protein (EGFP) was measured as a function of temperature for pH values of less than 6 in a temperature window between 10 and 60 °C.[151] Based on the same approach, simultaneous measurements of pH and temperature have been reported using pyranine as a fluorescent probe.[152] Nonetheless, the use of such an approach in physiological environments may be limited by the strong dependence of pH probes on buffer composition and ionic concentration.

Fluorescence polarization anisotropy (FPA) is directly linked to rotational diffusion of fluorescent molecules induced by Brownian thermal motion and can therefore be used for temperature measurements. The maximum sensitivity of this technique, based on the ratio of polarized intensities, is reached when the rotational and fluorescence lifetimes are of the same order. As a consequence, the increase of parameters, like the viscosity of the medium or the hydrodynamic volume of the fluorophore, is often used to

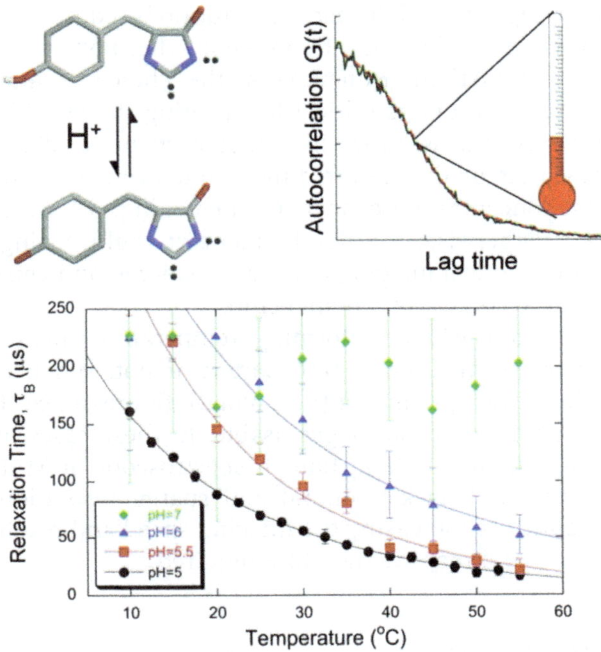

Figure 15.19   (Top) Schematic representation of the reversible protonation reaction responsible for the blinking of EGFP. (Bottom) The characteristic time associated with this phenomenon is strongly temperature dependent at low pH. The blinking characteristic time was detected by FCS. Reproduced with permission from ref. 151. © American Chemical Society 2007.

obtain reliable thermal measurements.[153] FPA as a function of the temperature for fluorescein dissolved in a glycerol/water mixture has been performed to map the temperature variations induced by nanosized heat sources with 300 nm spatial resolution and an accuracy of 0.1 °C.[154] GFP has been used as a nanothermometer for intracellular temperature mapping by measuring its FPA.[153] The temperature variations of the fluorescence anisotropy of Rhodamine 6G in glycerol have been measured using fluorescence anisotropy correlation spectroscopy (FACS) between 200 and 350 K with a time resolution of a few microseconds.[155]

15.4   Conclusion and Outlook

In this chapter, devoted to thermometry in the context of micro and nanofluidics, we proposed to introduce the field by considering the different techniques that are commonly used for temperature measurements. The selection of an appropriate method should be done with respect to the targeted application. Important parameters mainly concern the sensitivity, the accuracy, the temperature range and the spatial and temporal resolutions. Contact methods are too intrusive to be relevant especially when 2D temperature imaging with a high spatial resolution is required. Among semi-invasive measurement techniques, fluorescence thermometry has emerged as a very powerful method. Nonetheless, the choice of an appropriate fluorescent molecular probe is critical for meeting the specific needs of a given application. As discussed in this chapter, when possible, ratiometric measurements should be the preferred method to overcome artefacts due to photobleaching, concentration change, O_2 quenching, pH variation, optical fluctuations, *etc.* Otherwise, the development of multisensing probes for temperature and chemical imaging may be considered and could open new opportunities in the context of lab-on-a-chip.

A strong constraint when performing temperature measurements in liquid is whether the medium is transparent or not. When dealing with strongly diffusive or opaque media (*e.g.*, blood), fluorescence thermometry can be strongly altered and no longer usable. In these cases, non-invasive spectroscopic measurements (*e.g.*, Raman spectroscopy or MRI) can be alternatives, but they still lack temporal and spatial resolution. Technical improvements aimed at increasing the imaging speed and resolution with a better temperature accuracy are therefore required.

Acknowledgements

Part of this work was financially supported by the ANR project 10-NANO-012. Support from the French RENATECH network is also gratefully acknowledged.

References

1. M. A. Burns, B. N. Johnson, S. N. Brahmasandra, K. Handique, J. R. Webster, M. Krishnan, T. S. Sammarco, P. M. Man, D. Jones, D. Heldsinger, C. H. Mastrangelo and D. T. Burke, *Science*, 1998, **282**, 484.
2. J. Khandurina, T. E. McKnight, S. C. Jacobson, L. C. Waters, R. S. Foote and J. M. Ramsey, *Anal. Chem.*, 2000, **72**, 2995.
3. F. E. Torres, P. Kuhn, D. D. Bruyker, A. G. Bell, M. V. Wolkin, E. Peeters, J. R. Williamson, G. B. Anderson, G. P. Schmitz, M. I. Recht, S. Schweizer, L. G. Scott, J. H. Ho, S. A. Elrod, P. G. Schultz, R. A. Lerner and R. H. Bruce, *Proc. Natl. Acad. Sci. U. S. A.*, 2004, **101**, 9517.
4. J. Lerchner, A. Wolf, G. Wolf and I. Fernandez, *Thermochim. Acta*, 2006, **446**, 168.
5. T. K. Hakala, J. J. Toppari and P. Törmä, *J. Appl. Phys.*, 2007, **101**, 034512.
6. J. Xu, R. Reiserer, J. Tellinghuisen, J. P. Wikswo and F. J. Baudenbacher, *Anal. Chem.*, 2008, **80**, 2728.
7. W. Lee, W. Fon, B. W. Axelrod and M. L. Roukes, *Proc. Natl. Acad. Sci.*, 2009, **106**, 15225.

8. C. Hany, H. Lebrun, C. Pradere, J. Toutain and J.-C. Batsale, *Chem. Eng. J.*, 2010, **160**, 814.
9. G. Tang, D. Yan, C. Yang, H. Gong, J. C. Chai and Y. C. Lam, *Electrophoresis*, 2006, **27**, 628.
10. H. Mao, J. Ricardo Arias-Gonzalez, S. B. Smith, I. Tinoco Jr. and C. Bustamante, *Biophys. J.*, 2005, **89**, 1308.
11. M. Geissler, B. Voisin, L. Clime, B. Le Drogoff and T. Veres, *Small*, 2013, **9**, 654.
12. D. Koirala, J. A. Punnoose, P. Shrestha and H. Mao, *Angew. Chem., Int. Ed.*, 2014, **53**, 3470.
13. D. Ross, M. Gaitan and L. E. Locascio, *Anal. Chem.*, 2001, **73**, 4117.
14. V. Viasnoff, U. Bockelmann, A. Meller, H. Isambert, L. Laufer and Y. Tsori, *Appl. Phys. Lett.*, 2010, **96**, 163701.
15. U. Seger-Sauli, M. Panayiotou, S. Schnydrig, M. Jordan and P. Renaud, *Electrophoresis*, 2005, **26**, 2239.
16. J. J. Shah, S. G. Sundaresan, J. Geist, D. R. Reyes, J. C. Booth, M. V. Rao and M. Gaitan, *J. Micromech. Microeng.*, 2007, **17**, 2224.
17. D. D. Young, J. Nichols, R. M. Kelly and A. Deiters, *J. Am. Chem. Soc.*, 2008, **130**, 10048.
18. O. H. Elibol, B. Reddy, P. R. Nair, B. Dorvel, F. Butler, Z. Ahsan, D. E. Bergstrom, M. A. Alam and R. Bashir, *Lab Chip*, 2009, **9**, 2789.
19. D. Ross and L. E. Locascio, *Anal. Chem.*, 2002, **74**, 2556.
20. M. S. Munson, J. M. Meacham, L. E. Locascio and D. Ross, *Anal. Chem.*, 2008, **80**, 172.
21. M. S. Munson, J. M. Meacham, D. Ross and L. E. Locascio, *Electrophoresis*, 2008, **29**, 3456.
22. T. Huang and J. Pawliszyn, *Electrophoresis*, 2002, **23**, 3504.
23. B. Kates and C. L. Ren, *Electrophoresis*, 2006, **27**, 1967.
24. P. Baaske, C. J. Wienken, P. Reineck, S. Duhr and D. Braun, *Angew. Chem., Int. Ed.*, 2010, **49**, 2238.
25. L. H. Thamdrup, N. B. Larsen and A. Kristensen, *Nano Lett.*, 2010, **10**, 826.
26. M. Braun and F. Cichos, *ACS Nano*, 2013, **7**, 11200.
27. C. J. Wienken, P. Baaske, U. Rothbauer, D. Braun and S. Duhr, *Nat. Commun.*, 2010, **1**, 100.
28. V. Vandelinder, A. C. M. Ferreon, Y. Gambin, A. A. Deniz and A. Groisman, *Anal. Chem.*, 2009, **81**, 6929.
29. R. Narayanan, L. Zhu, Y. Velmurugu, J. Roca, S. V. Kuznetsov, G. Prehna, L. J. Lapidus and A. Ansari, *J. Am. Chem. Soc.*, 2012, **134**, 18952.
30. D. M. Sagar, S. Aoudjane, M. Gaudet, G. Aeppli and P. A. Dalby, *Sci. Rep.*, 2013, **3**, 2130.
31. W. N. Vreeland and L. E. Locascio, *Anal. Chem.*, 2003, **75**, 6906.
32. A. N. Hellman, K. R. Rau, H. H. Yoon, S. Bae, J. F. Palmer, K. S. Phillips, N. L. Allbritton and V. Venugopalan, *Anal. Chem.*, 2007, **79**, 4484.
33. P. García-Sánchez, A. Ramos and F. Mugele, *Phys. Rev. E*, 2010, **81**, 015303.

34. E. Verneuil, M. Cordero, F. Gallaire and C. N. Baroud, *Langmuir*, 2009, **25**, 5127.
35. X. Yao, J. Ju, S. Yang, J. Wang and L. Jiang, *Adv. Mater.*, 2014, **26**, 1895.
36. H. Mao, T. Yang and P. S. Cremer, *J. Am. Chem. Soc.*, 2002, **124**, 4432.
37. F. Closa, C. Gosse, L. Jullien and A. Lemarchand, *J. Chem. Phys.*, 2013, **138**, 244109.
38. K. Zrelli, T. Barilero, E. Cavatore, H. Berthoumieux, T. Le Saux, V. Croquette, A. Lemarchand, C. Gosse and L. Jullien, *Anal. Chem.*, 2011, **83**, 2476.
39. J. R. Adleman, D. A. Boyd, D. G. Goodwin and D. Psaltis, *Nano Lett.*, 2009, **9**, 4417.
40. M. P. Jonsson and C. Dekker, *Nano Lett.*, 2013, **13**, 1029.
41. E. Boulais, R. Lachaine, A. Hatef and M. Meunier, *J. Photochem. Photobiol., C*, 2013, **17**, 26.
42. B. Desiatov, I. Goykhman and U. Levy, *Nano Lett.*, 2014, **14**, 648.
43. G. Baffou, J. Polleux, H. Rigneault and S. Monneret, *J. Phys. Chem. C*, 2014, **118**, 4890.
44. J. Qiu and W. D. Wei, *J. Phys. Chem. C*, 2014, **118**, 20735.
45. P. Gründler, A. Kirbs and L. Dunsch, *ChemPhysChem*, 2009, **10**, 1722.
46. G.-U. Flechsig and A. Walter, *Electroanalysis*, 2012, **24**, 23.
47. A. J. Green, A. A. Alaulamie, S. Baral and H. H. Richardson, *Nano Lett.*, 2013, **13**, 4142.
48. C. Y. Jin, Z. Li, R. S. Williams, K.-C. Lee and I. Park, *Nano Lett.*, 2011, **11**, 4818.
49. D. Erickson, D. Sinton and D. Li, *Lab Chip*, 2003, **3**, 141.
50. B. Cetin and D. Li, *Electrophoresis*, 2008, **29**, 994.
51. P. M. Celliers and J. Conia, *Appl. Opt.*, 2000, **39**, 3396.
52. E. J. G. Peterman, F. Gittes and C. F. Schmidt, *Biophys. J.*, 2003, **84**, 1308.
53. T. Kang, S. Hong, Y. Choi and L. P. Lee, *Small*, 2010, **6**, 2649.
54. T. B. Jones, *J. Electrost.*, 2001, **51–52**, 290.
55. M. S. Jaeger, T. Mueller and T. Schnelle, *J. Phys. Appl. Phys.*, 2007, **40**, 95.
56. M. Kumemura, D. Collard, S. Yoshizawa, B. Wee, S. Takeuchi and H. Fujita, *ChemPhysChem*, 2012, **13**, 3308.
57. A. Kale, S. Patel, G. Hu and X. Xuan, *Electrophoresis*, 2013, **34**, 674.
58. W. B. McNamara, Y. T. Didenko and K. S. Suslick, *Nature*, 1999, **401**, 772.
59. S. L. Gac, E. Zwaan, A. van den Berg and C.-D. Ohl, *Lab Chip*, 2007, **7**, 1666.
60. Tandiono, S.-W. Ohl, D. S. W. Ow, E. Klaseboer, V. V. Wong, R. Dumke and C.-D. Ohl, *Proc. Natl. Acad. Sci.*, 2011, **108**, 5996.
61. P. A. Quinto-Su, M. Suzuki and C.-D. Ohl, *Sci. Rep.*, 2014, **4**, 5445.
62. F. Lemoine and G. Castanet, *Exp. Fluids*, 2013, **54**, 1.
63. S. C. Gibson, C. S. Feigerle and K. D. Cook, *Anal. Chem.*, 2014, **86**, 464.
64. P. R. N. Childs, J. R. Greenwood and C. A. Long, *Rev. Sci. Instrum.*, 2000, **71**, 2959.

65. C. Gosse, C. Bergaud and P. Loew, in *Thermal Nanosystems and Nano-materials*, ed. S. Volz, Springer-Verlag Berlin, Berlin, 2009, vol. 118, pp. 301–341.

66. T. Barilero, T. Le Saux, C. Gosse and L. Jullien, *Anal. Chem.*, 2009, **81**, 7988.

67. C. D. S. Brites, P. P. Lima, N. J. O. Silva, A. Millán, V. S. Amaral, F. Palacio and L. D. Carlos, *Nanoscale*, 2012, **4**, 4799.

68. A. Sode, M. Nebel, P. Pinyou, S. Schmaderer, J. Szeponik, N. Plumeré and W. Schuhmann, *Electroanalysis*, 2013, **25**, 2084.

69. H. Zhang, X. Xiao, T. Su, X. Gu, T. Jin, L. Du and J. Tang, *Electrochem. Commun.*, 2014, **47**, 71.

70. C. Wang, R. Xu, W. Tian, X. Jiang, Z. Cui, M. Wang, H. Sun, K. Fang and N. Gu, *Cell Res.*, 2011, **21**, 1517.

71. P. D. Tovee and O. V. Kolosov, *Nanotechnology*, 2013, **24**, 465706.

72. J. Khandurina, T. E. McKnight, S. C. Jacobson, L. C. Waters, R. S. Foote and J. M. Ramsey, *Anal. Chem.*, 2000, **72**, 2995.

73. E. T. Lagally, I. Medintz and R. A. Mathies, *Anal. Chem.*, 2001, 73, 565.

74. T. Yamamoto, T. Fujii and T. Nojima, *Lab Chip*, 2002, **2**, 197.

75. H. F. Arata, Y. Rondelez, H. Noji and H. Fujita, *Anal. Chem.*, 2005, 77, 4810.

76. H. F. Arata, P. Löw, K. Ishizuka, C. Bergaud, B. Kim, H. Noji and H. Fujita, *Sens. Actuators, B*, 2006, **117**, 339.

77. K. L. Davis, K. L. K. Liu, M. Lanan and M. D. Morris, *Anal. Chem.*, 1993, **65**, 293.

78. M. Becucci, S. Cavalieri, R. Eramo, L. Fini and M. Materazzi, *Appl. Opt.*, 1999, **38**, 928.

79. S. H. Kim, J. Noh, M. K. Jeon, K. W. Kim, L. P. Lee and S. I. Woo, *J. Micromech. Microeng.*, 2006, **16**, 526.

80. A. Ewinger, G. Rinke, A. Urban and S. Kerschbaum, *Chem. Eng. J.*, 2013, **223**, 129.

81. J. D. Smith, C. D. Cappa, W. S. Drisdell, R. C. Cohen and R. J. Saykally, *J. Am. Chem. Soc.*, 2006, **128**, 12892.

82. A. V. Kargovsky, *Laser Phys. Lett.*, 2006, **3**, 567.

83. R. J. Hopkins, C. R. Howle and J. P. Reid, *Phys. Chem. Chem. Phys.*, 2006, **8**, 2879.

84. R. Kuriyama and Y. Sato, *Microfluid. Nanofluidics*, 2013, **14**, 1031.

85. C. Heinisch, J. B. Wills, J. P. Reid, T. Tschudi and C. Tropea, *Phys. Chem. Chem. Phys.*, 2009, **11**, 9720.

86. V. Rieke and K. B. Pauly, *J. Magn. Reson. Imaging JMRI*, 2008, **27**, 376.

87. J. Yuan, C.-S. Mei, L. P. Panych, N. J. McDannold and B. Madore, *Quant. Imaging Med. Surg.*, 2012, **2**, 21.

88. A. S. Rathore, *J. Chromatogr. A*, 2004, **1037**, 431.

89. M. E. Lacey, A. G. Webb and J. V. Sweedler, *Anal. Chem.*, 2000, 72, 4991.

90. M. E. Lacey, A. G. Webb and J. V. Sweedler, *Anal. Chem.*, 2002, 74, 4583.

91. J. Liu, M. Enzelberger and S. Quake, *Electrophoresis*, 2002, **23**, 1531.

92. J. Noh, S. W. Sung, M. K. Jeon, S. H. Kim, L. P. Lee and S. I. Woo, *Sens. Actuators, A*, 2005, **122**, 196.

93. A. Iles, R. Fortt and A. J. de Mello, *Lab Chip*, 2005, 5, 540.

94. A. M. Chaudhari, T. M. Woudenberg, M. Albin and K. E. Goodson, *J. Microelectromech. Syst.*, 1998, 7, 345.

95. M. Basson and T. S. Pottebaum, *Exp. Fluids*, 2012, **53**, 803.

96. V. Hohreiter, S. T. Wereley, M. G. Olsen and J. N. Chung, *Meas. Sci. Technol.*, 2002, **13**, 1072.

97. C.-M. Cheng, M.-C. Chang, Y.-F. Chang, W.-T. Wang, C.-T. Hsu, J.-S. Tsai, C.-Y. Liu, C.-M. Wu, K.-L. Ou and T.-S. Yang, *Jpn. J. Appl. Phys.*, 2012, **51**, 067002.

98. E. A. Abbondanzieri, J. W. Shaevitz and S. M. Block, *Biophys. J.*, 2005, **89**, L61.

99. K. Chung, J. K. Cho, E. S. Park, V. Breedveld and H. Lu, *Anal. Chem.*, 2009, **81**, 991.

100. J. Millen, T. Deesuwan, P. Barker and J. Anders, *Nat. Nanotechnol.*, 2014, **9**, 425.

101. K. Swinney and D. J. Bornhop, *Electrophoresis*, 2001, **22**, 2032.

102. K. Swinney and D. J. Bornhop, *Electrophoresis*, 2002, **23**, 613.

103. C. J. Easley, L. A. Legendre, M. G. Roper, T. A. Wavering, J. P. Ferrance and J. P. Landers, *Anal. Chem.*, 2005, 77, 1038.

104. G. Baffou, P. Berto, E. Bermúdez Ureña, R. Quidant, S. Monneret, J. Polleux and H. Rigneault, *ACS Nano*, 2013, 7, 6478.

105. P. Bon, N. Belaid, D. Lagrange, C. Bergaud, H. Rigneault, S. Monneret and G. Baffou, *Appl. Phys. Lett.*, 2013, **102**, 244103.

106. B. Valeur and M. N. Berberan-Santos, *Molecular Fluorescence: Principles and Applications*, Wiley-VCH, Chichester, West Sussex U.K., 2nd edn, 2013.

107. R. Fu, B. Xu and D. Li, *Int. J. Therm. Sci.*, 2006, **45**, 841.

108. J. J. Shah, M. Gaitan and J. Geist, *Anal. Chem.*, 2009, **81**, 8260.

109. H. Mao, M. A. Holden, M. You and P. S. Cremer, *Anal. Chem.*, 2002, **74**, 5071.

110. M. Yoda and M. Kim, *Heat Transf. Eng.*, 2013, **34**, 101.

111. E. V. Keuren, D. Littlejohn and W. Schrof, *J. Phys. Appl. Phys.*, 2004, **37**, 2938.

112. O. Filevich and R. Etchenique, *Anal. Chem.*, 2006, **78**, 7499.

113. V. Augusto, C. Baleizão, M. N. Berberan-Santos and J. P. S. Farinha, *J. Mater. Chem.*, 2010, **20**, 1192.

114. R. Samy, T. Glawdel and C. L. Ren, *Anal. Chem.*, 2008, **80**, 369.

115. W. Jung, Y. W. Kim, D. Yim and J. Y. Yoo, *Sens. Actuators, A*, 2011, **171**, 228.

116. J. Wu, T. Y. Kwok, X. Li, W. Cao, Y. Wang, J. Huang, Y. Hong, D. Zhang and W. Wen, *Sci. Rep.*, 2013, **3**, 3321.

117. L. Liu, S. Creten, Y. Firdaus, J. J. A. F. Cuautle, M. Kouyaté, M. V. der Auweraer and C. Glorieux, *Appl. Phys. Lett.*, 2014, **104**, 031902.

118. L. Gui and C. L. Ren, *Appl. Phys. Lett.*, 2008, **92**, 024102.

119. T. Glawdel, Z. Almutairi, S. Wang and C. Ren, *Lab Chip*, 2009, **9**, 171.

120. E. J. McLaurin, L. R. Bradshaw and D. R. Gamelin, *Chem. Mater.*, 2013, **25**, 1283.

121. H. J. Kim, K. D. Kihm and J. S. Allen, *Int. J.* Heat Mass Transf., 2003, **46**, 3967.

122. V. K. Natrajan and K. T. Christensen, *Meas. Sci. Technol.*, 2009, **20**, 015401.

123. T. Kan, H. Aoki, N. Binh-Khiem, K. Matsumoto and I. Shimoyama, *Sensors*, 2013, **13**, 4138.

124. I. Węgrzyn, A. Ainla, G. D. M. Jeffries and A. Jesorka, *Sensors*, 2013, **13**, 4289.

125. J. S. Guasto and K. S. Breuer, *Exp. Fluids*, 2008, **45**, 157.

126. V. Deprédurand, A. Delconte and F. Lemoine, *Exp. Fluids*, 2011, **50**, 561.

127. G. Castanet, A. Labergue and F. Lemoine, *Int. J. Therm. Sci.*, 2011, **50**, 1181.

128. P. Lavieille, F. Lemoine, G. Lavergne and M. Lebouché, *Exp. Fluids*, 2001, **31**, 45.

129. R. Chen, V. D. Ta, F. Xiao, Q. Zhang and H. Sun, *Small Weinh. Bergstr. Ger.*, 2013, **9**, 1052.

130. G. Ke, C. Wang, Y. Ge, N. Zheng, Z. Zhu and C. J. Yang, *J. Am. Chem. Soc.*, 2012, **134**, 18908.

131. V. Viasnoff, A. Meller and H. Isambert, *Nano Lett.*, 2006, **6**, 101.

132. M. T. Albelda, E. García-España, L. Gil, J. C. Lima, C. Lodeiro, J. Seixas de Melo, M. J. Melo, A. J. Parola, F. Pina and C. Soriano, *J. Phys. Chem. B*, 2003, **107**, 6573.

133. J. Lou, T. A. Hatton and P. E. Laibinis, *Anal. Chem.*, 1997, **69**, 1262.

134. H. E. Gossage and L. A. Melton, *Appl. Opt.*, 1987, **26**, 2256.

135. K. Sun, A. Yamaguchi, Y. Ishida, S. Matsuo and H. Misawa, *Sens. Actuators, B*, 2002, **84**, 283.

136. M. Quintanilla, E. Cantelar, F. Cussó, M. Villegas and A. C. Caballero, *Appl. Phys. Express*, 2011, **4**, 022601.

137. F. Vetrone, R. Naccache, A. Zamarrón, A. Juarranz de la Fuente, F. Sanz-Rodríguez, L. Martinez Maestro, E. Martín Rodriguez, D. Jaque, J. García Solé and J. A. Capobianco, *ACS Nano*, 2010, **4**, 3254.

138. L. H. Fischer, G. S. Harms and O. S. Wolfbeis, *Angew. Chem., Int. Ed.*, 2011, **50**, 4546.

139. P. Haro-González, L. M. Maestro, M. Trevisani, S. Polizzi, D. Jaque, J. G. Sole and M. Bettinelli, *J. Appl. Phys.*, 2012, **112**, 054702.

140. L. Aigouy, L. Lalouat, M. Mortier, P. Löw and C. Bergaud, *Rev. Sci. Instrum.*, 2011, **82**, 036106.

141. J. Dong and J. I. Zink, *ACS Nano*, 2014, **8**, 5199.

142. M. I. J. Stich, L. H. Fischer and O. S. Wolfbeis, *Chem. Soc. Rev.*, 2010, **39**, 3102.

143. M. Srisa-Art, A. J. deMello and J. B. Edel, *Phys. Rev. Lett.*, 2008, **101**, 014502.

144. X. Casadevall i Solvas, M. Srisa-Art, A. J. deMello and J. B. Edel, *Anal. Chem.*, 2010, **82**, 3950.
145. R. K. P. Benninger, Y. Koç, O. Hofmann, J. Requejo-Isidro, M. A. A. Neil, P. M. W. French and A. J. deMello, *Anal. Chem.*, 2006, **78**, 2272.
146. S. Jeon, J. Turner and S. Granick, *J. Am. Chem. Soc.*, 2003, **125**, 9908.
147. T. Robinson, Y. Schaerli, R. Wootton, F. Hollfelder, C. Dunsby, G. Baldwin, M. Neil, P. French and A. deMello, *Lab Chip*, 2009, **9**, 3437.
148. D.-A. Mendels, E. M. Graham, S. W. Magennis, A. C. Jones and F. Mendels, *Microfluid. Nanofluidics*, 2008, **5**, 603.
149. M. A. Bennet, P. R. Richardson, J. Arlt, A. McCarthy, G. S. Buller and A. C. Jones, *Lab Chip*, 2011, **11**, 3821.
150. E. M. Graham, K. Iwai, S. Uchiyama, A. P. de Silva, S. W. Magennis and A. C. Jones, *Lab Chip*, 2010, **10**, 1267.
151. F. H. C. Wong, D. S. Banks, A. Abu-Arish and C. Fradin, *J. Am. Chem. Soc.*, 2007, **129**, 10302.
152. F. H. C. Wong and C. Fradin, *J. Fluoresc.*, 2011, **21**, 299.
153. J. S. Donner, S. A. Thompson, M. P. Kreuzer, G. Baffou and R. Quidant, *Nano Lett.*, 2012, **12**, 2107.
154. G. Baffou, M. P. Kreuzer, F. Kulzer and R. Quidant, *Opt. Express*, 2009, **17**, 3291.
155. R. Zondervan, F. Kulzer, H. van der Meer, J. A. J. M. Disselhorst and M. Orrit, *Biophys. J.*, 2006, **90**, 2958.

Multifunctional Luminescent Platforms for Dual-sensing

MARIO N. BERBERAN-SANTOS

CQFM - Centro de Química-Física Molecular and IN - Institute of
Nanoscience and Nanotechnology, Instituto Superior Técnico,
Universidade de Lisboa, 1049-001 Lisboa, Portugal
Email: berberan@tecnico.ulisboa.pt

16.1 Introduction

Many non-luminescent chemical and biochemical analytes can be detected
by luminescence (often fluorescence, but also phosphorescence) methods:
cations (H^+, alkali metal cations, alkaline earth cations, Zn^{2+}, Pb^{2+}, Al^{3+},
Cd^{2+}, Hg^{2+}, *etc.*); anions (halide ions, citrates, carboxylates, phosphates,
etc.); neutral molecules (H_2O_2, glucose, *etc.*); gases (O_2, CO_2, NO, *etc.*); biomolecules (amino acids, coenzymes, nucleosides, nucleotides, ATP, *etc.*);
and biological macromolecules (proteins, DNA, *etc.*).[1] Furthermore, luminescence can provide accurate information about physical parameters such
as temperature, pressure and viscosity.[1]

A clear distinction is made here between luminescence, fluorescence and
phosphorescence. *Luminescence* is the general word for the spontaneous
emission of radiation from an electronically excited species not in thermal
equilibrium with its environment. *Phosphorescence* and *fluorescence* are
special types of luminescence, respectively involving and not involving a
change in electron spin multiplicity in the radiative transition. In some
cases (*e.g.*, quantum dots) the spin multiplicity is not well defined, and it is

RSC Nanoscience & Nanotechnology No. 38
Thermometry at the Nanoscale: Techniques and Selected Applications
Edited by Luís Dias Carlos and Fernando Palacio

Published by the Royal Society of Chemistry, www.rsc.org

better to use the term *luminescence*, which is always appropriate.[1] The most common type of luminescence in sensing applications is *photoluminescence*, which is luminescence produced by light absorption.

Sensing is defined as the continuous monitoring of an analyte or of a physical parameter in real time.[1,2] In this chapter we will deal exclusively with sensing based on luminescence (often fluorescence). If the analyte is by itself luminescent, sensing is *direct* (also called *passive*). However, most analytes of interest are non-luminescent, and sensing is *indirect* (also called *active*), based on the response of a luminescent *probe* to the analyte. Temperature or pressure sensing by luminescence (as opposed to the universal thermal radiation) always requires a probe.

The probes can be small organic molecules, metal complexes, inorganic nanocrystals, fluorescent proteins, *etc.* They may contain not only a luminescent moiety (*luminophore*) but also an adjacent moiety for fast and reversible binding of the analyte (a *recognition unit* or *receptor*).[1] The luminophore acts as a signal transducer converting the information (presence/absence of an analyte) into a luminescence signal. The extent of change in the luminescence signal, usually resulting from an ensemble of individual probes, reflects the analyte concentration. The object to be sensed—where the analyte is located, or whose physical property (*e.g.*, temperature) should be known or mapped—can be a homogeneous 3D medium (gas, solution), a 2D medium (surface) or a complex micro/ nanostructured object (*e.g.*, a nanodevice or a living cell). The probe can be either inside the object (*e.g.*, dissolved in a liquid medium or encapsulated in nanoparticles suspended in the liquid) or at the interface with it (*e.g.*, dispersed in a solid support permeable to a gas, or in a film coating). If the probe is supported in some way, either by a nanoparticle or by a surface film, the global object is called the *transduction platform*. This platform can thus be micro/nanoscopic and *mobile* (*e.g.*, a micro/nanoparticle) or micro/ macroscopic and *fixed* (*e.g.*, in a solid or on a surface). Mobile platforms can be tailored to target specific parts of a heterogeneous sample, *e.g.*, specific organelles of a living cell or a particular site in a biomacromolecule, by surface functionalization.

The luminescent probe, signalling the presence of the analyte or responding to a physical parameter by changing its luminescence characteristics, is at the heart of the sensing process. Nevertheless, quantitative luminescence sensing requires a full optical *sensing apparatus* made of four distinct elements (Scheme 16.1):

1. A light source (with steady or time-dependent intensity);
2. The analyte-responsive species either inside the medium (*e.g.*, in a nanoparticle) or immobilized (*e.g.*, in a polymer or sol–gel matrix);
3. The optical system (eventually involving an optical fibre or a waveguide); and
4. The light detector (photomultiplier or photodiode) coupled to appropriate electronics for analog-to-digital conversion of the signal.[1]

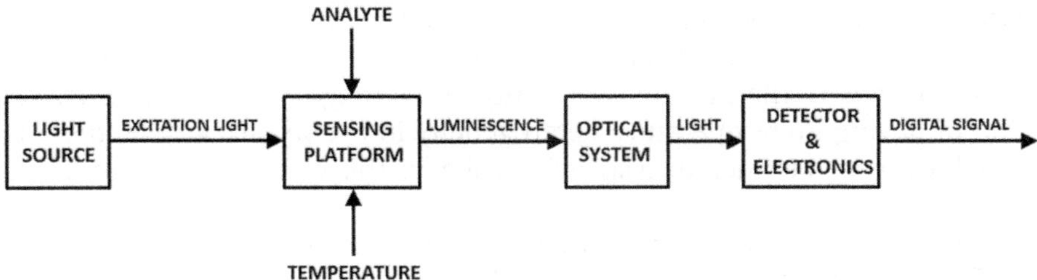

Scheme 16.1 The photoluminescence sensing process, considering the sensing of one analyte and the effect of temperature, which can also be sensed by a second probe.

Miniaturized and self-contained *sensing devices* designed for specific sensing applications are called *sensors* (this designation should not be used for the probe or the transduction platform alone, sometimes also called the *molecular sensor* or *nanosensor*, depending on its nature).[1–3] However, sensing for scientific or industrial purposes can be performed with sensing apparatuses or devices that are not sensors, for instance: (1) a fluorimeter or a fluorescence microscope allows the monitoring of analytes or physical parameters in specific research objects, *e.g.*, microreactors, microdevices, and living cells; and (2) sensor paints are used to map pressure and temperature on the surface of automobiles and aircraft.[4]

Dual-sensing is defined as the simultaneous monitoring of two analytes or of one analyte and a physical parameter, such as temperature, in real time and at the same point.[2] The meaning of "same point" in the previous sentence remains, however, open to interpretation. We view it here as the maximum area or volume in the sample where the two sensed quantities are still homogeneous.

Knowledge of temperature when using dual-sensors based on luminescence is almost mandatory, as in the overwhelming number of situations, the luminescence response to the analyte is temperature dependent[2] and therefore a precise knowledge of this physical parameter is needed. Ideally, temperature should be measured simultaneously and in the near vicinity of the probe whose reading is to be converted into a concentration, hence one of the probes of the dual-sensor should measure temperature, whenever temperature gradients are expected (see the "same point" definition in the previous paragraph). Provided these gradients are moderate (*e.g.*, they do not contain nanosized sources of heat, known as "hot spots") distances between the two probes (in the common platform) from a few tens of nanometres up to hundreds of micrometres (or even more) may be adequate. The probes can still be said to be *co-localized*. On the other hand, for linear scales shorter than tens of nanometres the very concept of temperature starts to lose a precise meaning,[5] and a finer mesh for temperature mapping is probably not realistic. One may also wonder if the very act of sensing may not modify the flow of heat (and possibly of mass and momentum, *e.g.*, for a liquid near a surface) and thus the temperature pattern, when the target and sensing platforms have comparable sizes and are close to each other. Furthermore, the present typical resolution (not the highest resolution) of fluorescence microscopes is in the hundreds of nanometres range.[1] In the extreme situation of having no temperature heterogeneity in the whole sample, there is probably no need to include a specific probe for this quantity in the multiple sensor.

Knowledge of temperature is by itself useful or necessary in many scientific, biomedical and industrial applications. In any case, luminescence dual-sensing including temperature should allow measuring of the temperature and the concentration of an analyte in real time, with a temporal resolution down to hundreds of microseconds. Imaging with a spatial resolution down to the hundreds of nanometres is possible in principle, although it has not yet been reported for luminescence dual-sensing. In several applications, the spatial scale of interest is indeed macroscopic.[1,2,4,6]

In this chapter the basic aspects and selected applications of luminescence dual-sensing (temperature included) will be presented, taking into account the subject matter of other chapters and the existence of several recent and detailed reviews: multiple sensing and imaging was reviewed with respect to fluorescence by Wolfbeis and co-workers[2] and fluorescence imaging by Schäferling.[7] The measurement of temperature was recently reviewed according to two partially overlapping viewpoints. One, covering the nanoscale aspects, was addressed by Brites and co-workers.[8] The other, by Wang, Wolfbeis and Meier, dealt with the use of luminescence in thermometry.[9] A third review, by Jaque and Vetrone, focused on the overlapping domain: luminescence nanothermometry.[10] Most of the temperature-related topics are of course treated afresh in several chapters of the present book. Three other reviews relevant to the subject of this chapter are those on optical chemical sensors by McDonagh, Burke and MacCraith,[11] on probes for oxygen sensors by Quaranta, Borisov and Klimant[12] and on oxygen sensing and imaging by Wang and Wolfbeis.[6] Luminescent probes, luminescence measuring methods, imaging and sensing are reviewed in ref. 1, and fluorescence sensing in the book by Demchenko.[13]

16.2 Sensing Formats

Temperature and analyte concentration are obtained from the luminescence parameters (especially intensity and lifetime) of emitting probes that are sensitive to those quantities.

In most applications, sensing platforms are preferred over the bare probe (which can be considered the simplest platform) owing to the higher selectivity and stability, both photochemical and chemical.

As mentioned, sensing platforms can be either mobile (*e.g.*, probes encapsulated in nanoparticles) or fixed (*e.g.*, probes in polymer films), see Scheme 16.2.

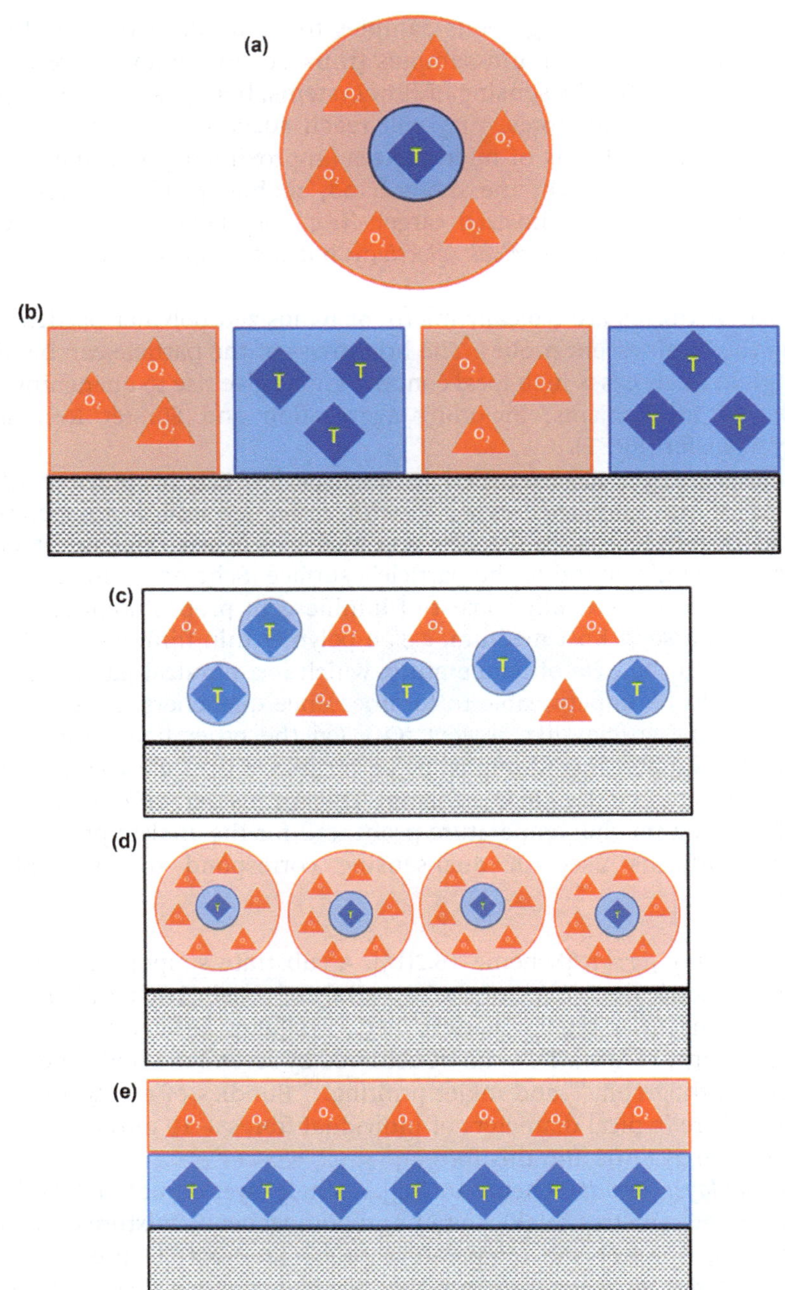

Scheme 16.2 Possible types of dual-sensing platforms: (a) mobile (depicted as a core–shell nanoparticle, with a temperature-sensitive core and a shell containing an oxygen-sensitive probe); (b) dual array, with alternating temperature- and oxygen-sensitive dots deposited on a substrate; (c) single layer, with a mixture of probes; (d) single layer, with dispersed bi-functional core–shell nanoparticles; and (e) two-layer format, the upper layer containing the oxygen-sensitive probes.

Fixed platforms allow gaseous samples to be used; also, they do not irreversibly blend with liquid samples (thus do not contaminate them), and are adequate for the sensing of flow systems, both gaseous and liquid. On the tip of optical fibres, they can reach specific locations but have a lower size limit of tens of micrometres (approximately the diameter of a thin human hair).[9] On the other hand, mobile platforms can reach microsized and even nanosized targets (*e.g.*, organelles of living cells), thus offering not only *chemical selectivity* but also superior *site selectivity* in 3D.

Mobile platforms are typically micro- or nanosized polymer particles enclosing the luminescent probes. The brightness of the particle can be high, although in many cases high local concentrations give rise to quenching by a number of mechanisms, including aggregation and Förster Resonance Energy Transfer (FRET).

Mobile sensing platforms used in dual-sensing can be polymer nanoparticles of the core–shell type,[7,14] with reference and/or temperature-sensitive probes located in the core and analyte-sensitive ones in the outer layer or covalently bound to the particle's surface [Scheme 16.2(a)].

Fixed platforms generally consist of luminescent probes homogeneously dispersed in a solid host medium, *e.g.*, a polymer thin film deposited on a substrate.[2,7] In the case of temperature, which is a physical parameter, the matrix should be impermeable to luminescence quenchers, thus avoiding interference by species like oxygen (O_2). On the other hand, for analyte sensing, the support medium must be permeable to it, as close contact with the probe is required (*e.g.*, in O_2 sensing). This means that different matrices should be used for the temperature probe and for the analyte probe. Fixed sensing platforms used in dual-sensing correspond to three planar formats:[2,7]

1. *Dual array format* [Scheme 16.2(b)]. A substrate supports an array of micrometric dots made of the two kinds of sensing material arranged according to a regular pattern (*e.g.*, a square lattice) and in close proximity (<1 mm). This can be achieved by several methods, including photolithography[15] and inkjet printing.[16] Bundles of two types of optical fibre (typical diameters of individual fibres 3–10 μm) may also be considered to fit into this format;[17]

2. *Single-layer format* [Scheme 16.2(c)]. A single layer of material (typically a few micrometres thick) contains an homogeneous mixture of the two probes, however the temperature probe (at least) is previously incorporated in oxygen-impermeable beads or nanoparticles, which are then dispersed in the final matrix (*binder*) together with the analyte probe[18] [Scheme 16.2(d)]; and

3. *Two-layer (sandwich) format* [Scheme 16.2(e)]. The sensing material consists of two adjacent layers (both typically a few micrometres thick), each with a different probe, and made of the appropriate material for the respective probe.[19]

16.3   Luminescence Measurement Methods

16.3.1   Intensity Based Sensing

In photoluminescence sensing, the probe is excited with light. Two different methods exist, depending on the time-dependence (or independence) of the light source intensity.

In the simplest method, *intensity based sensing*,[11] steady-state luminescence is used. In this technique,[1] the light source is continuous and the excitation intensity constant. The analyte concentration (or temperature, pressure, *etc.*) is inferred from the luminescence intensity, after appropriate calibration, as this is a relative method. The intensity is proportional to the probe concentration, and may not reflect in a simple way the analyte concentration. It is indeed subject to several possible errors, including photobleaching, chemical degradation and leaching of the probe, and competitive quenching by species other than the analyte. Background luminescence, inner filter effects (including scattering) and solid angle effects (in imaging) are also of concern. One of the best approaches is the *ratiometric method*,[1] however several potential sources of error remain. The ratiometric method has two main variants, *external reference* and *self-reference*: in the first, the luminescence intensity of a non-responsive compound is used as the reference signal and the probe emission intensity is compared to it. In the second, it is the relative intensity of two bands or peaks of the emission spectrum of the probe that are used. This method requires the emission spectrum of the probe to change upon interaction with the analyte (or to change with temperature, *etc.*). Again, the cause for the spectral change must be well understood, in order to discard effects other than the analyte concentration (or the sought-for temperature, pressure, *etc.*), and a calibration curve is still needed.

16.3.2   Lifetime-based Sensing

In the second photoluminescence sensing method, *lifetime-based sensing*,[11] a time-dependent source (pulsed or modulated)[1] is used, allowing the measurement of the luminescence lifetime of the probe. For sensing purposes (and in many other instances), lifetime measurements are preferable, as the measured decay times depend neither on the excitation intensity nor on the probe concentration (within broad limits). Furthermore, they are much less affected, or not affected at all, by inner filter effects, background luminescence, solid angle effects for emission collection, and light scattering.

Two general methods exist in time-resolved luminescence, according to the type of time-dependence: pulsed and modulated.[1,20] In the first case (*time domain*) the luminescence response is recorded after a short (ps–ns) laser or LED excitation impulse; in the second case (*frequency domain*) the optics and detection system are similar to those of the time domain method, except for the light source, which consists of an LED or CW laser and an acousto-optical modulator. The excitation intensity is modulated, usually with a sinusoidal time dependence, and the relative amplitude and phase of the response are measured at a single frequency or as a function of the excitation frequency (*multifrequency phase modulation*). A time domain determination allows eliminating (experimentally, with a delay) or identifying and discarding (upon data treatment) the contribution of fast decaying background luminescence. Identification and characterization of a short-lived background is also possible in the frequency domain but only with recourse to multifrequency measurements.

In many sensing applications, even if the luminescence decay is not single-exponential, knowledge of an average decay time suffices. For this, the *rapid lifetime determination* (RLD) method is used.[1,2,11,20] It basically approximates the experimental complex decay with an exponential function of time. The exponential is a two-parameter (pre-exponential and lifetime) function, hence is defined by two points: two intensities, taken at two different times (in practice, two narrow time windows).

A situation of practical relevance is a probe with two different locations in a solid but permeable matrix, each with its own response to a given analyte, *e.g.*, O_2. This model[12,21,22] gives a two-exponential decay, with lifetimes τ_1 and τ_2, related to the common lifetime in the absence of quenching, τ_0, by the Stern–Volmer relation[1]

$$\frac{\tau_0}{\tau_i} = 1 + K_{SVi}[Q] \quad (i=1,2), \tag{16.1}$$

where $[Q]$ is the quencher concentration in the sensed medium and K_{SVi} is the effective Stern–Volmer constant for location i. The average decay time τ (often obtained by the RLD method or by single-frequency phase-modulation measurements) is given by

$$\tau = f_1\tau_1 + f_2\tau_2, \tag{16.2}$$

where f_1 and f_2 are the fractional contributions of the two sites $(f_1 + f_2 = 1)$, hence

$$\frac{\tau}{\tau_0} = \frac{f_1}{1 + K_{SV1}[Q]} + \frac{1 - f_1}{1 + K_{SV2}[Q]}. \tag{16.3}$$

This three-parameter relation is frequently used to account for the non-linearity in Stern–Volmer plots (τ_0/τ *vs.* $[Q]$). The real heterogeneous structure of solid matrices used in O_2 sensing was studied by Lopez-Gejo and co-workers at a microscopic level.[23]

In the case of dual-sensing, at least two lifetimes must be determined, one for each probe. This can be achieved by selecting probes that emit at different wavelengths (but that are preferably excited at the same wavelength). However, this is not mandatory if the two probes have sufficiently different lifetimes, in which case the decay is measured at a common emission wavelength and subsequently subject to a two-exponential (or multi-exponential) analysis.

An extension of the RLD method, called *dual lifetime determination* (DLD) is also possible, by measuring the intensity at four different times.[2,20,24] For accurate determinations, or complex decays, a full mathematical analysis is nevertheless necessary. For this purpose, decay functions beyond the exponential exist, such as the Kohlrausch (stretched exponential)[1,25] and the Becquerel (compressed hyperbola) decay functions.[1,26,27]

16.3.3 Imaging

Luminescence imaging uses a CCD array and can be based on intensity or time-resolved data.[1,4] Simultaneous detection at two emission wavelengths is viable and common. Lifetime imaging offers all the advantages described above. However, collecting a number of photons similar to that used with macroscopic samples at each pixel is not possible, because: (1) acquisition of a full image would lead to prohibitive acquisition times; and (2) owing to photobleaching, as the same small sub-set of immobile luminophores is repeatedly excited. For these reasons, usually no more than a few thousand photons per pixel are collected. The reduced number of counts per pixel imposes limits on the accuracy of the results: as little as 200 counts are satisfactory in the single-exponential case, but this number is insufficient for defining complex decays. The small number of counts also restricts the number of channels in the decay histogram, which are typically between 32 and 128, compared to 1024 or 2048 for macroscopic samples. In time-domain lifetime imaging, laser or LED pulses periodically illuminate the field of view *via* an optical fibre and a lens of large numerical aperture, and two time windows at two different delay times are defined on a gated CCD. This allows the application of the RLD method to each pixel. Luminescence imaging is also possible in the frequency domain. The phase shift and modulation depth are measured relative to a known fluorescence standard or to scattering of the excitation light. A single frequency is often used, leading to accuracies similar to the RLD method, provided no background luminescence exists.

16.4 Probes and Analytes

16.4.1 Probes

A supported probe can respond reversibly to the analyte in three main ways:[1]

1. *Reaction of the analyte with the luminophore.* This is the case for protonation/deprotonation in fluorescent pH indicators;
2. *Binding of the analyte to the recognition unit.* The binding affects at least one luminescence property (*e.g.*, lifetime). This class of probes usually offers high selectivity; and
3. *Collisional quenching of the luminophore by the analyte.* Some analytes are able to quench the luminescence of probes by a physical process.

This method is well suited for the detection of small molecules in the gas phase (in some cases also in solution) such as O_2, SO_2, and Cl_2.

The response to a physical parameter may also rely on the last process: the common photoluminescent *pressure sensors* respond in fact to the concentration of gaseous oxygen, hence measuring its partial pressure, and not the total pressure. They do not work in pure nitrogen. The response to temperature, on the other hand, usually depends on intra- or intermolecular processes that are significantly temperature dependent. Probes for luminescence temperature sensing are discussed in detail in Chapters 5–9 and in ref. 1, 2, 6, 8–10, and 12. In dual-sensors they are typically ruthenium(II) and europium(III) complexes,[2,6] see Scheme 16.3.

Optical thermometers based on thermally activated delayed fluorescence (TADF)[1] are especially interesting, as the emission intensity increases with temperature, whereas almost all other probes show the opposite behaviour, owing to thermally activated quenching processes. A strong TADF effect has been observed in the fullerene C_{70} (Scheme 16.3),[28,29] which shows a very intense delayed fluorescence and an unusually long delayed fluorescence lifetime (tens of ms, up to 100 ms in $^{13}C_{70}$). For C_{70} dispersed in a polymer matrix [poly(*tert*-butyl methacrylate)], the working range extends from -80 to $+140\,^{\circ}C$.[30] This range is even wider when monitoring the lifetime of the delayed fluorescence, with an upper limit of several hundred degrees, only limited by the thermal stability of the matrix, and with a maximum sensitivity of about 2% K^{-1}. Owing to the strong quenching effect of oxygen on the long-lived triplet state, measurements must be carried out in a vacuum or inert atmosphere (less than 10 ppbv of oxygen). Encapsulation of fullerene-containing polymer nanoparticles within an oxygen-impermeable polymer like poly(acrylonitrile) enables the measurements to be made in air,[31] and simultaneously opens the possibility of temperature measurements with sub-micrometric resolution. The fullerene C_{70} can also be used to measure oxygen in very low concentrations (ppbv–ppmv range) and at high temperatures, again owing to the exceptional TADF.[19,32,33] Typical probes for

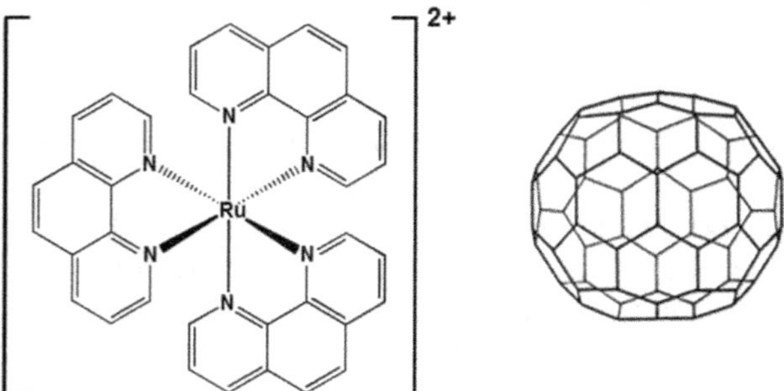

Scheme 16.3 Two temperature-sensing probes (*molecular thermometers*): Ru-phen and C_{70}.

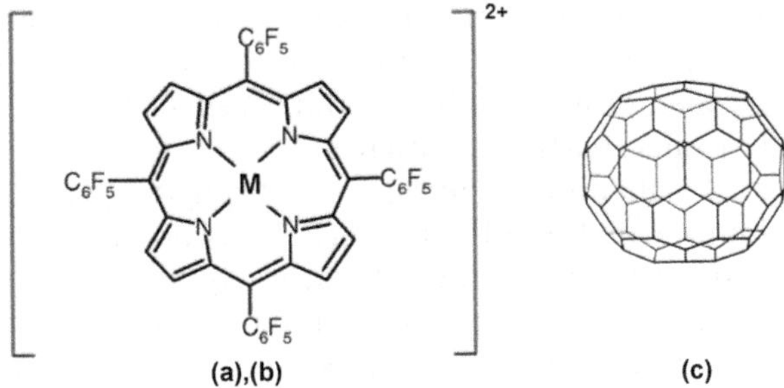

Scheme 16.4 Three oxygen-sensing probes: (a) and (b) MTFPP (M = Pt, Pd) and (c) C_{70}.

oxygen in dual-sensors are palladium(II) and platinum(II) porphyrins, and ruthenium(II) and iridium(III) complexes,[2,6] see Scheme 16.4, with intrinsic lifetimes in the range 10 μs to 1 ms, depending on the compound.

Control of the oxygen permeability of the medium allows selection of the range of measurable oxygen concentrations for a given probe, however this may affect both the response time and the shape of the Stern–Volmer plot.

16.4.2 Analytes

Owing to its importance, one of the most common analytes in photo-luminescence sensing is oxygen,[6] and this is especially true in photo-luminescence dual-sensing, where the temperature sensor always comes in tandem with an oxygen sensor.[16,18,19,24,34–54] Indeed, the main reason for the temperature sensor is not so much knowledge of the temperature *per se*, but mainly the calibration of the oxygen sensor,[2,34,35] as occurs with *pressure sensitive paints* (PSPs).[4,7,34] Quenching of excited singlet and triplet states by oxygen is collisional (induced intersystem crossing)[1] and thus temperature dependent, as is, in general, the intrinsic lifetime of a probe.

To date, only two multiple (two- or three-analyte + temperature) sensing platforms have been reported, both fixed, and both including the measurement of oxygen and temperature: a triple platform also measuring pH,[53] and a quadruple platform (with three layers) also measuring pH and CO_2.[54]

16.5 Selected Platforms

Although mobile platforms with dual-sensing capability seem increasingly relevant, only fixed planar platforms have been reported in the literature thus far.[16,18,19,24,34–54] Three of these fixed platforms, with different characteristics, are described next.

One of the first dual-sensors, developed by Wolfbeis and co-workers in 2006,[18] is of the single-layer type (*ca.* 100 μm thickness), and uses a composite material containing two indicators incorporated in a porous polymer (polyurethane hydrogel) matrix [Scheme 16.2(c)], allowing simultaneous determination of oxygen partial pressure and temperature. The temperature-sensitive dye is ruthenium tris-1,10-phenanthroline (Scheme 16.3), chosen for its highly temperature-dependent luminescence. It is incorporated in poly(acrylonitrile) microparticles, which are virtually impermeable to oxygen. A fluorinated palladium(II) tetraphenylporphyrin (palladium(II) *meso*-tetrakis(pentafluorophenyl)porphyrin, PdTFPP) (Scheme 16.4) incorporated in poly(styrene-*co*-acrylonitrile) microparticles is the oxygen probe. The binder used is biocompatible, and the material is adequate for biomedical applications. The luminescence of the dyes can be separated both spectrally (due to the different absorption and emission spectra of the probes) and by luminescence decay time (*ca.* 680 nm and 1 ms for the phosphorescence of PdTFPP, and *ca.* 580 nm and a few μs for the luminescence of the Ru complex). The material allows temperature-compensated oxygen sensing in the range between 0 and 60 °C. The response time of the phosphorescent probe to changes in oxygen composition (O_2 replacement by N_2) was nevertheless long, 620 s. Simultaneous imaging of pressure and temperature was also achieved.

A sensing platform of the single-layer type, made of doped core–shell nanoparticles embedded in a sol–gel matrix for the dual-sensing of temperature and oxygen [Scheme 16.2(d)] was reported in 2012 by Sung and Lo.[49] The nanoparticles (*ca.* 25 nm diameter) comprise a CdSe quantum dot (QD) core (*ca.* 4.2 nm diameter) and a silica shell doped with a fluorinated platinum(II) tetraphenylporphyrin (PtTFPP), Scheme 16.4. The QD core is the temperature-sensing probe (emission at 532 nm) while the platinum compound (phosphorescence emission at 648 nm) is the oxygen-sensing probe. Both probes are excited using a common wavelength, and sensing is based on steady-state luminescence intensities. The doped nanoparticles are embedded in an ormosil xerogel and coated on the end of an optical fibre. The temperature range is 30 to 100 °C. The response time when replacing oxygen with nitrogen was shorter than 5 s.

A dual luminescent sensor for a precise mapping of pressure on a solid surface, consisting of discrete dot arrays of pressure- and temperature-sensitive paints (PSPs and TSPs), Scheme 16.2(b), was recently (2014) developed by Kameya and co-workers.[16] The sensor arrays were produced by inkjet printing of PSP and TSP solutions. Given that pressure- and temperature-sensitive luminophores are isolated from each other, interaction between the two luminophores, observed in a mixed sensing material, is avoided. In this dual-array sensor, use of an optimal solvent and of an optimal binder for each luminophore is possible: a 2-propanol solution of PtTFPP (Scheme 16.4) and a toluene solution of ZnS–AgInS$_2$ nanoparticles were employed as PSP and TSP solutions, respectively. The diameter of the dots was *ca.* 300 μm, and the centre-to-centre distance 600 μm, in order to avoid overlap. The pressure distribution on the surface with a non-uniform temperature distribution was successfully measured by the dual-array sensor from steady-state intensities recorded at 530 nm (TSP) and 690 nm (PSP), with excitation at 395 nm.

16.6 Perspectives

Luminescence dual-sensing, including temperature measurement, has to date been essentially restricted to macroscopic planar platforms whose analyte is O_2. Temperature measurement is mostly used only for calibrating the oxygen probe. Maximum spatial resolution is currently on the order of a few micrometres, although the probes used are nanosized or even molecular. On the other hand, temperature measurement and mapping at the nanoscale using photoluminescence has already been carried out.[8–10] There are thus plenty of opportunities for progress in the area of mobile nanoscopic platforms (scaffolds), especially having in view biological targets, *e.g.*, intracellular ones, where spatial resolution at the nanoscale is relevant. For this purpose, fluorescence microscopy (with one- and two-photon excitation and also with possible recourse to superresolution methods[1] in some cases) will probably become the key technique. The recent work by Wolfbeis and co-workers[55] on intracellular oxygen imaging using polystyrene nanoparticles carrying a ruthenium probe, and the review by Nocera and co-workers[56] on the use of QDs incorporating a single luminescent probe (sensing pH, O_2, or glucose, depending on the probe) for metabolic tumour profiling strongly suggest that multiple sensing with the use of nanoparticles of different types and other scaffolds is a promising tool for the study of biological and artificial micro/nanostructured systems, in particular those with time-dependent temperature and/or temperature inhomogeneities coupled with chemical/biochemical reactions, whose reactants, intermediates and products are potential analytes.

Acknowledgements

The research leading to this work has received funding from the European Union Seventh Framework Programme (FP7/2007–2013) under grant agreement no. 314032 and from project RECI/CTM-POL/0342/2012 (FCT, Portugal).

References

1. B. Valeur, M. N. Berberan-Santos, Molecular Fluorescence. Principles and Applications, Wiley-VCH, Weinheim, 2nd edn, 2012.
2. M. I. J. Stich, L. H. Fischer and O. S. Wolfbeis, *Chem. Soc. Rev.*, 2010, **39**, 3102.
3. O. S. Wolfbeis, *Angew. Chem. Int. Ed.*, 2013, **52**, 9864.
4. O. S. Wofbeis, *Adv. Mater.*, 2008, **20**, 3759.
5. M. Wautelet and D. Duvivier, *Eur. J. Phys.*, 2007, **28**, 953.
6. X. Wang and O. S. Wolfbeis, *Chem. Soc. Rev.*, 2014, **43**, 3666.
7. M. Schäferling, *Angew. Chem. Int. Ed.*, 2012, **51**, 3532.
8. C. D. S. Brites, P. P. Lima, N. J. O. Silva, A. Millán, V. S. Amaral, F. Palacio and L. D. Carlos, *Nanoscale*, 2012, **4**, 4799.

9. X. Wang, O. S. Wolfbeis and R. J. Meier, *Chem. Soc. Rev.*, 2013, **42**, 7834.

10. D. Jaque and F. Vetrone, *Nanoscale*, 2012, **4**, 4301.

11. C. McDonagh, C. S. Burke and B. D. MacCraith, *Chem. Rev.*, 2008, **108**, 400.

12. M. Quaranta, S. M. Borisov and I. Klimant, *Bioanal. Rev.*, 2012, **4**, 115.

13. A. P. Demchenko, *Introduction to Fluorescence Sensing*, Springer, Berlin, 2009.

14. R. G. Chaudhuri and S. Paria, *Chem. Rev.*, 2012, **112**, 2373.

15. A. Revzin, R. J. Russell, V. K. Yadavalli, W.-G. Koh, C. Deister, D. D. Hile, M. B. Mellott and M. V. Pishko, *Langmuir*, 2001, **17**, 5440.

16. T. Kameya, Y. Matsuda, Y. Egami, H. Yamaguchi and T. Niimi, *Sens. Actuators, B*, 2014, **190**, 70.

17. J. R. Epstein and D. R. Walt, *Chem. Soc. Rev.*, 2003, **32**, 203.

18. S. M. Borisov, A. S. Vasylevska, C. Krause and O. S. Wolfbeis, *Adv. Funct. Mater.*, 2006, **16**, 1536.

19. C. Baleizão, S. Nagl, M. Schäferling, M. N. Berberan-Santos and O. S. Wolfbeis, *Anal. Chem.*, 2008, **80**, 6449.

20. B. B. Collier and M. J. McShane, *J. Lumin.*, 2013, **144**, 180.

21. J. N. Demas, B. A. DeGraff and W. Xu, *Anal. Chem.*, 1995, **67**, 1377.

22. G. A. Baker, B. R. Wenner, A. N. Watkins and F. V. Bright, *J. Sol-Gel Sci. Technol.*, 2000, **17**, 71.

23. J. Lopez-Gejo, D. Haigh and G. Orellana, *Langmuir*, 2010, **26**, 2144.

24. S. Nagl, M. I. J. Stich, M. Schäferling and O. S. Wolfbeis, *Anal. Bioanal. Chem.*, 2009, **393**, 1199.

25. M. N. Berberan-Santos, E. N. Bodunov and B. Valeur, *Chem. Phys.*, 2005, **315**, 171.

26. M. N. Berberan-Santos, E. N. Bodunov and B. Valeur, *Chem. Phys.*, 2005, **317**, 57.

27. F. Menezes, A. Fedorov, C. Baleizão, B. Valeur and M. N. Berberan-Santos, *Methods Appl. Fluoresc.*, 2013, **1**, 015002.

28. M. N. Berberan-Santos and J. M. M. Garcia, *J. Am. Chem. Soc.*, 1996, **118**, 9391.

29. T. Palmeira, A. Fedorov and M. N. Berberan-Santos, *Methods Appl. Fluoresc.*, 2014, **2**, 035002.

30. C. Baleizão, S. Nagl, S. M. Borisov, M. Schäferling, O. S. Wolfbeis and M. N. Berberan-Santos, *Chem. – Eur. J.*, 2007, **13**, 3643.

31. V. Augusto, C. Baleizão, M. N. Berberan-Santos and J. P. Farinha, *J. Mater. Chem.*, 2010, **20**, 1192.

32. S. Nagl, C. Baleizão, S. M. Borisov, M. Schäferling, M. N. Berberan-Santos and O. S. Wolfbeis, *Angew. Chem. Int. Ed.*, 2007, **46**, 2317.

33. S. Kochmann, C. Baleizão, M. N. Berberan-Santos and O. S. Wolfbeis, *Anal. Chem.*, 2013, **85**, 1300.

34. L. M. Coyle and M. Gouterman, *Sens. Actuators, B*, 1999, **61**, 92.

35. J. Hradil, C. Davis, K. Mongey, C. McDonagh and B. D. MacCraith, *Meas. Sci. Technol.*, 2002, **13**, 1552.

36. K. Mitsuo, K. Asai, M. Hayasaka and M. Kameda, *J. Visualization*, 2003, **6**, 213.
37. B. Zelelow, G. E. Khalil, G. Phelan, B. Carlson, M. Gouterman, J. B. Callis and L. R. Dalton, *Sens. Actuators, B*, 2003, **96**, 304.
38. M. E. Köse, A. Omar, C. A. Virgin, B. F. Carroll and K. S. Schanze, *Langmuir*, 2005, **21**, 9110.
39. M. E. Köse, B. F. Carroll and K. S. Schanze, *Langmuir*, 2005, **21**, 9121.
40. S. M. Borisov and O. S. Wolfbeis, *Anal. Chem.*, 2006, **78**, 5094.
41. A. S. Kocinkova, S. M. Borisov, C. Krause and O. S. Wolfbeis, *Anal. Chem.*, 2007, **79**, 8486.
42. C.-S. Chu and Y.-L. Lo, *IEEE Photon. Technol. Lett.*, 2008, **20**, 63.
43. P. A. S. Jorge, C. Maule, A. J. Silva, R. Benrashid, J. L. Santos and F. Farahi, *Anal. Chim. Acta*, 2008, **606**, 223.
44. M. I. J. Stich, S. Nagl, O. S. Wolfbeis, U. Henne and M. Schäferling, *Adv. Funct. Mater.*, 2008, **18**, 1399.
45. L. H. Fischer, M. I. J. Stich, O. S. Wolfbeis, N. Tian, E. Holder and M. Schäferling, *Chem. – Eur. J.*, 2009, **15**, 10857.
46. L. H. Fischer, S. M. Borisov, M. Schäferling, I. Klimant and O. S. Wolfbeis, *Analyst*, 2010, **135**, 1224.
47. H. Lam, G. Rao, J. Loureiro and L. Tolosa, *Talanta*, 2011, **84**, 65.
48. L. H. Fischer, C. Karakus, R. J. Meier, N. Risch, O. S. Wolfbeis, E. Holder and M. Schäferling, *Chem. – Eur. J.*, 2012, **18**, 15706.
49. T.-W. Sung and Y.-L. Lo, *Sens. Actuators, B*, 2012, **173**, 406.
50. S. M. Borisov and I. Klimant, *Anal. Bioanal. Chem.*, 2012, **404**, 2797.
51. C.-S. Chu and T.-H. Lin, *Sens. Actuators, B*, 2014, **202**, 508.
52. C.-S. Chu and T.-H. Lin, *Sens. Actuators, B*, 2014, **195**, 259.
53. M. I. J. Stich, M. Schäferling and O. S. Wolfbeis, *Adv. Mater.*, 2009, **21**, 2216.
54. S. M. Borisov, R. Seifner and I. Klimant, *Anal. Bioanal. Chem.*, 2011, **400**, 2463.
55. X.-D. Wang, D. E. Achatz, C. Hupf, M. Sperber, J. Wegener, S. Bange, J. M. Lupton and O. S. Wolfbeis, *Sens. Actuators, B*, 2013, **188**, 257.
56. C. M. Lemon, P. N. Curtin, R. C. Somers, A. B. Greytak, R. M. Lanning, R. K. Jain, M. G. Bawendi and D. G. Nocera, *Inorg. Chem.*, 2014, **53**, 1900.

Subject Index